Foundation Chemistry
Question & Answer

Badruddin Khan
Department of Chemistry
University of Kashmir
Srinagar - 190 006

2012

Studium Press (India) Pvt. Ltd.

Foundation Chemistry
Question & Answer

ISBN: 978-93-80012-34-6

Published by:

Studium Press (India) Pvt. Ltd.
4735/22, 2nd Floor, Prakash Deep Building (Near PNB),
Ansari Road, Darya Ganj, New Delhi-110002
Phone: +91-11-43240200-15 (15 lines); Fax: +91-11-43240215
E-mail: studiumpress@gmail.com

Printed at:
Thomson Press (India) Ltd.

About the Author

Born at Gorsara in Ghazipur District of Uttar Pradesh, Dr. Badruddin Khan passed High School and Intermediate examinations from U.P. Board, Allahabad and obtained B.Sc. (Hons), M.Sc., M.Phil and Ph.D. degrees in Chemistry from A. M. U. Aligarh. He joined the University of Kashmir in 1982 as a lecturer and is today, the senior most faculty in the Department of Chemistry which he also headed during 1994 - 97 & 2003 - 2004. Besides remaining stuck with his parent institution with utmost dedication and sincerity, Dr. Khan also enriched the University of Jimmah, Ethiopia, as an Associate Professor and worked as Principal of S.G.B. Podar Post Graduate College, Nawalgarh, Rajasthan.

While Dr. Khan has written five books; contributed over 270 articles; supervised 14 M.Phil and Ph.D. students and published many research papers in journals of national and international repute, he has been pronounced as an excellent teacher by the Internal Quality Assurance Cell of the University of Kashmir on the basis of assessment conducted by it. Having become part and parcel of the prevalent educational set up of the country for more than 40 years in different capacities, Dr. Khan stands well posted about both the need of the students in respect of reading materials and dearth of appropriate books, keeping which in view he has started authoring books on Foundation Chemistry not only for the benefit of learners and competitors but also for the assistance of teachers who are custodians of knowledge bank of the students.

Preface

While the basic chemistry courses are not only the main cause of concern but also the real nightmares among students as well as teachers in schools and colleges, there are several such topics in chemistry which may need types of studying tailored to the topics concerned. More over students often find themselves confused in respect of probable questions and their appropriate answers when they are to take up terminal or final examinations due to which they often run after sample questions, guesses and keys etc. This book is out come of my experience of teaching and examining students in chemistry for more than 30 years. In this treatise, I have tried to educate students in the basic principles of chemistry through questions and their answers so that instead of going through an entire chapter, they may concentrate on questions & answers of their interest. I have also tried to present the material in a clear and easy to understand manner under the intention to provide greater clarity and present things differently than usual chemistry text books do.

In fact this basic chemistry treatise is a survey course in chemistry that covers a little bit of every aspect with emphasis on historical back grounds, common terminology and study techniques. My aim has been to give students some chemical and general scientific literacy besides training them to be good chemists. In this book efforts have been made to include all fundamental items together with all necessary backgrounds so that besides understanding the underlying principles and vocabulary students can understand auxiliary topics such as Units & Measurement with mathematical back grounds. The most useful advice to learners is that they should strive at least to stay up with if not be ahead of the class for if they fall significantly behind in basic chemistry, it will be hard to catch up, while catching up is the only way to continue. The basic facts which need to be learned or memorized at the beginning have been used well in this treatise keeping in view the established fact that fluency in the basic material is necessary for learners to be able to understand and learn the more complex ideas.

Dr. Badruddin Khan

Contents

Chapter 1

INTRODUCTION TO CHEMISTRY

1. WHAT DO WE STUDY IN CHEMISTRY?

Chemistry is the subdivision of science which specializes in the composition, structure and properties of substances and the transformations, these substances go through. This might sound confusing to some, but really it is quite simple to understand. Chemistry can be found all over the world, even in such basic things as cooking; for instance, when we boil water and dissolve salt in it, chemistry explains why water evaporates and why salts dissolve in it. Basically, Chemistry studies chemicals, to which its name may be attributed and applies this knowledge to the society. To understand how to make good use of the fruits of the knowledge of chemistry, one must first understand the underlying principles of Chemistry.

In this branch of science we study about the atomic building blocks of nature, how they combine and what is outcome of their combinations which form the solids, liquids and gases that make up most forms of matter. Atomic theory, the theory which states that all matter is composed of a set of very small units, called atoms, is basic to Chemistry. One of the very first laws to be discovered leading to the establishment of Chemistry as a science, is the Law of Conservation of Matter. The law states that there is no detectable change in the quantity of matter during an ordinary chemical reaction. Modern Physics now teaches that atoms and energy can be neither created nor destroyed. On a superficial level this means that if we start off with 10,001 atoms and proceed with many chemical reactions, we will be left with 10,001 atoms. Even if we started off with something green and gooey and ended up with something black and hard there will still be the same number of atoms. The mass will be the same too if the energy gained or lost is accounted for.

Chemistry studies the interactions of these atoms, sometimes alone but more often combined with (bonded to) other atoms to form ions and molecules. These atoms interact with other atoms (*e.g.* a wood fire is the combination of oxygen atoms from the air with the carbon and hydrogen atoms in the wood) and they also interact with light (a photograph is formed from the changes that light causes to the chemicals on a film) and other types of radiation. One surprisingly early finding was that these atoms almost always combine in definite ratios or proportions (*e.g.* silica sand (SiO_2) is a structure where the ratio of silicon atoms to oxygen atoms is 1:2). We now know that there are exceptions to the Law of Definite Proportions.

Another key discovery in chemistry was that when a change is made, the amount of energy gained or lost will always be the same. This leads to the important concepts of equilibrium, thermodynamics and kinetics. The most interesting theory describing all of chemistry is Quantum Mechanics. This theory is complex, non-intuitive and difficult to master. Often, simpler concepts are used to predict the results of experiments. These concepts (*e.g.* acid/base theories) are limited in scope, but much easier to understand and apply. College students typically study chemistry under various subheadings such as: Analytical Chemistry, Physical Chemistry, Organic Chemistry, Inorganic Chemistry, Polymer Chemistry and Biochemistry. Often, discoveries in chemistry are made by physicists, biologists, chemical engineers or pharmacists."

2. WHAT HAS BEEN THE EARLY HISTORY OF CHEMISTRY AND THE SCIENTIFIC APPROACH?

Chemistry as we know it today has its roots in the earliest history of humankind. The ancients were proficient in the art of metallurgy and dyeing, both of which are chemical in nature. The structure of matter concerned the philosophers of Greece and Rome. The alchemists of the Middle Ages practiced chemistry as they searched for the philosopher's stone that would change matter "baser metals" into gold. During the eighteenth century, science became a popular hobby of the rich. It was common for men (like Lavoisier) to have laboratories in their homes where they did experiments, considered the implications of the experimental findings and formulated theories that could be tested by new experiments. These experimentalists met with one another to discuss their work and formulate theories on the nature of matter. This approach to science formed the basis for the pattern of experimentation that was illustrated by Lavoisier's experiment; we call it the scientific method. According to the scientific method, new knowledge and an understanding of the world around us are most reliably gained if the observers organize their work around the following steps:

- The investigators first define the event or situation they wish to explain. The event may be one of which no studies have been made, or it may be one for which our investigators have hypothesized a new explanation.
- Careful observations are made about this event. These may be direct observations of nature or observations that others have made.
- A hypothesis or model is constructed that explains or consolidates these observations.
- New experiments to test the hypothesis are planned and carried out.
- The original hypothesis is modified to be consistent with both the new and the original observations and capable of predicting the results of further investigations.

A hypothesis that survives extensive testing becomes accepted as a theory. Although our present hypotheses and theories are the best we have devised so far, we have no guarantee that they are final. Regardless of how many experiments have been done to test a given theory and how much data has been accumulated to support it, a single experiment that can be repeated by other scientists and whose results contradict the theory forces its modification or rejection. Some of our currently accepted theories on the nature of matter may in the future have to be modified or even rejected on the basis of data from new experiments. We must keep an open mind and be ready to accept new data and new theories.

One possible series of the steps of the scientific method may be summarized as under:

Definition of problem or Intuitive hypothesis based on previously collected data ↓ Collection of data specifically aimed at solution of problem or testing intuitive hypothesis ↓ Tentative statement of hypothesis ↓ Collection of more data to test hypothesis ↓ Restatement of hypothesis ↓ Collection of more data to test restatement of hypothesis ↓ Hypothesis becomes theory, which continues to be tested using new data.

Note: Hypothesizing and data collecting continue to alternate for some time before the hypothesis earns the right to be called theory. Although all scientists collect data and form hypotheses, it is sometimes difficult to describe when each step is taken.

In spite of the vast amounts of new data being collected and the number of new theories proposed, the understanding of scientific events does not increase at a steady rate. Its forward movement is less like that of a smoothly flowing river than it is like that of a mountain stream, which sometimes rushes ahead and at other times scarcely moves or even wanders off into dead-end swamps. Although there is little doubt that chemical knowledge is expanding, we still have far to go before our understanding of the chemistry of life and of the chemical world in which we live, is complete.

An inherent part of the scientific method is the element of creativity. The scientist assembles all of the observations that have been made in a particular area and combines this knowledge in a new way, out of which comes an original and unique hypothesis. For some scientists it is a new concept; for others it is the refinement and clarification of an existing concept. We see many examples of creativity in science as we move through texts. Certainly Einstein's equation: $E = mc^2$, relating mass and energy, is an example of such creativity.

It is desirable to note the important distinctions between scientific fact and scientific hypothesis and between a scientific law and a scientific theory. While a scientific fact is an observed phenomenon, such as the decomposition of sodium hydrogen carbonate on heating, a scientific hypothesis is an attempt to explain a fact, such as an explanation of why heat causes decomposition. A scientific law is the compilation of the observations of many scientific facts, such as the law that all carbonates decompose on heating. A scientific theory explains a law. The scientific theory that would accompany the law that all carbonates decompose on heating would explain the relationship between the atoms of a carbonate and how heating changes this relationship so that the carbonate decomposes.

3. WHAT ARE PARTS OF A SCIENTIFIC METHOD?

In all sciences, there is a system that describes how to go about solving problems. This system is called the Scientific Method. Basically, there are four parts to this system that are very logical in order. First, an observation must be made. Observations can be either in the form of qualitative observations or quantitative ones. While qualitative observations have to do with a physical characteristic of something, quantitative observations have to do with numbers and concentration. For example, when an observer observes a two winged red butterfly, the fact that the butterfly has two wings is a quantitative observation and the fact that the wings are red is a qualitative observation. This leads to the second part of the scientific method.

In the third step, after making many observations, the observer formulates a question, which is normally, why did that happen? From there, the observer looks for possible reasons why, or what is called formulates theories. A theory is the observer's interpretation of the data that he/she has collected. The first set of theories is called a hypothesis. A hypothesis is simply an educated guess on the part of the observer to explain the question at hand. The fourth and the final step for the observer is to test his theory by experimentation. If the theory is wrong the observer goes back and formulates a new theory and if he/she is right, he/she goes back and retests his/her theory. However, science is in no way perfect. In accordance with human nature, people do make mistakes, but as technology improves and more people test the previous theories, the wrong ones get sorted out from the right ones and science corrects itself.

4. WHAT IS THE UNDERLYING PHILOSOPHY OF EXPERIMENTAL SCIENCE?

It is impossible to discuss the concepts of chemistry without an overview of the history of natural philosophy and science, since the emergence of chemistry as a clearly separate discipline is only two or three centuries old. Much of the thought and knowledge we now use in chemistry, date back well beyond it. In fact, they lie beyond the dawn of recorded history. We begin by making a distinction between science, or more precisely pure science (whose driving inquiry is the desire to understand nature) and applied science or technology (whose driving inquiry is the desire to control nature). However, the distinction is somewhat artificial, because separation of pure and applied chemistry is perhaps possible in principle but not possible in practice. One cannot control effectively something one does not understand, so understanding must precede control or science precedes technology. On the other hand, only technological control permits the use of studies under controlled conditions, we call experiments. Experiments are necessary to advance our understanding of nature and so control must precede understanding. Hence, Science and technology are as inseparable as chickens and eggs.

Experiments often give rise to a general statement about the nature of the subjects of the experiments which is called a scientific law. A scientific law is an empirical statement which describes the results of many experiments. A law is a description, not an explanation; a law states how the subjects of the experiments behave but does not state why they behave in that way. A description which explains why the experimental results are as they are found to be is called a scientific theory. Once a theory to explain the experimental results has been devised, the empirical laws are seen to be consequences of the theory. They follow logically from the theoretical explanation. Thus, theories explain laws.

5. WHAT ARE IMPORTANT BRANCHES OF THE MODERN DAY CHEMISTRY?

Since chemistry is the study of matter and its changes, it includes everything in the universe from a simple hydrogen atom to very large replicating molecules in life processes. Chemistry is involved with the development of medicines that control and cure diseases; food production through specific and safe agricultural chemicals; consumer products such as cleaners, plastics and clothing; new methods of energy production, transfer and storage; new materials for electronic components and new methods for protection and cleaning up of the environment. Chemists are needed to help solve some of society's most difficult technological problems through research, development and teaching. Chemistry is not only a broad and exciting field that contains numerous other branches, but in recent years it has become a discipline that intrudes on our lives from all directions. Accordingly it has become unavoidable to study different aspects of chemistry and chemical phenomena as different disciplines.

Broadly Chemistry can be divided into branches according to either the substances studied or the types of study conducted. While the primary division of the first type is between organic chemistry and inorganic chemistry, divisions of the second type are physical chemistry and analytical chemistry. The original distinction between organic and inorganic chemistry arose as chemists gradually realized that compounds of biological origin were quite different in their general properties from those of mineral origin. Organic chemistry was defined as the study of substances produced by living organisms. However, when it was discovered in the 19th century that organic molecules can be produced artificially in the laboratory, this definition had to be abandoned. Now while Organic chemistry is most simply defined as the study of the compounds of carbon, inorganic chemistry is the study of chemical elements and their compounds (with the exception of carbon compounds). Generally chemistry has been divided into many sub-branches such as:

Theoretical Chemistry

Theoretical chemistry is the study of chemistry via fundamental theoretical reasoning (usually within mathematics or physics). In particular the application of quantum mechanics to chemistry is called quantum chemistry. Since the end of the Second World War, the development of computers has allowed a systematic development of computational chemistry, which is the art of developing and applying computer programs for solving chemical problems. Theoretical chemistry has large overlap with (theoretical and experimental) condensed matter physics and molecular physics.

Stoichiometry

Stoichiometry is the branch of chemistry and chemical engineering that deals with the quantities of substances that enter into and are produced by, chemical reactions. Stoichiometry provides the quantitative relationship between reactants and products in a chemical reaction. For example, when methane unites with oxygen in complete combustion, 16g of methane require 64g of oxygen. At the same time 44g of carbon dioxide and 36g of water are formed as reaction products. Every chemical reaction has its characteristic proportions. The method of obtaining these from chemical formulas, equations, atomic weights and molecular weights and determination of what and how much is used and produced in chemical processes, is the major concern of Stoichiometry.

Inorganic Chemistry

Inorganic chemistry is the major branch of chemistry that is generally considered to embrace all chemical substances except compounds of carbon, though inclusive of some compounds of carbon such as carbonates, carbon dioxide, carbon monoxide, etc. It covers a broad range of subjects, among which are atomic structure, crystallography, chemical bonding, coordination compounds, acid-base reactions, ceramics and the various subdivisions of electrochemistry (electrolysis, battery science, corrosion, semi conduction, etc.). It is important to mention that inorganic and organic chemistry often overlap. For example, chemical bonding applies to both disciplines, electrochemistry and acid-base reactions have their organic counterparts and catalysts and coordination compounds may be either organic or inorganic.

Regarding the importance of inorganic chemistry, R.T. Sanderson has written: "All chemistry is the science of atoms, involving an understanding of why they possess certain characteristic qualities and why these qualities dictate the behavior of atoms when they come together. All properties of material substances are the inevitable result of the kind of atoms and the manner in which they are attached and assembled. All chemical changes involve a rearrangement of atoms. Inorganic chemistry is the only discipline within the chemistry that examines specifically the differences among all the different kinds of atoms".

Organic Chemistry

Organic chemistry is a major branch of chemistry which embraces all compounds of carbon except such binary compounds as the carbon oxides, the carbides, carbon disulfide, etc.; such ternary compounds as the metallic cyanides, metallic carbonyls, phosgene ($COCl_2$), carbonyl sulfide (COS), etc.; and the metallic carbonates, such as calcium carbonate and sodium carbonate. The total number of organic compounds is indeterminate, but some 6,000,000 have been identified and named. Important areas of organic chemistry include polymerization, hydrogenation, isomerization, fermentation, photochemistry and stereochemistry. There is no sharp dividing line between organic and inorganic chemistry, for the two often tend to overlap. It is concerned with the study of the structure, properties, composition, mechanisms and reactions of organic compounds. An organic compound is defined as any compound based on a carbon skeleton. Organic chemistry once was defined as the chemistry of substances derived from living matter but that definition is no longer valid. We can say only that the substances, organic chemists work with, usually contain a great deal of carbon and not many metals. Chemists who work with polymers, petroleum and rubber are organic chemists.

Physical Chemistry (also called Physicochemistry)

Physical chemistry is the study of the physical and fundamental basis of chemical systems and processes. In particular, the energetics and dynamics of such systems and processes are of interest to physical chemists. Important areas of physical chemistry include chemical thermodynamics, chemical kinetics, electrochemistry, statistical mechanics and spectroscopy. Physical chemistry has large overlap with molecular physics. Physical chemistry involves the use of infinitesimal calculus in deriving equations. It is usually associated with quantum chemistry and theoretical chemistry. Physical chemistry is a distinct discipline from chemical physics.

In fact physical chemistry is the application of physics to macroscopic, microscopic, atomic, subatomic and particulate phenomena in chemical systems within the field of chemistry traditionally using the principles, practices and concepts of thermodynamics, quantum chemistry, statistical mechanics and kinetics. It is mostly defined as a large field of chemistry, in which several sub-concepts are applied. The inclusion of quantum mechanics is used to illustrate the application of physical chemistry to atomic and particulate chemical interaction or experimentation.

Physical chemistry is mostly referred to as a macromolecular doctrine, as majority of the principles on which physical chemistry was founded are composed entirely of macromolecular concepts, such as colloids. The relationships that physical chemistry tries to resolve include the effects of: intermolecular forces on the physical properties of materials (plasticity, tensile strength, surface tension in liquids), reaction kinetics on the rate of a reaction and the identity of ions on the electrical conductivity of materials.

Analytical Chemistry

Analytical Chemistry is the subdivision of chemistry concerned with identification of materials (qualitative analysis) and with determination of the percentage composition of mixtures or the constituents of a pure compound (quantitative analysis). The gravimetric and volumetric (or "wet") methods (precipitation, titration and solvent extraction) are still used for routine work and new titration methods have been introduced *e.g.* cryoscopic, pressure-metric (for reactions that produce a gaseous product), redox methods and use of a ion-sensitive electrode, etc. However, faster and more accurate techniques (collectively called instrumental) have been developed in the last few decades. Among these are infrared, ultraviolet and x-ray spectroscopy where the presence and amount of a metallic element is indicated by lines in it's emission or absorption spectrum, colorimetry by which the percentage of a substance is determined by the intensity of it's colour, chromatography of various types by which the components of a liquid or gaseous mixture are determined by passing it through a column of porous material or on thin layers of finely divided solids, separation of mixtures in ion exchange columns and radioactive tracer analysis. Optical and electron microscopy, mass spectrometry, microanalysis, Nuclear Magnetic Resonance (NMR) and Nuclear Quadrupole Resonance (NQR) spectroscopy all fall within the area of analytical chemistry. New and highly sophisticated techniques have been introduced in recent years, in many cases replacing traditional methods.

Biochemistry (Biological Chemistry)

Originally a subdivision of chemistry but now an independent branch of science, biochemistry includes all aspects of chemistry that apply to living organisms. Thus, photochemistry is directly involved with photosynthesis and physical chemistry with osmosis, two phenomena that underlie all plant and animal life. Other important chemical mechanisms that apply directly to living organisms are catalysis, which takes place in biochemical systems by the agency of enzymes; nucleic acid and protein constitution and behavior, which is known to control the mechanism of genetics; colloid chemistry, which deals in part with the nature of cell walls, muscles, collagen, etc; acid-base relations, involved in the pH of body fluids; and such nutritional components as amino acids, fats, carbohydrates, minerals, lipids and vitamins, all of which are essential to life. The chemical organization and reproductive behavior of microorganisms (bacteria and viruses) and a large part of agricultural chemistry are also included in biochemistry. Particularly active areas of biochemistry are nucleic acids, cell surfaces (membranes), enzymology, peptide hormones, molecular biology and recombinant DNA.

Nuclear Chemistry

Nuclear chemistry is the division of chemistry dealing with changes in or transformations of the atomic nucleus. It is the study of how subatomic particles come together and make nuclei. It includes spontaneous and induced radioactivity, the fission or splitting of nuclei and their fusion, or union; also the properties and behavior of the reaction products and their separation and analysis. The reactions involving nuclei are usually accompanied by large energy changes, far greater than those of chemical reactions; that are carried out in nuclear reactors for electric power production and manufacture of radioactive isotopes for medical use, also (in research work) in cyclotrons. Modern Transmutation is a large component of nuclear chemistry and the table of nuclides is an important result and tool for this field.

Materials Chemistry

Material chemistry is the preparation, characterization and understanding of substances with a useful function. The field is a new breadth of study in graduate programs and it integrates elements from all classical areas of chemistry with a focus on fundamental issues that are unique to materials. Primary systems of study include the chemistry of condensed phases (solids, liquids, polymers) and interfaces between different phases. In the recent past many other branches of Chemistry have emerged. There are many other fields of chemistry. There are also several main cross-disciplinary and more specialized fields of chemistry.

6. WHY SHOULD STUDENTS STUDY CHEMISTRY?

"Man is slightly nearer to the atom than the stars. From his central position he can survey the grandest works of Nature with the astronomer, or the minutest works with the chemist." ***Arthur Eddington (a British astronomer)***

While there may be many answers to this simple question, the simplest answer will be as simple as "Because it's related to materials, hence we want to explore it". Since students at foundation level stand well posted with the concepts of Atoms and Bonding, we may start with brief introductory remarks to both, the atom and the bonding so that instead of looking for answer to this question outside the domain of Chemistry, we may feel the fragrance of chemistry from within.

Perhaps people take chemistry for granted because nature makes everything look so easy. We know that everyday, exploding stars eject new elements into the vastness of space and extreme temperature and pressure create precious gemstones deep in Earth's belly. Further on its surface, green plants transform carbon dioxide and sunlight into food for a living planet. Yet no one blinks an eye. We are living in the world of the atom, ruled by the tiny particles, called electrons. They were discovered barely 100 years ago and modern chemistry now knows enough about them to explain why glass breaks whereas rubber bounces. Understanding the properties of electrons fuels rockets into deep space, powers lightning-fast computer chips and targets medical laser beams that destroy cancer cells. But it has been a long road.

For thousands of years, scientists labored under the misguided theories of the ancient Greek natural philosophers. Pseudo scientists, called alchemists, appeared in the Middle Ages, promising chemical elixirs that would cure all disease and lead to immortality. And as late as the 1850s, respected chemists experimented with the invisible (and nonexistent) substance called phlogiston, which was thought to have negative weight and caused materials to burn. Yet the longest road is often the most enlightening. As we retrace this journey marked with countless wrong turns and dead ends, what jumps out of the story are not the mistakes, but, rather, how often the right answers were there all along. Beginning with the very nature of matter, we may examine the forces at work in the unseen chemical world.

These forces are powerful indeed. Atoms, the building blocks of all matter, are more than 99% empty space. Yet we cannot thrust a hand through a solid wall because the force between electrons causes them to repel each other. Sometimes, though, matter does interact. Chemical reactions occur and bonds form and break all around us. To understand how and why this happens, one will have to learn about the nature of the elements, the rules that govern chemical behavior and some of the broader laws and theories of modern chemistry.

Atoms are the building blocks of all substances. But what is it that keeps atoms connected together? They are held together by chemical bonds, strong attractive forces, between atoms. Without these ties, the universe would be nothing more than a mass chaos of individual atoms. So we need to know what constitutes a chemical bond. A bond is formed when electrons from two atoms interact with each other and their atoms become bonded. The electrons that interact with each other are valence electrons, the ones that reside in the outermost electron shell of an atom.

As we sit reading this sentence, billions of atoms surround us. When any two of them approach each other, their outermost electrons come into close contact. We know that electrons are negatively charged particles, so a chance of meeting between atoms usually results in the two repelling each other. Occasionally, however, atoms will stick together by forming a chemical bond. There are two main types of bonding processes. A covalent bond results when two atoms "share" valence electrons between them. An ionic bond occurs when one atom gains a valence electron from a different atom, forming a negative ion (Anion) and a positive ion (Cation), respectively. These oppositely charged ions are attracted to each other, forming an ionic bond.

One may ask as to why chemical bonds are important. The type of chemical bond that occurs in a molecule or substance in part defines its properties. For example, let us consider sodium chloride (NaCl) and hydrogen chloride (HCl). Both these substances contain chlorine, but NaCl is the white solid crystalline substance sprinkled on eatables and HCl is a foul smelling gas. When this gas is dissolved in water, it forms a solution known as hydrochloric acid. This is the acid that our stomach uses to digest food. How can this be if both materials have chlorine in them? The chemical bonding that takes place in NaCl is different than that in HCl. While the former is an ionic bond, the later is a polar covalent bond. This gives NaCl and HCl very different structures, appearances and properties.

What other differences are noticeable among molecules that result from different types of chemical bonding? Let us think about what happens when a bunch of sodium and chlorine ions join together to form rock salt. If we hit this with a hammer, it shatters into tiny pieces. It does this because the bonds between the atoms in rock salt are ionic. The particles are arranged in such a way that they line up along rows of positive and negative charges. Under enough stress, the salt crystals break along those lines into much smaller pieces.

The bonds that hold the carbon and hydrogen atoms in rubber together, on the other hand, are not ionic but covalent. Each carbon atom shares four of its outermost electrons with its immediate neighbors. Under stress, the bonds stretch and then snap back as each atom pulls on the shared electrons. And that's the way the ball bounces. There is a third type of bonding, called Metallic bonding. As the name implies, metallic bonding usually occurs in metals, such as copper. A piece of copper metal has a certain arrangement of copper atoms. The valence electrons of these atoms are free to move about the piece of metal and are attracted to the positive cores of copper, thus holding the atoms together.

Essential to understanding all types of chemical bonding is realizing that all bonds use electron "glue." Every substance is made up of atoms and all atoms are surrounded by the negatively charged particles, called electrons. Electrons in the outermost shell can and do bond with those of other atoms to create everything from quartz and quarters to plastic and pennies. In large part, the difference between materials as diverse as diamonds and pencils is how they're glued together. We can conclude our notion by revealing that once we understand about the nature of the forces operating at molecular level, we can have a closer look at the various types of ties that bind the universe and the things within. It is obvious that without understanding chemistry, students can not understand them and without studying chemistry, they can not understand the underlying chemistry. Hence not only students but we all need to study chemistry.

7. WHY PEOPLE NEED KNOWLEDGE OF CHEMISTRY?

Whereas definitions of chemistry such as 'Chemistry is a subdivision of science which specializes in the composition, structure and properties of substances and the transformations through which these substances undergo or chemistry is broadly defined as the study of matter and its properties, are acceptable to everyone, these definitions do not clarify why one needs to know about chemistry and chemistry of the matter. This might sound confusing to some, but really it is quite simple and interesting to understand. While the one line answer to this query will be that the world in which we live is a chemical world, chemistry can be found all over the world, even in very basic things such as cooking, cleansing, playing, bathing, etc. For instance, when we boil water and dissolve salt in it, we may use our knowledge of Chemistry to explain as to why water evaporates and salts dissolve in it. Our body itself is a complex chemical factory that uses chemical processes to change the food we eat and the air we breathe, into bones, muscle, blood and tissue and even into the energy that we use in our daily living. When illness prevents some part of these processes from functioning correctly, the doctor may prescribe, as a medicine, a chemical compound, either isolated from nature or prepared in a chemical laboratory by a chemist. The world around us is also a vast chemical laboratory. The daily news bulletins and dailies are filled with reports of acid rain, toxic wastes, the risks associated with nuclear power plants and the derailment of trains carrying often toxic and corrosive chemicals such as vinyl chloride, sulfuric acid, ammonia etc. However, not all chemical news is of disasters. The daily news also carries stories (often in smaller headlines) of new drugs that cure chronic diseases; of fertilizers, insecticides and herbicides designed by chemists to help the farmers grow more in order to feed our growing populations and of other new products to make our lives more pleasant.

The packages we buy at the provision stores, list their contents, including what chemicals a package contains, such as preservatives and the nutritional contents in terms of vitamins, minerals, fats, carbohydrates and proteins. Since our everyday life is besieged with chemicals and their products, it is both desirable and advisable to know about their chemistry as many of them are bound to change or affect our body chemistry. Basically, Chemistry studies chemicals, from which its most probable name originates and we apply this knowledge for the betterment of society. In order to understand how to apply the knowledge of Chemistry prudently, one must first understand the basic principles behind Chemistry. But in the beginning of the study of chemistry it will be unwise to start with topics as complex as the latest miracle drugs. Rather it is better to begin with the study of composition of matter and the different kinds of matter. We can then talk about the properties of the different types of matter and the changes that each can undergo. Later on we may make people learn that each of these changes is accompanied by an energy change and then they may be made to realize the significance of these energy changes.

8. WHAT COMES TO OUR MIND WHEN WE THINK OF CHEMISTRY?

Undoubtedly it sounds nice and simple that Chemistry is the science of substances and matters and the ways they interact in our world, but in reality it's a broad and complex discipline. In terms of the important major sciences, chemistry builds upon and uses a lot of the theory from physics and in turn is used and applied extensively in biology, so it serves as a bridge between the general world and the real life. This makes chemistry cool and important and earned it the nickname of "the central science". When we think of Chemistry, we are lost in imaginations such as the following:

"Chemistry involves substances and the most basic substance in nature is called an element. In fact elements are the basic building blocks of all chemical substances. All elements are made of atoms and an atom is the tiniest piece that retains the properties of its element. An atom has a nucleus, which contains protons, positive particles and neutrons, which are neutral. It also has orbiting electrons, tiny negative particles. The number of protons of an atom gives us the atomic number, which is unique for each element. For example, the element Hydrogen has an atomic number of 1, so it has 1 proton. The element Oxygen has atomic number 8, so it has 8 protons. The element Mendelevium has atomic number 101, so a Mendelevium atoms has 101 protons in its nucleus. The number of neutrons in the nucleus is combined with the number of protons to give us the atomic mass of an atom. Atomic mass is not measured in kilograms or milligrams since atoms are ridiculously small, so instead atoms are measured relative to each other's mass, with Hydrogen being the smallest element of atomic mass 1 amu (atomic mass unit). So for example, as Phosphorus has an atomic number of 15 (*i.e.* 15 protons) and an atomic mass of 31, it would have 31–15=16 neutrons in its nucleus.

Some elements have isotopes, which are atoms that have the same number of protons but a different number of neutrons (and thus a different atomic mass). These isotopes generally share the same chemical properties, because they have the same number of protons and electrons in a neutral state. The number of electrons of an atom determines its overall charge and facilitates compounds. So, these are our elements. The best way of presenting all chemical elements would be to use the Periodic Table. This table, created by Mendeleev, uses rows and columns to divide elements into distinct groups that share similar properties. There are various ways to group elements.

Elements can combine to form compounds. A compound has 2 or more types of elements in it that are bonded together in a fixed proportion. An example of a compound would be water, in which every oxygen atom is bounded to exactly 2 hydrogen atoms. Compounds can be classified in many ways, such as by the type of bonds they include. A compound is generally represented by a chemical formula that explains the types of elements involved and their proportions, for example, water is H_2O. Using water as our example, a glass of water contains billions of molecules. One water molecule is simply one oxygen atom bonded with 2 hydrogen atoms. Splitting a molecule into separate atoms will cause it to lose the properties of the compound, it used to represent. Molecules have specific sizes and geometrical shapes.

Molecular mass denotes the relative mass of a molecule compared to other molecules and atoms. It is calculated simply as the sum of the masses of the atoms involved. For example, a water molecule with 1 oxygen atom (16 amu) and 2 hydrogen atoms (1 amu each) has a molecular mass of 18 amu. Chemical substances are either compounds or elements. They have constant composition and their atoms or molecules cannot be broken down without a chemical reaction. Mixtures, on the other hand, can contain several substances and can be separated using simpler means. Homogeneous mixtures are ones that have definite and consistent properties. We may think of them like a veggie-shake, which has the vegetables cut so small and mixed up so that the shake has a specific consistency. Heterogeneous mixtures have inconsistent and non-uniform composition, like a salad, wherein it's easy to see and separate the different vegetables."

Chapter 2

MATTER

1. WHAT ARE CHARACTERISTICS OF MATTER?

In the beginning of Chemistry, there are always a few terms which must be defined clearly because either their definitions are very close to each other or they are very important. One of such general terms is 'matter'. It is important to understand the meaning of matter because everything in the universe is made up of matter. So much so everything we can hold, taste, or smell is made of matter. In fact matter makes up not only everything we can see, such as clothes, water, food, plants and animals etc but also it makes up some things we cannot see, such as air or the smell of perfume. We can describe the kind of matter by its material properties such as its color or hardness. Matter is made up of particles so tiny that only the most powerful microscope can see them. Matter can broadly be divided into two groups: non-living matter and living matter. While non-living matter neither moves on its own, nor grows, or reproduces, the living things do so. The rocks that make up the Earth are examples of non-living matter and all living things, including animals and plants, are examples of living matter. However, everything is not made of matter. Non-matter includes the light from a torch, the heat from a fire and the sound of an ambulance siren as we cannot hold, taste, or smell these things. Though they are not types of matter, they are certainly forms of energy. Hence everything that exists can be classed as either a type of matter or a form of energy.

In fact matter is any thing that occupies space and can be felt by our senses. In 1804, the English scientist John Dalton formulated the atomic theory, which set out some fundamental characteristics of matter and which is still used today. According to this theory, matter is composed of extremely small particles called atoms, which can neither be created nor destroyed. Atoms can, however, attach themselves (bond) to each other in various arrangements to form molecules. Matter can exist in three phases: solid, liquid and gas inclusive of plasma and condensate forms. Solids are rigid structures which have definite volume and definite shape. Due to the strength of solids, they are not easily compressed. Liquids, the second state of matter, have definite volume, but lack definite shape. This is why liquids fill containers with odd shapes, but are not easily compressible. Gases are the third and final normal state in which matter can exist. Since gases do not have either definite volume or definite shape, they fill containers with odd shapes and are easily compressible. Most kinds of matter can exist in all the three states.

Every kind of matter possesses a number of properties by which it can be identified. Broadly these properties fall into two large categories: (1) physical properties: those that can be observed without changing the composition of the sample and (2) chemical properties: those whose observation involves a change in composition. We know that baking soda dissolves readily in water. If water is evaporated from a solution of baking soda, the baking soda is recovered unchanged; thus, solubility of a substance is a physical property. The decomposition of baking soda on heating is a chemical property. We can observe the decomposition of baking soda, but, after we make this observation, we no longer have baking soda. Instead we have carbon dioxide, water and sodium carbonate. While a physical change alters only physical properties, such as size and shape, a chemical change alters chemical properties, such as composition.

We are quite familiar with water as a solid (ice), a liquid and a gas (steam). We have seen wax as a solid at room temperature and a liquid when heated. We have heard of existence of carbon dioxide as a solid (dry ice) and are aware of it as a colorless gas at higher temperatures but we have not seen solid carbon dioxide. The temperatures at which a given kind of matter changes from a solid to a liquid (its melting point) or from a liquid to a gas (its boiling point) are physical properties. For example, the melting point of ice (0°C) and the boiling point of water (100°C) are physical properties of the substance water. Like pure substances, mixtures too can exist in the three physical states of solid, liquid and gas. For example, air is a gaseous mixture of approximately 78% nitrogen, 21% oxygen and varying percentages of several other gases; rubbing alcohol is a liquid mixture of approximately 70% isopropyl alcohol and 30% water and steel is a solid mixture of iron and other pure substances.

2. WHAT IS A 'SUBSTANCE'?

Science as it generally is understood today is of Western origin and much of it still uses the forms first clearly worked out in ancient Greece. Chemistry as a practical art is far older and goes back to the first smiths of the Bronze Age and to early potters and flintknappers before that, yet prior to the civilization of Greece no attempt was made to understand the nature of chemical materials in a form which has survived. The concept of substance dates back to this pre-theoretical and even prehistoric period. By the concept of substance we mean the existence, not of matter itself, but of different kinds of matter which have different properties. These different kinds of matter are different substances. Even in the Stone Age, the selection of stone for spear points and wood for spear shafts or bow staves was done with care and skill; considerable attention was paid to the properties of different woods and stones. The modern distinction between pure substances and mixtures was not drawn, but the idea of different substances, having markedly different properties, was both well known and well used. Homogeneous substances, those substances which appear to be uniform throughout, were distinguished very early from heterogeneous substances, which have parts that are visibly different, although the terms date from Greek and Roman times. Common homogeneous substances include sea water, wine and some common rocks and minerals. Most natural materials are obviously heterogeneous substances. Homogeneous substances were then, as they still are, described in terms of their properties such as color, hardness, melting point, size, shape and mass. A modern scientist draws a distinction between two types of properties.

The different substances known in the ancient world were divided into the three classes we now call the three states of matter: solids, liquids and gases. Again, no distinction was made between pure solid substances and solid mixtures, between pure liquid substances and solutions, or between pure gases and smokes or fogs. Nevertheless, the concept of different substances classified under these general headings was carried into the natural philosophy of classical Greece in the concept of the four elements. The modern definitions of the three states of matter are the same as the ancient ones.

3. WHAT IS DIFFERENCE BETWEEN A PURE SUBSTANCE AND A MIXTURE?

While Chemistry is defined as the study of matter and the process of bond breaking and bond making, students can not study all types of matter in the beginning. Hence it is better to concentrate on simple substances, the properties that identify them, and the changes they undergo. The discussion of properties of matter takes us to the point related to the differences between pure substances and mixtures.

Pure Substances

A pure substance consists of a single kind of matter. It always has the same composition and the same set of properties. For example, baking soda is a single kind of matter, known chemically as sodium hydrogencarbonate or sodium bicarbonate. A sample of pure baking soda, regardless of its source or size, will be white solid containing 57.1% sodium, 1.2% hydrogen, 14.3% carbon and 27.4% oxygen. The sample will dissolve in water. When heated to 270°C the sample will decompose, giving off carbon dioxide and water vapor and leaving a residue of sodium carbonate. Thus, by definition, baking soda is a pure substance because it has a constant composition and a unique set of properties, some of which we have mentioned. The properties we have described hold true for any sample of baking soda. These properties are the kinds which we are interested in.

Here we may note that the term pure means a single substance, not a mixture of substances. As used by the U.S. Food and Drug Administration (USFDA), the term pure means "fit for human consumption." Though milk, whether whole, 2% fat, or skimmed, may be pure (fit for human consumption) by public health standards, it is not pure in the chemical sense. Milk is a mixture of a great many substances, including water, butterfat, proteins and sugars. Each of these substances is present in different amounts in each of the different kinds of milk.

A pure substance can either be an element or a compound. Elements are those pure substances that are made of only one kind of atoms and cannot be decomposed by ordinary chemical means such as heating, electrolysis, or reaction. Gold, silver and oxygen are among examples of elements. Compounds are pure substances formed by the combination of elements; they can be decomposed by ordinary chemical means. Baking soda is a compound; it contains the elements sodium, hydrogen, carbon and oxygen and it decomposes on heating. Mercury (II) oxide is another compound; it contains the elements mercury and oxygen and on heating it decomposes to those elements.

Mixtures

In Chemistry and the real world, substances are not always pure. While pure substances are substances that have constant composition, mixtures are substances or a group of substances which are variable in composition and can be separated by physical means. An example of a mixture might be collection of sand and gravel. A mixture consists of two or more pure substances. Most of the matter we see around us is composed of mixtures. For example, seawater contains dissolved salts; river water contains suspended mud; and hard water contains salts of calcium, magnesium and iron. Both seawater and river water also contain dissolved oxygen, without which fish and other aquatic life could not survive. Unlike the constant composition of a simple substance, the composition of a mixture can be changed.

The properties of the mixture depend on the percentage of each pure substance in it. Steel is an example of a mixture. All steel starts with the pure substance iron. Refiners then add varying percentages of other elements such as carbon, nickel, chromium, vanadium, etc to obtain steels of a desired hardness, tensile strength, corrosion resistance and so on. The properties of a particular type of steel depend not only on the substances that are mixed with the iron but also on the relative percentage of each. One type of chromium-nickel steel contains 0.6% chromium and 1.25% nickel. Its surface is easily hardened, a property that makes it valuable in the manufacture of automobile gears, pistons and transmissions. The stainless steel used in the manufacture of surgical instruments, food-processing equipments and kitchenware is also a mixture of iron, chromium and nickel; it contains 18% chromium and 8% nickel. Steel with this composition can be polished to a very smooth surface and is very resistant to rusting.

There are two types of mixtures. One type is a homogeneous mixture. In this type of mixture, the substance is the same throughout. The second type of mixture is heterogeneous. A heterogeneous mixture is a mixture containing regions of different properties. An example of heterogeneous mixture is obtained when we take a bucket of water and dump dirt in it. Some of the dirt stays suspended in the water while most of the dirt falls to the bottom, creating regions of different properties. We define pure substances as substances that have constant composition and because of this, pure substances can be broken down into small groups called compounds and elements. A compound is a substance with constant composition which can be broken down into its elements by chemical processes. Chemical processes are processes where the atoms in a substance are reorganized to form a new substance. After substances are broken down into their basic elements, elements can no longer be broken up. It is worth noting that physical processes, such as filtration, can be reversed, but chemical processes, such as combustion, can not be.

Very often one can tell from the appearance of a sample whether it is a mixture or a pure substance. For example, if river water is clouded with mud or silt particles, we know it is a mixture. If a layer of brown haze lies over a city, we say that the atmosphere is mixed with pollutants. However, the appearance of a sample is not always sufficient evidence by which we can judge its composition. A sample of matter may look pure without being so. For instance, air looks like a pure substance but it is actually a mixture of oxygen, nitrogen and other gases. Rubbing alcohol is a clear, colorless liquid that looks pure but is actually a mixture of isopropyl alcohol and water, both of which are clear and colorless liquids. As another example, we cannot look at a piece of metal and know whether it is pure iron or a mixture of iron with some other substance such as chromium or nickel.

A mixture can be separated into its components by differences in their physical properties. A mixture of salt and sand can be separated because salt dissolves in water but sand does not. If we add water to a salt-sand mixture, the salt will dissolve, leaving the sand at the bottom of the container. If we pour off the water, the sand will remain. If we boil off the water from the salt solution, we will get the salt by itself. We have separated the two components of the mixture by a difference in their ability to dissolve in water. Solubility is a physical property. Pure substances, on the other hand, can be separated into their components only by chemical changes. When added to water, the pure substance sodium bicarbonate does not separate into sodium, hydrogen, carbon and oxygen, although these components of sodium bicarbonate differ greatly in their solubilities in water.

Compounds differ from mixtures in that the elements in a compound are held together by chemical bonds and cannot be separated by differences in their physical properties. The components of a mixture are not joined together by any chemical bonds and they can be separated from one another by differences in their physical properties. We may note that mixtures can be separated into their components by differences in physical properties but compounds can be separated into their components only by chemical changes.

4. WHAT IS MEANT BY THE LAW OF CONSERVATION OF MASS AND WHY LAVOISIER HAS BEEN CALLED 'THE FATHER OF CHEMISTRY'?

The Law of Conservation of Mass is an established scientific law which states that matter can be changed from one form into another, mixtures can be separated or made and pure substances can be decomposed, but the total amount of mass remains constant. We can state this important law in another way. The total mass of the universe is constant within measurable limits; whenever matter undergoes a change, the total mass of the products of the transformation is, within measurable limits, the same as the total mass of the reactants.

The formulation of this law near the end of the eighteenth century marked the beginning of modern chemistry. By that time many elements had been isolated and identified, most notably oxygen, nitrogen and hydrogen. It was also known that, when a pure metal was heated in air, it became what was then called a calx (which we now call an oxide) and that this change was accompanied by an increase in mass. The reverse of this reaction was also known. Many calxes on heating lost mass and returned to pure metals. Many imaginative explanations of these mass changes were proposed. Antoine Lavoisier (1743–1794), a French nobleman who was later guillotined in the revolution for reasons beyond description at this stage, was an amateur chemist with a remarkably analytical mind. He considered the properties of metals and then carried out a series of experiments designed to allow him to measure not just the mass of the metal and the calx but also the mass of the air surrounding the reaction. Results of his experiments showed that the mass gained by the metal in forming the calx was equal to the mass lost by the surrounding air.

With this simple experiment, in which accurate measurement was critical to the correct interpretation of the results, Lavoisier established the Law of Conservation of Mass and chemistry became an exact science, based on careful measurement. For his pioneering work in the establishment of that law and his analytical approach to experimentation, Lavoisier has been called the father of modern chemistry. We may note that this step forward, like so many others in science, depended on technology, in this instance, on the development of an accurate and precise balance.

5. WHAT IS CORRELATION BETWEEN MATTER AND ENERGY?

A study of the properties of matter must include a study of energy. Energy is the ability or capacity to do work or to produce change. Forms of energy include heat, light, sound, electricity and chemical energy. Energy and work are measured in the same units: foot-pounds, joules, ergs, or some other, depending on the system of measurement being used. When a force acts on a body, the work performed (and the energy expended) is the product of the force and the distance over which it is exerted. Energy is a state function commonly defined as the capacity to do work. Since work is defined as the movement of an object through a distance, energy can also be described as the ability to move an object through a distance. It is the ability of a system to do work. Some forms of energy include kinetic (motion), potential (location), thermal (heat), gravitational and electrical. In chemistry, energy is required to form and break bonds of molecules. Common units of energy are the calorie and joule. According to the Law of Conservation of Energy, energy can neither be created nor destroyed, though it can be transformed from one form to another.

Potential energy is the capacity for doing work that a body possesses because of its position or condition. For example, a stone resting on the edge of a cliff has potential energy due to its position in the earth's gravitational field. If it falls, the force of gravity (which is equal to the stone's weight) will act on it until it strikes the ground; the stone's potential energy is equal to its weight times the distance it can fall. A charge in an electric field also has potential energy because of its position; a stretched spring has potential energy because of its condition. Chemical energy is a special kind of potential energy; it is the form of energy involved in chemical reactions. The chemical energy of a substance is due to the condition of the atoms of which it is made; it resides in the chemical bonds that join the atoms in compound substances.

Kinetic energy is energy a body possesses because it is in motion. The kinetic energy of a body with mass m moving at a velocity v is one half the product of the mass of the body and the square of its velocity, *i.e.*, $KE = 1/2mv^2$. Even when a body appears to be at rest, its atoms and molecules are in constant motion and thus have kinetic energy. The average kinetic energy of the atoms or molecules is measured by the temperature of the body. The difference between kinetic energy and potential energy and the conversion of one to the other, is demonstrated by the falling of a rock from a cliff, when its energy of position is changed to energy of motion. Another such example is provided in the movements of a simple pendulum. As the suspended body moves upward in its swing, its kinetic energy is continuously being changed into potential energy; the higher it goes the greater becomes the energy that it owes to its position. At the top of the swing the change from kinetic to potential energy is complete and in the course of the downward motion that follows the potential energy is in turn converted to kinetic energy.

6. WHAT IS MEANT BY 'CONVERSION AND CONSERVATION OF ENERGY'?

It is common for energy to be converted from one form to another; however, the law of conservation of energy, a fundamental law of physics, states that although energy can be changed in form it can be neither created nor destroyed. The theory of relativity shows, however, that mass and energy are equivalent and thus that one can be converted into the other. As a result, the law of conservation of energy includes both mass and energy. Many transformations of energy are of practical importance. Combustion of fuels results in the conversion of chemical energy into heat and light. In the electric storage battery chemical energy is converted to electrical energy and vice versa. In the photosynthesis of starch, green plants convert light energy from the sun into chemical energy. Hydroelectric facilities convert the kinetic energy of falling water into electrical energy, which can be conveniently carried by wires to its place of use. The force of a nuclear explosion results from the partial conversion of matter to energy.

One of the characteristics of energy is that one form of energy can be converted to another. When wood is burned, some of its potential energy is changed to radiant energy (heat and light). Some is changed to kinetic energy as the water and carbon dioxide formed move away from the burning log. Some remains as potential energy in the composition of the water and carbon dioxide produced by the burning. Throughout all these changes, the total amount of energy remains constant. All changes must obey the

Law of Conservation of Energy, which states that energy can neither be created nor destroyed. An alternative statement is that the total amount of energy in the universe remains constant.

The Law of Conservation of Mass and the Law of Conservation of Energy are interrelated principles. Mass can be changed into energy and energy into mass according to the equation: $E=mc^2$, where E = energy (rather energy change), m = mass change (in grams) and c = speed of light (3.00×10^8 m/sec, or 186,000 mi/sec). This relationship allows us to state the two laws as a single law, called the Law of Conservation of Mass/ Energy: Energy and mass may be interconverted, but together they are conserved. This law was first stated by Albert Einstein (1879–1955). In most changes, the amount of matter converted to energy is much too small to be detected by even the most sensitive apparatus and we can say "in this change both mass and energy are separately conserved." It is nevertheless important to be aware of this relationship between mass and energy, because nuclear energy is obtained by just such a conversion of mass to energy.

7. WHAT IS DIFFERENCE BETWEEN MASS AND WEIGHT?

The first set of words that might pose a problem to a beginner in science in general and chemistry in particular, is mass and weight. In today's society, mass and weight may be used interchangeably, but in Chemistry they are used for very different notions. Mass is the quantity of matter in a body regardless of its volume or of any forces acting on the body. Newton defined the mass of an object as the quantity of matter it possessed. A small rock, for example, has a mass, a fixed, unchanging quantity of matter. If we were to take that rock along with us on a trip to the moon, it would have the same quantity of matter (the same mass) that it had on Earth. Its weight, however, would be less on the moon. The term should not be confused with weight, which is the measure of the force of gravity acting on a body. An example of the difference between mass and weight is that a person on earth may have a mass of 220 kilograms but may weigh 100 pounds and on the moon that same person will have a mass of 220 kilograms and weight of 16.7 pounds. The reason why the weight changed was because the moon's gravitational pull is much less than earth's, but one must take note of the fact that the mass never changes no matter what the weight does.

Under ordinary conditions the mass of a body can be considered to be constant; its weight, however, is not constant, since the force of gravity varies from place to place. There are two ways of referring to mass, depending on the law of physics defining it: gravitational mass and inertial mass. The gravitational mass of a body may be determined by comparing the body on a beam balance with a set of standard masses; in this way the gravitational factor is eliminated. The inertial mass of a body is a measure of the body's resistance to acceleration by some external force. One body may have twice as much inertial mass as another body if it offers twice as much force in opposition to the same acceleration.

All evidence seems to indicate that the gravitational and inertial masses of a body are equal, as demanded by Einstein's equivalence principle of relativity; so that at the same location equal (inertial) masses have equal weights. Because the numerical value for the mass of a body is the same anywhere in the world, it is used as a basis of reference for many physical measurements, such as density and heat capacity. According to the special theory of relativity, mass is not strictly constant but increases with the speed according to the formula $m=m_0/$, where m_0 is the rest mass of the body, v is its speed and c is the speed of light in vacuum. This increase in mass, however, does not become appreciable until very great speeds are reached. The rest mass of a body is its mass at zero velocity. The special theory of relativity also leads to the Einstein mass-energy relation, $E=mc^2$, where E is the energy and m and c are the (relativistic) mass and the speed of light, respectively. Because of this equivalence of mass and energy, the law of conservation of energy was extended to include mass as a form of energy.

8. WHAT IS A CHEMICAL ELEMENT?

An element, in chemistry, is a substance that cannot be decomposed into simpler substances by chemical means. It is a substance made up of only one kind of atom (atoms having the same atomic number). The atomic number of an element is the number of protons found in the nucleus of each atom of that element; the number of protons in the nucleus equals the number of electrons. A substance such as a compound can be decomposed into its constituent elements by means of a chemical reaction, but no further simplification can be achieved. An element can, however, be decomposed into simpler substances, such as protons and neutrons or various combinations of them, by the methods of particle physics, *e.g.*, by bombardment of the nucleus.

The smallest unit of a chemical element that has the properties of that element is called an atom. Many elements (*e.g.*, helium) occur as single atoms. Other elements occur as molecules made up of more than one atom. Elements that ordinarily occur as diatomic molecules include hydrogen, nitrogen, oxygen and the halogens, but oxygen also occurs as a triatomic form called ozone. Phosphorus usually occurs as a tetratomic molecule and crystalline sulfur occurs as molecules containing eight atoms.

Regardless of how many atoms, the element is composed of, each atom has the same number of protons in its nucleus and this is different from the number in the nucleus of any other element. Thus this number, called the atomic number (at. no.), defines the element. For example, the element carbon consists of atoms all with atomic No. 6, *i.e.*, all having 6 protons in the nucleus; any atom with atomic no. 6 is a carbon atom. By 2011, 118 elements are known, ranging from hydrogen with an atomic No. of 1 to an as yet unnamed element (temporarily known as ununoctium) with an atomic No. of 118. The nuclei of most atoms also contain neutrons. The total number of protons and neutrons in the nucleus of an atom is called the mass number. For example, the mass number of a carbon atom with 6 protons and 6 neutrons in its nucleus is 12.

Although all atoms of an element have the same number of protons in their nuclei, they may not all have the same number of neutrons. Atoms of an element with the same atomic but different mass number make up an isotope of the element. All known elements have isotopes; some have more than others. Hydrogen, for example, has only 3 isotopes, while xenon has 16. Approximately 300 naturally occurring isotopes are known and more than 2,500 radioactive isotopes have been artificially produced. There are 13 isotopes of carbon, having from 2 to 14 neutrons in the nucleus and therefore mass numbers from 8 to 20.

Not all of the elements have stable isotopes. Some have only radioactive isotopes, which decay to form other isotopes, usually of other elements. In some cases all the isotopes of an element are very unstable and the element is therefore not found in nature. Only 94 or so of the elements are known to occur naturally on earth. Of these, 6 occur in minute amounts produced by the decay of other elements. These 6 extremely scarce elements and those that do not occur at all naturally were discovered when they were produced in the laboratory; they are often called the man-made, artificially produced, or synthetic elements.

Although some of the elements were discovered in prehistoric times, they were not recognized as elements. Only 12 elements were known before 1700 and only about twice that many by 1800, but by 1900 over 80 elements had been identified. In 1919 Ernest Rutherford found that hydrogen was given off when nitrogen was bombarded with alpha particles. This first transmutation encouraged further study of nuclear reactions and eventually led to the discovery in 1937 of technetium, the first synthetic element. Neptunium (atomic number 93) was the first transuranium element to be synthesized (1940). Its discovery prompted the search that led to the discovery of other trans-uranium elements.

9. WHAT ARE CHARACTERISTICS OF ELEMENTS?

Properties of an element are sometimes classed as either chemical or physical. Chemical properties are usually observed in the course of a chemical reaction, while physical properties are observed by examining a sample of the pure element. The chemical properties of an element are due to the distribution of electrons around the atom's nucleus, particularly the outer, or valence, electrons; it is these electrons that are involved in chemical reactions. A chemical reaction does not affect the atomic nucleus; the atomic number therefore remains unchanged in a chemical reaction. Some properties of an element can be observed only in a collection of atoms or molecules of the element. These properties include color, density, melting point, boiling point and thermal and electrical conductivity. While some of these properties are due chiefly to the electronic structure of the element, others are more closely related to properties of the nucleus, *e.g.*, mass number. The elements are sometimes grouped according to their properties. One major classification of the elements is as metals, nonmetals and metalloids. Elements with very similar chemical properties are often referred to as families; some families of elements include the halogens, the inert gases and the alkali metals. In the periodic table the elements are arranged in order of increasing atomic number in such a way that the elements in any column have similar properties. A transuranium (beyond uranium) element is any of the chemical elements with atomic numbers higher than 92, which is the atomic number of uranium. Ever since the eighteenth century when chemists began to recognize certain substances as chemical elements, uranium had been the element with the highest atomic weight; it had the heaviest atoms of all the elements that could be found on Earth.

10. WHAT IS MEANT BY 'FAMILIES OF ELEMENTS'?

A family of chemical elements usually consists of elements that are in the same group (the same column) on the periodic table. The term is also applied to certain closely related elements within the same period (row). Just as the individual members in a human family are all different but have some common characteristics. Similarly the elements in a chemical family have certain common properties. Elements are identified by the nuclei of the atoms of which they are made. For example, an atom having six protons in its nucleus is carbon and one having 26 protons is iron. There are about 90 naturally occurring elements, with uranium (92 protons) being the heaviest (heavier nuclei have been produced in reactors on Earth). Nuclei also contain certain neutrons, usually in numbers greater than the number of protons.

Each element is assigned an official symbol by the International Union of Pure and Applied Chemistry (IUPAC). For example, the symbol for carbon is C which comes from carbo and the symbol for silver is Ag which comes from its Latin name argentum. There are several ways of designating an isotope. One designation consists of the name or symbol of the element followed by a hyphen and the mass number of the isotope; thus the isotope of carbon with mass number 12 can be designated carbon-12 or C-12. The mass number is often written as a superscript, *e.g.*, ^{12}C; sometimes the atomic number is written as a subscript preceding the symbol, *e.g.*, $_6{}^{12}C$. The IUPAC rules for nomenclature of inorganic chemistry state that the subscript atomic number and superscript mass number should both precede the symbol, *e.g.*, $^{12}_{6}C$.

Many isotopes were given special names and symbols when they were first discovered in natural radioactive decay series (*e.g.*, uranium-235 was called actinouranium and represented by the symbol AcU). This practice is discouraged in the modern nomenclature except in the case of hydrogen. The isotopes hydrogen-2 and hydrogen-3 are usually called deuterium and tritium, respectively. Hydrogen-1, the most abundant isotope, has the name protium but is usually simply called hydrogen. Newly discovered elements that have been synthesized by one laboratory and not yet confirmed by a second are given a provisional name based on Greek and Latin roots; when the discovery is confirmed, the laboratory that first made it may suggest a name for the element.

11. HOW ELEMENTS HAVE BEEN NAMED AND ASSIGNED SYMBOLS?

We know that every element has a name and many of these names such as gold, silver, copper; chlorine, platinum, carbon, oxygen and nitrogen are already familiar to us.

Many names refer to a property of the element, *e.g.* the Latin name for gold is aurum, meaning "shining dawn" and the Latin name for mercury, hydrargyrum, means "liquid silver." The practice of naming an element after one of its properties continues. Cesium was discovered in 1860 by the German chemist Bunsen (the inventor of the Bunsen burner). Because this element imparts a blue color to a flame, Bunsen named it cesium from the Latin word caesius, meaning "sky blue." Many elements have been named for people. Curium is named for Marie Curie (1867–1934), a pioneer in the study of radioactivity. Marie Curie, a French scientist of Polish origin, was awarded the Nobel Prize in Physics in 1903 for her studies of radioactivity. She was also awarded the Nobel Prize in Chemistry in 1911 for her discovery of the elements polonium which she preferred to name after her native country Poland and radium (Latin, radius, "ray").

Some elements are named for places. The small town of Ytterby in Sweden has four elements, namely terbium, yttrium, erbium and ytterbium named for it. Californium is another example of an element named for the place where it was first observed. This element does not occur in nature. It was first produced in 1950 in the Radiation Laboratory at the University of California, Berkeley, by a team of scientists headed by Glenn Seaborg who was also the first to identify curium at the metallurgical laboratory at the University of Chicago (now Argonne National Laboratory) in 1944. Seaborg became a Nobel laureate in 1951 for his pioneering work in the preparation of other unknown elements.

Every element has a symbol too, one or two English alphabets that represent the element much as our initials represent us. The symbol of an element also represents one atom of that element. For 14 of the elements, the symbol consists of one letter. With the possible exceptions of yttrium (Y) and vanadium (V), we are familiar with the names of all elements having one-letter symbols. For 12 of these elements, the symbol is the first letter of the name. Potassium was discovered in 1807 and named for potash, the substance from which potassium was first isolated. Potassium's symbol, K, comes from kalium, the Latin word for potash. Tungsten, discovered in 1783, has the symbol W, for wolframite, the mineral from which it was first isolated. Most other elements have two-letter symbols. In these two-letter symbols, the first letter is always capitalized and the second is always lowercased. Eleven elements have names (and symbols) beginning with the letter C. One of these, carbon, has a one-letter symbol, C. The other ten namely, Cadmium, Calcium, Californium, Cerium, Cesium, Chlorine, Chromium, Cobalt, Copper and Curium have two-letter symbols.

While we study chemistry, we often need a list of the elements. This list must include the symbols, the atomic numbers and the atomic weights of the elements. Each element has a number between 1 and 118 called its atomic number. This number is as unique to the element as its name or symbol. The second list, called the periodic table, arranges the elements in order of increasing atomic number in rows of varying length. Throughout the chemistry texts we refer to the periodic table, because it contains an amazing amount of information. For now we need only be aware that elements in the same column have similar properties and that the heavy stair-step line that crosses the table diagonally from boron (B) to astatine (At) separates the metallic elements from the nonmetallic elements. The characteristic properties of a metal may be summarized as:

- It is shiny and lustrous.
- It conducts heat and electricity.
- It is ductile and malleable; that is, it can be drawn into a wire and can be hammered into a thin sheet.
- It is a solid at 20°C. Mercury is the only exception to this rule as it is a liquid at room temperature. Two other metals, gallium and cesium, have melting points close to room temperature.

Nonmetals vary more in their properties than do metals. Some of them may even have one or more of the metallic properties listed. Some nonmetals are gaseous; chlorine and nitrogen are gaseous nonmetals. At 20°C one nonmetal, bromine is a liquid and others such as carbon, sulfur and phosphorus are solids.

12. HOW ARE ELEMENTS DISTRIBUTED NATURALLY?

The known elements are not equally distributed throughout the world. Only 92 or so (the number varies between 88 and 92) are found in the Earth's crust, oceans, or atmosphere; the others have been produced in laboratories. Traces of some but not all of these elements have been found on Earth or in the stars. The search for the others continues. One might read of its success or of the isolation of new elements as he/she studies this text. Distribution of Elements in the Earth's crust, oceans and atmosphere may be summarized and tabulated as under:

Table 2.1: Distribution of elements as percent of total mass

Element	*Percent of total mass*	*Element*	*Percent of total mass*
Oxygen	49.2	Chlorine	0.19
Silicon	25.7	Phosphorous	0.11
Aluminum	7.50	Manganese	0.09
Iron	4.71	Manganese	0.09
Calcium	3.39	Sulfur	0.06
Sodium	2.63	Barium	0.04
Potassium	2.40	Nitrogen	0.04
Magnesium	1.93	Fluorine	0.03
Hydrogen	0.87	All others	0.49
Titanium	0.58		

The above table (Table 2.1) lists the 18 elements that are most abundant in the Earth's crust, oceans and atmosphere, along with their relative percentages of the Earth's total mass. One of the most striking points about this list is the remarkably uneven distribution of the elements. Oxygen is by far the most abundant element. It makes up 21% of the volume of the atmosphere and 89% of the mass of water. Oxygen in air, water and elsewhere constitutes 49.2% of the mass of the Earth's crust, oceans and atmosphere. Silicon is the Earth's second most abundant element (25.7% by mass). Silicon is not found free in nature but occurs in combination with oxygen, mostly as silicon dioxide (SiO_2), in sand, quartz, rock crystal, amethyst, agate, flint, jasper and opal, as well as in various silicate minerals such as granite, asbestos, clay and mica.

Aluminum is the most abundant metal in the Earth's crust (7.5%). It is always found combined in nature. Most of the aluminum used today is obtained by processing bauxite, an ore that is rich in aluminum oxide. These three elements (oxygen, silicon and aluminum) plus iron, calcium, sodium, potassium and magnesium make up more than 97% of the mass of the Earth's crust. Another surprising feature of the distribution of elements is that several of the metals that are most important to our civilization are among the rarest; these metals include lead, tin, copper, gold, mercury, silver and zinc.

The distribution of elements in the cosmos is quite different from that on Earth. According to our present knowledge, hydrogen is by far the most abundant element in the universe, accounting for as much as 75% of its mass. Helium and hydrogen together make up almost 100% of the mass of the universe. The following table (Table 2.2) lists the biologically important elements, those found in a normal, healthy body. The first four of these elements: oxygen, carbon, hydrogen and nitrogen, make up about 96% of total body weight. The other elements listed, although present in much smaller amounts, are nonetheless necessary for good health.

Table 2.2: Biologically important elements (amounts given per 70-kg body weight)

Major elements	*Approximate amount (kg)*	*Major elements*	*Approximate amount (kg)*
Oxygen	45.5	Carbon	12.6
Hydrogen	7.0	Nitrogen	2.1
Calcium	1.0	Phosphorous	0.70
Magnesium	0.35	Potassium	0.24
Sulfur	0.18	Sodium	0.10
Chlorine	0.10	Iron	0.003
Zinc	0.002		

***Elements present in less than 1-mg amounts listed alphabetically*:** Arsenic, Chromium, Cobalt, Copper, Fluorine, Iodine, Manganese, Molybdenum, Nickel, Selenium, Silicon and Vanadium.

***The distribution of elements (by mass) in the human body is as*:** Oxygen–65%, Carbon–18%, Hydrogen–10%, Nitrogen–03% and Calcium–025, Others–02%.

13. HOW ELEMENTS OCCUR IN NATURE?

Elements occur in nature as single atoms or as groups of atoms chemically bonded together. Groups of atoms bonded together chemically are called molecules or formula units. Molecules may contain atoms of a single element, or they may contain atoms of different elements (in which case the molecule is of a compound.) Just as an atom is the smallest unit of an element, a molecule is the smallest unit of a compound that is, the smallest unit having the chemical identity of that compound. Let us consider how the elements might be categorized by the way they are found in the universe.

The Noble Gases

Only a few elements are found as single, uncombined atoms. Along the extreme right-hand column of the periodic table of elements is a group of elements known as the noble gases: helium, neon, argon, krypton, xenon and radon. Also known as the rare gases, they once were called inert gases, because scientists believed them incapable of reacting with other elements. Rare though they are, these gases are a part of everyday life, as evidenced by the helium in balloons, the neon in signs and the harmful radon in some homes and offices. Under normal conditions, all of these elements are gases and collectively, they are known as the noble gases. They are also called monatomic gases, meaning that they exist uncombined as single atoms. The molecular formula for each of the noble gases is simply its symbol. When the formula of helium is required, the symbol 'He' is used wherein the subscript 1 is understood and not mentioned.

Metals

Almost three-quarters of all elements are metals, such as gold and silver. There are also some elements we may not think of as metals, such as the calcium in our bones and the sodium in table salt (sodium chloride). Metals are defined by their metallic properties, such as high melting points. Mixtures of metals are called alloys. Solder is an alloy that is used to join metals in plumbing and electrical wiring. It is mainly tin with lead or silver. Pure metals are treated as though they existed as single, uncombined atoms even though a sample of pure metal is an aggregate of billions of atoms. Thus, when the formula of copper is required, its symbol, Cu, is used to mean one atom of copper. Metals are classified according to their position in the periodic table. Each group has a set of properties that make its metals useful for different purposes.

Alkali metals: lithium, sodium, potassium, rubidium, cesium and francium form Group I of the periodic table. They are extremely reactive metals. They react strongly with water to form strong alkalis.

Alkaline-earth metals make up Group II of the periodic table. They combine with many elements in the Earth's crust. Their oxides react with water to form alkalis.

Transition metals' groups include copper, silver, gold and many others. They are hard and shiny, have high melting points and are good conductors of heat and electricity.

There are also other metals called poor metals, these metals are fairly soft and melt easily. They include tin and aluminium and are often used in alloys. Bronze is an alloy of tin and copper.

Nonmetals

The majority of elements on the periodic table are metals: solids (along with one liquid, mercury) which are lustrous or shiny in appearance. Nonmetals, as their name implies, are elements that display properties quite different from those of metals. Some nonmetals exist, under normal conditions of temperature and pressure, as molecules containing two, four, or eight atoms. Some of the nonmetals that occur as diatomic (two-atom) molecules include O_2 and N_2 as the formula for oxygen and nitrogen respectively. Among the nonmetals, sulfur exists as S_8 and phosphorus is found as P_4. For other nonmetals a monatomic formula is used, for example, 'As' for arsenic and Se for selenium.

14. WHAT ARE ATOMS?

Atom [Greek = uncuttable (indivisible)], is the basic unit of matter; more properly, the smallest unit of a chemical element having the properties of that element. An atom consists of a central, positively charged core, the nucleus and negatively charged particles called electrons that are found in orbitals around the nucleus. Almost the entire mass of the atom is concentrated in the nucleus, which occupies only a tiny fraction of the atom's volume. The nucleus of an atom consists of neutrons and protons, the neutron being an uncharged particle and the proton a positively charged one. Their masses are almost equal. Atoms containing the same number of protons but different numbers of neutrons represent different forms, or isotopes, of the same element.

Surrounding the nucleus of an atom are its electrons; for a neutral atom, the number of electrons is equal to the atomic number. The outermost electrons of an atom determine its chemical and electrical properties. An atom may combine chemically with another atom in various ways, either by giving up or receiving electrons, thus setting up an electrical attraction between the atoms, or by sharing one or more pairs of electrons. Because metals have few outermost electrons and tend to give them up easily, they are good conductors of electricity or heat.

The electrons are often described as revolving about the nucleus as the planets of solar system revolve about the sun. This picture, however, is misleading. The quantum theory has shown that all particles in motion also have certain wave properties. For a particle of the size of an electron, these properties are of considerable importance. As a result the electrons in an atom cannot be pictured as localized in space; rather they should be viewed as smeared out over the entire orbit so that they form a cloud of charge. The electron clouds around the nucleus represent regions in which the electrons are most likely to be found. The shapes of these clouds can be very complex, in marked contrast to the simple elliptical orbits of planets. Surprisingly, the sizes of all atoms are comparable, in spite of the large differences in the number of electrons they contain.

15. WHAT IS THE BASIS OF MODERN ATOMIC THEORY?

By the end of the eighteenth century, experimenters had well established that each pure substance had its own characteristic set of properties such as density, specific heat, melting point, boiling point etc. Also established was the fact that certain quantitative relationships, such as the Law of Conservation of Mass, governed all chemical changes. But there was still no understanding of the nature of matter itself. For example, it was not clear whether matter was continuous, like a ribbon from which varying amounts could be snipped, or was it granular, like a string of beads from which only whole units or groups of units could be removed? Whereas some scientists believed strongly in the continuity of matter, others believed equally strongly in granular matter. However, both views were based solely on speculations. In 1803, an English humble and poor schoolmaster, named John Dalton (1766–1844), summarized and extended the then-current theory of matter. The postulates of his theory, changed only slightly from their original statement, form the basis of modern atomic theory. Today, we express these four postulates as under:

1. Matter is made up of tiny particles called atoms. (A typical atom has a mass of approximately 10^{-23}g and a radius of approximately 10^{-10}m.)
2. Over 100 different kinds of atoms are known; each kind is an element. All the atoms of a particular element are alike chemically but can vary slightly in mass and other physical properties. Atoms of different elements have different masses.
3. Atoms of different elements combine in small, whole-number ratios to form compounds. For example, hydrogen and oxygen atoms combine in a ratio of 2:1 to form the compound water, H_2O. Carbon and oxygen atoms combine in a ratio of 1:2 to form the compound carbon dioxide, CO_2. Iron and oxygen atoms combine in a ratio of 2:3 to form the familiar substance rust, Fe_2O_3.
4. The same atoms can combine in different whole-number ratios to form different compounds. As just noted, hydrogen and oxygen atoms combined in a 2:1 ratio to form water; but when they combine in a 1:1 ratio they form hydrogen peroxide, H_2O_2. Carbon and oxygen atoms combined in a 1:2 ratio form carbon dioxide and when combined in a 1:1 ratio, they form carbon monoxide, CO.

16. WHY DO WE USE THE TERM 'ATOMIC MASS UNIT (AMU)'?

Because the actual masses of atoms and subatomic particles are very very small, we often describe their masses by comparison rather than in SI units, hence the term relative mass is used. If a proton is assigned a mass of 1.007, then a neutron will have a relative mass of 1.008 and an electron's relative mass will be 5.45×10^{-4}. When talking about relative masses, we use the term atomic mass unit (amu). Using this unit, a proton has a mass of 1.007 amu, a neutron a mass of 1.008 amu and an electron's mass will be 5.45×10^{-4} amu. Charges, too, are given relative to one another. If a proton has a charge of +1, then an electron has a charge of –1. It should be kept in mind that the negative charge of an electron is in comparison to the assigned positive charge of a proton.

17. WHY ABSOLUTE VALUES OF ATOMIC MASSES ARE LESS CERTAIN?

Atomic mass is the mass of a single atom, usually expressed in atomic mass units (amu). Most of the mass of an atom is concentrated in the protons and neutrons contained in the nucleus. Each proton or neutron weighs about 1 amu and thus the atomic mass is always very close to the mass number (total number of protons and neutrons in the nucleus). Atoms of an isotope of an element all have the same atomic mass. Atomic masses are usually determined by mass spectrograph. Though they have been determined with great relative accuracy, their absolute value is less certain.

18. WHAT DOES ATOMIC WEIGHT OF AN ELEMENT REFLECT?

The atomic weight (or atomic mass) of an element is the average relative mass of the naturally occurring atoms of that element. The atomic weight of an element is based on the variety of naturally occurring isotopes of that element and the relative abundance of each. A collection of naturally occurring carbon atoms contains 98.89% carbon-12 atoms and 1.11% carbon-13 atoms, along with a trace percentage of carbon-14 atoms. The atomic weight of carbon (12.01) reflects the relative abundance of these three isotopes. The atomic weight of chlorine (35.45) reflects the fact that 75.53% of naturally occurring chlorine is chlorine-35 and 24.47% is chlorine-37. Atomic weight is the average mass of the atoms of an element taking into account the masses of the various isotopes and their relative abundance. For example, oxygen has an atomic weight of 15.9994. This means that the average of the atoms' masses is 15.9994 atomic mass units (amu). The atomic weight is a weighted average of all of the atomic masses of the naturally occurring isotopes. This means that, instead of each mass counting equally toward the average, the mass is scaled according to each isotope's abundance. To find the atomic weight multiply each isotope's atomic mass by its percent abundance in decimal form then find the sum of these weighted masses.

19. WHY DOES NO ELEMENT HAS AN ATOMIC WEIGHT LESS THAN 1 AND WHAT IS DIFFERENCE BETWEEN ATOMIC WEIGHT AND MASS NUMBER?

The atomic weight of some elements is given as a whole number enclosed in parentheses. These elements are unstable; that is, their nuclei decompose radioactively. The number in parentheses is the mass number of the most stable or best-known isotope of that element. Atomic weights are measured in atomic mass units. One atomic mass unit is defined as 1/12 the mass of an atom of carbon-12. With this reference standard, no element has an atomic weight less than one. The approximate atomic weight of an element can be calculated if the relative abundance of its isotopes is known. If the identity of the naturally occurring isotopes and the atomic weight of an element are known, it is possible to estimate which of the isotopes is most abundant.

Oxygen has an atomic weight of 15.9994. This means that the average of the atoms' masses is 15.9994 atomic mass units (amu). A proton has a mass of slightly more than one amu as does a neutron. An electron's mass is so small that it would close to two thousand electrons to be as massive as a proton. Most of the naturally occurring atoms of oxygen have eight protons, eight neutrons and eight electrons giving a mass of close to 16 amu. Many students assume that the mass number (always an integer) is a rounded form of the atomic weight (never an integer). The atomic weight is an average while the mass number is a particle count. Though the atomic weight can be found on the periodic table, it is never written as part of an atomic symbol.

20. HOW DO WE CALCULATE ATOMIC WEIGHTS?

Atomic weight is the average mass of the atoms of an element taking into account the masses of the various isotopes and their relative abundance. For example, oxygen has an atomic weight of 15.9994. This means that the average of the atoms' masses is 15.9994 atomic mass units (amu). The atomic weight is a weighted average of all of the atomic masses of the naturally occurring isotopes. This means that, instead of each mass counting equally toward the average, the mass is scaled according to each isotope's abundance. To find the atomic weight multiply each isotope's atomic mass by its percent abundance in decimal form then find the sum of these weighted masses. For example, Chromium has four naturally occurring isotopes: ^{50}Cr, ^{52}Cr, ^{53}Cr and ^{54}Cr. The atomic masses and abundance data are:

Table 2.3: Isotopic abundances of chromium

Isotope	*Mass*	*Abundance*
^{50}Cr	49.9460464 amu	4.345 %
^{52}Cr	51.9405098 amu	83.789 %
^{53}Cr	52.9406513 amu	9.501 %
^{54}Cr	53.9388825 amu	2.365 %

21. WHAT ARE CHARACTERISTICS OF MOLECULES?

A molecule is the smallest particle of a compound that has all the chemical properties of that compound. A single atom is usually not referred to as a molecule and ionic compounds such as common salt are not made up of molecules. Unlike ions, molecules carry no electrical charge. Molecules are made up of two or more atoms, either of the same element or of two or more different elements, joined by one or more covalent chemical bonds. According to the kinetic-molecular theory, the molecules of a substance are in constant motion. The state (solid, liquid, or gaseous) in which matter appears depends on the speed and separation of the molecules in the matter. Substances differ according to the structure and composition of their molecules. A molecular compound is represented by its molecular formula; for example, water is represented by the formula H_2O. A more complex structural formula is sometimes used to show the arrangement of atoms in the molecule. The terms atom and molecule were used interchangeably until the early 19th century. Initial experimental work with gases led to what is essentially the modern distinction between them. J.A.C. Charles and R. Boyle had shown that all gases exhibit the same relationship between a change in temperature or pressure and the corresponding change in volume. J.L. Gay-Lussac had shown that gases always combine in simple whole-number volume proportions and had rediscovered the earlier findings of Charles, which had not been published.

22. WHAT IS MEANT BY MOLECULAR WEIGHT AND HOW IS IT DETERMINED EXPERIMENTALLY?

Molecular weight is the weight of a molecule of a substance expressed in atomic mass units (amu). The molecular weight may be calculated from the molecular formula of the substance; it is the sum of the atomic weights of the atoms making up the molecule. For example, water has the molecular formula H_2O, indicating that there are two atoms of hydrogen and one atom of oxygen in a molecule of water. Rounded to three decimal places, the atomic weight of hydrogen is 1.008 amu and that of oxygen is 15.999 amu. The molecular weight of water is thus $(2 \times 1.008) + (1 \times 15.999) = 2.016 + 15.999 = 18.015$ amu. Since atomic weights are average values, molecular weights are also average values. On the average, a molecule of ordinary water weighs 18.015 amu. Both hydrogen and oxygen are made up of several isotopes. One isotope of hydrogen is deuterium, or heavy hydrogen. Atoms of deuterium are about twice as massive as the average for all hydrogen atoms in ordinary water. Therefore water that contains only atoms of deuterium, called heavy water, has a higher molecular weight than ordinary water. Some substances, especially ionic compounds such as common salt, are not made up of molecules and thus have neither a molecular formula nor a molecular weight.

Molecular weights of substances may be determined experimentally in various ways, the method employed usually depending on the state (solid, liquid, or gas) of the substance. Methods for determining the molecular weights of gaseous substances are based on Avogadro's law, which states that under given conditions of temperature and pressure a given volume of any gas contains a specific number of molecules of the gas; thus a comparison of the weights of equal volumes of different gases under the same conditions of temperature and pressure is equivalent to a direct comparison of the weights of molecules of the gases. The molecular weights of substances that are not normally gaseous and do not evaporate without decomposition are sometimes determined from their effects on the melting point, boiling point, vapor pressure, or osmotic pressure of some solvent. However, if the substance ionizes or does not completely separate into molecules, the molecular weight so determined will be erroneous. Highly accurate molecular weights are sometimes determined by using the mass spectrograph.

Some substances, *e.g.*, proteins, viruses and certain synthetic polymers, have very high molecular weights. Their molecular weights may be determined by measurement of sedimentation rate in an ultracentrifuge, by light-scattering photometry, or by other methods. The methods may give different results, since usually the molecules of a substance such as a polymer do not all have exactly the same molecular weight. These methods determine an average molecular weight for the molecules in the sample. The number-average molecular weight determined by the ultracentrifuge method gives a value that is equal to the weight of the sample divided by the number of molecules in the sample. This number-average molecular weight can also be determined by other methods based on measurement of colligative properties. The light-scattering method determines what is called the weight-average molecular weight. Although this may be the same value as the number-average molecular weight if all the molecules have nearly the same weight, it will be higher if some of the molecules are heavier than others.

23. WHAT IS SIGNIFICANCE OF FORMULA WEIGHT AND HOW SHOULD WE CALCULATE IT?

Formula weight, in chemistry, is a quantity computed by multiplying the atomic weight (in atomic mass units) of each element in a formula by the number of atoms of that element present in the formula and then adding all of these products together. For example, the formula weight of water (H_2O) is two times the atomic weight of hydrogen plus one times the atomic weight of oxygen. Numerically, this is $(2\times1.00797) + (1 \times 15.9994) = 2.01594 + 15.9994 = 18.01534$. If the formula used in computing the formula weight is the molecular formula, the formula weight computed is the molecular weight. The percentage by weight of any atom or group of atoms in a compound can be computed by dividing the total weight of the atom (or group of atoms) in the formula by the formula weight and multiplying by 100. For example, the weight percentage of hydrogen in water is determined by taking two times the atomic weight of hydrogen, dividing it by the formula weight of water and multiplying by 100. Numerically, this is $100\times (2 \times 1.00797)/18.01534 = 11.19\%$ hydrogen in water by weight. Formula weights are especially useful in determining the relative weights of reagents and products in a chemical reaction. For example, it is known that two molecules of hydrogen gas, H_2, react with one molecule of oxygen gas, O_2, to form two molecules of water, H_2O. This reaction may be represented by the chemical equation: $2H_2 + O_2 \rightarrow 2H_2O$. The formula weight of hydrogen gas is 2.01594 and that of oxygen gas 31.9998 and that of water is 18.01534. Our chemical equation is numerically equivalent to $2 \times 2.01594 + 31.9998 = 2 \times 18.01534$ or $4.03188 + 31.9998 = 36.03068$ if the formula weight of each reactant is substituted for the formula of that reactant. From this equation we know, for example, that 4.03188 grams of hydrogen gas will react with 31.9998 grams of oxygen gas to yield 36.03068 grams of water. The relative proportions by weight of these reactants are the same in any reaction of hydrogen and oxygen to form water. These relative weights computed from the chemical equation are sometimes called equation weights.

24. WHY MOLE HAS BEEN INCLUDED IN THE INTERNATIONAL SYSTEM OF UNITS?

In fact Mole, in chemistry, is a quantity of particles of any type equal to Avogadro's number, or 6.02×10^{23} particles. One gram-molecular weight of any molecular substance contains exactly one mole of molecules. The term mole is often used in place of gram-molecular weight; *e.g.*, one speaks of 18 grams of water as one mole of water rather than as one gram-molecular weight of water. Mole, the standard unit for counting numbers of particles, is needed in chemistry because atoms and molecules react with one another particle by particle. The amount of a chemical reaction that is how much of the chemicals are used up or produced in a reaction, is determined by the numbers of particles that are reacting. The mole has been included in the International System of Units (SI). Weighing the chemicals wouldn't tell us anything very meaningful unless we knew how to count the number of species in the given amount.

25. WHAT IS CORRELATION BETWEEN AVOGADRO'S NUMBER AND THE MOLE?

Generally we talk about moles of atoms and know that one mole of a particular element has a mass in grams equal to the element's atomic weight. We can also have moles of compounds or ions. One mole of a compound or an ion will have a mass in grams equal to the formula weight of that compound or ion. A mole is a collection of atoms with a mass equal to the atomic weight in grams. The number of atoms in a mole is: 6.02×10^{23}. We can calculate the formula weight of sulfuric acid which comes to be 98.09 amu, thus it follows that one mole of sulfuric acid has a mass of 98.09 g. The formula of a compound also tells how many moles of a particular element are contained in one mole of the compound. For example, one mole of sulfuric acid contains: 2 mol hydrogen atoms, weighing 2(1.008 g) or 2.016 g, 1 mol sulfur atoms, weighing 1(32.07 g) or 32.07 g and 4 mol oxygen atoms, weighing 4(16.00 g) or 64.00 g. These individual masses add up to our calculated formula weight for the acid. This relationship between mass and moles of a compound is often used as a conversion factor in solving problems.

This works because one amu is 1.66×10^{-24} gram, so it takes 6.02×10^{23} of them to make one gram. This is one of the most useful concepts in chemistry. The atomic weight of lithium is 6.941amu. This means that a sample with a mass of 6.941 grams has 6.02×10^{23} atoms of lithium in it. Note that this works for molecules and ions, too. The number 6.02×10^{23} is known as Avogadro's number in honor of the chemist who first proposed a law relating the volume of a gas to the number of gaseous particles it contains. Note that the law works for mass too, but it was easier for chemists in the 17th century to measure the volume of a gas than to weigh it.

26. WHY AVOGADRO'S NUMBER IS GENERALLY SHOWN TO THREE SIGNIFICANT FIGURES?

We know that a mole is also the amount that contains Avogadro's Number of species. Avogadro's number is generally shown to three significant figures, which is the degree of accuracy usually required in calculations. Actually, the number of items in a mole has been determined to six or more significant figures, the exact number depending on the method by which the number was determined. Equally important is the fact that one mole of a substance has a mass in grams numerically equal to the formula weight of that substance. Thus, one mole of an element has a mass in grams equal to the atomic weight of that element and contains 6.023×10^{23} atoms of the element. For those elements that do not occur as single atoms, that is, the diatomic gases inclusive of polyatomic elements sulfur and phosphorus, it is important to be certain that we specify what we are talking about. One mole of atoms of oxygen has a mass of 16 g, as 16 is the atomic weight of oxygen and contains 6.02×10^{23} atoms of oxygen. One mole of oxygen gas, which has the formula O_2, has a mass of 32 g and contains 6.02×10^{23} molecules of oxygen but 12.04×10^{23} $(2 \times 6.02 \times 10^{23})$ atoms, because each molecule of oxygen contains two oxygen atoms. These definitions allow a new definition of atomic weight: The atomic weight of an element is the mass in grams of one mole of naturally occurring atoms of that element. Formula weight = grams/ mole. Using these relationships, we can calculate the number of atoms in a given mass of an element or the mass of a given number of atoms.

27. WHAT ARE COMPOUNDS?

Although many elements can occur in the uncombined state, all elements except some of the noble gases are also found combined with other elements in compounds. We define a compound as a substance that can be decomposed by ordinary chemical means. A compound is a chemical species in which atoms combine in such a way that the compound always has the same composition, unless it is chemically altered in some way. A compound can also be defined as a pure substance that contains two or more elements. The composition of a compound is expressed by a formula that uses the symbols of all the elements in the compound. Each symbol is followed by a subscript, a number that shows how many atoms of the element occur in one molecule (the simplest unit) of the compound; the subscript 1 is not shown. Water is a compound with the formula H_2O, meaning that one molecule (or formula unit) of water contains two hydrogen atoms and one oxygen atom.

The compound sodium hydrogen carbonate has the formula $NaHCO_3$, meaning that a single formula unit of this compound contains one atom of sodium, one atom of hydrogen, one atom of carbon and three atoms of oxygen. Notice that the symbol of the metal in sodium hydrogen carbonate is written first, followed by the nonmetals and that of oxygen is written last. This order is customary. Sometimes a formula will contain a group of symbols enclosed in parentheses as, for example, $Cu(NO_3)_2$. The parentheses imply that the group of atoms they enclose act as a single unit. The subscript following the parenthesis means that the group is taken two times for each copper atom. The properties of a compound are quite unlike those of the elements from which it is formed. This fact is apparent if we compare the properties of carbon dioxide, CO_2 (a colorless gas used in fire extinguishers), with those of carbon (a black, combustible solid) and oxygen (a colorless gas necessary for combustion). Chemistry is a subdivision of science which specializes in the composition, structure and properties of substances and the transformations in which these substances go through.

28. WHAT ARE GENERAL CHARACTERISTICS OF COMPOUNDS?

A compound is a chemical species in which the constituent atoms combine in such a way that the compound concerned always has the same composition, unless it is chemically altered in some way. Atoms of the elements make up compounds and although there are only about 90 elements that occur in nature, there are literally many millions of compounds. When atoms of one element bond to atoms of another, they form chemical species quite different from either element. Sugar, for instance, is made up of carbon, the material in coal and graphite, along with two gases, hydrogen and oxygen. None of these is sweet, yet when they are brought together in just the right way and correct ratios, they make the compound that sweetens everything from breakfast cereals to colas and tasty sweets. A compound is not the same as a mixture, which has a variable composition, but until chemists understood the atomic and molecular substructure of compounds, the distinction was not always clear.

A compound cannot be understood purely in terms of its constituent atoms unless an awareness of the structure is also made available. Likewise, it is important to know just how to name a compound, using a uniform terminology, since there are far too many substances in the world to give each an individual name. Compounds have constant compositions and a unique set of properties. The composition of a compound is represented by its formula, which lists the symbols of the elements it contains, with each symbol followed by a subscript which tells how many atoms of that element are contained in the simplest unit of the compound. Some compounds exist as molecules. These molecular compounds usually contain only nonmetals.

29. WHAT ARE CHARACTERISTICS OF MOLECULAR COMPOUNDS?

Molecular compounds are composed of molecules in which atoms share electrons via bonds. The chemistry of molecular compounds is varied and complicated, far more so than ionic compounds. For this reason together with some tradition and a pinch of inertia, most of the content of a general chemistry course involves ionic compounds. Most molecular compounds contain two or more non-metals (though it is not true with the ammonium ion). While there are exceptions, the absence of metals is good signal for the beginning chemistry student to note that the compound concerned is a molecular one. While many molecular compounds contain hydrogen, if the formula of a compound has hydrogen written first, one should assume that it is an acid rather than a molecular compound.

The only type of molecular compounds a beginning student is expected to name, are the binary molecular compounds. These contain two non-metallic elements (but not carbon with hydrogen which are hydrocarbons and have their own special nomenclature). Many binary molecular compounds have common names. For example, H_2O is named as water and NH_3 as ammonia. One just has to memorize these. There is also an older system of naming binary compounds which contain oxygen. For example, N_2O is nitrous oxide and NO is nitric oxide. While students may need to memorize a list of these compounds early in the course, at later stages they will be expected to know as to how binary molecular compounds are named. Unlike ionic compounds, the elements in a molecular compound can combine with a variety of ratios. We use prefixes to indicate the number of atoms of each element in a molecule. To name the compound using the formula, one should first name the first element then the second element giving it an -ide ending. For example, the compound PCl_3 is phosphorus trichloride and the compound SO_2 is sulfur dioxide.

30. WHAT ARE CHARACTERISTICS OF IONIC COMPOUNDS?

These compounds are formed by the combination of ions. Ionic compounds include hydroxides and salts. A salt is an ionic compound that contains a cation other than hydrogen ion and an anion other than hydroxide ion. Many acids exist as ions in solution but as molecules in the pure state. Ionic compounds are composed of positively and negatively charged ions and are held together by electrostatic attractions. Ionic compounds do not contain molecules. Instead, because opposite charges attract one another, the ions stack themselves in crystals such that positive ions are surrounded by negative ions and vice versa. Students can identify a compound as being ionic from either the name or the formula. Metals form positive ions relatively easily and non-metals tend to form negative ions. Hydrogen, H^+ and ammonium ion, NH_4^+, are the only common positive ions that do not contain a metal. To identify a compound as being ionic, scan the name or the formula and identify the elements. If the compound contains a metal (or the ammonium ion) and at least one non-metal, it is almost always an ionic compound. Compounds that contain the hydrogen ion are acids, which we will discuss separately under acid-base chapter.

To name ionic compounds, first name the cation and then the anion. Monatomic cations have the same name as the element from which they are formed. For example, Ca^{+2} is calcium ion. The names of the simple monatomic anions are similar to the names of their element but end in an -ide suffix. S^{-2} is sulphide ion. The name of the ionic compound containing Ca^{2+} and S^{-2} is calcium (II) sulfide. Some metals can form more than one ion. The names of these ions include a Roman numeral in parenthesis to indicate the charge. For example, Cu^{+1} is copper (I) ion and Cu^{+2} is copper (II) ion. The name of the ionic compound containing Cu^{+1} and O^{-2} is copper (I) oxide. There is an older style of nomenclature which uses the Latin form of the metal name and a suffix, the higher charged ion has an -ic ending and the lower charged ion has an -ous ending. This means that Cu^{+1} is also called cuprous ion. This leads to some difficulties (not just learning the Latin), because the higher/lower charges might be +2/+1 or +2/+3 or some other combination. The Roman numeral designation always represents the charge of the ion (except for the polyatomic mercury ion Hg_2^{+2} which is mercury (I), the exceptions are annoying). The metals that form only one type of ion do not need a Roman numeral in their names, but, when in doubt, put one in. It is not wrong, merely redundant and hence obsolete.

An ion may have more than one atom joined together by bonds to form a single charged unit. These are called polyatomic ions. While there are certain rules governing the names of polyatomic anions, there are so many exceptions that the beginning student is better off just memorizing them. We see subscripts in the formulas for many polyatomic ions. These represent the number of atoms of the element immediately preceding the subscript: NO_3^{1-} is nitrate ion and it has one nitrogen atom and three oxygen atoms bonded together to form an ion with a negative one charge. Note that one must learn the formulae, charges and names of the common ions.

The name of an ionic compound is the name of the cation and anion. For example, $CuSO_3$ is copper (II) sulfite because, for the compound to be neutral, the copper ion must have a charge of 2+ to match the charge of the sulfite ion, 2–. Cu_2S is copper(I) sulphide because, for the compound to be neutral, the copper ion must have a charge of 1+ for two of them to balance the charge of the sulphide ion, 2–. $Ca(NO_3)_2$ is calcium nitrate; here a Roman numeral is not needed because calcium has only one possible ion. Note that the nitrate is enclosed in parenthesis and given a subscript of 2. This indicates that there are two nitrate ions needed for every one calcium ion, for the compound to be neutral.

31. HOW DO WE FIND EMPIRICAL FORMULAS?

The empirical formula of a compound expresses a ratio between the numbers of atoms of different elements present in a molecule of the compound. This ratio is a mole ratio as well as a ratio between numbers of atoms. From the formula it is possible to calculate the percent composition of a compound. Going in the opposite direction from the composition of a compound, it is possible to calculate its empirical formula. Consider the compound chloroform. The percent composition of chloroform is 10.06% carbon, 0.85% hydrogen and 89.09% chlorine. We know then that 100 g chloroform contains 10.06 g carbon, 0.85 g hydrogen and 89.09 g chlorine. This weight relationship can be converted to a mole ratio by the following calculations:

$$\text{Carbon} = 10.06 \text{ g C} \times 1 \text{ mol C} / 12.01 \text{ g C} = 0.8376 \text{ mol C}$$
$$\text{Hydrogen} = 0.85 \text{ g H} \times 1 \text{ mol H} / 1.008 \text{ g H} = 0.84 \text{ mol H}$$
$$\text{Chlorine} = 89.09 \text{ g Cl} \times 1 \text{ mol Cl} / 35.45 \text{ g Cl} = 2.513 \text{ mol Cl}$$

These calculations show that the mole ratio between the elements in chloroform is: 0.84 mol C to 0.84 mol H and 0.84 mol C to 2.51 mol Cl. This ratio can be expressed by the formula: C0.8376H0.84Cl2.513. However, formulas by definition can contain only whole numbers of atoms. The ratio can be changed to whole numbers by dividing each subscript by the smallest subscript, giving the formula of chloroform as: C0.8376/0.8376H0.84/0.8376Cl2.513/0.8376 or $CHCl_3$.

32. HOW ARE EMPIRICAL AND MOLECULAR FORMULAS INTERRELATED?

The formulas we have calculated in the preceding section express the simplest atomic ratio between the elements in the compound. Such formulas are called empirical formulas. An empirical formula does not necessarily represent the actual numbers of atoms present in a molecule of a compound; it represents only the ratio between those numbers. The actual numbers of atoms of each element that occur in the smallest freely existing unit or molecule of the compound is expressed by the molecular formula of the compound. The molecular formula of a compound may be the empirical formula, or it may be a multiple of the empirical formula. For example, the molecular formula of butene, C_4H_8, shows that each freely existing molecule of butene contains four atoms of carbon and eight atoms of hydrogen. Its empirical formula is CH_2. One molecule of ethylene (molecular formula C_2H_4) contains two atoms of carbon and four atoms of hydrogen. Its empirical formula is CH_2. Both have the same empirical formula, yet they are different compounds with different molecular formulas. Butene is C_4H_8, or four times the empirical formula and ethylene is C_2H_4, or twice the empirical formula.

The following table (Table 2.4) shows three groups of compounds.Within each group, the compounds have the same empirical formula and percent composition but different molecular formulas. The fact that they are different compounds is shown by their different boiling points.

Table 2.4: Compounds with the same empirical formula but different molecular formulas

Empirical formula	*Compound*	*Molecular formula*	*Boiling point, °C*
CH (92.2% C; 7.8% H)	Acetylene	C_2H_2	–84
	Benzene	C_6H_6	80
CH_2 (85.6% C; 14.4% H)	Ethylene	C_2H_4	–103
	Butene	C_4H_8	–6.3
	Cyclohexane	C_6H_{12}	80.7
CH_2O (40.0% C; 6.7% H; 53.3% O)	Formaldehyde	CH_2O	–21
	Acetic acid	C_2H_4O	117
	Glyceraldehyde	$C_3H_6O_3$	140

The molecular formula of a compound can be determined from the empirical formula if the formula weight, or molecular weight, is known.

33. HOW CAN STUDENTS WRITE FORMULAE OF MOLECULAR COMPOUNDS?

A student can write the formula of a binary molecular compound from its name by translating the prefixes into subscripts. Given the name, write the symbol for the first element followed by the symbol for the second element. Translate the prefixes into subscripts. For example, $SeCl_2$ is selenium dichloride and PF_5 is phosphorus pentafluoride. It is very common for beginning chemistry students to mix up the system for naming ionic compounds with the system for naming binary molecular compounds. Some students write the formula for binary molecular compound then attempt to balance the charges. They then get frustrated because both elements seem to have negative charges and they can't get the compound as a neutral one. Another common mistake is to become fond of the prefixes used for binary molecular compounds and use that system with ionic compounds. Some of the polyatomic ions have names that seem similar to these prefixes. Chromate ion is CrO_4^{-2} and dichromate ion is $Cr_2O_7^{2-}$ but $Sn(CrO_4)_2$ is tin(IV) chromate and not tin dichromate. There are two tin dichromates: tin(II) dichromate is $SnCr_2O_7$ and tin(IV) dichromate is $Sn(Cr_2O7)_2$.

34. HOW DO WE WRITE FORMULAS OF IONIC COMPOUNDS?

To write the formula of an ionic compound from its name, identify the ions, write their symbols and use subscripts to make the compound neutral. Iron (III) sulfate contains the iron (III) ion and the sulfate ion. The symbol for iron (III) is Fe^{3+} and the symbol for sulfate is SO_4^{2-}. These charges do not balance in a one to one ratio, so the compound must have multiples of the ions in order to be neutral. If there are two iron (III) ions for every three sulfate ions, the charges will be six positive and six negative, which balance. This gives the ionic formula $Fe_2(SO_4)_3$. The parentheses are needed to make it clear that this compound has three parts sulfate ion rather than 43 parts oxygen.

The formula of an ionic compound is neutral. The ratio of the cations and anions it contains is such that there is no excess charge. Thus, the combination of a sodium ion, Na^+, with a sulfate ion, SO_4^{2-}, to form sodium sulfate must be in a 2:1 ratio so that the resulting compound is neutral, Na_2SO_4. The combination of an aluminum ion, Al^{3+}, with a chloride ion, Cl^-, to form aluminum chloride must be in the 1:3 ratio, giving the formula $AlCl_3$, which is neutral. The combination of an ammonium ion, NH_4^+, with a phosphate ion, PO_4^{3-}, to form ammonium phosphate must be in a 3:1 ratio, giving the neutral formula $(NH_4)_3PO_4$. Here we notice that in this last formula the ammonium ion is enclosed in parentheses and followed by the subscript 3. This notation means that the whole ion is taken three times. When a polyatomic ion is taken more than once in a formula, it is enclosed in parentheses and the number of ions contained in the formula is indicated by a subscript following the parentheses. Monatomic ions and polyatomic ions taken only once (for example, the sulfate ion in sodium sulfate) are not enclosed in parentheses.

Let us have a look of the cases of potassium sulphide, magnesium nitrate and aluminum sulfate produced below as proto type examples:

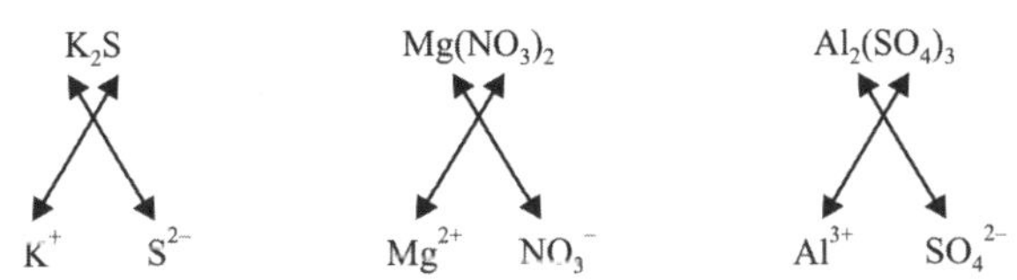

We may note two things about the examples shown above:

1. In both magnesium nitrate and aluminum sulfate, the polyatomic ions are enclosed in parentheses because they are taken more than once; the monatomic ion is not enclosed in parentheses regardless of its subscript.

2. If the charge on the two ions differs in magnitude, the number of times the cation is taken equals the magnitude of the charge on the anion. Similarly, the number of times the anion is taken equals the magnitude of the charge on the cation.

Note: Each compound is given a name. Ideally, this name should indicate the composition of the compound and perhaps something of its properties. Such names are called systematic names and are based on a set of rules drawn up by IUPAC. Although all compounds have systematic names, many also have trivial, or common, names.

35. HOW ARE COMPOUNDS NAMED?

Compounds fall into two large groups: molecular, in which the smallest unit of the compound is the molecule and ionic that are formed by the union of ions in a ratio such that the compound and its formula have no net charge. Each compound has a systematic name that identifies its composition. Many compounds also have common names. Compounds are given systematic names based on certain rules, many of which depend on the oxidation numbers of the elements in the compound. Another set of rules is used to assign oxidation numbers.

The formula of a compound states which elements and how many atoms of each are present in the smallest unit of the compound. It also represents the mole ratio between the elements in the compound. The formula weight of a compound can be calculated from the formula and the atomic weights of elements present in the compound. The formula weight in grams is the mass of one mole of the compound. Using formula weights, the percent composition of a compound can be calculated. If the percent composition of a compound is known, its empirical formula can be calculated. If both the molecular weight and the percent composition of a compound are known, the molecular formula of the compound can be calculated. Several compounds may have the same empirical formula but different molecular formulas.

36. WHAT IS SIGNIFICANCE OF OXIDATION NUMBER IN NAMING OF COMPOUNDS?

Many of the rules by which names of compounds are assigned are based on the concept of oxidation numbers. The oxidation number of an element represents the positive or negative character (nature) of an atom of that element in a particular bonding situation. Oxidation numbers are assigned according to the following rules:

1. The oxidation number of an uncombined element is 0. In the equation:

   $$Zn + 2HCl \rightarrow H_2 + ZnCl_2$$

   For example, the oxidation number of zinc (Zn) as an uncombined atom is 0 and the oxidation number of hydrogen in H_2 is 0.
2. The oxidation number of a monatomic ion is the charge on that ion. In $ZnCl_2$, the oxidation number of chlorine as Cl^- is –1 and that of zinc as Zn^{2+} is +2. In Ag_2S, the oxidation number of silver as Ag^+ is +1 and that of sulfur as S^{2-} is –2.

3. Hydrogen in a compound usually has the oxidation number +1. An exception to this rule occurs when hydrogen is bonded to a metal in hydrides.
4. Oxygen in a compound usually has the oxidation number –2. Peroxides are an exception to this rule: In hydrogen peroxide, H_2O_2, for example, the oxidation number of oxygen is –1.
5. The sum of the oxidation numbers of the atoms in a compound is 0. For example, in the compound $ZnCl_2$, the oxidation number of the zinc ion is +2 and that of each chloride ion is –1. The sum of these oxidation numbers (+2 for zinc and – 2 for the two chloride ions) is 0.
6. In a polyatomic ion, the net charge on the ion is the sum of the oxidation numbers of the atoms in the ion. We can use this rule to calculate the oxidation number of nitrogen in the nitrate ion, NO_3^-.

The charge on the ion as a whole is –1 and oxidation number of Oxygen = –2. By substituting all these, we get: Oxidation number of nitrogen +3(–2) = –1. By rearranging, them we get. Oxidation number of nitrogen = –1 – 3(–2) = –1 + 6 = + 5

37. HOW ARE BINARY COMPOUNDS NAMED?

Many chemical compounds are binary; that is, they contain only two elements. Binary compounds are of several varieties.

Binary Compounds Containing a Metal and a Nonmetal

Binary compounds of a metal and a nonmetal contain a metallic cation and a nonmetallic anion. The alkali metals form ions with only a +1 charge, the alkaline earth metals form ions with only a +2 charge and aluminum forms only the ion Al^{3+}. For these ions, the name of the element followed by the term ion is an unambiguous name. For example, the sodium ion can only be Na^+, the calcium ion can only be Ca^{2+}. According to IUPAC rules, the names of all other metallic cations contain the name of the element followed by its oxidation state (in parentheses) in that ion. This rule rules out any ambiguity. The name chromium ion does not say whether the ion is Cr^{2+} or Cr^{3+}; the proper names for these ions are chromium(II) and chromium(III). The anions in binary compounds are named by using the root name of the element,followed by the suffix ide;for example, bromide ion is Br^-, the sulfide ion is S^{2-} and the oxide ion is O^{2-}. In these examples, the root name of the element is italicized. For binary compounds, the cation is named first and the anion second. Thus, $NiCl_2$ is nickel(II) chloride, K_2S is potassium sulfide, $CaBr_2$ is calcium bromide and ZnO is zinc(II) oxide.

Before leaving this group of compounds, we should once again mention the second and less-preferred method of naming cations of the same element in different oxidation states. This older method gives the ending -ous to the ion of lower oxidation state and the ending -ic to the ion of higher oxidation state.Often this system also uses the Latin root of the name of the element. Thus, in this system, Fe^{2+} is ferrous and Fe^{3+} is ferric;

Pb^{2+} is plumbous and Pb^{4+} is plumbic. The elements for which Latin roots are generally used may be shown as under:

Table 2.5: Some elements with non-English root names

Element	*Latin name*	*Element*	*Latin name*
Copper	Cuprum	Lead	Plumbum
Gold	Aurum	Silver	Argentum
Iron	Ferrum	Tin	Stannum

Binary Compounds Containing Two Nonmetals but not Hydrogen

Binary compounds of two nonmetals, neither of which is hydrogen, are molecular rather than ionic. They do not contain cations and anions. Carbon dioxide (CO_2) and phosphorus trichloride (PCl_3) are two examples of such compounds. They are named using prefixes to state how many atoms of an element are in one molecule of the compound. The name of the second element is modified to the root of its name followed by the ending -ide. In both the formula and the name of these compounds, the most nonmetallic element comes first. The prefix mono is often omitted for the first element but never omitted for the second. Thus, CO is carbon monoxide, SF_6 is sulfur hexafluoride and N_2O is dinitrogen monoxide.

38. HOW DO WE NAME PSEUDO-BINARY COMPOUNDS?

Several polyatomic ions act so much like monatomic ions that they are classified as such. These ions are called pseudo-binary ions. They include the ammonium ion, NH^{4+}, the hydroxide ion, OH^-, the cyanide ion, CN^- and several others. Compounds containing these ions are pseudo-binary compounds. The properties of the ammonium ion are much like those of the alkali-metal ions. Compounds containing the hydroxide ion are bases. A general definition of a base is that its aqueous solution contains more hydroxide than hydrogen ions. The cyanide ion behaves very much like a halogen ion. Many compounds containing the cyanide ion are extremely toxic.

39. HOW ARE TERNARY COMPOUNDS NAMED?

Ternary compounds are compounds containing three elements. Ionic ternary compounds are formed by the combination of a monatomic cation with a polyatomic (containing several atoms) anion, as in sodium nitrate, $NaNO_3$. A polyatomic anion is derived from a ternary acid.

40. HOW DO WE NAME SALTS CONTAINING MORE THAN ONE CATION?

Occasionally we encounter a salt containing more than one cation. If both cations are metals, they are named in the order in which they are written according to the rules already given. If one of the cations is hydrogen, the salt can be named either by calling the cation-hydrogen or by adding bi as a prefix to the name of the anion. Thus, $NaHCO_3$ can be named sodium hydrogen carbonate or sodium bicarbonate. Salts with more than one cation, one of which is hydrogen, are sometimes called acid salts.

41. HOW CAN WE SUMMARIZE THE PROCEDURE FOR NOMENCLATURE OF COMPOUNDS?

Choose a formula to name. Begin at the top of the chart and ask yourself each of the questions. If the answer is 'yes,' follow the instructions and ask the next question. If the answer is 'no,' follow the instructions and skip to the appropriate question. If you are instructed to skip to another number, ignore all of the questions in between. Eventually you will get to a statement that will tell you which rule to apply. Stop at that point and choose a new compound to name.

Start

1. Does the formula contain a metal? If yes, go to question 2; if no, skip to question 5.
2. Does the formula contain square brackets? If yes, the compound is a coordination compound. You will need more information about coordination compounds than this flow chart can provide. If the answer is no, go to question 3.
3. Does the metal have only one oxidation state (only one possible charge)? If yes, you can name the compound by naming the positive ion (same as the metal name) then the negative ion. If the answer is no, go to question 4.
4. What is the charge of the metal ion in this particular case? Use the subscripts and the charge of the negative ion to determine this. Write the charge of the metal ion as a Roman numeral following the metal name then write the negative ion name.
5. Does the formula contain the ammonium ion, NH_4^+? If yes, the name of the ionic compound is ammonium followed by the name of the negative ion. If the answer is no, the compound is not ionic. Go on to question 6.
6. Does the formula start with H? If yes, go to question 7. If no, the compound is not an acid (at least for naming purposes). Skip to question 10.
7. Identify the anion and its root word. Does the name of the anion end with the suffix "-ate?" If yes, the acid name follows the format " — ic acid" with the root word in the blank. If the answer is no, go on to question 8.
8. Does the name of an anion end with the suffix "-ite?" if yes, the acid name follows the format " — ous acid" with the root word in the blank. If the answer is no, go on to question 9.
9. The anion ends with the suffix "_ide." Is the compound in an aqueous solution? If the answer is yes, the acid name follows the format "hydro — ic acid" with the root word in the blank. If the answer is no, go on to question 12.
10. The compound is molecular. Does it contain carbon? If the answer is yes, go on to question 11. If the answer is no, skip to question 12.

11. Does the compound have only one carbon atom and one other element (not hydrogen)? If yes, go to question 12. If the answer is no, the compound is organic. You will need more information about organic compounds than this flow chart can provide. If the answer is no, go on to question 13.
12. Is the compound one of these: CO_2, CO, CCl_4? If yes, go on to question 13. If no, this organic compound is not one that is easy to name. You need more information about organic nomenclature.
13. Does the compound contain only two non-metals? If the answer is yes, name the compound by naming the first element with a prefix that matches the subscript in the formula then name the second element with a prefix that matches its subscript. End the name of the second element with the suffix "-ide." If the answer is no, the name is more complicated than the rules in this flow chart and you will need more detailed nomenclature rules.

42. WHAT IS AN ION?

An ion is an atom or molecule that has a charge because it has lost or gained electrons (note that protons are not lost or gained). Electrons are negatively charged, so an atom that has lost electrons is a positively charged ion. Another name for a positively charged ion is a cation. An atom that has gained electrons is a negatively charged ion. Another name for a negatively charged ion is an anion. The charge is always written as part of the symbol and can be found at the upper right. Examples are Mg^{+2} and Br^{-1}.

43. WHY DOWE SAY THAT CHARGE IS A CONSERVED QUANTITY?

Atoms are neutral species yet they are made up of positively and negatively charge particles (protons and electrons). The charge on the proton is exactly the same strength as the charge on the electron, so they cancel one another out. When equal numbers of protons and electrons are present, the atom is electrically neutral and more generally, any physical system containing equal numbers of positive and negative charges are neutral. Charge is a conserved quantity; the net electric charge in a closed physical system is constant. Whenever charges are created, as in the decay of a neutron into a proton, an electron and an antineutrino, equal amounts of positive and negative charge must be created. Although charge is conserved, it can be transferred from one body to another. Electric current, on which much of modern technology is dependent, is a flow of charge through a conductor electrolyte. Although current is usually treated as a continuous quantity, it actually consists of the transfer of millions of individual charges from atom to atom, typically by the transfer of electrons. A precise description of the behavior of electric charge in crystals and in systems of atomic and molecular dimensions requires the use of the quantum theory.

Chapter 3

SUBATOMIC PARTICLES

1. WHAT SUBATOMIC PARTICLES ARE OF INTEREST TO CHEMISTS?

Atom, the smallest unit of a chemical element having the properties of that element, consists of a central, positively charged core, the nucleus and negatively charged species called electrons that revolve in orbitals around the nucleus. Almost the entire mass of the atom is concentrated in the nucleus, which occupies only a tiny fraction of the atom's volume. The nucleus of an atom consists of neutrons and protons, the neutron being an uncharged particle and the proton a positively charged one. Their masses are almost equal. Atoms containing the same number of protons but different numbers of neutrons represent different forms, or isotopes, of the same element. An atom is too tiny to be visualized. Its mass is between 10^{-21} and 10^{-23}g. A row of 10^7 atoms (10,000,000 atoms) extends only 1.0 mm. Besides electrons, protons and neutrons, atoms contain many different subatomic particles such as mesons, neutrinos and quarks, but the atomic model used by chemists requires knowledge of only electrons, protons and neutrons, so discussion of subatomic particles for chemists is generally limited to these three species alone.

2. WHAT ARE HISTORICAL ASPECTS AND CHARACTERISTICS OF ELECTRONS?

Electron is an elementary particle carrying a unit charge of negative electricity. It is one of the basic constituents of matter. Ordinary electric current is the flow of electrons through a conductor. An atom consists of a small, dense, positively charged nucleus surrounded by electrons that whirl about it in orbits, forming a cloud of charge. Ordinarily there are just enough negative electrons to balance the positive charge of the nucleus and the atom is neutral. As the atoms are electrically neutral, the positive electrical charge on the nucleus has the same magnitude as the negative charge due to all the electrons. The electrons are held in the atom by the attractive force exerted on them by the positively charged nucleus.

If electrons are added to or removed from an atom, a net +ve or –ve charge results and the atom is said to be ionized. Atomic electrons, especially the valence shell electrons, are responsible for the chemical properties of matter. The name electron was first used for a unit of negative electricity by the English physicist G.J. Stoney in the late 19th century. The actual discovery of the particle, however, was made in 1897 by J.J. Thomson, who showed that cathode rays are composed of electrons and who measured the ratio of charge to mass for the electron. In 1909, R.A. Mullikan measured the charge of the electron. Combining these two results gives the mass of the electron equal to about 1/1,840 of the mass of the proton. Ernest Rutherford, in 1903, showed that beta rays are high-energy electrons. In 1927, Davisson and Germer, working with high-speed electron beams, discovered that electrons sometimes exhibit the wave property of diffraction. This confirmed L.V. de Broglie's hypothesis that electrons, which had previously been thought of as particles, also possess certain wave properties.

The wavelike properties of electrons are utilized in the electron microscope and other devices. The electron is the lightest particle having a non-zero rest mass. It belongs to the lepton class of particles and, together with its antiparticle, the positron and its associated neutrino and antineutrino, constitutes a subfamily of the leptons. In any particle reaction involving any of the four members of the electron family, the total electron family number (+1 for ordinary particles, –1 for antiparticles) must be conserved). As a consequence, an electron and a positron (total electron family number equals zero) can annihilate each other to yield two or more photons or a neutrino-antineutrino pair, but not two neutrinos (total electron family number equals two).

3. HOW WAS ELECTRON DISCOVERED?

The electron was discovered and its properties defined during the last quarter of the nineteenth century. The experiments that proved its existence were studies of the properties of matter in gas-discharge or cathode-ray tubes. This apparatus consists of a glass tube sealed at both ends. Within the tube are two metal plates called electrodes, which were connected to an outside power supply. If the tube was full of air or some other gas, no current flew between the electrodes, regardless of how large a voltage was applied from the power source. If the tube had been evacuated before sealing (that is, almost all the gas had been pumped out of it), the application of a high voltage from the power source across the two electrodes gave rise to a glow inside the tube and simultaneously, a current began to flow between the electrodes.

We need not discuss in detail the various experiments performed with this apparatus; we may only state the conclusions drawn from them. The current is carried by streams of tiny particles given off by the negative electrode, called the cathode and the positive electrode is called the anode. In these experiments, the presence of these electrons and their properties did not change if the metal of the electrode was changed, nor were any changes observed in their properties when different gases were used in the tube. Eventually, the experimenters became convinced that all matter contains electrons. Each electron carries a single, negative electric charge and has a mass of 9.108×10^{-28}g. Because the mass of an atom is approximately 10^{-23}g, the mass of an electron is negligible compared to that of an atom.

4. HOW WAS EXISTENCE OF THE NUCLEUS AND PROTONS ESTABLISHED?

Gas-discharge tubes of slightly different design were used to identify small, positively charged particles that moved from the positive electrode (anode) to the negative electrode (cathode). The mass and charge of these particles varied but were always a simple multiple of the mass and charge of the positive particle observed when the gas-discharge tube contained hydrogen. The particle formed from hydrogen is called the proton. Proton is an elementary particle having a single positive electrical charge and constituting the nucleus of the ordinary hydrogen atom. The positive charge of the nucleus of any atom is due to its protons. Every atomic nucleus contains one or more protons; the number of protons, called the atomic number, is different for every element. The mass of the proton is about 1,840 times the mass of the electron and slightly less than the mass of the neutron. The total number of nucleons, as protons and neutrons are collectively called; in any nucleus is the mass number of the nucleus.

The existence of the nucleus was postulated by Ernest Rutherford in 1911 to explain his experiments on the scattering of alpha particles; in 1919 he discovered the proton as a product of the disintegration of the atomic nucleus. The proton and the neutron are regarded as two aspects or states of a single entity, the nucleon. The proton is the lightest of the baryon class of elementary particles. The proton and other baryons are composed of triplets of the elementary particle called the quark. A proton, for instance, consists of two quarks called up and one quark called down; a neutron consists of two down quarks and an up quark. The antiparticle of the proton, the antiproton, was discovered in 1955; it has the same mass as the proton but a unit negative charge and opposite magnetic moment. Protons are frequently used in a particle accelerator either as the bombarding (accelerated) particle or the target nucleus, or both. The possibility that the proton may have a finite lifetime has recently come under examination. If the proton does indeed decay into lighter products, however, it takes an extremely long time to do so. All atoms contain one or more protons.

5. WHY DOES PREPONDERANCE OF NEUTRONS BECOME MORE MARKED FOR VERY HEAVY NUCLEI?

The neutron is the third subatomic particle of interest to chemists. Its mass of 1.675×10^{-24}g is very close to mass of the proton. Hence neutron is an uncharged elementary particle of slightly greater mass than the proton. It was discovered by James Chadwick in 1932. The stable isotopes of all elements except hydrogen and helium contain a number of neutrons equal to or greater than the number of protons. The preponderance of neutrons becomes more marked for heavy nuclei as a nucleus with an excess of neutrons is radioactive due to which the extra neutrons convert to protons by beta decay.

6. WHAT IS A NUCLEON?

Nucleon is a general term applying to the proton and the neutron, the two constituents of atomic nuclei. The nucleon may be considered a single entity, of which the proton and the neutron are two different states. In fact nucleon is a generic word for the heavy particles that make up the atomic nucleus: the protons and neutrons. It is like the generic word fruit, used to include apples, oranges and many others, except that the class of nucleons contains only two members. The use of the ending -on for the names of subatomic particles, began with the discovery and naming of electron.

7. WHY DO PHYSICISTS REGARD THE NEUTRON AND THE PROTON AS TWO STATES OF A SINGLE ENTITY?

In a nucleus the neutron can be stable, but a free neutron decays with a half-life of about 17 min, into a proton, an electron and an antineutrino. The fact that the neutron possesses a magnetic moment suggests that it has an internal structure of electric charge, although the net charge is zero. The electron-scattering experiments of Robert Hofstadter indicate that the neutron, like the proton, is surrounded by a cloud of pions; protons and neutrons are bound together in nuclei by the exchange of virtual pions. The neutron and the proton are regarded by physicists as two aspects or states of a single entity, the nucleon. The antineutron, the neutron's antiparticle, was discovered in 1956. The neutron, like other particles, also possesses certain wave properties, as explained by the quantum theory. The field of neutron optics is concerned with such topics as the diffraction and polarization of beams of neutrons. The formation of images using the techniques of neutron optics is known as neutrography.

8. WHAT ARE CHARACTERISTICS OF SUBATOMIC PARTICLES?

The subatomic particles of most interest to a chemist are: electrons, protons and neutrons. While electrons have a relative charge of –1 and a relative mass of 5.45×10^{-4} amu and protons have a relative charge of +1 and a relative mass of 1.007 amu, neutrons have a relative mass of 1.008 amu and no charge. Each atom is an aggregate or mix of these three particles. The atomic number of an atom is equal to the number of protons in the nucleus of that atom and also to the number of electrons of the neutral atom. Rutherford's gold-foil experiment, carried out in 1911, confirmed the existence of the nucleus. Most of the mass of an atom is concentrated in the nucleus, where neutrons and protons are found. The total number of particles in the nucleus (neutrons plus protons) is equal to the mass number of the atom. Atoms of the same element always contain the same number of protons but can differ slightly in the number of neutrons. This variation explains the existence of isotopes, which are atoms of the same element varying slightly in mass.

The atomic weight of an element is the average of the relative masses of the naturally occurring atoms of that element. To three significant figures, one mole of an element contains 6.02×10^{23} atoms of that element. The mole is a counting unit; a mole is 6.02×10^{23} things. One mole of an element has a mass in grams numerically equal to the atomic weight of the element. The nuclei of some atoms decompose radioactively to produce alpha particles, beta particles, or gamma rays. These reactions can be shown by equations. The rate of decay of radioactive isotopes is measured in half-lives. Radioactivity has important applications in the health and biological sciences. Nuclear reactions can be used to produce energy.

9. WHAT IS A NUCLIDE?

Nuclear Chemistry is the study of atomic nuclei, especially of radioactive nuclei and their reactions with neutrons and other nuclei. The number of protons in a nucleus is also the atomic number, which defines the chemical element concerned. All nuclei with 11 protons for example, are nuclei of sodium (Na) atoms. An element can have various isotopes the nuclei of which have differing numbers of neutrons. For example, stable sodium nuclei contain 12 neutrons, whereas those with 13 are radioactive. These isotopes are symbolized as ${}^{23}_{11}Na_{12}$ and ${}^{24}_{11}Na_{13}$, where the left-hand subscript indicates the atomic number; the right-hand indicates the number of neutrons. The superscript represents the total number of nucleons, or neutrons and protons. Any species of nucleus designated by certain atomic and neutron numbers is called a nuclide.

10. WHY DO DIFFERENT ELEMENTS HAVE DIFFERENT ATOMIC NUMBERS?

Each element is assigned a unique number known as atomic number which equals the number of protons in the nucleus. Since each proton carries a single positive charge, the atomic number is also equal to the total positive charge of the atomic nucleus of an element. The atomic numbers are listed along with the names and symbols of the elements in the periodic table of the chemical elements. It also equals the number of protons in the nucleus and numerically it is the number of electrons in the neutral atom. The atomic number has the symbol Z. Different elements have different atomic numbers because atoms of different elements contain different numbers of protons (and electrons). For example, oxygen has the atomic number 8 as its atoms contain 8 protons and 8 electrons; Uranium has the atomic number 92 as its atoms contain 92 protons and 92 electrons. All of the atoms of the element iron have 26 protons, so the atomic number of iron is 26. During nuclear reactions, an element may change into another. For clarity and as an aid to balancing nuclear reaction equations, the atomic number may be written as part of the atomic symbol. If so, it appears at the bottom left, as ${}_{26}Fe$. The relationship between atomic number and the number of protons or electrons can be stated as: Atomic number = number of protons per atom = number of electrons per neutral atom.

11. HOW ARE ATOMIC NUMBER AND MASS NUMBER SYMBOLIZED?

Mass number is the number of protons and neutrons in an atom of a specific isotope of an element. For example, oxygen has an atomic number of eight. The isotope of oxygen having ten neutrons has a mass number of 18. The atomic symbol may include the mass number, if it is relevant. The mass number is written at the top left, as ${}^{18}O$. Every atom also has a mass number, denoted by the symbol A. since the mass number of an atom is equal to the number of protons plus the number of neutrons that it contains, the number of neutrons in any atom is its mass number minus its atomic number. Hence number of neutrons = mass number – atomic number Or Mass number = number of protons + number of neutrons. The atomic number and the mass number of an atom of an element can be shown by writing, in front of the symbol of the element, the mass number as a superscript and the atomic number as a subscript:

Mass number Atomic number	Symbol of element	or	A Z	X

For example, an atom of gold (symbol Au), with an atomic number 79 and mass number of 196 is denoted as:

196 79	Au

12. WHY DO WE USE SUBSCRIPTS WHEN DEALING WITH ATOMIC SYMBOLS?

Since we deal with atomic symbols, we should include subscripts for the sake of clarity. The smallest particle of an element is the atom. The atomic symbol identifies which of roughly one hundred elements, it is. It also may include information about the number of subatomic particles in that particular atom/ion. The smallest particle of the naturally occurring form of some elements is a group of atoms joined together. This group of atoms is known as a molecule. A number at the lower right of the atomic symbol tells how many atoms are in that molecule. For example, there are two naturally occurring forms of oxygen, one called simply oxygen and one called ozone. Oxygen molecule has two oxygen atoms joined together. Ozone has three oxygen atoms joined together. Their molecular symbols are O_2 and O_3, respectively. If a substance consists of atoms that are joined together, its formula must include the subscript. If a particle has lost or gained electrons and has a charge, its formula must include that charge. The mass numbers and atomic numbers are only included in the symbol if that information is pertinent to the discussion.

13. WHAT ARE ISOTOPES AND HOW ARE THE VARIOUS ISOTOPES OF AN ELEMENT DESIGNATED?

Although all atoms of a given element must have the same atomic number, they need not all have the same mass number. For example, some atoms of carbon (atomic number 6) have a mass number of 12, others have a mass number of 13 and still others have a mass number of 14. These different kinds of the same element are called isotopes. In chemistry and physics, isotopes are two or more forms of an element that have the same atomic number but different numbers of neutrons in the atomic nucleus and therefore a different atomic weight. Most elements occur naturally as mixtures of isotopes. These atoms have the same number of protons in their nuclei, but different numbers of neutrons and therefore different mass numbers. The term isotope comes from the Greek isos topos/isotopos, which means same place, because isotopes of the same element have the same atomic number and therefore occupy the same place in the periodic table. The concept of isotope was introduced by F. Soddy in explaining aspects of radioactivity; the first stable isotope (of neon) was discovered by J.J. Thomson. The nuclei of isotopes contain identical numbers of protons, equal to the atomic number of the atom and thus represent the same chemical element, but do not have the same number of neutrons. Thus isotopes of a given element have identical chemical properties but slightly different physical properties and very different half-lives, if they are radioactive. The symbol for an isotope usually consists of the mass number placed as a leading superscript before the elemental symbol. The various isotopes of an element can be designated by using superscripts and subscripts to show the mass number and the atomic number respectively. They can also be identified by the name of the element with the mass number of the particular isotope. For example, as an alternative to

$$^{12}_{6}\text{C}, \quad ^{13}_{6}\text{C}, \quad \text{and} \quad ^{14}_{6}\text{C}$$

We can write carbon-12, carbon-13 and carbon-14 respectively.

14. WHAT USES RADIOACTIVE ISOTOPES MAY BE PUT TO?

Though some elements have only one naturally occurring isotope, for most elements, both stable and radioactive isotopes are known. About 350 isotopes occur naturally on Earth and another 1500 have been produced artificially. The isotopes of a given element are by no means equally abundant. For example, 98.89% of all carbon occurring in nature is carbon-12, 1.11% is carbon-13 and only a trace is carbon-14. Radioactive isotopes of many common elements, such as carbon and phosphorus, are used as tracers in medical, biological and industrial research. Their radioactive nature makes it possible to follow the substances in their paths through a plant or animal body and through many chemical and mechanical processes; thus a more exact knowledge of the processes under investigation can be obtained. The very slow and regular transmutations of certain radioactive substances, notably carbon-14, make them useful as "nuclear clocks" for dating archaeological and geological samples. By taking advantage of the slight differences in their physical properties, the isotopes may be separated. The mass spectrograph uses the slight difference in mass to separate different isotopes of the same element. Depending on their nuclear properties, the isotopes thus separated have important applications in nuclear energy. For example, the highly fissionable isotope uranium-235 must be separated from the more plentiful isotope uranium-238 before it can be used in a nuclear reactor or atomic bomb. In atmospheric chemistry, long lived isotopes such as carbon-14 (^{14}C) are used to determine the age of objects that contained living matter. Other isotopes are used to track/trace the routes of air parcels in the atmosphere.

15. WHAT FORCES PROVIDE CONCLUSIVE EVIDENCES REGARDING THE INNER STRUCTURE OF THE ATOM?

Generally we discuss about electrons, protons, neutrons and some related species and their various aspects along with the ways to determine how many of each of the subatomic particles a particular atom contains. However, the question still remains: Are these particles randomly distributed inside the atom like blueberries in a muffin, or does an atom have some organized inner structure? At the beginning of the twentieth century, scientists were trying to answer this question. Various theories were proposed, but none was verified by experiment. It has rightly been suggested that at various points in its development, science has marked time until someone performed a key experiment that provided new insights. In the history of the study of atoms, a key experiment was performed in 1911 by Ernest Rutherford and his colleagues.

Our understanding of the conclusions drawn from Rutherford's experiment depends on knowledge of the forces acting between bodies. Therefore, before discussing his experiment, a brief review of these forces will be in order. Force, commonly, is a "push" or "pull," more properly defined in physics as a quantity that changes the motion, size, or shape of a body. Force is the term used for an outside influence exerted by one body on another which produces a change in state of motion or state of configuration. This limited meaning in science compared to our everyday usage is most important because of the specific results of this out-side influence. Force producing a change in state of motion gives the body concerned acceleration.

Force is a vector quantity, having both magnitude and direction. The magnitude of a force is measured in units such as the pound, dyne and newton, depending upon the system of measurement being used. Unbalanced force acting on a body free to move will change the motion of the body. The quantity of motion of a body is measured by its momentum, the product of its mass and velocity. According to Newton's second law of motion, the change in momentum is directly proportional to the applied force. Since mass is constant at ordinary velocities, the result of the force is a change in velocity, or an acceleration, which may be a change either in the speed or in the direction of the velocity.

Two or more forces acting on a body in different directions may balance, producing a state of equilibrium. For example, the downward force of gravity on a person weighing 200 lb when standing on the ground is balanced by an equivalent upward force exerted by the earth on the person's feet. If the person were to fall into a deep hole, then the upward force would no longer be acting and the person would be accelerated downward by the unbalanced force of gravity. If a body is not completely rigid, then a force acting on it may change its size or shape. Scientists study the strength of materials to anticipate how a given material may behave under the influence of various types of force.

There are four basic types of force in nature. Two of these are easily observed; the other two are detectable only at the atomic level. First is the force of gravity that exists between all bodies. Its magnitude depends on the respective masses and the distance between the centers of gravity of the two interacting bodies. We are familiar with gravity; it acts to keep our feet on the ground and the moon in orbit. Although gravitational force is the weakest of the four forces, it is the most easily observed because it affects all matter, is always attractive and because its range is theoretically infinite, *i.e.*, the force decreases with distance but remains measurable at the largest separations. Thus, a very large mass, such as the sun, can exert over a distance of many millions of miles a force sufficient to keep a planet in orbit.

The electromagnetic force, which can be observed between electric charges, is stronger than the gravitational force and also has infinite range. Electrical forces also exist between charged particles. The magnitude of the electrical force between two charged bodies depends on the charge on each body and on the distance between their centers. If the charges are of the same sign (either positive or negative), the bodies repel each other; if the charges are of opposite sign, the bodies attract each other. Magnetic forces are similar to electrical forces. Each magnet has two poles: a north pole and a south pole. When two magnets are brought together, a repulsive force exists between the like poles and an attractive force between the unlike poles. The magnetic and electrical forces can interact in the charged body. Both electric and magnetic forces are ultimately based on the electrical properties of matter; they are propagated together through space as an electromagnetic field of force. These three forces were known at the end of the nineteenth century when the structure of the atom came under intensive study. At the atomic level, two additional types of force exist, both having extremely short range. The strong nuclear force, or strong interaction, is associated with certain reactions between elementary particles and is responsible for holding the atomic nucleus together. The weak nuclear force, or weak interaction, is associated with beta particle emission and particle decay; it is weaker than the electromagnetic force but stronger than the gravitational force.

16. WHAT OUTCOMES OF RUTHERFORD'S EXPERIMENT ASTOUNDED HIM AND HIS CO-WORKERS?

In 1911, it was generally accepted that the atom contained electrons and protons but that they were probably not arranged in any set pattern. Rutherford wished to establish whether a pattern existed. He hoped to gain this information by studying how the protons in the atom deflected the path of another charged particle shot through the atom. For his second particle, he chose alpha particles. An alpha particle contains two protons and two neutrons, giving it a relative mass of 4 amu and a charge of +2 and it is sufficiently close in mass and charge to a nucleus of helium atom. Its path would be changed if it passed close to the proton. In the experiment, a beam of alpha particles was directed at a piece of gold foil, so thin as to be translucent and, more importantly for Rutherford, only a few atoms thick. The foil was surrounded by a zinc sulfide screen that flashed each time it was struck by an alpha particle. By plotting the location of the flashes, it would be possible to determine how the path of the alpha particles through the atom was changed by the protons in the atom. The three paths shown in the following figure (Fig. 3.1) (as paths A, B and C) are representative of those, observed. Most of the alpha particles followed path A; they passed directly through the foil as though it were not there. Some were deflected slightly from their original path, as in path B; and an even smaller number bounced back from the foil as though they had hit a solid wall (path C).

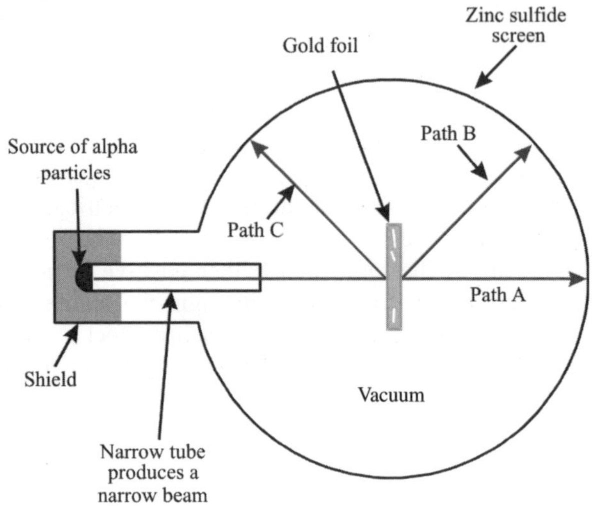

Fig. 3.1(a): Cross-section of Rutherford's apparatus

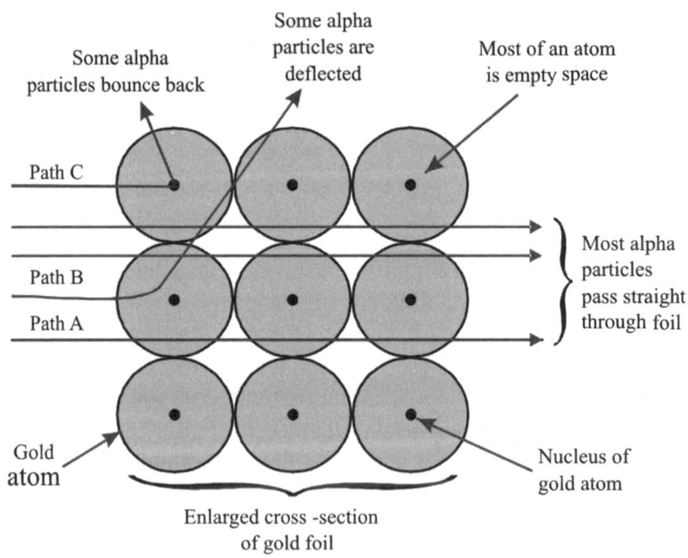

Fig. 3.1(b): Enlarged cross-section of the gold foil in the apparatus, showing the deflection of alpha particles by the nuclei of the gold atoms.

Although we may be surprised that any alpha particles passed through the gold foil, Rutherford was not. He had expected that many would pass straight through (path A). He had also expected that, due to the presence in the atom of positively charged protons, some alpha particles would follow a slightly deflected path (path B). The fact that some alpha particles bounced back (path C) is what astounded Rutherford and his co-workers. Path C suggested that the particles had smashed into a region of dense mass and had bounced back. To use Rutherford's analogy, the possibility of such a bounce was as unlikely as a cannonball bouncing off a piece of tissue paper.

17. WHAT ARE IMPORTANT OUTCOMES OF THE RESULTS OF RUTHERFORD'S EXPERIMENT?

Careful consideration of the results of his experiment, particularly the fact that some alpha particles bounced back, convinced Rutherford (and the scientific community) that an atom contains a very small, dense nucleus and a large amount of extra nuclear space. According to Rutherford's theory, the nucleus of an atom contains all the mass of the atom and therefore all the protons. The protons give the nucleus a positive charge. Because like charges repel each other, positively charged alpha particles passing close to the nucleus are deflected (path B). The nucleus, containing all the protons and neutrons, is more massive than an alpha particle; therefore, an alpha particle striking the nucleus of a gold atom bounces back from the collision, as did those following path C.

Outside the nucleus, in the relatively enormous extra nuclear space of the atom, are the tiny electrons. Because electrons are so small relative to the space they occupy, the extra nuclear space of the atom is essentially empty. In Rutherford's experiment, alpha particles encountering this part of the atoms in the gold foil passed through the foil undeflected. If the nucleus contains virtually all the mass of the atom, it must be extremely dense. Its diameter is about 10^{-12} cm, about 1/10,000 that of the whole atom. Given this model, if the nucleus was the size of a marble, the atom with its extra nuclear electrons would be 300 m in diameter. If a marble had the same density as the nucleus of an atom, it would weigh 3.3×10^{10} kg.

This model of the nucleus requires the introduction of a force other than those discussed earlier, one that will allow the protons, with their mutually repelling positive charges, to be packed close together in the nucleus, separated only by the uncharged neutrons. These nuclear forces seem to depend on interactions between protons and neutrons. Some are weak and some are very strong. Together they hold the nucleus together, but they are not yet understood. The model of the atom based on Rutherford's work is, of course, no more than a model; we cannot see these subatomic particles or their arrangement within the atom. However, this model does give us a way of thinking about the atom that coincides with observations made about its properties. We can now determine not only what subatomic particles a particular atom contains but also whether or not they are in its nucleus. For example, an atom of carbon-12 contains 6 protons and 6 neutrons in its nucleus and 6 electrons outside the nucleus.

We have two distinct parts of an atom, the nucleus and the extra nuclear space. Here two points are to be noted. First that the nucleus of an atom does not play any role in chemical reactions, but it does participate in radioactive reactions and second that the chemistry of an atom depends on its electrons, the valence electrons (how many there are and how they are arranged in the extra nuclear space).

18. WHAT FORCE HOLDS PROTONS AND ELECTRONS TOGETHER IN ATOMS?

By charge we mean an electrical charge which is the characteristic of a sample of matter and determines how that sample will behave in an electro-magnetic field. Charge is the property of matter that gives rise to all electrical phenomena. The basic unit of charge, usually denoted by e, is that on the proton or the electron; that on the proton is designated as positive (+e) and that on the electron is designated as negative (–e). All other charged elementary particles have charges equal to +e, –e, or some whole number times one of these, with the exception of the quark, whose charge could be 1/3e or 2/3e. There are three states of charge: positive, negative and neutral (none). Every charged particle is surrounded by an electric field of force such that it attracts any charge of opposite sign brought near it and repels any charge of like sign, the magnitude of this force being described by Coulomb's law. This force is much stronger than the gravitational force between two particles and is responsible for holding protons and electrons together in atoms and for chemical bonding. While two objects with the same charge will repel one another, two objects that have opposite charges (one positive and one negative) will be attracted to one another.

19. WHAT IS QUANTUM MECHANICS?

Quantum mechanics is a theory that describes the behavior of the electrons in an atom. Though Quantum mechanics is fascinating and very rewarding to study, unfortunately, it is also rather difficult to understand. At the foundation level chemistry teachers will not expect students to understand quantum mechanics. However, they will expect their students to be familiar with the terms and to be able to apply the major conclusions of the theory. One should remember that theories are just tools. This is a descriptive overview of the theory rather than a rigorous treatment. Quantum mechanics is based on the probability of finding an electron in a particular volume of space. This probability can be stated by a mathematical function. This function has four index numbers designated by the letters n, l, ml and s. The principal quantum number, n, gives us an idea of how far, on an average; the electron is likely to be from the nucleus and how much energy it has. The azimuthal quantum number, l, gives the shape of the volume of space the electron is likely to occupy and more specific information about its energy. The magnetic quantum number, ml, identifies which one of a set of volumes the electron occupies. The spin quantum number, ms, gives the magnetic properties of the electron in question and pinpoints which of two electrons is the concerned one. Together, the set of index numbers serve to locate the most probable volume that is likely to contain an electron with specific characteristics.

The set of quantum numbers is like an electron's address. We may imagine that electrons live in tiny little houses, two electrons per house. A set of quantum numbers would be an electron's mailing address. The principal number, n, would be the state, but a state is a pretty big place. We can use l and ml to specify the town, then the street number of the electron's address. This gives us a house where two opposite (boy and girl) electrons live. The ms quantum number, either up or down, identifies one of the two electrons (Mr. or Mrs.). This gives us the most probable place to find the electrons, but just because we know the address doesn't mean that the electrons are there. One is not always at home, either. This is just an analogy and it isn't a very good one because the electrons are not unique individuals with a set address.

20. WHAT ARE CHARACTERISTICS OF NUCLEAR DECAY?

We know that the atoms of any element have two distinct parts: the nucleus, which contains the protons and neutrons and the extranuclear space, which contains the electrons. The electrons in the atom, particularly those farthest from the nucleus, determine the chemical properties of the element. Here we may describe properties of the nucleus and, in particular, the characteristics of nuclear decay, which is also called radioactivity or radioactive decay of the nucleus. In nuclear decay, the nuclei of radioactive atoms decay spontaneously to form other nuclei, a process that always results in a loss of energy and often involves the release of one or more small particles. Some atoms are naturally radioactive. Others that are normally stable can be made radioactive by bombarding them with subatomic particles. Often, one isotope of an element is radioactive and others of the same element are stable. A radioactive isotope is called a radioisotope.

Radioactivity is a common phenomenon. Of the 350 isotopes known to occur in nature, 67 are radioactive. Over a thousand radioactive isotopes have been produced in the laboratory. Every element, from atomic number 1 through number 109, has at least 1 natural or artificially produced radioactive isotope. Of the 3 known isotopes of hydrogen, namely hydrogen-3, more commonly known as tritium, is radioactive. Oxygen, the Earth's most abundant element, has 8 known isotopes, 5 of which namely oxygen-13, -14, 15, -19 and -20, are radioactive. Iodine, an element widely used in nuclear medicine, has 24 known isotopes ranging in mass from 117 to 139 amu. Of these, only iodine-127 is stable; this isotope is the only naturally occurring one. Uranium has 14 known isotopes, all of which are radioactive.

21. WHAT ARE RADIO WAVES?

Radio waves are a form of electromagnetic radiation with long wavelengths and low frequencies. The radio section of the electromagnetic spectrum covers a fairly wide band and includes waves with frequencies ranging from about 10 kilohertz to about 60,000 megahertz (which correspond to wavelengths between 30,000 m and 0.5 cm. The term 'Radio' is used for the technology and practice that enables the transmission and reception of information carried by long-wave electromagnetic radiation. Radio waves make it possible to establish wireless two-way communication between individual pairs of transmitter and receiver and it is used for one-way broadcasts to many receivers.

22. WHAT IS MEANT BY RADIOCHEMISTRY?

Radiochemistry is the chemistry of radioactive substances. Radioactive isotopes are very useful as tracers to study the mechanisms of complex organic reactions, since even minute amounts of these isotopes are easily detected by means of a Geiger counter or photographic film. For example, by feeding plants carbon dioxide that contains the radioisotope carbon-14 and by monitoring the carbon compounds through the plants' life cycle, the intermediate stages of the photosynthetic process can be determined. A method developed by W. F. Libby uses carbon-14 to date archaeological discoveries and other samples containing organic matter. Radioactive substances have also been used to investigate the properties of artificially produced elements and to measure the rates of electron transfers.

23. WHAT ARE CHARACTERISTICS OF NUCLEAR REACTIONS?

Radioactivity is the decay or disintegration of the nucleus of an atom. During the process, either alpha or beta particles may be emitted. Energy, in the form of gamma rays, may also be released by this process and a different atom is formed. This new atom may be of a different element, or a different isotope of the same element. All of these characteristics and more can be shown by using an equation to describe the radioactive process. Like a chemical equation, a nuclear equation must be balanced. First, the total mass of the products must equal the total mass of the reactants. Second, the total charge of the reactants (the sum of their atomic numbers) must equal the total charge of the products. Let us consider the equation for the decay of radium-226 to radon-222, with the simultaneous loss of an alpha particle and energy in the form of a gamma ray. Radium-226 is the reactant; radon, an alpha particle and a gamma ray are the products. The equation is:

$$\underset{\text{radium-226}}{{}^{226}_{88}\text{Ra}} \longrightarrow \underset{\text{radon-222}}{{}^{222}_{86}\text{Rn}} + \underset{\text{alpha particle}}{{}^{4}_{2}\text{He}} + \underset{\text{energy}}{{}^{0}_{0}\gamma}$$

In the notation for particles, the superscript shows the mass of the particle and the subscript shows the charge which on each of these particles is its atomic number. The equation is balanced with respect to mass because the sum of the masses of the reactants (226) equals the sum of the masses of the products (222 + 4 + 0). The equation is balanced with respect to charge because the sum of the atomic numbers of the reactants (88) equals the sum of the atomic numbers of the products (86 + 2 + 0). The energy change accompanying the reaction is shown by the release of gamma rays. A similar equation can be written for nuclear decay by beta emission. Iodine-131 is a beta emitter commonly used in nuclear medicine. The equation for its decay is:

$${}^{131}_{53}\text{I} \longrightarrow {}^{131}_{54}\text{Xe} + {}^{0}_{-1}\beta + {}^{0}_{0}\gamma$$

In this case we may note that both the charge and the mass are balanced and that iodine-131 emits both a gamma ray and a beta particle. For this reason, iodine-131 is known as a beta-gamma emitter. Carbon-14, the isotope widely used in radiodating of archaeological artifacts containing carbon, is also a beta emitter:

$$^{14}_{6}C \longrightarrow ^{14}_{7}N + ^{0}_{-1}\beta$$

One may ask as to how can nuclei give off beta particles (high-energy electrons) if the nucleus has no electrons? The process is not yet clearly understood, but it may occur through the disintegration of a neutron to form a proton and the emitted electron:

$$\text{neuron} \longrightarrow \text{proton} + \text{electron}$$

$$^{1}_{0}n \longrightarrow ^{1}_{1}H + ^{0}_{-1}e$$

The electron is ejected and the proton remains in the nucleus. In beta emission, the atomic number of the product nucleus is one greater than that of the reactant nucleus because the nucleus now contains one more proton. The mass of the product nucleus is approximately the same as that of the reactant nucleus because an electron's mass is negligible with respect to that of a proton. Emission of a gamma ray changes neither the mass nor the charge of the nucleus. It accompanies the rearrangement of a nucleus from a less stable, more energetic nuclear configuration to a more stable, less energetic form. The identity and mass of the nucleus stay the same. Given the atomic number and mass number of a radioactive isotope and the type of radiation emitted during its decay, we can easily predict the mass number, atomic number and identity of the new element formed.

24. WHAT IS SIGNIFICANCE OF HALF-LIFE?

The rate of decay of a radioactive isotope (also called a radioisotope) is measured in terms of its half-life. Half-life is defined as the length of time required for half of the sample to decay. The length of half-lives varies from fractions of a second for some isotopes to billions of years for others. The half-life of a process is an indication of how fast that process proceeds, a measure of the rate or rapidity of the process. Specifically, the half-life is the length of time that it takes for a substance involved in that process to diminish to one-half of its initial amount. The faster is a process, the less time it will take to use up one-half of the substance. In fact half-life is a measure of the average lifetime of a radioactive substance or an unstable subatomic particle and one half-life is the time required for one half of any given quantity of the substance to decay. For example, the half-life of a particular radioactive isotope of thorium is 8 minutes. If 100 grams of the isotope are originally present, then only 50 grams will remain after 8 minutes, 25 grams after 16 minutes (2 half-lives), 12.5 grams after 24 minutes (3 half-lives) and so on. Of course the 87.5 grams are no longer present as the original substance after 24 minutes and have not disappeared but remain in the form of one or more other substances in the isotope's radioactive decay series. Similarly Iodine-131 has a half-life of 8.1 days. If we start today with a 25-mg sample of iodine-131 then after 8.1 days that sample will be left with only 12.5 mg of iodine-131. At the end of 16.2 (2 × 8.1) days, the sample will contain only 6.25 mg iodine-131. Of course, the matter in the sample does not disappear; it changes to another element, the product of the radioactive decay of iodine-131. Individual decays are random and cannot be predicted, but this statistical measure of the great number of atoms in the sample is very accurate. The half-life of a radioactive isotope is a characteristic of that isotope and is not affected by any change in physical or chemical conditions. Knowing the identity of a radioisotope, its half-life and the type of radiation it emits, we can determine the identity of the product and calculate the amount formed in a given period of time.

25. HOW IMPORTANT ARE RADIOACTIVE ISOTOPES?

The use of radioisotopes is widespread in chemistry, biology, medicine and other areas of science and industry. In all applications of radioactivity, the following characteristics of nuclear decay must be considered:

1. The chemical properties and reactions of a radioisotope are exactly the same as those of a non radioactive isotope of the same element.
2. Radiation can be detected some distance from its source.
3. Each radioisotope has a characteristic half-life.
4. Radioactive emissions interfere with normal cell growth.

Several uses of radioisotopes in the health and biological sciences illustrate ways in which they can provide information that would be difficult or impossible to obtain by any other means.

26. WHY RADIOACTIVE ISOTOPES HAVE IMPORTANT USES IN MEDICAL THERAPY AND DIAGNOSIS?

A radioactive isotope or radioisotope is a natural or artificially created isotope of a chemical element having an unstable nucleus that decays, emitting alpha, beta, or gamma particles and rays until stability is reached. The stable end product is a nonradioactive isotope of another element, *i.e.*, radium-226 decays finally to lead-206. Very careful measurements show that many materials contain traces of radioactive isotopes. For a time it was thought that these materials were all members of the actinide series; however, exacting radiochemical research has demonstrated that certain of the light elements also have naturally occurring isotopes that are radioactive. Since minute traces of radioactive isotopes can be sensitively detected by means of the Geiger counter and other methods, they have various uses in medical therapy, diagnosis and research. In therapy, they are used to kill or inhibit specific malfunctioning cells. For example, radioactive phosphorus is used to treat abnormal cell proliferation, *e.g.*, polycythemia (increase in red cells) and leukemia (increase in white cells). Radioactive iodine can be used in the diagnosis of thyroid function and in the treatment of hyperthyroidism. Since the iodine taken into the body concentrates in the thyroid gland, the radio action can be confined to that organ. In research, radioactive isotopes as tracer agents make it possible to follow the action and reaction of organic and inorganic substances within the body, many of which could not be studied by any other means. They also help to ascertain the effects of radiation on the human organism. In industry, radioactive isotopes are used for a number of purposes, including measuring the thickness of metal or plastic sheets by the amount of radiation they can stop, testing for corrosion or wear and monitoring various processes.

27. WHAT IS THE UNDERLYING PRINCIPLE OF RADIO DATING?

In the nineteenth century, prominent scientists such as Charles Lyell, Charles Darwin, Sir William Thomson (Lord Kelvin) and Thomas Huxley, were in continual debate about the age of the earth. The discovery of the radioactive properties of uranium in 1896 by Henri Becquerel subsequently revolutionized the way scientists measured the age of artifacts. Radioactive dating is the way of establishing the relative age of various materials with the use of carbon-14. This involves measuring the amount of ^{14}C and of ^{12}C and comparing the measured ratio to the one established by the production of ^{14}C in the upper atmosphere by cosmic rays. When an organism is alive the $^{14}C/ ^{12}C$ ratio in its biomass is constant (because of constant atmospheric ^{14}C production, diffusion to the lower atmosphere, absorption by organisms and because of constant ^{14}C radioactive decay) but when the organism dies the ratio begins to change at a very predictable rate. Therefore knowing the carbon-14/carbon-12 ratio now in some artifact gives a very good measure of how long the carbon has been "dead." The object however must obviously have organic material either in it or on it for this method to work. Also the length of time one can "look back" is limited because the amount of carbon-14 must be detected with some certainty for the age to be known with confidence and the longer the sample has been dead the less ^{14}C is present because it is continually decaying.

28. WHAT IS MEANT BY RADIOACTIVE FALLOUT?

The nucleus of each atom has a specific number of protons and neutrons and is either stable or unstable, depending on the relative number of each. The most stable atoms are those that have an equal number of protons and neutrons. Atoms that are unstable are radioactive. Radioactive decay is the spontaneous breakdown of certain kinds of atomic nuclei into one or more nuclei of different elements, involving the release of energy and subatomic particles. An atom that is radioactive can also be called a radionuclide. Radioactive fallout is material produced by a nuclear explosion or a nuclear reactor accident that enters the atmosphere and eventually falls to Earth. This fallout consists of minute, radioactive particles of dust, soil and other debris. While some fallout results from natural sources, the term is usually used in reference to radioactive particles that were released into the atmosphere by nuclear explosions.

29. WHAT ARE CHARACTERISTICS OF ALPHA AND BETA PARTICLES AND GAMMA RAYS?

Nuclei undergoing nuclear decay release various kinds of emissions. Three of these emissions are: alpha particles, beta particles and gamma rays. All three are forms of ionizing radiation, so called because their passage through matter leaves a trail of ions and molecular debris.

Alpha (α) particles

An alpha particle is identical to a helium atom that has been stripped off its two electrons; thus, an alpha particle contains two protons and two neutrons. Because an alpha particle has no electrons to balance the positive charge of the two protons, it has a charge of +2 and can be represented as He^{2+}. If a particle has a charge, whether negative or positive, it can be shown as a superscript. Thus He^{2+} means a helium atom that has lost two electrons and has a +2 charge. The symbol O^{2-} means an oxygen atom that has gained two electrons and thus has a charge of –2. Atoms that have acquired a charge by losing or gaining electrons are called ions. Besides He^{2+}, other symbols for this particle are:

$$^{4}_{2}\text{He and } ^{4}_{2}\alpha$$

When ejected from a decaying nucleus, alpha particles interact with all matter in their path, whether it is a photographic film, lead shielding, or body tissue, stripping electrons from other atoms as they go. In their way, they leave a trail of positive ions and free electrons. A single alpha particle, ejected at high speed from a nucleus, can create up to 100,000 ions along its path before it gains two electrons to become a neutral helium atom. In air, an alpha particle travels about 4 cm before gaining the two electrons. Within body tissue, its average path length is only a few thousandths of a centimeter. Hence an alpha particle is unable to penetrate the outer layer of human skin. Because of this limited penetrating power, external exposure to alpha particles is not nearly as serious as internal exposure. If a source of alpha emissions is taken internally, the alpha radiation can do massive damage to the surrounding tissue; therefore alpha emitters are never used in nuclear medicine.

Beta (β) particles

Beta particles are one of the three types of radiations (alpha, beta and gamma) resulting from natural radioactivity. Beta radiation (or beta rays) was identified and named by E. Rutherford, who found that it consists of high-speed electrons. A beta particle is a high-speed electron ejected from a decaying nucleus; it carries a charge of –1. A beta particle is represented as:

$$^{0}_{-1}e \quad \text{and} \quad ^{0}_{-1}\beta$$

Unlike alpha and gamma particles, whose energy can be explained as the difference of the energies of the radioactive nucleus before and after emission, beta particles emerge with a variable energy. This apparent violation of the law of conservation of energy led to the hypothesis that a second undetected particle, the neutrino, is emitted along with the electron and shares the total available energy. In some forms of induced, or artificial, radioactivity, the electron's antiparticle, the positron, is emitted from the excited nucleus; the positron in this case is also called a beta particle and denoted by $\beta+$ (the ordinary beta particle is $\beta-$).

Like alpha particles, beta particles cause the formation of ions by interacting with whatever matter is in their path. Beta particles are far less massive than alpha particles and carry a charge with only half the magnitude of that of the alpha particle. (This property depends only on the size of the charge, not its sign). Thus beta particles produce less ionization and travel farther through matter before combining with a positive ion to become a neutral particle. The path of a beta particle in air can be 100 times that of an alpha particle. About 25 cm of wood, 1 cm of aluminum, or 0.5 cm of body tissue will stop a beta particle. Because beta particles cause less ionization than alpha particles, beta particles are more suitable for use in radiation therapy, since the likelihood of damage to healthy tissue is greatly reduced. Beta emitters such as calcium-46, iron-59, cobalt-60 and iodine-131 are widely used in nuclear medicine.

Gamma (γ) rays

Gamma radiation is high-energy photons emitted as one of the three types of radiation resulting from natural radioactivity. The release of either alpha or beta particles from a decaying nucleus is generally accompanied by the release of nuclear energy in the form of gamma rays, represented as:

$$^{0}_{0}\gamma$$

It is not only a highly energetic form of electromagnetic radiation but also the most energetic form of electromagnetic radiation, with a very short wavelength (high frequency). Wavelengths of the longest gamma radiation are less than 10^{-10} m, with frequencies greater than 10^{18} hertz (cycles per sec). The wavelength of a gamma ray is very short, less than the radius of an atom, the energy they carry can be millions of electron volts. Gamma rays originate in the nucleus of an atom and are created when cosmic rays collide with atoms in molecules of gas.

Gamma rays have no charge or mass and are similar to X-rays. Even though they bear no charge, gamma rays are able to produce ionization as they pass through matter. The degree of penetration of gamma rays through matter is much greater than that of either alpha or beta particles. The path length of a gamma ray can be as much as 400 m in air and 50 cm through tissue. Because of their penetrating power, gamma rays are especially easy to detect. Virtually all radioactive isotopes used in diagnostic nuclear medicine are gamma emitters. Additional gamma emitters commonly used in nuclear medicine include chromium-51, arsenic-74, technetium-99 and gold-198. Gamma rays are essentially very energetic X-rays; the distinction between the two is not based on their intrinsic nature but rather on their origins. X-rays are emitted during atomic processes involving energetic electrons. Gamma radiation is emitted by excited nuclei or other processes involving subatomic particles; it often accompanies alpha or beta radiation, as a nucleus emitting those particles may be left in an excited (higher-energy) state.

The applications of gamma radiation are much the same as those of X-rays, both in medicine and in industry. In medicine, gamma ray sources are used for cancer treatment and for diagnostic purposes. Some gamma-emitting radioisotopes are also used as tracers. In industry, principal applications include inspection of castings and welds. Data from artificial satellites and high-altitude balloons have indicated that a flux of gamma radiation is reaching the earth from outer space, thus opening up the field of research known as gamma-ray astronomy. Gamma ray bursts are brief, seconds-long, blasts of radiation of mysterious origin that, in nature, seem to come from the depths of interstellar space. Bursts of gamma radiation also have been measured coming from severe thunderstorms and are a component of nuclear bomb detonation. The accidental discovery of cosmic gamma ray bursts was confirmed in 1973.

30. HOW DOES A RADIOMETER WORK?

A radiometer is an instrument for detection or measurement of electromagnetic radiation; the term is applied in particular to devices used to measure infrared radiation. One of the earliest experiments in radiometry was performed by W. Herschel, c.1800, who observed the heating of a mercury thermometer by sunlight; he was also able to detect heat radiated from hot but not incandescent bodies. E. Becquerel was able (c.1843) to detect near-infrared radiation by photographic means. Radiometers that function by an increase in the temperature of the device, such as Herschel's thermometer, are called thermal detectors. Commonly used thermal detectors include the thermocouple, which produces a voltage when heated and the bolometer, which changes in electrical resistance when heated. Devices that can, in principle, detect a single quantum of radiant energy, such as Becquerel's photographic plate, are called quantum detectors. Many current quantum detectors are based on the photoelectric cell. The term radiometer is often used to refer specifically to a type of thermal detector invented by Sir William Crookes (c.1874). Because his device was somewhat insensitive and not readily calibrated, it is rarely used today as a scientific instrument.

31. HOW DOES CROOKE'S RADIOMETER WORK?

A Crooke's radiometer consists essentially of two parts. The first part is a glass bulb from which most of the air has been removed, creating a partial vacuum. The second part is a rotor that is mounted on a vertical support inside the bulb. The rotor consists of four light, horizontal arms mounted at right angles to one another on a central pivot; the rotor can turn freely in the horizontal plane. At the outer end of each arm is mounted a metal vane, placed vertically. Each vane has one side polished and the other blackened; the vanes are arranged so that the polished side of one faces the blackened side of the next. When radiant energy strikes the polished surfaces, most of it is reflected away, but when it strikes the blackened surfaces, most of it is absorbed, raising the temperature of the surfaces. The air near a blackened surface thus becomes hotter, exerts a greater pressure on the blackened surface and causes the rotor to turn. The rate of rotation provides an indication of the intensity of the radiation.

32. WHY RADIOACTIVE WASTE IS DANGEROUS?

Certain atoms are radioactive, meaning they exhibit radioactivity during spontaneous transformation from an unstable isotope to a more stable one. Radioactive pollution results from contamination of the environment with such substances, which may represent a significant health risk to humans and other organisms. Radioactive pollution differs from conventional pollution in that it cannot be detoxified. Radioactive waste is generated during the production of electricity by nuclear power plants, by the eventual disposal of those facilities and during the manufacturing and disposal of nuclear weapons and machines used in medical diagnosis and treatments, academic and industrial research and certain industrial applications. Radioactive waste produces ionizing radiation.

33. HOW TRACERS WERE RECOGNIZED IN BIOCHEMISTRY AND HOW DO THEY WORK?

Because the chemical properties and reactions of a radioisotope are exactly the same as those of a nonradioactive isotope of the same element, a radioisotope can be substituted for a stable isotope of the same element in a molecule or compound, without changing the chemical properties of the compound. Such a compound is said to be "tagged" or "labeled." Because the radiation emitted by the radioisotope can be detected some distance from the radiating atom, the progress of those atoms through the body can be followed, or traced. Hence, such labeled compounds are called tracers. A tracer is an identifiable substance used to follow the course of a physical, chemical, or biological process. In chemistry the ideal tracer has the same chemical properties as the molecule it replaces and undergoes the same reactions but can at all times be detectible and quantitatively assessed. In biochemistry tracers have been in use since the beginning of the 20th century. Using synthetic methods, Franz Knoop in 1904 made various derivatives of fatty acids, the degradation of which he studied by feeding the derivatives to dogs and by monitoring the appearance of unusual products in the dogs' urine. From these studies were obtained the first descriptions of the metabolic pathway for fatty acid catabolism. About these sorts of experiments, however, the argument could always be made that the derivatives were "unphysiological," that is, did not occur naturally and might be handled by the enzymes of the body differently than "physiological" compounds. This difficulty was overcome in 1935 when Rudolf Schoenheimer and David Rittenberg described the use of the isotope, deuterium (identical to the hydrogen atom except that it contains an extra neutron) in following biochemical reactions. They argued persuasively that deuterium-labeled compounds (those having a deuterium atom substituted for a hydrogen atom) were essentially indistinguishable from non-labeled compounds as far as metabolic processes were concerned but that the amount of deuterium in any given sample could be quantitatively determined by the properties of the water produced upon combustion of the sample. Although this was the first declaration of the general usefulness of the approach, George Hevesy in 1923 was the first investigator to use an isotope in metabolic studies; he explored lead transport in the bean plant using radioactive thorium.

Radioactive isotopes are more easily detected than nonradioactive ones, such as deuterium; therefore, when the radioactive isotopes of various atoms commonly occurring in organic molecules became widely available after World War II, metabolic studies proliferated. Isotopes in common use today include carbon-14, iodine-131, nitrogen-15, oxygen-17, phosphorus-32, sulfur-35, tritium (hydrogen-3), iron-59 and sodium-24. Iodine-131 is the most commonly used radioisotope for the study of iodine metabolism in humans. The thyroid gland has a remarkable ability to extract iodine from the bloodstream and use it to produce the thyroid hormones: thyroxin and triiodothyronine. These two hormones have a direct effect on the body's metabolism. Very small amounts of iodine-131, in the form of sodium iodide, can be injected into the bloodstream and, within minutes, will begin to concentrate in the thyroid gland. By monitoring the accumulation of radioactivity, it is possible to estimate the size and shape of the thyroid gland and to determine whether any part of it is functioning abnormally. The uptake of iodine by the thyroid gland can be measured by tracing atoms of radioisotope iodine-131.

34. WHAT ARE BIOLOGICAL EFFECTS OF RADIATION?

Radiation is a term applied to the emission and transmission of energy through space or through a material medium and also to the radiated energy itself. In its widest sense the term includes electromagnetic, acoustic and particle radiation and all forms of ionizing radiation. The word radiation comes from the Latin for "ray of light," and is used in a general sense to cover all forms of energy that travel through space from one place to another as "rays." Radiation may be in the form of a spray of subatomic particles, like miniature bullets from a machine gun, or in the form of electromagnetic waves, which are nothing but pure energy.

Commonly radiation refers to the electromagnetic spectrum, which, in order of decreasing wavelength, includes radio, microwave, infrared, visible-light, ultraviolet, X-ray and gamma-ray emissions. All of these travel through space at the speed of light (c.300,000 km/186,000 mi per sec) but differ in wavelength and frequency. According to the quantum theory, the energy carried in the form of electromagnetic radiation may be viewed as made up of tiny bundles or packets, each bundle being known as a photon.

The sun is the source of much radiant energy in the form of sunlight and heat. Heat radiation is infrared radiation. All types of electromagnetic radiation can be reflected and absorbed in the same manner as is visible light. Acoustic radiation, propagated as sound waves, may be sonic (in the frequency range from 16 to 20,000 cycles per sec), infrasonic, or subsonic (frequency less than 16 cycles per sec) and ultrasonic (frequency greater than 20,000 cycles per sec). Examples of particle radiation are: alpha and beta rays in radioactivity and many kinds of atomic and subatomic particles such as electrons, mesons, neutrons, protons and heavier nuclei. Radiation is usually considered to travel from a source in straight lines, but its path may be affected by external factors; for instance, charged particles travel in curved paths in magnetic fields. The Van Allen radiation belts consist of charged particles trapped in the earth's magnetic field.

Although the exact manner in which radiation damages tissues and cells is not fully understood, cellular damage clearly can occur any time alpha or beta particles or other ionizing radiation passes through the cell. The effects of ionizing radiation range from minor damage of specific molecules to the death of cells. Ionizing radiation is especially damaging to the cell nucleus, particularly nuclei undergoing rapid division and nuclei of younger, less mature cells. Many types of cancer cells are especially sensitive to gamma radiation because they are growing rapidly and are therefore less mature than cells of surrounding noncancerous tissue. This sensitivity is the reason behind the use of radiation to destroy cancer cells. The apparatus used for such treatments is designed to focus the radiation sharply on the cancer cells, thus minimizing the damage to nearby healthy cells.

35. WHAT ARE RADIATION DETECTORS?

Radiation exposure occurs any time when energy in the form of electromagnetic rays or particles interacts with biological tissue. Ionizing radiation is particularly energetic; examples include: x rays, gamma radiation and subatomic particles. Biological damages caused by exposure to ionizing radiation range from mild tissue burns to cancer, genetic damage and ultimately, death. Radiation detectors are devices which sense and relay information about incoming radiation. Though the name brings to mind images of nuclear power plants and science fiction films, radiation detectors have found homes in such fields as medicine, geology, physics and biology. The term radiation refers to energies or particles given off by radioactive matter.

36. HOW DO SCIENTISTS USE NUCLEAR REACTIONS TO PRODUCE ENERGY?

We know that there are four characteristics of radioactivity and nuclear decay that form the basis for the use of radioisotopes in the health and biological sciences. A fifth characteristic of nuclear reactions is that they release enormous amounts of energy.

The first nuclear reactor to achieve controlled nuclear disintegration was built in the early 1940s by Enrico Fermi and his colleagues at the University of Chicago. Since that time, a great deal of effort and expense has gone into developing nuclear reactors as a source of energy. The nuclear reactions presently used or studied by the nuclear power industry fall into two categories: fission reactions and fusion reactions. Nuclear power is any method of doing work that makes use of either nuclear fission or nuclear fusion reactions. In its broadest sense, the term refers to both the uncontrolled release of nuclear energy, as in fission or fusion weapons and to the controlled release of energy, as in nuclear power plants. Most commonly, however, the expression "nuclear power" is reserved for the latter. A nuclear reactor is a device by which energy is produced as the result of a nuclear reaction, either fission or fusion. At the present time, all commercially available nuclear reactors make use of fission reactions, in which the nuclei of large atoms such as uranium (the fuel) are broken apart into smaller nuclei, with the release of energy. Nuclear weapons are explosive devices that release nuclear energy. An individual nuclear device may have an explosive force equivalent to millions of tons (megatons) of trinitrotoluene (TNT, the chemical explosive traditionally used for such comparisons) and is more than enough to inflict devastating physical damage to a big city.

37. WHY DO NUCLEAR POWER PLANTS DEPEND PRIMARILY ON THE FISSION OF URANIUM-235 AND PLUTONIUM-239?

Nuclear fission is a process in which the nucleus of an atom splits, usually into two pieces. This reaction was discovered when a target of uranium was bombarded by neutrons. Fission fragments were shown to fly apart with a large release of energy. The fission reaction was the basis of the atomic bomb, which was developed by the United States and used during World War II against Japan. In nuclear fission a large nucleus is split into two medium-sized nuclei. Only a few nuclei are known to undergo fission. Nuclear power plants currently in use depend primarily on the fission of uranium-235 and plutonium-239. When a nucleus of uranium-235 undergoes fission, it splits into two smaller atoms and, at the same time, releases neutrons (n) and energy. Some of these neutrons are absorbed by other atoms of uranium-235. In turn, these atoms split apart, releasing more energy and more neutrons. A typical reaction is:

$$^{235}_{92}U + ^{1}_{0}n \longrightarrow [^{236}_{92}U] \longrightarrow ^{139}_{56}Ba + ^{94}_{36}Kr + 3\,^{1}_{0}n + \text{energy}$$

The brackets around U indicate that it has a highly unstable nucleus. Under proper conditions, the fission of a few nuclei of uranium-235 sets in motion a chain reaction that can proceed with explosive violence if not controlled. In fact, this reaction is the source of energy in the atomic bomb.

In nuclear power plants, the energy released by the controlled fission of uranium-235 is collected in the reactor and used to produce steam in a heat exchanger. The

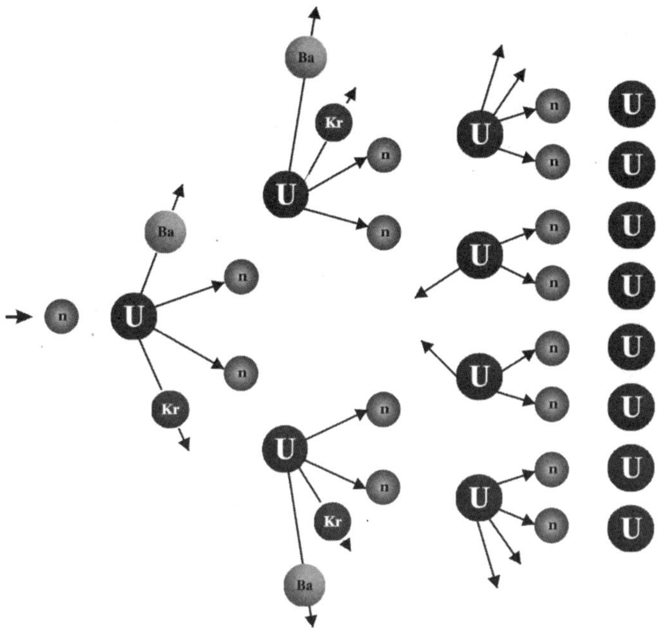

Fig. 3.2: Diagram of a nuclear fission chain reaction. Every fission process results in two (or more) neutrons that can react with other uranium atoms so that the number of nuclear fissions occurring soon reaches an enormous number.

steam then drives a turbine to produce electricity. Energy generation can be regulated by inserting control rods between the fuel rods in the reactor to absorb excess neutrons, thereby controlling the rate of the chain reaction. A typical nuclear power plant in operation today uses about 2 kg uranium-235 to generate 1000 megawatts of electricity. About 5600 tons (5.1×10^6 kg) of coal are required to produce the same amount of electricity in a conventional power plant.

Uranium-235 (natural abundance 0.71%) is very scarce and difficult to separate from uranium-238 (natural abundance 99.28%). The much more abundant uranium-238 does not undergo fission and therefore cannot be used as a fuel for nuclear reactors. However, if uranium-238 is bombarded with neutrons (from uranium-235, for example), it absorbs a neutron and is transformed into uranium-239. This isotope undergoes beta emission to generate neptunium-239, which, in turn, undergoes another beta emission to produce plutonium-239:

$$^{235}{}_{92}U + {}^{1}{}_{0}n \rightarrow {}^{239}{}_{92}U$$

$$^{239}{}_{92}U \rightarrow {}^{239}{}_{93}Np + {}^{0}{}_{-1}\beta$$

$$^{239}{}_{93}Np \rightarrow {}^{239}{}_{94}Pu + {}^{0}{}_{-1}\beta$$

Plutonium-239 also undergoes fission, with the production of more energy and more neutrons. These neutrons can then be used to breed more plutonium-239 from uranium-238. Thus, a so-called breeder reactor can produce its own supply of fissionable material. Several breeder reactors are now functioning in Europe and some other countries.

38. WHAT LACUNAS LEAD TO THE THREE MILE ISLAND ACCIDENT?

Nuclear reactors using fissionable materials pose several serious risks to the environment. First is the ever-present danger that leaks, accidents, or acts of sabotage will release radioactive materials from the reactor into the environment. This problem has been a continuing concern since the accident at the Three-Mile Island nuclear plant in March 1979, a concern that has increased in the wake of the disaster at the Chernobyl reactor in the former Soviet Union in April 1986, where radioactive fallout spread within days across the globe. Three Mile Island is site of a nuclear power plant 10 mi (16 km) south of Harrisburg. On Mar. 28, 1979, failure of the cooling system of the No. 2 nuclear reactor led to overheating and partial melting of its uranium core and production of hydrogen gas, which raised fears of an explosion and dispersal of radioactivity. Thousands living near the plant left the area before the 12-day crisis ended, during which time some radioactive water and gases were released. A federal investigation, assigning blame to human, mechanical and design errors, recommended changes in reactor licensing and personnel training, as well as in the structure and function of the Nuclear Regulatory Commission. The accident also increased public concern over the dangers of nuclear power and slowed construction of other reactors.

Second, many of the products of nuclear fission are themselves radioactive. The radioactivity from spent nuclear fuel and from the products of nuclear fission will remain lethal for thousands of years; the safe disposal of these materials is a problem that has not yet been solved. Third, obsolete generating plants also present a problem to future generations, for they contain much radioactive material. One suggestion has been to encase such plants in concrete for 100 years or more. Although it may become necessary, this solution is hardly simple or permanent.

39. WHAT IS MAGNETIC BOTTLE APPROACH AND WHY SCIENTISTS HAVE NOT YET FOUND A WAY TO PRODUCE AND CONTROL NUCLEAR FUSION ON EARTH?

Nuclear fusion, in contrast to nuclear fission, is the process by which two light atomic nuclei combine to form one heavier atomic nucleus. As an example, a proton and a neutron will, under the proper circumstances, combine to form a deuteron (the nucleus of an atom of "heavy hydrogen"). In general, the mass of the heavier product nucleus is less than the total mass of the two lighter nuclei. Nuclear fusion, although it was known theoretically in the 1930s as the process by which the sun and most other stars radiate their great output of energy, was not achieved by scientists until the 1950s. Fusion reactions are also known as thermonuclear reactions because the temperatures required to initiate them are more than 1,000,000°C. In the hydrogen bomb, such temperatures are provided by the detonation of a fission bomb. The energy released during fusion is even greater than that released during fission. Moreover, the fuel for fusion reactions, isotopes of hydrogen, is readily available in large amounts and there is no release of radioactive byproducts.

In stars ordinary hydrogen, whose nucleus consists of a single proton, is the fuel for the reaction and is fused to form helium through a complex cycle of reactions. This reaction takes place too slowly, however, to be of practical use on the earth. The heavier isotopes of hydrogen, deuterium and tritium, have much faster fusion reactions. For sustained, controlled fusion reactions, a fission bomb obviously cannot be used to trigger the reaction. The difficulties of controlled fusion center on the containment of the nuclear fuel at the extremely high temperatures necessary for fusion for a time long enough to allow the reaction to take place. For deuterium-tritium fusion, this time is about 0.1 sec. At such temperatures the fuel is no longer in one of the ordinary states of matter but is instead a plasma, consisting of a mixture of electrons and charged atoms. Obviously, no solid container could hold such a hot mixture; therefore, containment attempts have been based on the electrical and magnetic properties of a plasma, using magnetic fields to form a "magnetic bottle." In 1994 U.S. researchers achieved a fusion reaction that lasted about a second and generated 10.7 million watts, using deuterium and tritium in a magnetically confined plasma. The use of tritium lowers the temperature required and increases the rate of the reaction, but it also increases the release of radioactive neutrons. Another method has used laser beams aimed at tiny pellets of fusion fuel. If practical controlled fusion is achieved, it could have great advantages over fission as a source of energy. Deuterium is relatively easy to obtain, since it constitutes a small percentage of the hydrogen in water and can be separated by electrolysis, in contrast to the complex and expensive methods required to extract uranium-235 from its sources. In 2005 a six-member consortium (China, the European Union, Japan, Russia, South Korea and the United States) agreed to build an experimental fusion reactor at Cadarache in France that would use the "magnetic bottle" approach. One of the most promising fusion reactions generates energy by the fusion of two deuterium (hydrogen-2) atoms to form an atom of helium-3:

$$^{2}{}_{1}H + {}^{2}{}_{1}H - energy \longrightarrow {}^{3}{}_{2}He + {}^{1}{}_{0}n + energy$$

Such reactions require enormously high energy to force the two positively charged nuclei close together enough to fuse. Once the nuclei fuse, however, much more energy is released than is required for the reaction. Nuclear fusion occurs in the core of the sun, where the temperature is approximately 40 million degrees Celsius. Unfortunately, scientists have not yet found a way to produce and control nuclear fusion on Earth. Controlled nuclear fusion produces almost no radioactive wastes and would therefore be a nonpolluting source of energy. A massive effort is being made in many countries to find ways to harness this energy source.

40. WHAT IS MEANT BY 'NUCLEAR WINTER'?

Nuclear winter is a theory estimating the global climatic consequences of a prolonged and worldwide cooling and darkening caused by sunlight-blocking smoke and soot entering the atmosphere due to nuclear war. The term nuclear winter was first defined and used by American astronomer Carl Sagan (1934–1996) and his group of colleagues in their 1983 article (later referred to as the TTAPS-article). In the event of a significant nuclear war, researchers (Paul Crutzen and John Birks initially) predicted that a barrier of smoke (from incinerated urban centers and forests) would fill tropospheric skies and eventually surround the earth, reducing the tropospheric temperature and causing damage to ecosystems and atmospheric components such as stratospheric ozone.

Chapter 4

RADIANT ENERGY, ELECTRONIC CONFIGURATION AND PERIODIC PROPERTIES

1. WHAT ARE FORMS AND CHARACTERISTICS OF ENERGY?

A study of the properties of matter must include a study of energy. Energy, defined as the capacity to do work, has many forms. Potential energy is stored energy; it may be due to composition (the composition of a battery determines the energy it can release), to position (a rock at the top of a cliff will release energy if it falls to lower ground), or to condition (a hot stone will release heat energy if it is moved to a cooler place). Kinetic energy is energy of motion. We are undoubtedly aware that the faster a car is moving, the more damage it will do on crashing into an object. Because it is moving faster, it has more kinetic energy and has a greater capacity to do work (in this case, damage).

One of the characteristics of energy is that one form of energy can be converted to another. When wood is burned, some of its potential energy is changed to radiant energy (heat and light). Some is changed to kinetic energy as the water and carbon dioxide formed move away from the burning log. Some remains as potential energy in the composition of the water and carbon dioxide produced by the burning. Throughout all these changes, the total amount of energy remains constant. All changes must obey the Law of Conservation of Energy, which states that energy can neither be created nor destroyed. An alternative statement is that the total amount of energy in the universe remains constant. The Law of Conservation of Mass and the Law of Conservation of Energy are interrelated principles. Mass can be changed into energy and energy into mass according to the equation: $E = mc^2$, where E = energy change, M = mass change (in grams) and C = speed of light (3.00×10^8 m/sec, or 186,000 mi/sec).

This relationship allows us to state the two laws as a single law, called the Law of Conservation of Mass/Energy: Energy and mass may be interconverted, but together they are conserved. This law was first stated by Albert Einstein (1879–1955). In most changes, the amount of matter converted to energy is much too small to be detected by even the most sensitive apparatus and we can say "in this change both mass and energy are separately conserved." It is nevertheless important to be aware of this relationship between mass and energy, because nuclear energy is obtained by just such a conversion of mass to energy.

2. WHAT ARE CHARACTERISTICS OF RADIANT ENERGY?

To some extent, the properties of electrons can be compared with the properties of a closely related phenomenon, radiant energy or light. Radiant energy, also known as electromagnetic radiation, travels through a vacuum in waves at a constant speed of 3.0×10^8 m/sec. In many ways the waves of electromagnetic radiation are like waves in water. We have seen closely spaced, choppy waves on a small lake. We also have seen, on larger bodies of water like the ocean, waves that are farther apart. The difference between these two types of waves is in their wavelength λ (lambda), which is the distance from crest to crest or trough to though. A wave can also be characterized by its frequency, that is, the number of wave crests that pass a given point in a unit time. The frequency ν(nu) of a wave is related to its wavelength λ by the equation. $C = \lambda\nu$ where c is the constant speed of light, 3.0×10^8 m/sec. From this equation we can see that as wavelength increases, the frequency of the wave decreases. The energy associated with a wave is directly proportional to its frequency. Hence the higher is the frequency, the shorter is the wavelength and the higher is the energy of the wave.

We may consider the wide range of electromagnetic radiation from AM radio waves with a wavelength of 10^4 m to gamma waves with a wavelength of 10^{-12} m. This range is called the electromagnetic spectrum. The common names of the other kinds of electromagnetic radiation and their wavelengths are also given. Notice that visible light, electromagnetic radiation with wavelengths between 4×10^{-7} and 7×10^{-7} m, comprises only one small part of the electromagnetic spectrum. We can see that red light has a longer wavelength than blue light. Red light, then, has a lower frequency and is associated with less energy than blue light. Infrared light, microwaves, television waves and radio waves are invisible forms of electromagnetic radiation; their wavelengths are greater than that of visible light and thus their energies are lower than that of visible light.

Ultraviolet light, X rays and gamma rays, all of which have wavelengths shorter and energies higher than those of visible light, are also invisible forms of radiant energy. When an object is heated, it radiates energy, often in the form of visible light. Our sun is probably the most familiar example of a heated body giving off light. The "white" light from the sun is a collection of light of all wavelengths and is called continuous light. When light passes through a prism, it is separated into its various wavelengths. We have seen how sunlight, when passed through a prism, separates out into all the colors and wavelengths of the rainbow. However, if a gaseous sample of a single element is heated, the light emitted is not continuous and is only of a few wavelengths. When this light is passed through a prism, instead of a rainbow we see a series of brightly colored lines, each line corresponding to a particular wavelength of the emitted light. The pattern of wavelengths (or lines) of light thus produced is unique for each element and is called its emission spectrum (plural, spectra). Because the pattern is characteristic, it can be used to show the presence of that element in even the tiniest amounts. The light from sodium-vapor street lamps is that of the yellow-orange lines of the sodium spectrum. Neon signs use the red lines of the neon spectrum to produce their color. The spectra of these elements, as well as those of all the other elements, show other lines in the invisible parts of the electromagnetic spectrum.

Atoms can also absorb energy. If continuous light is passed through the vapor of an element, some wavelengths of light are absorbed. Analysis of the emerging light shows a rainbow interspersed with some black lines. This spectrum, called an absorption spectrum, shows that some energy (measured by the wavelength at which the black lines appear) has been absorbed by the vaporized element. If the vaporized element is hydrogen, the black lines will appear at exactly the same places in the absorption spectrum as did the bright lines in the emission spectrum of hydrogen. Similarly for sodium, neon and other elements, the bright lines in their emission spectra are of the same wavelengths as the black lines in their absorption spectra. We may conclude from these observations that, first, atoms can lose or gain energy in amounts similar to the energy of light and, second, the energy of an electron can change only by certain increments and not by any amount randomly.

3. WHAT CONCLUSIONS HAVE BEEN DRAWN ABOUT THE BEHAVIOR OF ELECTRONS IN ATOMS FROM STUDY OF SPECTRA?

The energy of an electron is of the same order of magnitude (is in the same range) as the energy of light. The lines in the spectrum of an element represent changes in the energy of electrons within the atoms of that element. By studying these spectra, scientists have drawn some conclusions, as stated below, about the behavior of electrons in atoms.

1. The energy of an electron in an atom depends on its location with respect to the nucleus of the atom. The higher the energy of an electron in an atom, the farther is its most probable location from the nucleus. It is to be noted that we say probable location. Because of the electron's small size and high energy, we are limited in how precisely we can mark its position at any instant. We can only describe regions around the atom's nucleus within which the electron may be found.
2. In describing these regions of space, we also recognize that the energy of an electron is quantized. A property is quantized if it is available only in multiples of a specific amount. If we are pouring a soft drink from a can, we can pour out as much or as little as we like. However, if we are buying a soft drink from a store's machine, we can buy only a certain specified amount. We cannot buy a half or a third of a can of soda; we can buy only a whole can or several cans. Thus soft drinks dispensed by a machine are available only in multiples of a set volume, or quantum. hence, the dispensing of soft drinks by machine has been quantized. Whenever we call our people from STD booths or perform journey by a hired taxi, the payment we made, is on quantized basis.

Energy can also be quantized. If we are climbing a ladder, we can stop only on the rungs and not in between them. The energy needed to climb the ladder is used in finite amounts to lift our body from one rung to the next. To move upward, we must use enough energy in order to move our feet to the next higher rung. If the available energy is enough only to move partway up to the next rung, we cannot move at all because we cannot stop between rungs. Thus, in climbing the ladder, our expenditure of energy is quantized. If we are going up a hill instead of climbing a ladder, our energy expenditure is not quantized. We can go straight up the hill or can zigzag back and forth, going up gradually. Further we can take big steps or little steps; no limitations are placed on where we can stop or on how much energy we must use.

Now let us apply the analogy of the ladder and its rungs to an atom and its electrons. In climbing the ladder, we can place our feet only on the rungs. Similarly, an atom has only certain places, set distances from the nucleus called energy levels, where electrons may be found. Unlike a ladder, which has a limited length, the energy levels of an atom extend infinitely out from the nucleus and the energy levels are not evenly spaced. As the distance from the nucleus increases, the levels get closer together and contain more-energetic electrons. The energy of an electron in one of the levels at a considerable distance from the nucleus is greater than that of an electron in a closer level. Therefore for an electron to move from one energy level to the next higher level, it must gain the right amount of energy. If less than that amount is available, the electron stays where it is. Electrons always move from one level to another; they cannot stop in between. Thus there are certain regions of space within an atom where an electron can be and other regions where an electron cannot be.

4. WHAT IS THE BASIS OF THE PRESENT DAY ATOMIC MODEL?

The present model of the atom is based on the concept of energy levels for electrons within an atom and on the mathematical interpretation of detailed atomic spectra. The requirements for this model are:

1. Each electron in a particular atom has a unique energy that depends on the relationship between the negatively charged electron and both the positively charged nucleus and the other negatively charged electrons in the atom.
2. The energy of an electron in an atom can increase or decrease, but only by specific amounts, known as quanta.

We picture an atom as having a small nucleus surrounded by a much larger volume of space containing the electrons. This space is divided into regions called principal energy levels, numbered 1, 2, 3, 4 . . ., ∞ outward from the nucleus. Each principal energy level can contain up to $2n^2$ electrons, where n is the number of the level concerned. Thus, the first level can contain up to 2 electrons, $2(1^2) = 2$; the second up to 8 electrons, $2(2^2) = 8$; the third up to 18, $2(3^2) = 18$; and so on. Only seven energy levels are needed to contain all the electrons in an atom of any of the elements known so for. As stated earlier, the energy associated with an energy level increases as the distance of that level from the nucleus increases. An electron in the seventh energy level has more energy associated with it than does one in the first energy level. The lower the number of the principal energy level, the closer the negatively charged electron in it is to the positively charged nucleus and the more difficult it is to remove this electron from the atom.

When an electron at a particular moment is in a particular energy level, it is more likely to be found in some parts of that level than in others. These parts are called orbitals. Orbitals of equivalent energy are grouped in sublevels. Each orbital can contain a maximum of two electrons. When in a magnetic field, the two electrons in a particular orbital differ very slightly in energy because of a property called electron spin. The theory of electron spin states that the two electrons in a single orbital spin in opposite directions on their axes, causing an energy difference between them. (Like many models, this explanation is an oversimplification, but for the purpose of a foundation course, it is a useful description.)

Each principal energy level has at least one sublevel containing one orbital, known as orbital that can contain a maximum of two electrons. Electrons in this orbital are called 's' electrons and have the lowest energy of any electrons in that principal energy level. The first principal energy level contains only an 's' sublevel; therefore, it can hold a maximum of two electrons only. The next principal energy level above the first contains one 's' orbital and three 'p' orbitals. A set of three 'p' orbitals, called the 'p' sublevel, can hold a maximum of six electrons. Therefore, the second level can contain a maximum of eight electrons, 2 in the s orbital and 6 in the three p orbitals.

Each principal energy level above the second contains, in addition to one s orbital and three p orbitals, a set of five d orbitals, called the d sublevel. The five d orbitals can hold up to 10 electrons. Thus, the third level holds a maximum of 18 electrons: 2 in the s orbital, 6 in the three p orbitals and 10 in the five d orbitals. The fourth and higher levels also have an f sublevel, containing seven f orbitals, which can hold a maximum of 14 electrons. Thus, the fourth level can hold up to 32 electrons: 2 in the s orbital, 6 in the three p orbitals, 10 in the five d orbitals and 14 in the seven f orbitals.

To distinguish which s, p, d, or f sublevel we are talking about, we prefix the letter by the number of the principal energy level concerned. For example, the s sublevel of the second principal energy level is designated as 2s; the s sublevel of the third principal energy level is designated as 3s; and so on. The number of electrons occupying a particular sublevel is shown by a superscript after the letter of the sublevel. The notation $4p^5$ means that five electrons are contained in the p sublevel of the fourth energy level that can be sketched as under.

Principal energy level ⟶ $4p^5$ ⟵ Number of electrons

↑

Sublevel

Table 4.1: Sublevels of the first four energy levels

Principal energy level	*Sublevel*	*Number of orbitals in sub level*	*Total possible occupying electrons*	
1	s	1	2	2
2	s	1	2	8
	p	3	6	
	s	1	2	
3	p	3	6	18
	d	5	10	
	s	1	2	
4	p	3	6	32
	d	5	10	
	f	7	14	

5. WHAT ARE SHAPES AND SIZES OF VARIOUS ORBITALS?

Each orbital has a unique shape and size. The shapes of s and p orbitals are shown in the following figure. In these diagrams, the nucleus is at the origin of the axes. The s orbitals are spherically symmetrical about the nucleus and increase in size as distance from the nucleus increases. The 2s orbital is a larger sphere than the 1s orbital; the 3s orbital is larger than the 2s orbital and so on.

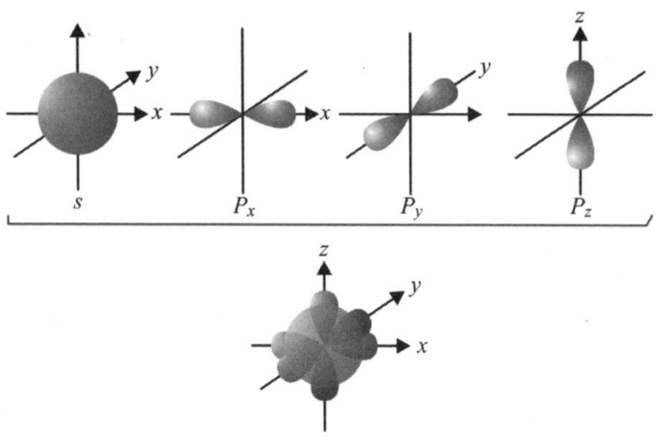

Fig. 4.1: Perspective representations of the s and the three p orbitals of a single energy level. The clouds show the space within which the electron is most apt to be. The lower sketch shows how these orbitals overlap in the energy level.

The three p orbitals are more or less dumbbell-shaped, with the nucleus at the center of the dumbbell. They are oriented at right angles to one another along the x, y and z axes; hence we denote them as p_x, p_y and p_z. Like the s orbitals, the p orbitals increase in size as the number of the principal energy level increases; thus a 4p orbital is larger than a 3p orbital.

The shapes of d orbitals are shown in the proceeding figure (Fig. 4.2a). The five d orbitals are denoted by dxy, dyz, dxz, dx^2-y^2 and dz^2. We notice that these shapes are

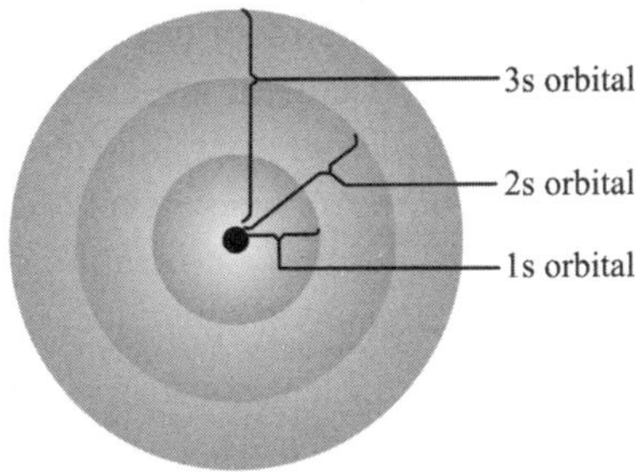

Fig. 4.2(a): Cross-sectional views of the s orbitals of an atom showing their relative sizes and overlap.

more complex than those of p orbitals and recall that the shapes of p orbitals are more complex than those of s orbitals. Clearly, the shape of an orbital becomes more complex as the energy associated with that orbital increases. We can predict that the shapes of f orbitals will be even more complex than those of the d orbitals.

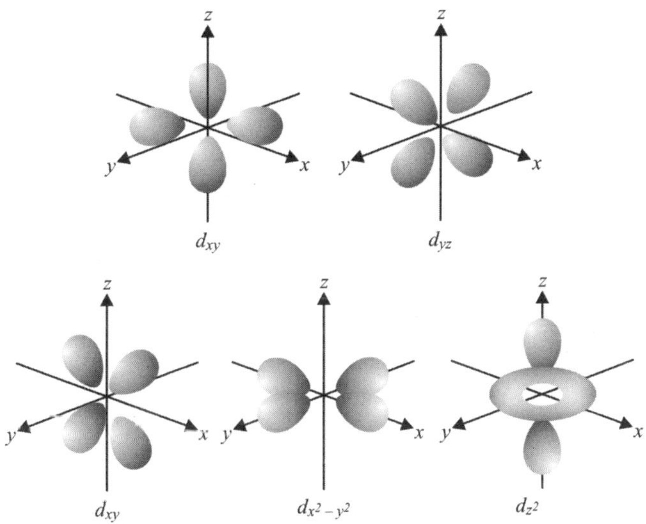

Fig. 4.2(b): The shapes and orientations of the d orbitals

There is one further, important note about orbital shapes. These shapes do not represent the path of an electron within the atom; rather, they represent the region of space in which an electron of that sublevel is most apt to be found. Thus, a p electron is most apt to be within a dumbbell-shaped space in the atom, but we make no pretense of describing its path.

6. HOW DOES ENERGY OF AN ELECTRON VARY IN ORBITALS?

Within a given principal energy level, electrons in p orbitals are always more energetic than those in s orbitals, those in d orbitals are always more energetic than those in p orbitals and electrons in f orbitals are always more energetic than those in d orbitals. For example, within the fourth principal energy level, we have:

$$\xrightarrow[\text{Increasing energy}]{4s < 4p < 4d < 4f}$$

In addition, the energy associated with an orbital, increases as the number of the principal energy level of the orbital increases. For instance, the energy associated with a 3p orbital is always higher than that associated with a 2p orbital and the energy of a 4d orbital is always higher than that associated with a 3d orbital. The same is true of s orbitals:

$$\xrightarrow[\text{Increasing energy}]{1s < 2s < 3s < 4s < 5s}$$

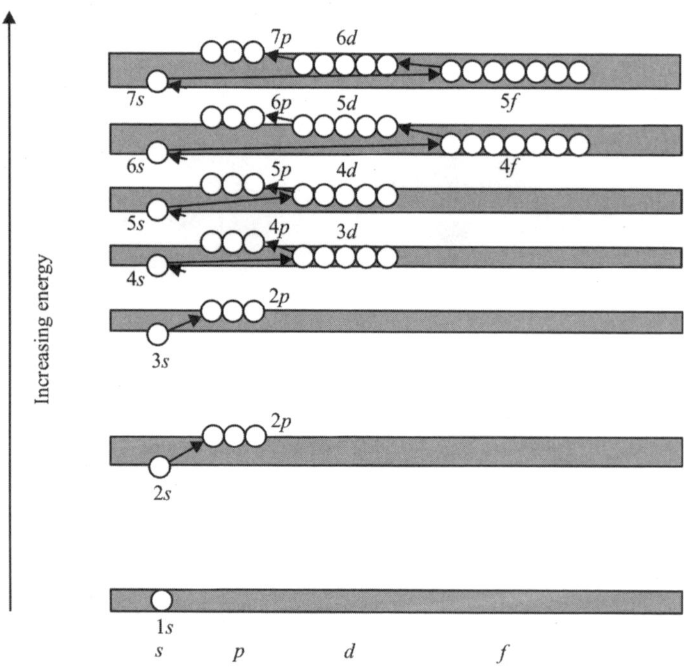

Fig. 4.3: The principal energy levels of an atom and the sublevels and orbitals each contains. The arrows show the order in which the sublevels fill

Each orbital is not a region of space separate from the space of other orbitals. If all those orbitals were superimposed on one another, we would see that a great deal of space is included in more than one orbital. For example, a 3p electron can be within the space assigned to a 3d or 3s orbital as well as within its own 3p space. There is also an interweaving of energy levels. Figure 4.3 shows, in order of increasing energy, all the orbitals of the first four energy levels. Here we notice that the energy of a 3d orbital is slightly higher than that of a 4s orbital and that of a 4d orbital is a little higher than that of a 5s orbital. We may note especially the overlap of orbitals in the higher principal energy levels.

7. WHAT CAUSES VARIATION IN ENERGY AMONG THE ELECTRONS IN AN ATOM AND WHY DOES EACH ELEMENT MUST HAVE A UNIQUE SPECTRUM?

According to the present day model of the atom, electrons are distributed among the energy levels and orbitals of the atom according to certain rules. Moreover each electron has a unique energy that is determined by the position of its orbital. When an atom absorbs the right amount of energy, one of its electron moves from its original orbital to a higher-energy orbital that has a vacancy. Similarly, when an atom emits energy, an electron drops to a lower-energy orbital that has a vacancy. For example, an electron in a 3s orbital can drop to the 2p orbital, or the 2s orbital, or the 1s orbital. The energy emitted by an electron in dropping to a lower-energy orbital is released in the form of radiation and determines the lines in the spectrum of the element.

When all the electrons of an atom are in the lowest possible energy states (meaning that the energy levels have been filled in order of increasing energy), the atom and its electrons are said to be in the ground state. If one of these electrons moves to a higher energy level, the atom is in an excited state. We know that each element has a unique spectrum. These spectra show that the energy differences among the electrons in an atom vary from one element to another. In order to know as to what causes this variation; we may recall that the nucleus of an atom is positively charged, that electrons carry a negative charge and that oppositely charged bodies attract one another. The atoms of one element differ from those of another element in the number of protons in the nucleus and, consequently, in the charge on the nucleus.

The attraction for an electron and therefore its energy, will differ from one element to the next according to differences in nuclear charge. In addition, the atoms of one element contain a different number of electrons than do atoms of any other element. The energy of each electron within the atom depends not only on its interaction with the positively charged protons in the nucleus, but also on its interaction with the other electrons in the atom. Therefore, the energies of the electrons of one element will differ from the energies of the electrons of another element. Considering these two variables, viz nuclear charge and number of electrons, we can see that each element must have a unique spectrum derived from its unique set of electron energy levels.

8. HOW DO WE DETERMINE THE ELECTRON CONFIGURATION OF A PARTICULAR ATOM?

The electron configuration of an atom shows the number of electrons in each sublevel of every energy level of the ground-state atom. To determine the electron configuration of a particular atom, we should start at the nucleus and add electrons one by one until the number of electrons becomes equal to the number of protons in the nucleus. Each added electron is assigned to the lowest-energy sublevel available. The first sublevel filled will be the 1s sublevel, then the 2s sublevel, the 2p sublevel, the 3s, 3p, 4s, 3d and so on. This order is not only difficult to remember but also often hard to determine from energy-level diagrams.

A more convenient way to remember the order is to use the following scheme. The principal energy levels are listed in columns, starting at the left with the 1s level. To use this scheme, read along the diagonal lines in the direction of the arrow. The order is summarized under the diagram.

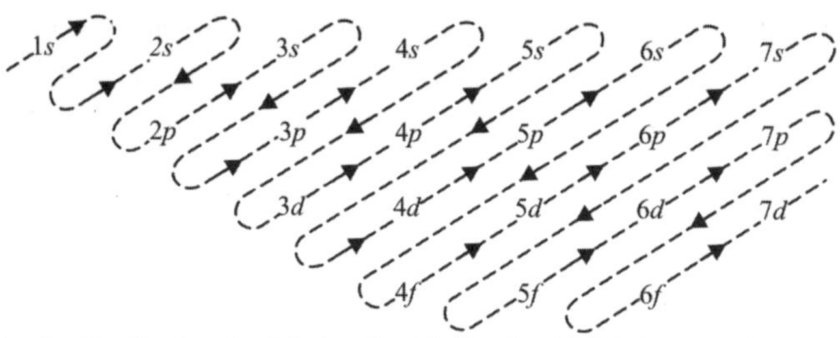

1s 2s 2p 3s 3p 4s 3d 4p 5s 4d 5p 6s 4f 5d 6p 7s 5f 6d 7p

Note: The arrow shows a way of remembering the order in which sublevels fill

Fig. 4.4: The energy level diagram for ground state electron configuration

An atom of hydrogen (atomic number 1) has one proton and one electron. The single electron is assigned to the 1s sublevel, the lowest-energy sublevel in the lowest-energy level. Therefore, the electron configuration of hydrogen is written as:

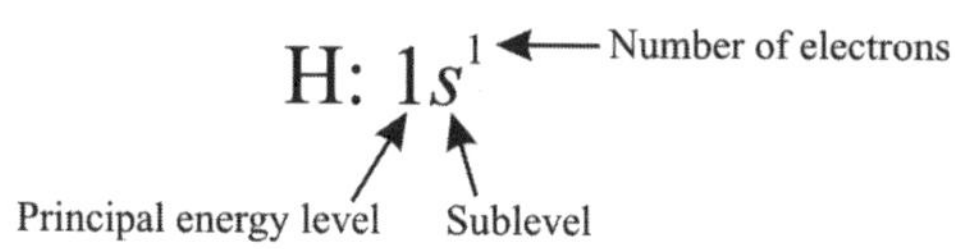

For helium (atomic number 2), which has two electrons, the electron configuration is: He: $1s^2$

Two electrons completely fill the first energy level. Because the helium nucleus is different from the hydrogen nucleus, neither of the helium electrons will have exactly the same energy as the single hydrogen electron, even though all are in the 1s sublevel. The element lithium (atomic number 3) has three electrons. In order to write its electron configuration, we must first determine (from the above scheme) that the 2s sublevel is next higher in energy after the 1s sublevel. Therefore, the electron configuration of lithium is: $1s^2 2s^1$

Boron (atomic number 5) has five electrons. Four electrons fill both the 1s and 2s orbitals. The fifth electron is added to a 2p orbital, the sublevel next higher in energy. The electron configuration of boron is: $1s^2 2s^2 2p^1$. The electron configurations of the elements with atomic numbers 1 through 18. May be depicted in a tabular form and the electron configurations of elements with higher atomic number can be written by following the orbital-filling chart in the above scheme.

Table 4.2: The electronic configuration of elments with atomic number 1 through 18.

Element	*Atomic number*	*Electron configuration*
Hydrogen	1	$1s^1$
Helium	2	$1s^2$
Lithium	3	$1s^22s^1$
Beryllium	4	$1s^22s^2$
Boron	5	$1s^22s^22p^1$
Carbon	6	$1s^22s^22p^2$
Nitrogen	7	$1s^22s^22p^3$
Oxygen	8	$1s^22s^22p^4$
Fluorine	9	$1s^22s^22p^5$
Neon	10	$1s^22s^22p^6$
Sodium	11	$1s^22s^22p^63s^1$
Magnesium	12	$1s^22s^22p^63s^2$
Aluminum	13	$1s^22s^22p^63s^23p^1$
Silicon	14	$1s^22s^22p^63s^23p^2$
Phosphorus	15	$1s^22s^22p^63s^23p^3$
Sulfur	16	$1s^22s^22p^63s^23p^4$
Chlorine	17	$1s^22s^22p^63s^23p^5$
Argon	18	$1s^22s^22p^63s^23p^6$

9. HOW THE ELECTRONS OF A PARTICULAR SUBLEVEL ARE DISTRIBUTED AMONG THE ORBITALS?

If an atom has a partially filled sublevel, it is be important to know as to how the electrons of that sublevel are distributed among the orbitals. Research has shown that unpaired electrons (a single electron in an orbital) are in a lower energy configuration than are paired electrons (two electrons in an orbital). The energy of the electrons in a sublevel would then be lower with half-filled orbitals than with some filled and some empty orbitals. We can show the distribution of electrons by using box diagrams, where each box represents an orbital and the arrows within the boxes represent the electrons in that orbital. The direction of the arrow represents the spin of the electron. (We know that two electrons in an orbital spin in opposite directions on their axes). Therefore, if an orbital contains two electrons, its box will contain two arrows, one pointing up and the other down. Using a box diagram, we show the electron configuration of nitrogen as:

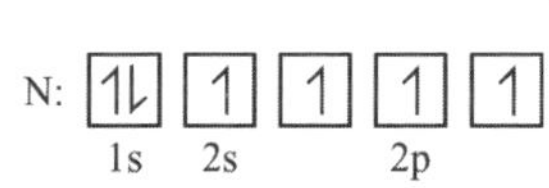

Notice that the 2p electrons are shown as

Rather than

which would mean that, of the three p orbitals, one is filled, one is half-filled and one is empty.

10. WHAT IS THE UNDERLYING HISTORY BEHIND CLASSIFICATION OF THE ELEMENTS AND WHAT IS MEANT BY 'FAMILIES OF ELEMENTS'?

Generally it is implied from our discussion that the arrangement of elements in the periodic table followed knowledge of electron configuration. That implication is incorrect. In the late eighteenth and early nineteenth centuries, many elements were discovered and their properties described; those elements with similar properties were grouped in families and the families arranged in a periodic table. The following shows the location of these families in the periodic table.

One family of elements is that of the alkali metals: lithium, sodium, potassium, rubidium, cesium and francium. These elements, found in column 1 of the periodic table, have a single valence electron. They are all soft, silvery gray solids with a clearly metallic luster. They are all very reactive and their reactions with water are vigorous. These elements do not occur free in nature but are found combined with other elements, often with chlorine. The most common of these compounds is sodium chloride (table salt), the substance that makes seawater salty and it is also found in huge underground deposits (salt mines).

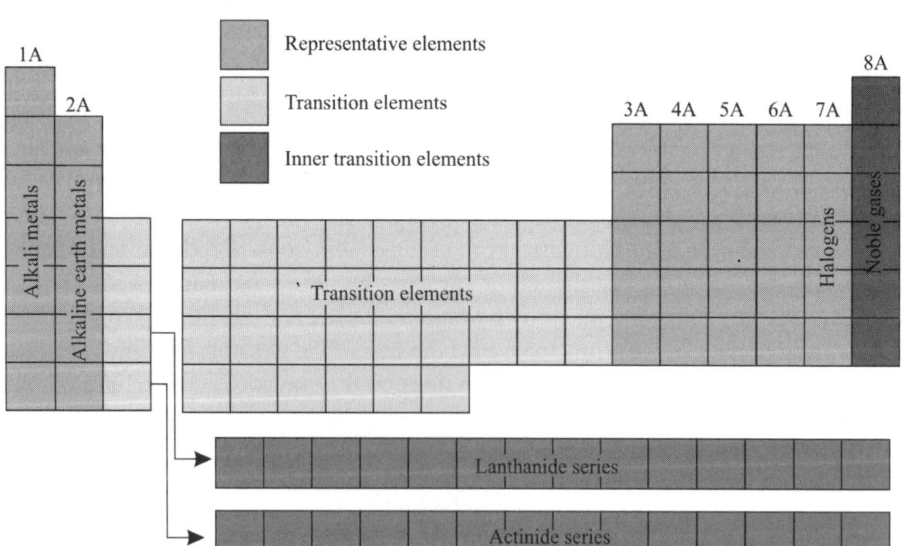

Fig. 4.5: An old version of the periodic table, showing the location of the representative, transition and inner transition elements.

The alkaline earth metals, another family include all the elements in column 2 of the periodic table. Their atoms have two valence electrons. These metals are harder and stronger than the alkali metals. They, too, are found in combination with other elements. Although all alkaline earth metals have similar chemical reactivity, they have a wider range of reactivity than do the alkali metals. Beryllium and magnesium are unaffected by water, calcium reacts slowly with boiling water and barium reacts violently with cold water.

The elements of Group 7/17, with seven valence electrons, are known as the halogens. Chemically, these elements are very similar; physically, they are less alike. The lightest halogen, fluorine, is a pale yellow gas; iodine is a shiny, black solid. The heaviest halogen, astatine, is quite rare and is found in uranium ores. The total amount of astatine in the Earth's crust is probably less than 1g. The longest-lived isotope of this element, astatine-210, has a half-life of only 8.3 hours. Several characteristic properties of halogens are shown in the proceeding table. We may examine as to how these properties change as the atomic number increases.

Table 4.3: Properties of halogens

Halogen	*Symbol*	*Atomic number*	*Atomic weight*	*Radius (10^{-8} cm)*	*Melting point, °C*
Fluorine	F	9	19	0.72	–219.6
Chlorine	Cl	17	35.45	0.99	–101
Bromine	Br	35	79.9	1.14	–7.2
Iodine	I	53	126.9	1.33	113.7
Astatine	At	85	(210)	1.45	302

The elements in column 8/18 are the noble gases. All these elements are monatomic gases and occur free and uncombined. They are all singularly unreactive, so much so that they were known earlier as the "inert gases." Only in 1960 were any of them shown to take part in chemical reactions. Even now, only krypton and xenon are known to form chemical compounds with other elements.

The elements in other columns of the table do not show striking similarities of properties. These columns are crossed by the stair-step line that separates metals from nonmetals, so they have nonmetals at the top and metals at the bottom of the column. For example, Group 4/14 is headed by the nonmetal carbon and has lead, a typical metal, as its heaviest member. Group 5/15 starts with the nonmetal nitrogen and ends with the metal bismuth.

11. WHAT CHRONOLOGICAL DEVELOPMENTS HAVE TAKEN PLACE IN THE HISTORY OF PERIODIC TABLE?

By the middle of the nineteenth century, scientists knew the atomic weights of all the then-known elements and had observed that these weights progressed from 1 amu (hydrogen) to over 200 amu (lead). Efforts were made to find a systematic arrangement of all the elements that would place them in order of increasing atomic weight and in groups with similar properties. Such an arrangement, to be known as the periodic table, was conceived in 1869 both by Julius Lothar Meyer (1830–1395) in Germany and by Dmitri Mendeleev (1834–1907) in Russia. Mendeleev is usually given most of the credit for constructing the periodic table because he not only described the ordering of the known elements but also used this arrangement to predict the existence of other elements not yet discovered.

Mendeleev's arrangement of the elements is shown in the following table. It resembles today's periodic table but has fewer columns and a different overall shape. Our current Group 8/18 is missing because the noble gases had not yet been discovered. The shorter columns of the d block are missing; instead, the transition elements are placed as subgroups in the long columns. Nevertheless, Mendeleev's table is the ancestor of our periodic table.

Table 4.4: Medeleev's periodic table

Series	*Group I*	*Group II*	*Group III*	*Group IV*	*Group V*	*Group VI*	*Group VII*	*Group VIII*
1	H=1							
2	Li=7	Be=9.1	B=11	C=12	N=14	O=16	F=19	
3	Na=23	Mg=24.4	Al=27	Si=28	P=31	S=32	Cl=35.5	Fe=56, Ni=58.5
4	K=39.1	Ca=40	–=44	Ti=48.1	V=51.2	Cr=52.3	Mn=55	Co 59.1, Cu63.3
5	(Cu)=63.3	Zn=65.4	–=69	–=72	As=75	Se=79	Br=80	Rh=103, Ru=103.8
6	Rb=85.4	Sr=87.5	Y=89	Zr=90.7	Nb=94.2	Mo=95.9	–=100	Pd=108, Ag=107.9
7	(Ag)=107.9	Cd=112	In=113.7	Sn=118	Sb=120.3	Te=125.2	1=126.9	–
8	Cs=132.9	Ba=137	La=138.5	Ce=141.5	Di=145	–	–	–
9	(–)	–	–	–	–	–	–	Ir=193.1, Pt=194.8
10	–	–	Yb=173.2	–	Ta=182.3	W=184	–	Os=200, Au=196.78
11	(Au)=196.7	Hg=200.4	TI=204.1	Pb=206.9	Bi=208	–	–	–
12	–	–	–	Tb=233.4	–	U=239	–	–

***Note*:** Mendeleev's periodic table (1871). The formulas of the oxide and the hydride of an element determined which column it would be in. We clearly notice the spaces left empty but filled later as new elements were discovered. The atomic weights shown are those accepted in 1871; many have since been changed.

In making this table, Mendeleev followed most closely his aim to put similar elements in the same column. This aim had two important results. First, if the elements did not match, spaces were left and properties predicted for the elements yet to be discovered. Germanium, one of those for which a space was left, was discovered in 1886 by using Mendeleev's predicted properties as a guide. Second, if two neighboring elements were close in atomic weight and did not match their respective columns when placed in consecutive order, the order was reversed. Iodine and tellurium are two such elements. Although tellurium has a higher atomic weight, it is clearly not a halogen, but iodine is; thus, tellurium was placed in Group 6/16 and iodine in Group 7/17.

Mendeleev's arrangement of the elements held until about 1912 as an illustration of his statement of the periodic law: The properties of the elements are periodic functions of their atomic weights. In 1912 H. G. J. Moseley, (1887–1915), a student of Rutherford, assigned to each element an atomic number that was equal to the charge on its nucleus and showed that this charge increased by one from one element to the next. Moseley's work suggested a restatement of the periodic law to read: the properties of the elements are periodic functions of their atomic numbers. This restatement resolved the tellurium-iodine problem as well as other similar problems. It recognized that atomic weights are based on isotopic distribution and that the real distinction between elements is the difference in the number of protons in their atoms. The periodic table we use today came into wide acceptance during the 1940s as the importance of electron configuration in predicting properties was recognized. Today we use the table as a summary of knowledge about the elements, whereas Mendeleev's table was held to be an interesting but not particularly useful artifact.

12. WHAT PROPERTIES OF ELEMENT CAN BE PREDICTED FROM THE KNOWLEDGE OF THE PERIODIC TABLE?

The properties that can be predicted from the knowledge of periodic table are as under:

Atomic Radius

While comparing the relative sizes of the atoms of the representative elements, we notice that atoms' size increases from top to bottom in a column and from right to left across a row. This trend is related to electron configuration. As we look at the elements in column 1, for example, we see that the single valence electron for each successive element is in a higher principal energy level than the last one and the electron is thus farther away from the nucleus; hence, the atomic radius increases going from top to bottom. This same regular increase in size can be observed in each column of the periodic table.

Atoms decrease in size while going across a period from left to right as for elements within a period, electrons are being added one by one to the same principal energy level. At the same time, protons are also being added one by one to the nucleus, increasing the number of positively charged species. This increasing positive charge increases the attraction of the nucleus for all electrons and pulls them all closer to the nucleus, decreasing the atom's radius. Thus, atomic size is a periodic property that increases from top to bottom within a column and from right to left across a period.

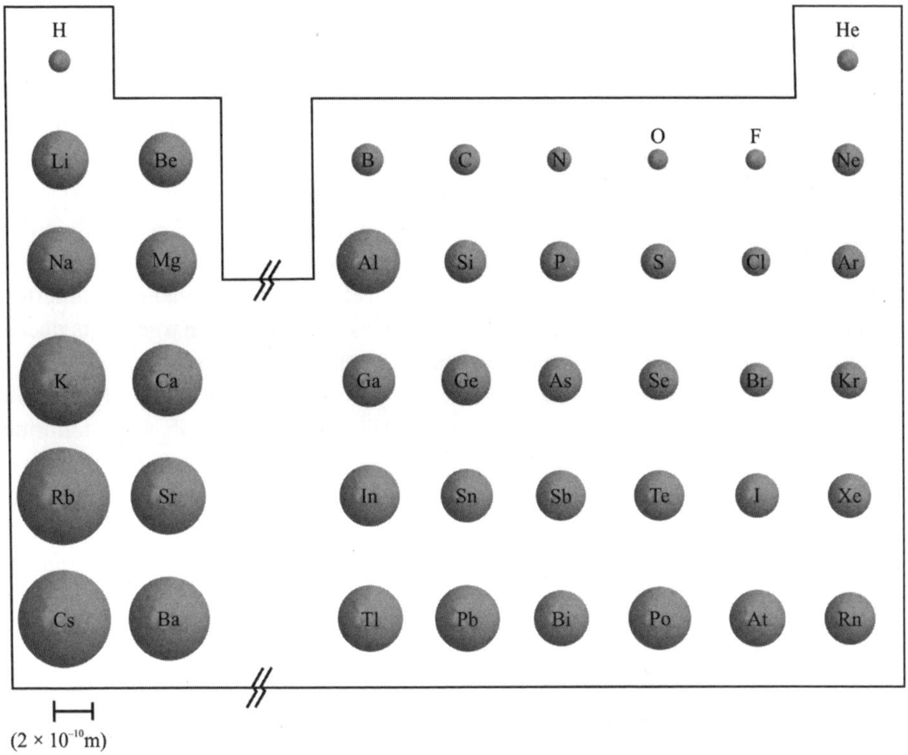

Fig. 4.6: The relative sizes of the atoms of the representative elements.

Ionization Energy

The ionization energy of an element is the minimum energy required to remove an electron (the most looselybound) from a gaseous atom of that element, leaving a positive ion. An equation expressing the ionization of sodium would be:

$$Na \rightarrow Na^+ + e^-$$

Electrons are held in the atom by the attractive force of the positively charged nucleus. The farther the outermost electrons are from the nucleus, the less tightly they are held. Thus, the ionization energy within a group of elements decreases as the elements increase in atomic number. Among the atoms of naturally occurring alkali metals, the single valence electron of cesium is farthest from the nucleus (in the sixth principal energy level) and we can correctly predict that the ionization energy of cesium is the lowest of all the alkali metals. (Recall that francium is not naturally occurring).

From left to right across a period, the ionization energy of the elements tends to increase. The number of protons in the nucleus (the nuclear charge) increases, yet the valence electrons of the elements are in the same energy level. It becomes increasingly more difficult to remove an electron from the atom. The ionization energy of chlorine is much greater than

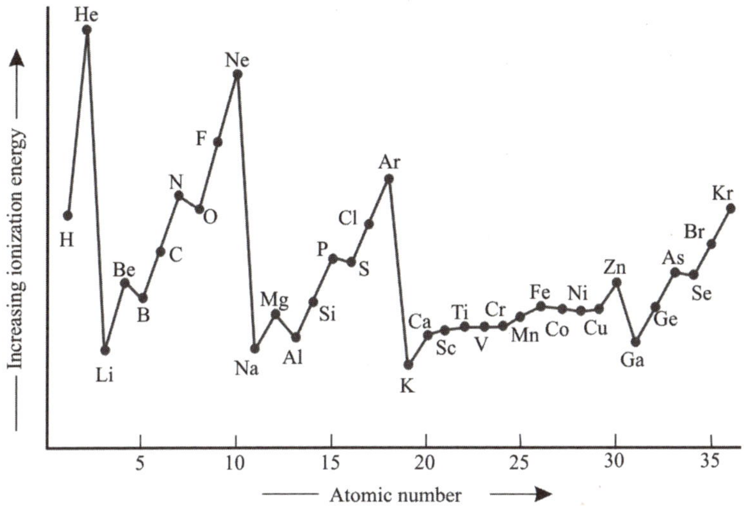

Fig. 4.7: The ionization energies of elements 1 through 36 plotted against their atomic numbers.

that of sodium, an element in the same period. The ionization energies of elements 1 through 36 are plotted against their atomic numbers in the following figure.

The peaks of the graph are the high ionization energies of the noble gases. The height of the peaks decreases as the number of the highest occupied energy level increases. The low points of the graph are the ionization energies of the alkali metals, which have only one electron in their valence shell. These points, too, decrease slightly as the number of the highest occupied energy level increases. The graph shows that ionization energy is periodically related to atomic number. Even within a row of the periodic table, the variations in ionization energy are closely related to electron configuration. In fact the ionization energy of an element is a measure of its metallic nature. From the figure, we see that each alkali metal has the lowest ionization energy of the elements in its period. Therefore, alkali metals are the most metallic elements. From bottom to top in the periodic table and from left to right across it, the metallic nature of the elements decreases.

Nonmetals, located in the upper-right section of the periodic table, have high ionization energies. Except for the noble gases, fluorine has the highest ionization energy. Therefore, excluding the noble gases, fluorine is the least metallic (or most nonmetallic) element. From top to bottom in a column or to the left of fluorine, elements become more metallic. In summary, ionization energy increases from bottom to top of a column and from left to right across a period.

Electron Affinity

Electron affinity is closely but inversely related to the ionization energy. Electron affinity is the energy change that occurs when an electron is added to a neutral atom. For a nonmetal this change is usually a release of energy. The equation showing this reaction for chlorine is: $Cl_2 + 2e^- \rightarrow 2Cl^-$

Electron affinities are fairly difficult to measure. Their accurate values have been determined for only a few elements. In general the values become increasingly negative from left to right across a period. Consequently, a halogen will have the most-negative electron affinity of all the representative elements in its period (remember that the noble gases are not representative elements). Electron affinity does not change with the same regularity as does atomic radius or ionization energy and relative values cannot be predicted as easily. The proceeding figure summarizes how atomic radii, ionization energies and metallic properties change within the periodic table.

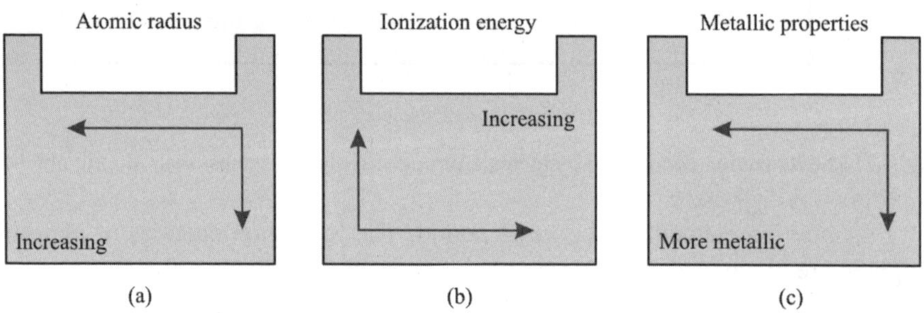

Fig. 4.8: Trends of various atomic properties as related to position in the periodic table

13. HOW ARE ELEMENTS CATEGORIZED IN THE PERIODIC TABLE?

The elements on the periodic table are categorized as under:

The Representative Elements

Elements in the s and p blocks are collectively known as representative elements or main group elements. The term representative dates from early times, when chemists believed that the chemistry of these elements was representative of all elements. Group 8/18 is never included in the representative elements because the chemistry of the noble gases is unique to them. In period 7 there are no elements in the p block. The p block of period 7 would contain elements with atomic numbers greater than 112; such elements have not yet been found in the Earth's crust. In the s and p blocks, the period in which the element occurs has the same number as the highest energy level that contains electrons in a ground-state atom. The number of the column in which an element is found is the same as the number of s and p electrons in that level. Sodium is a representative element with 11 electrons. Its electron configuration is: $1s^2 2s^2 2p^6 3s^1$. Sodium is in column 1 of the third period. In a sodium atom, the highest-energy principal energy level containing electrons is the third energy level and that energy level contains one electron.

The Transition Elements

The transition elements (or transition metals, for they are all metals) are the elements found in the short columns of the d block. Many of these elements are familiar to us. The coinage metals: gold, silver and copper,-are here. So is iron, the principal ingredient of steel, as well as those elements that are added to iron to make particular kinds of steel: chromium, nickel and manganese. Many of the properties of the transition elements are related to the fact that, in their electron structures, the occupied s and d sublevels of highest energy are very close in energy.

The Inner Transition Elements

The inner transition elements are those found in the f block of the periodic table (in the two rows below the main body of the table). The elements in this block are chemically very much alike, which will seem reasonable when we consider that they have the same electron configurations in the two outermost energy levels. The differences occur in the next further-in energy level. For example, the electron configuration of cerium (Ce, atomic number-58) is:

$$1s^22s^22p^63s^23p^64s^23d^{10}4p^65s^24d^{10}5p^66s^24f^2$$

and that of praseodymium (Pr, atomic number -59) is:

$$1s^22s^22p^63s^23p^64s^23d^{10}4p^65s^24d^{10}5p^66s^24f^3$$

The only difference between these two configurations is in the number of 4f electrons. Both the fifth and sixth energy levels contain electrons. The elements in the lanthanide series are also known as the rare earths. They are used extensively in producing monitors for color television. The elements in the actinide series are all radioactive and only three are found in appreciable concentration in the Earth's crust. Of the others, only some have been found in trace amounts in the Earth or in the stars. All have been produced in laboratories as products of nuclear reactions.

14. WHY THE ELECTRON CONFIGURATION OF THE NOBLE GASES IS TAKEN AS CORE NOTATION?

We know that there is a relationship between the electron configuration of an element and its location in the periodic table. Let us look closer now at the electron configurations of the noble gases, the elements in Group 8/18 of the periodic table. The electron configurations of these elements are shown in the proceeding table (Table 4.5).

A careful examination of these configurations shows that none has any partially filled sublevels. The symbol of a noble gas enclosed in brackets is used to represent those filled sublevels. As an example, let us consider the electron configuration of bromine: Br: $1s^22s^22p^63s^23p^63d^{10}4s^24p^5$ .The first 18 electrons are in the same orbitals as those of an atom of argon. If we use the symbol [Ar] to represent those 18 electrons, we can write the electron configuration of bromine as Br: [Ar] $3d^{10}4s^24p^5$

Table 4.5: Electron configurations of the noble gases (Group 8/18 elements)

Element symbol	*Atomic number*	*Electron configuration*
He	2	$1s^2$
Ne	10	$1s^22s^22p^6$
Ar	18	$1s^22s^22p^63s^23p^6$
Kr	36	$1s^22s^22p^63s^23p^63d^{10}4s^24p^6$
Xe	54	$1s^22s^22p^63s^23p^63d^{10}4s^24p^64d^{10}5s^25p^6$
Rn	86	$1s^22s^22p^63s^23p^63d^{10}4s^24p^64d^{10}4f^{14}5s^25p^65d^{10}6s^26p^6$

This device is useful because we can write electron configurations more quickly. More importantly, this notation emphasizes the electron configurations in the higher energy levels, where the differences are important in determining the chemistry of an element. This use of the noble gases to represent certain configurations is known as core notation. The symbol of a noble gas enclosed in brackets represents the inner, filled orbitals of an element. Additional electrons are shown outside the brackets in the standard way. Note that only the noble gases can be used in core notation. When using this method, remember that, even though the inner configuration of an element may be written the same as that of a noble gas, the energies of these inner electrons are slightly different. The following table shows, in core notation, the electron configurations of the elements in Groups 1 and 6/16 of the periodic table. Notice how this method emphasizes the similar structure of the elements in a single column.

Table 4.6: Electron configurations of elements in Groups I and VI/16, using core notations

Group 1		*Group 6/16*	
H	$1S^1$		
Li	$[He]2s^1$	O	$[He]2s^22p^4$
Na	$[Ne]3s^1$	S	$[Ne]3s^23p^4$
K	$[Ar]4s^1$	Se	$[Ar]4s^23d^{10}4p^4$
Rb	$[Kr]5s^1$	Te	$[Kr]5s^24d^{10}5p^4$
Cs	$[Xe]6s^1$	Po	$[Xe]6s^24f^{14}5d^{10}6p^4$
Fr	$[Rn]7s^1$		

15. WHAT IS OCTET RULE?

We have already observed that the noble gases are very unreactive. The lack of this reactivity is generally attributable to a stable electron configuration. Looking back to the electronic configuration, we can see that all the noble gases except helium have eight electrons (two s and six p) in the highest occupied energy level. When atoms of the other representative elements react, they lose, gain, or share enough electrons to attain the noble-gas electron structure, a complete octet, eight electrons, in their outer shell. This tendency is expressed by the octet rule: An atom generally reacts in ways that give it an octet of electrons in its outer shell. Hydrogen and lithium are exceptions; they react in ways that give them the same electron configuration as helium, with two outer-shell electrons. An atom with one, two, or three valence electrons usually reacts by losing these electrons to acquire the electron configuration of the noble gas next below it in atomic number. An atom with six or seven valence electrons will usually react by adding enough electrons to acquire the electron configuration of the noble gas next above it in atomic number. Other atoms may attain a complete octet by sharing electrons with a neighboring atom.

16. WHAT IS SIGNIFICANCE OF THE VALENCE ELECTRONS IN THE PERIODIC TABLE?

In discussing the chemical properties of an element, we often focus on electrons in the outermost occupied energy level. These outer-shell electrons are called valence electrons and the energy level they occupy is called the valence shell. Valence electrons participate in chemical bonding and chemical reactions. The valence electrons of an element are shown by using a representation of the element called an electron-dot structure or Lewis structure, named after G.N. Lewis, the twentieth-century American chemist who first pointed out the importance of outer-shell electrons. A Lewis structure shows the symbol of the element surrounded by a number of dots equal to the number of electrons in the outer energy level of the atoms of the element.

We have noticed in writing electron configurations that the s sublevel of a principal energy level n is always occupied before d electrons are added to the principal energy level numbered n – 1. Immediately after filling the d sublevel of principal level n – 1, the p sublevel of principal level n is filled and the next sublevel filled will be the s sublevel of the n + 1 principal energy level. The significance of these observations is that, in the electron configuration of any atom, the principal energy level with the highest number that contains any electrons cannot contain more than eight electrons. This also means that the valence electrons of an atom are the s and p electrons in the occupied principal energy level of highest number. Consequently, no atom can have more than eight valence electrons.

In drawing the Lewis structure of an atom, we imagine a four-sided box around the symbol of the atom and consider that each side of that box corresponds to an orbital. We represent each valence electron as a dot. The first two valence electrons will be s electrons; they would be represented by two dots on a side (it doesn't matter which side) of the symbol. The valence electrons that are in the p subshell are placed first, one on each of the remaining sides of the symbol and then a second one is added to each side. This method of filling is similar to the one used in drawing box diagrams of electron configurations. As an example, consider the Lewis structure of sodium. Looking back at Table 4.6, we see that the core notation for sodium is $[Ne]3s^1$. This tells us that a sodium atom has one electron in its outer shell, so its Lewis structure is Na•. The core notation for selenium is $[Ar]3d^{10}4s^24p^4$. Its Lewis structure is: •• Se •• • • . The ten 3d electrons of selenium are not shown because they are not in the outer shell, which is the principal energy level 4.

17. HOW VARIOUS FEATURES OF THE PERIODIC TABLE CORRELATE THE ELECTRON CONFIGURATION OF THE ELEMENTS AND THEIR POSITION IN THE TABLE?

We know that the periodic table is an arrangement of the elements wherein metals are separated from the nonmetals. In this section let us show how the various features of the table relate to the electron configuration of the different elements and to their position in the table. First let us point out those features using the complete periodic table in which the elements are placed in rows and columns of varying length. Seven rows are used to show all of the elements now known. These rows are called periods and each period is numbered (From 1-7). We may note that the elements labeled as "lanthanides", placed below the table, belong in period 6 between element 57 (lanthanum) and element 72 (hafnium). In some periodic tables, lanthanum is the first member of the lanthanide series. Similarly, the element labeled as "actinides" belongs to period 7 between element 89 (actinium) and element 104 (rutherfordium). Again, in some tables actinium is the first member of the actinide series. These two displays are customarily put below the table so that the table will fit into a reasonable space. The columns of the periodic table vary in length. Some are numbered, but the short columns, those in the middle of the table, have not been numbered in the older versions of the periodic table.

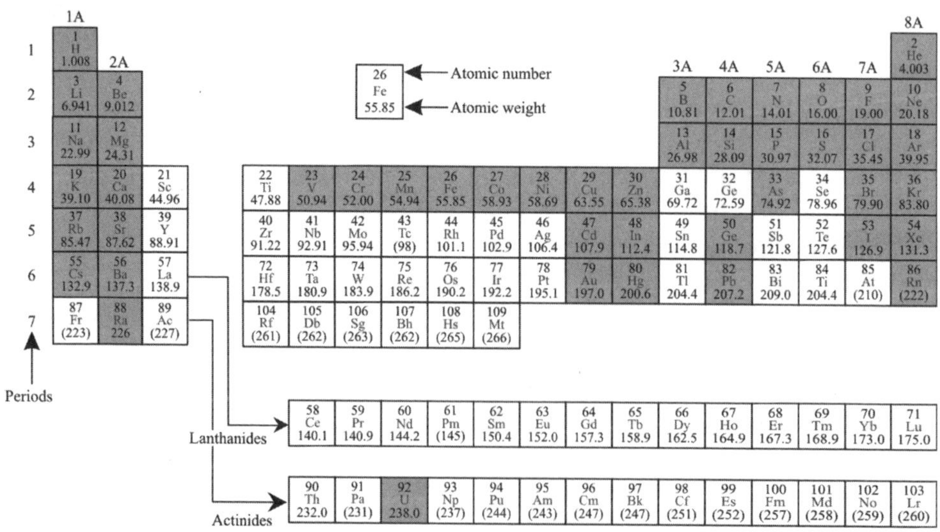

Fig. 4.9: Older version of the periodic table

The elements in a column make up a family of elements. A family is also known as a group. Thus the elements in column 8/18 are known as the family or group of noble gases. We may also show the periodic table without the symbols of the elements. Instead it may show the last sublevels filled in describing the electron configurations of the elements in each section. We can use these tables to relate the electron configuration of an element to its position in the periodic table.

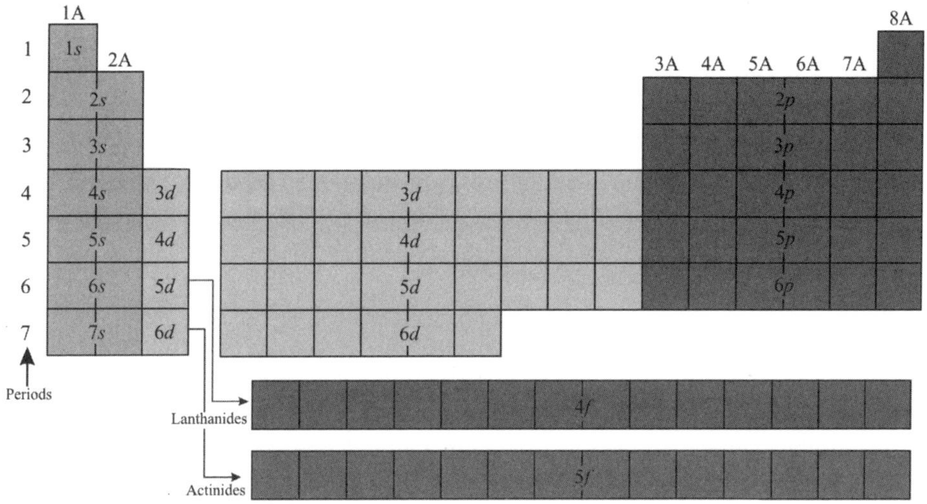

Fig. 4.10: The periodic table and the energy level subshells

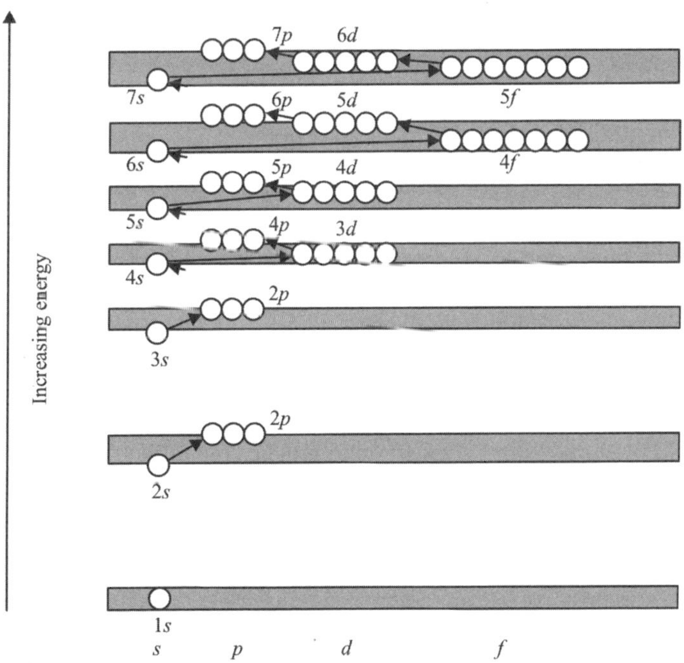

Fig. 4.11: The principal energy levels of an atom and the sublevels and orbitals, each contains. The arrows show the order in which the sublevels fill.

In period 1, there are two boxes. In the usual table, these boxes would contain the symbols for hydrogen and helium, the only elements in this period. We show instead the letter s indicating that the last added electron for the elements in these boxes is in the 1s sublevel. In period 2, there are eight boxes. Instead of symbols for eight elements, the figure concerned shows s in the first two boxes and p in the last six boxes, showing that the 2s and 2p sublevels are being filled as the electron configurations of the elements in these boxes are completed. Period 3 also has eight boxes, which would correspond to the electrons needed to fill the 3s and 3p sublevels.

Let us look back now to the figure that shows the order in which the sublevels fill. Notice that the 4s sublevel is filled immediately after the 3p sublevel. The figure shows that elements whose last added electron goes into an s sublevel are in columns 1 and 2. So we must start here a new period, period 4 and put boxes for the elements formed by filling the 4s sublevel in those columns. The figure shows that the next sublevel to fill is the 3d sublevel. These are the first d electrons added, so we start new columns for the elements formed by their addition. Ten electrons are needed to fill the five d orbitals, so we start ten columns in this fourth period, placing the columns next to column 2 and between it and column 3. The 4p sublevel is filled next, after the 3d sublevel. The boxes for the elements formed by filling the p orbitals are in place under the boxes for elements formed by adding the 3p electrons. By consulting the figure concerned, we see that the next sublevels filled are in the order: 5s, 4d and 5p. Boxes for the elements formed by filling the orbitals of these sublevels are arranged as were those in period 4. Just as period 4 contains more elements than period 3, period 6 contains more elements than period 5. Period 6 starts with elements whose last added electron is in the 6s sublevel. The next step is where period 6 differs from period 5. Looking again at the figure we note that the 4f sublevel is filled after the 6s sublevel and before the 5d sublevel. We will need 14 boxes to contain the electrons needed to fill the seven f orbitals. These are the boxes of the lanthanide series, shown below the table. There is some evidence that these orbitals do not fill before one electron is in a 5d orbital, so we have shown the lanthanide series coming after the first d column. After the 4f orbitals are filled, boxes are shown for the rest of the elements formed by adding 5d and 6p electrons. The seventh period contains boxes for the elements formed by filling the 7s, the 5f (the actinide series shown below the table) and finally the 6d sublevels. Periodic table thus shows the close relationship that exists between the electron configuration of an element and its location in the periodic table. This relationship is further expressed by the following names sometimes given to parts of the table:

Columns 1 and 2	s Block
Columns 3–8 (13–10)	p Block
Short columns	d Block
Lanthanides and actinides	f Block

The groups of elements found in these blocks are also known by other names.

18. WHY DO WE NEED KNOWLEDGE OF ELECTRON CONFIGURATION AND PERIODIC TABLE?

Our knowledge of the electron configuration of atoms is based on studies of the properties of light and spectra of atoms. Because of the electrons' negligible mass and high energies, their properties are similar to those of light. Each electron in a ground-state atom has unique energy properties. In describing these properties, we assume that the nucleus of an atom is surrounded by energy levels; the energy associated with these levels increases as their distance from the nucleus increases. Each energy level is divided into sublevels that differ slightly from one another in energy. The sublevels are divided into orbitals, each of which can contain two electrons spinning in opposite directions on their axes. The electron configuration of a particular atom is described by stating how many electrons are in each sublevel of a particular principal energy level. If electron spin is important, box diagrams are used. Core notation emphasizes the outer-level electrons. If only the valence electrons are of interest, Lewis structures are used.

Although the periodicity of elements was clearly recognized in 1869 when the periodic table was first proposed, it was only in last century (20th) that the relationship of an element's electron configuration to its location in the periodic table has been clearly defined. Elements in the same long columns of the table have the same valence-shell configuration and have quite similar chemical properties. Elements in the same row of the table show a reasonably regular progression of properties, particularly with respect to atomic radius, metallic nature and ionization potential. Elements lose or gain electrons to become ions. Metals become cations by losing electrons, thus gaining a positive charge. Nonmetals become anions by gaining electrons, thus attaining a negative charge. Polyatomic ions are groups of atoms with an excess or deficiency of electrons. The oxides of metals react with water to yield hydroxide ions; hydroxides are a subset of a larger group called bases. Oxides of nonmetals typically react with water to form acids. The solution of an acid contains an excess of hydrogen ions; and that of a base contains, an excess of hydroxide ions.

19. HOW ARE IONS FORMED?

Atoms are electrically neutral as the number of positively charged protons in the nucleus of an atom equals the number of negatively charged electrons outside the nucleus. If electrons are added to or lost from an atom, the atom concerned acquires a charge and becomes an ion. Thus an ion is an atom or group of atoms having a net electric charge. A neutral atom or group of atoms becomes an ion by gaining or losing one or more electrons or protons. Since the electron and proton have equal but opposite unit charges, the charge of an ion is always expressed as a whole number of unit charges and is either positive or negative. A simple ion consists of only one charged atom and a complex ion consists of an aggregate of atoms with a net charge. If an atom or group of atoms loses electrons or gains protons, it will have a net positive charge and is called a cation. If an atom or group of atoms gains electrons or loses protons, it will have a net negative charge and is called an anion.

Since ordinary matter is electrically neutral, ions normally exist as groups of cations and anions such that the sum total of positive and negative charges is zero. In the common table salt, sodium chloride, (NaCl), the sodium cations, Na^+, are neutralized by chlorine anions, Cl^-. In the salt sodium carbonate, Na_2CO_3, two sodium cations are needed to neutralize each carbonate anion, CO^{-2}_3, because its charge is twice that of the sodium ion.

Ionization of neutral atoms can occur in several different ways. Compounds such as salts dissociate in solution into their ions, *e.g.*, in solution sodium chloride exists as free Na^+ and Cl^- ions. Substances that ionize in solution are called electrolytes; those that do not ionize, like sugar and alcohol, are called nonelectrolytes. Ions in solution conduct electricity. Ionization can also be caused by the bombardment of matter with high-speed particles or radiation. Ultraviolet radiation and low-energy X-rays excite molecules in the upper atmosphere sufficiently to cause them to lose electrons and become ionized, giving rise to several different layers of ions in the earth's atmosphere. A gas can be ionized by passing an electron current through it; the ionized gas then permits the passage of a much higher current. Heating to high temperatures also ionizes substances; certain salts yield ions in their melts as they do in solution.

20. WHAT ARE CATIONS AND ANIONS?

A cation is an atom or group of atoms carrying a net positive charge. The charge results because there are fewer electrons than protons in the cation. Cations can be formed from metals by oxidation or from neutral acids or polar compounds by ionization. An anion is an atom or group of atoms carrying a negative charge. The charge results because there are more electrons than protons in the anion. Anions can be formed from nonmetals by reduction or from neutral acids or polar compounds by ionization. Anionic species include Cl^-, SO_4^{2-} and CH_3COO^-. Highly colored intermediates in organic reactions are often radical anions (anions containing an unpaired electron). Salts are made up of anions and cations.

Cations (Positive Ions)

When a neutral atom loses an electron, it forms a positively charged ion, called a cation. In general, metals lose electrons to form cations. The atom thereby attains the electron configuration of the noble gas next below it in atomic number. For example, an alkali metal loses one electron to form a cation with a single positive charge. Sodium loses its single 3s valence electron to form the ion Na^+, which has the electron configuration of neon:

$$Na^{\bullet} \rightarrow Na^+ + e^-$$

An alkaline earth metal loses two electrons to form a cation with a charge of +2. In forming the magnesium ion, Mg^{2+}, a magnesium atom loses its two valence electrons:

$$Mg: \rightarrow Mg^{2+} + 2\ e^-$$

Aluminum loses its three valence electrons to form a cation with a charge of +3:

$$Al : \rightarrow Al^{3+} + 3\ e^-$$

Transition elements and the metals to their right do not always follow the octet rule; frequently they form more than one cation. For example, iron forms Fe^{2+} and Fe^{3+}; cobalt forms Co^{2+} and Co^{3+}. The names of these ions must indicate the charge they carry. The preferred system of nomenclature (naming) is the one recommended by the International Union of Pure and Applied Chemistry (IUPAC). In this system, the name of the metal is followed by a Roman numeral (in parentheses) showing the charge on the ion. No extra space is left between the name and the number. Thus, Fe^{2+} is iron(II) (pronounced "iron two") and Fe^{3+} is iron(III). In the older system, the name of the cation of lower charge ends in ous and the name of the cation of higher charge ends in ic. When ions are formed, electrons are pulled off one by one from the atom/ion. Thus, the first electron is removed from a neutral atom, the second electron from an ion of charge +1, the third electron from an ion of charge +2 and so on. The amount of energy necessary to remove an electron increases dramatically as the positive charge of the ion increases. To remove a fourth electron and form an ion of charge +4 is energetically unlikely.

Anions (Negative Ions)

When a neutral atom gains an electron, it forms a negatively charged ion, called an anion. Typically, nonmetals form anions, gaining enough electrons to acquire the electron configuration of the noble gas of next higher atomic number. Elements of group 6/16, with six valence electrons, form anions by gaining two electrons; the halogens, with seven valence electrons, form anions by gaining one electron. The names of these anions include the root name of the element and the ending -ide.

The ions described in the preceding paragraphs are monatomic ions; that is, each contains only one atom. Many polyatomic ions are also known. Polyatomic ions are groups of atoms bonded together that carry a charge due to an excess or deficiency of electrons. The following table lists the formulas and names of several common polyatomic ions. The symbols in the formula show which elements are present. The subscripts tell how many atoms of each element are present in the ion ("1" is understood).

Table 4.7: Common polyatomic ions

Charge	*Formula*	*Name*
+1	NH_4^+	Ammonium ion
–1	OH^-	Hydroxide ion
	NO_3^-	Nitrate ion
	HCO_3^-	Bicarbonate ion
–2	CO_3^{-2}	Carbonate ion
	SO_4^{-2}	Sulfate ion
–3	PO_4^{-3}	Phosphate ion

So far we have shown that metals usually have one, two, or three valence electrons. They have low ionization energies and are found to the left in the periodic table. Nonmetals have four, five, six, or seven valence electrons, have high ionization energies and are in the upper-right section of the periodic table. All these properties are closely related to electron configurations and we have used electron configurations in discussing them. However, before electron configurations were known, people did knew the differences between metals and nonmetals. Early chemists observed how the physical properties of a metal (malleability, luster and conductivity), contrasted with those of nonmetals. These chemists also identified differences in chemical properties. The compounds formed when oxides of metals react with water are very much alike and very different from those formed when oxides of nonmetals react with water. When a metallic oxide reacts with water, a hydroxide is formed.

Chapter 5

PHYSICAL VERSUS CHEMICAL CHANGES

1. WHAT ARE PHYSICAL AND CHEMICAL PROPERTIES OF MATTER?

In general the variety of observable properties of matter is divided into two classes: Physical and chemical. Physical properties are those properties whose observation does not involve a change in the composition of the sample. For example, a sample of water does not change its composition if we pour it from a tall pitcher into a flat bowl. It can be frozen to a solid or vaporized to steam. Yet it remains water, with the formula H_2O. Through all these physical changes, the composition of water is unchanged. Another physical change is the crushing of limestone. Even though the limestone is crushed to smaller particles, the composition of the limestone does not change.

But the observation of a chemical property involves a change in the composition of the sample, that is, a chemical change. For example, when an electric current is passed through water which contains a few drops of sulfuric acid, the water decomposes to gaseous hydrogen and oxygen. Water molecules are no longer present; instead, we have two new substances: hydrogen and oxygen. Similarly, limestone (known chemically as calcium carbonate) when heated, is converted into two new substances: lime (calcium oxide) and carbon dioxide that have very different properties from those of limestone.

2. HOW DO WE DESCRIBE A CHEMICAL REACTION?

A chemical reaction can be described with an equation that shows the formula of each reactant and each product. The equation must be balanced to show that no mass is lost or gained during the reaction. An equation may also show the physical state of each component and the energy changes associated with the reaction. For example, a complete equation for the burning of propane, C_3H_8, is:

$$C_3H_8(g) + 5O_2(g) \rightarrow 3CO_2(g) + 4H_2O(l) \ \Delta H = -2220 \text{ kJ}$$

The symbol after each formula shows the physical state of that substance. The above equation can be read as: One molecule of gaseous propane reacts with five molecules of gaseous oxygen to yield three molecules of gaseous carbon dioxide and four molecules of liquid water.

An equation can also be read in terms of moles. This above said equation states: One mole of gaseous propane reacts with five moles of gaseous oxygen to yield three moles of gaseous carbon dioxide and four moles of liquid water. The ΔH (delta H) term following the equation describes the enthalpy change. For this equation, the enthalpy term states that the combustion of one mole of gaseous propane to gaseous carbon dioxide and liquid water is accompanied by the release of 2220 kJ of energy. The equation given for the burning of propane is a complete equation. It is to be noted that when a discussion centers on the reactants and products of a reaction, rather than on the associated energy changes, the physical states and the enthalpy change (ΔH) are often omitted.

3. WHAT ENERGY CHANGES ACCOMPANY CHEMICAL REACTIONS?

All changes, whether chemical or physical, are accompanied by a change in energy. Each reacting molecule possesses a certain amount of energy due to the nature of chemical bonds. So does each product molecule. As the bonds of the reacting molecules break and the new bonds of the products form, energy is released or absorbed, depending on whether the reactants have higher or lower energy than the products. We can measure energy changes in several ways. The two kinds of energy change of most interest to us are: (1) the change in free energy (ΔG), which is the energy available to do useful work and (2) the change in enthalpy (ΔH), which is the heat energy absorbed or released by the reaction and measured at constant pressure. Most chemical reactions take place under the constant pressure of the atmosphere. The energy released or absorbed by such reactions is the change in enthalpy, ΔH, which can be shown as: ΔHreaction = Hproducts - Hreactants

In reporting values of ΔH, a superscript is used to show the temperature at which the measurements were made. For example, the symbol ΔH 0°C shows that the change in enthalpy was measured at 0°C. If no temperature is shown, the enthalpy change was measured at 25°C. All changes are measured at one atmosphere pressure.

The value of ΔH given with an equation refers to that particular equation. When the enthalpy change was measured, the physical states of the components were those stated in the equation. If the physical states are different, there will be a different enthalpy change. This difference is illustrated by the next two equations for the formation of water. They differ in enthalpy change. In the first, gaseous water is formed and in the second, liquid water is formed; the difference between their enthalpy changes reflects the difference in energy content between a gas and a liquid.

$$H_2(g) + \frac{1}{2}O_2(g) \rightarrow H_2O(g)\ \Delta H = -241\text{ kJ}$$

$$H_2(g) + \frac{1}{2}O_2(g) \rightarrow H_2O(l)\ \Delta H = -286\text{ kJ}$$

The enthalpy change given for a reaction also depends on the coefficients used in the equation for the reaction. Thus, if the equation for the formation of water is written

$$2H_2(g) + O_2(g) \rightarrow 2H_2O(l)\ \Delta H = -482\text{ kJ}$$

The enthalpy change is twice what it was in the previous equation for the formation of gaseous water when the coefficient of water was 1. This last problem can be resolved by doing as we do in several equations where we report the enthalpy change per mole of one component of the reaction, thus removing any ambiguity in interpretation.

4. WHAT ARE ENDOTHERMIC AND EXOTHERMIC REACTIONS?

A reaction that absorbs energy is an endothermic reaction; its enthalpy change (ΔH) is positive. The enthalpy of the products of the reaction is greater than that of the reactants. Energy is absorbed from the surroundings. The following reactions are endothermic.

1. The formation of hydrogen iodide:

$$\frac{1}{2}H_2(g) + \frac{1}{2}I_2(g) \rightarrow HI(l)\ \Delta H = +25.9 \text{ kJ/mol HI} \qquad (eq.\ A)$$

2. The decomposition of water:

$$H_2O(l) \rightarrow H_2(g) + \frac{1}{2}O_2(g)\ \Delta H = +285.8 \text{ kJ/mol } H_2O \qquad (eq.\ B)$$

A reaction that releases energy is an exothermic reaction; its enthalpy change is negative. The enthalpy of the products is less than that of the reactants. Energy is released to the surroundings. The following reactions are exothermic.

The combustion of methane:

$$CH_4(g) + 2O_2(g) \rightarrow CO_2(g) + 2H_2O(l)\ \Delta H = -891 \text{ kJ/mol } CH_4 \qquad (eq.\ C)$$

The formation of water:

$$H_2(g) + \frac{1}{2}O_2(g) \rightarrow H_2O(l)\ \Delta H = -285.8 \text{ kJ/mol } H_2O \qquad (eq.\ D)$$

Notice that the decomposition of water (equation B) is endothermic and requires the input of 285.8 kJ of energy per mole of water decomposed. The reverse reaction, the formation of one mole of water from hydrogen and oxygen (equation D), is exothermic and releases 285.8 kJ energy. The amount of energy is the same, but the sign of the energy change is different.

Another example as the relationship between energy change and the direction of a reaction is the formation and decomposition of glucose. Glucose ($C_6H_{12}O_6$) is formed from carbon dioxide and oxygen in the cells of green plants in the process called photosynthesis. Photosynthesis is an endothermic reaction. The source of the energy for the formation of glucose is light (radiant energy), usually from the sun.

$$6CO_2(g) + 6H_2O(l) \xrightarrow{\text{photosynthesis}} C_6H_{12}O_6(s) + 6O_2\ (g)$$
$$\Delta H = +2.80 \times 10^3 \text{ kJ/mol glucose}$$

$$C_6H_{12}O_6(s) + 6O_2(g) \xrightarrow{\text{metabolism}} 6CO_2\ (g) + 6H_2O(l)$$
$$\Delta H = -2.80 \times 10^3 \text{ kJ/mol glucose}$$

Thus, green plants have the remarkable ability to trap the energy of sunlight and use that energy to produce glucose from carbon dioxide and water. The energy is stored in the glucose. Animal and plant cells have the equally remarkable ability to metabolize glucose and use the energy released to maintain body temperature or do biological work, such as contracting muscles or thinking.

5. HOW ARE CHEMICAL CHANGES CLASSIFIED?

One common way of classifying chemical reactions is to divide them into four categories: combination, decomposition, displacement and double displacement. We may give some examples of each type for two reasons: (1) to ensure that students understand the scope of each category and (2) to help them gain experience in interpreting and balancing equations. The following are parts of a chemical equation:

Reactants: The substances that combine in the reaction. (Their formulas must be correct.)

Products: The substances that are formed by the reaction. (Their formulas must also be correct.)

ΔH: The enthalpy (heat energy) change accompanying the reaction. Energy is released if $\Delta H < 0$ or negative and energy is absorbed if $\Delta H > 0$ or positive

Arrows →: Found between reactants and products, means "reactants react to form products."

An arrow of the type:/→: means that the equation is not balanced

↑

Is placed after the formula of a product that is a gas

↓

Is placed after the formula of a product that is an insoluble solid, that is, a precipitate Physical state indicates the physical state of the substance whose formula it follows.

(g) Indicates that the substance is a gas

(l) Indicates that the substance is a liquid

(s) Indicates that the substance is a solid

(aq) Means that the substance is in aqueous (water) solution

Coefficients: The numbers placed in front of the formulas to balance the equation.

Conditions: Words or symbols placed over the arrow (→) to indicate conditions used to make the reaction occur.

Δ (delta) means that Heat is added; hv (nu) means that light is added and Elec means that electrical energy is supplied.

6. WHAT IS THE COMPLETE MEANING OF THE FOLLOWING EQUATION?

$$2H_2O(l) \xrightarrow{elec} 2H_2(g) + O_2(g)\ \Delta H\ -591\ kJ$$

The equation shows that, when an electric current is passed through water, two molecules (or two moles) of liquid water will decompose to yield two molecules (or two moles) of gaseous hydrogen and one molecule (or one mole) of gaseous oxygen and energy of 591 KJ will be absorbed for every two moles of water decomposed. The equation also shows that a molecule of water contains two atoms of hydrogen and one atom of oxygen and that hydrogen and oxygen are diatomic gases.

7. WHAT HAPPENS IN 'COMBINATION REACTIONS'?

If two substances combine to form a single compound, it is a combination reaction. Some examples of such reactions are the reaction of solid magnesium with gaseous oxygen to form magnesium oxide, a solid and the reaction of a hydrogen gas with chlorine gas to form gaseous hydrogen chloride shown below as reactions a & b respectively:

$$2Mg(s) + O_2(g) \rightarrow 2MgO(s)\ (a)$$

$$H_2(g) + Cl_2(g) \rightarrow 2HCl(g)\ (b)$$

Note: The brilliant white light associated with some fireworks is due to the release of energy when magnesium reacts with oxygen.

Other combination reactions have compounds as reactants. For example, the reaction of gaseous carbon dioxide with solid calcium oxide to form solid calcium carbonate is an example of such a reaction.

$$CaO(s) + CO_2(g) \rightarrow CaCO_3(s)$$

8. WHAT HAPPENS IN DECOMPOSITION REACTIONS?

In a decomposition reaction, a compound is decomposed either to its component elements or to some other compounds. Although some compounds decompose spontaneously, usually light or heat is needed to initiate the decomposition. The antiseptic hydrogen peroxide is sold in opaque brown bottles because hydrogen peroxide decomposes in light. The equation for this decomposition is:

$$2H_2O_2(aq) \xrightarrow{hv} 2H_2O(l) + O_2(g)$$

Oxygen can be prepared by heating solid potassium chlorate in the presence of manganese dioxide, a catalyst. A catalyst is a chemical that, when added to a reaction mixture hastens the reaction but can be recovered unchanged after the reaction is complete.

$$2KClO_3(s)\ (\text{in presence of } MnO_2) \rightarrow 2KCl(s) + 3O_2(g)$$

Note: Light hastens the decomposition of hydrogen peroxide. The dark bottle in which hydrogen peroxide is usually stored keeps the light out, thus protects the hydrogen peroxide from decomposition.

When slaked lime, $Ca(OH)_2(s)$, is heated, lime (CaO) and water vapor are produced:

$$Ca(OH)_2(s) \rightarrow CaO(s) + H_2O(g)$$

9. WHAT ARE CHARACTERISTICS OF DISPLACEMENT REACTIONS?

In displacement reactions, an uncombined element reacts with a compound and displaces an element from that compound. For example, bromine is found in seawater as sodium bromide. When chlorine is bubbled through seawater, bromine gas is released and a solution of sodium chloride is formed:

$$2NaBr(aq) + Cl_2(g) \rightarrow 2NaCl(aq) + Br_2(g)$$

In another example, when an iron nail is dropped into a solution of copper(II) sulfate, iron(II) sulfate is formed in solution and metallic copper is deposited:

$$CuSO_4(aq) + Fe(s) \rightarrow FeSO_4(aq) + Cu(s)$$

As another displacement reaction, the reaction of metallic copper with silver nitrate may be mentioned in which copper displaces silver under the equation:

$$2AgNO_3 + Cu \rightarrow Cu(NO_3)_2 + 2Ag$$

10. WHAT HAPPENS IN DOUBLE-DISPLACEMENT REACTIONS?

In double displacement reactions that are sometimes called as metathesis or ion exchange process, two ionic compounds react to form two different compounds. These reactions fall into a pattern that can be expressed as: $AB + CD \rightarrow CB + AD$, in which A and C are cations and B and D are anions. These reactions are often called "exchanging-partner" reactions because the cations A and C exchange the anions with which they are associated. Double-displacement reactions fall into two categories: (1) those in which an acid reacts with a base to form a salt and water, which are known as neutralization reactions and (2) those in which one of the products is insoluble, which are usually precipitation reactions, although occasionally the insoluble product is a gas.

11. WHAT REACTIONS ARE CATEGORIZED AS 'NEUTRALIZATION REACTIONS'?

In neutralization reactions, an acid reacts with a base (alkali) to form a salt and water. We know that in the most general sense an acid is a compound that liberates hydrogen ions in aqueous solution and a base is a compound that liberates hydroxide ions in aqueous solution (we will center here on hydroxides, a subgroup of bases). A salt is defined as an ionic compound in which the cation is not hydrogen and the anion is not hydroxide ion. These reactions are called neutralization reactions because the base neutralizes the acid. Some examples of these reactions are:

a. The reaction of sodium hydroxide with hydrochloric acid to form sodium chloride and water:

$$NaOH(aq) + HCl(aq) \rightarrow NaCl(aq) + H_2O(l)$$

We may note that sodium chloride, the salt formed results from combination of the cation, Na^+, of the base with the anion, Cl^- of the acid. The formula of the salt is the neutral combination of these ions, here it is a 1:1 combination in sodium chloride, NaCl.

b. The reaction of magnesium hydroxide with phosphoric acid to form magnesium phosphate and water:

$$3Mg(OH)_2(aq) + 2H_3PO_4(aq) \rightarrow Mg_3(PO_4)_2(aq) + 6H_2O(l)$$

Here again the salt formed, magnesium phosphate, $Mg_3(PO_4)_2$, is a neutral combination of the cation, Mg^{2+}, of the base, with the anion, PO_4^{3-} of the acid. A polyprotic acid is one whose molecules ionize to yield more than one hydrogen ion. Sulfuric acid, H_2SO_4, phosphoric acid, H_3PO_4 and carbonic acid, H_2CO_3, are examples of polyprotic acids. When a polyprotic acid happens to be one of the reactants, neutralization may be incomplete and an acid salt may result. An example of this reaction is:

$$NaOH(aq) + H_2SO_4(aq) \rightarrow NaHSO_4(aq) + H_2O(l)$$

In this reaction, only one of the hydrogen atoms of the diprotic acid, H_2SO_4, has reacted and the product is an acid salt, sodium hydrogen sulfate. The addition of more sodium hydroxide neutralizes the second hydrogen of this diprotic acid:

$$NaHSO_4(aq) + NaOH(aq) \rightarrow Na_2SO_4(aq) + H_2O(l)$$

12. WHAT DOUBLE-DISPLACEMENT REACTIONS FORM INSOLUBLE IONIC PRODUCTS?

Precipitation reactions, the second group of double-displacement reactions, result in the formation of insoluble ionic compounds. Ionic compounds differ enormously in the extent to which they dissolve in water that is their solubility. We may illustrate this point by listing the solubilities of several ionic compounds in cold water. Several species, such as barium iodide and silver(I) nitrate, are very soluble in water, whereas others, such as lead(II) chloride, are only slightly soluble. Others, such as silver(I) chloride, are virtually insoluble. Generally, if more than 0.1 g of an ionic solid dissolves in 100 mL (0.1L) of water, the compound is said to be soluble (less than 0.1g calcium carbonate, barium sulfate and silver(I) chloride dissolve in 100 mL water). Therefore, they are classified as insoluble compounds. Following are the solubility rules by which the solubility of an ionic compound in water can be predicted.

1. All common salts of ammonium ion are soluble
2. All common salts of sodium and potassium are soluble
3. All nitrates are soluble
4. All acetates except Iron (III) acetate, Fe $(CH_3COO)_3$ are soluble
5. All chlorides, bromides and iodides are soluble except those of silver, mercury(I), lead(II). However $PbCl_2$ and $BrCl_2$ are slightly soluble in hot water
6. All sulfates except $CaSO_4$, $BaSO_4$, $PbSO_4$ and Ag_2SO_4 are soluble
7. Only ammonium phosphate and alkali metal phosphates and carbonates are soluble
8. Only ammonium hydroxide and alkali metal hydroxides are soluble (Hydroxides of calcium, barium and strontium are slightly soluble).

13. WHAT HAPPENS IN A PRECIPITATION REACTION?

In precipitation reactions, solutions of two ionic compounds are combined. If two of the ions in the resulting mixture combine to form an insoluble compound, a precipitation reaction is said to occur. If no insoluble product is produced, no reaction occurs For example when a clear colorless solution of lead nitrate [$Pb(NO_3)_2$] is added to a clear colorless solution of sodium iodide (NaI), a yellow precipitate of lead iodide (PbI_2) appears. Similarly if a solution of barium iodide is added to a solution of ammonium nitrate, no reaction takes place because the predicted products barium nitrate and ammonium iodide are both soluble.

$$BaI_2(aq) + 2NH_4NO_3(aq) \rightarrow Ba(NO_3)_2(aq) + NH_4I(aq)$$

A reaction would occur if a solution of barium iodide were added to a solution of silver nitrate, because one of the products, silver iodide, is insoluble.

$$BaI_2(aq) + 2AgNO_3(aq) \rightarrow Ba(NO_3)_2(aq) + 2AgI(\downarrow)$$

In the equations for these reactions, the physical state of the insoluble product, the precipitate, is indicated either by (s) or by an arrow pointing downward after its formula; the soluble components of the reaction are shown as (aq). As another example we may write the balanced equation for the reactions when solutions of lead(II) nitrate and sodium iodide react to form a yellow precipitate.

The formulas for the reactants are: $Pb(NO_3)_2$ and NaI. The formulas of the products of a reaction between these two compounds would have an interchange of anions, yielding PbI_2 and $NaNO_3$. Arranging these formulas in an unbalanced equation, we get:

$$Pb(NO_3)_2(aq) + NaI(aq) / \rightarrow PbI_2 + NaNO_3$$

Balancing of this equation requires two iodide ions and therefore 2NaI. Two molecules of sodium nitrate are formed: $Pb(NO_3)_2(aq) + 2NaI(aq) \rightarrow PbI_2(\downarrow) +$

$2NaNO_3(aq)$ Because all sodium salts are soluble, the precipitate must be lead(II) iodide; we place an arrow after that formula.

As yet another example we may mention the reaction between a solution of copper(II) nitrate and a solution of potassium sulfide yielding a heavy black precipitate. The formulas of the reactants are $Cu(NO_3)_2$ and K_2S respectively. The formulas of the products are CuS and KNO_3 respectively. We know that potassium nitrate is soluble, so the precipitate must be CuS, copper(II) sulfide. The unbalanced equation is:

$$Cu(NO_3)_2(aq) + K_2S(aq)/\!\rightarrow CuS(\downarrow) + KNO_3(aq)$$

Balancing of this equation requires two molecules of potassium nitrate. The balanced equation is:

$$Cu(NO_3)_2(aq) + K_2S(aq) \rightarrow CuS(\downarrow) + 2KNO_3(aq)$$

All nitrates are soluble, so the precipitate is copper(II) sulfide. A downward-pointing arrow is put after its formula.

Occasionally the insoluble product is a gas, as happens in the following examples:

1. Hydrogen chloride is released as a gas when concentrated sulfuric acid is added to solid sodium chloride. Although hydrogen chloride is very soluble in water, it is quite insoluble in concentrated sulfuric acid. The acid salt, sodium hydrogen sulfate, is the second product.

$$NaCl(s) + H_2SO_4(l) \rightarrow HCl(g) + NaHSO_4(s)$$

2. Acetic acid may be released as a gas in a reaction similar to that in the first example. The equation for the reaction of concentrated hydrochloric acid with sodium acetate is:

$$NaC_2H_3O_2(s) + HCl(aq) \rightarrow HC_2H_3O_2(g) + NaCl(s)$$

3. A carbonate reacts with an acid to form carbonic acid, which immediately decomposes to gaseous carbon dioxide and water.

$$Na_2CO_3(s) + 2HCl(aq) \rightarrow 2NaCl(aq) + CO_2(g) + H_2O(l)$$

14. WHAT HAPPENS IN OXIDATION-REDUCTION REACTIONS?

One way of classifying chemical reactions is to separate them into only two groups: (1) those of the reactions that do not involve a change in oxidation number but do result in a decrease in the number of ions in solution and (2) those of the reactions that involve a transfer of electrons and hence changes in oxidation number. Those reactions which result in a decrease in the number of ions in solution are usually double-displacement reactions. In a neutralization reaction, hydrogen ion combines with hydroxide ion to form the covalent, un-ionized compound water, thus decreasing the number of ions in solution. In a precipitation reaction, the insoluble product removes ions from the solution.

Other reactions in this category are those that form weak electrolytes, such as acetic acid or aqueous ammonia (often wrongly called ammonium hydroxide). Examples of these reactions are the reaction between sodium acetate and hydrochloric acid:

$$NaC_2H_3O_2(aq) + HCl(aq) \rightarrow NaCl(aq) + HC_2H_3O_2(aq)$$

and the reaction between sodium hydroxide and ammonium chloride:

$$NaOH(aq) + NH_4Cl(aq) \rightarrow NaCl(aq) + NH_3(aq) + H_2O(l)$$

Those reactions that involve a transfer of electrons include combination, displacement and decomposition reactions. Reactions that involve a transfer of electrons are known as oxidation-reduction or redox reactions. The reaction of sodium with chlorine is a redox reaction:

$$2Na + Cl_2 \rightarrow 2NaCl$$

Each chlorine atom gains an electron to form a chloride ion:

$$2e^- + :\ddot{\underset{..}{Cl}} - \ddot{\underset{..}{Cl}}: \longrightarrow 2\ Cl^-$$

Each sodium atom loses an electron to form a sodium ion:

$$\dot{Na} + \dot{Na} \longrightarrow 2\ Na^+ + 2e^-$$

The element that loses electrons is oxidized. In the reaction of sodium with chlorine, sodium is oxidized. The element that gains electrons is reduced. In this reaction, chlorine is reduced. Displacement reactions are usually oxidation-reduction reactions. A typical displacement reaction is that of copper with silver nitrate:

$$Cu(s) + 2AgNO_3(aq) \rightarrow 2Ag + Cu(NO_3)_2(aq)$$

In this reaction, the copper loses electrons (is oxidized):

$$Cu^0 \rightarrow Cu^{2+} + 2\ e\text{-}$$

and the silver ion gains electrons (is reduced):

$$Ag^+ + e^- \rightarrow Ag^0\ O$$

Oxidation without reduction is impossible. If an atom in a reaction loses electrons, another atom in the reaction must gain electrons.

15. HOW CAN WE IDENTIFY OXIDATION-REDUCTION REACTIONS?

Oxidation numbers have many uses, but the one that concerns us very often is their role in determining whether or not a particular reaction involves oxidation-reduction. In an oxidation-reduction reaction, at least two oxidation numbers change. The element that is oxidized increases its oxidation number and the element that is reduced decreases its oxidation number. In the reaction of sodium with chlorine, sodium atoms are oxidized to sodium ions; the oxidation number of sodium increases from 0 to +1. At the same time, chlorine is reduced to chloride ions; the oxidation number of chlorine decreases from 0 to –1. By assigning oxidation numbers to all the elements in the reactants and the products of a reaction, we can determine whether the reaction results in a change in oxidation number or not. If a change does occur, the reaction is an oxidation-reduction reaction. For example, let us consider the reaction between magnesium and oxygen:

$$\underset{0}{2Mg} + \underset{0}{O_2} \longrightarrow 2\,\underset{+2}{Mg}\,\underset{-2}{O}$$

Below each atom in every substance in the equation, we have written its oxidation number. The oxidation number of magnesium increases from 0 to +2; hence magnesium is oxidized. The oxidation number of oxygen decreases from 0 to –2; hence oxygen is reduced. Thus, the reaction of magnesium with oxygen is an oxidation-reduction reaction. A reaction that is not an oxidation-reduction reaction will cause no changes in oxidation numbers. Consider the reaction of sodium hydroxide with hydrochloric acid:

$$\underset{+1,}{Na}\,\underset{-2,}{O}\,\underset{+1,}{H} + \underset{+1,}{H}\,\underset{-1}{Cl} \longrightarrow \underset{+1,}{Na}\,\underset{-1}{Cl} + \underset{+1,}{H_2}\,\underset{-2}{O}$$

Below each atom in the equation, we have written its oxidation number. Because each element has the same oxidation number in the products as it does in the reactants, we know that this neutralization reaction is not an oxidation-reduction reaction.

As an illustrative example we may mention the reaction of sodium with chlorine to form sodium chloride and try to identify which substance is the oxidizing agent and which is the reducing agent?

First of all we should write the equation for the reaction and assign oxidation numbers:

$$2\,\underset{0}{Na(s)} + \underset{0}{Cl_2(g)} \longrightarrow 2\,\underset{+1-1}{NaCl(s)}$$

Because chlorine changes oxidation number form 0 to -1, it is reduced and is the oxidizing agent. Because sodium changes oxidation number from 0 to +1, it is oxidized and is the reducing agent.

The characteristics of oxidation and reduction may be summarized as:

'A substance that is oxidized loses electrons, attains a more positive oxidation number and is reducing agent while a substance that is reduced, gains electrons, attains a more negative oxidation number and is the oxidizing agent.'

16. WHAT REACTIONS ARE DUBBED AS 'COMBUSTION REACTIONS'?

Combustion reactions are a special type of oxidation-reduction reaction. We correctly associate combustion reactions with burning. In the usual combustion reaction, the elements in the reacting compound combine with oxygen to form oxides as happens in the combustion of propane to form carbon dioxide and water:

$$2\ C_3H_8(l) + 10\ O_2(g) \rightarrow 6\ CO_2(g) + 8\ H_2O(l)$$

The above reactions take place when an adequate supply of oxygen is present. In the absence of an adequate supply of oxygen, carbon monoxide may be formed instead of carbon dioxide.

$$2C_3H_8(l) + 7O_2(g) \rightarrow 6CO_2(g) + 8H_2O(l)$$

As illustrative examples we may write the balanced equations for the complete combustion in oxygen of: a. sulfur to form sulfur dioxide; b. butane, C_4H_{10} and c. ethyl alcohol, C_2H_5OH

We must note that the physical states of these substances are not given. They are therefore omitted from the equations.

a. We should write the formulas of the reactants, sulfur and oxygen and of the product, SO_2:

$$S_8 + 8O_2 \rightarrow 8SO_2$$

b. Since butane contains only carbon and hydrogen and the combustion is complete, the products are carbon dioxide and water. The unbalanced equation is:

$$C_4H_{10} + O_2 \not\rightarrow CO_2 + H_2O$$

Four atoms of carbon on the left give four molecules of carbon dioxide on the right and ten atoms of hydrogen on the left give 5 molecules of water on the right which requires thirteen atoms of oxygen (6½ molecules) on the left.

$$C_4H_{10} + 6\tfrac{1}{2}\ O_2 \rightarrow 4CO_2 + 5H_2O$$

To write the equation using only whole numbers of molecules, we must multiply throughout by 2 to get:

$$2C_4H_{10} + 13O_2 \rightarrow 8\ CO_2 + 10H_2O$$

c. Ethyl alcohol contains carbon, hydrogen and oxygen. The products of complete combustion will be carbon dioxide and water.

$$C_2H_5OH + O_2 \not\rightarrow CO_2 + H_2O$$

Initial balancing gives two atoms of carbon on the left, two molecules of carbon dioxide on the right; six atoms of oxygen on the right, 3½ molecules of oxygen on the left:

$$C_2H_5OH + 3\tfrac{1}{2}O_2 \rightarrow 2CO_2 + 3H_2O$$

We multiply by 2 to clear the fraction and give the balanced equation.

$$2C_2H_5OH + 7O_2 \rightarrow 4CO_2 + 6H_2O$$

17. HOW CAN WE SUMMARIZE THE VARIOUS ASPECTS OF A CHEMICAL REACTION?

Whereas reactions can broadly be classified into two types: exothermic and endothermic, one classification divides them into four categories: combination, decomposition, displacement and double displacement. A second classification separates reactions into those that involve a transfer of electrons and those that do not. If electrons are transferred, the reaction is an oxidation-reduction reaction. In such a reaction, one substance is oxidized and loses electrons and another substance is reduced and gains electrons. Oxidation numbers are useful in identifying oxidation-reduction reactions. A chemical reaction can be described by an equation that gives the correct formula of each reactant and product and the relative numbers of formula units that react or result. Symbols in parentheses after the formula of a substance may be used to show the physical state of the substances involved in the reaction. An equation must be balanced to be a quantitative description of a reaction. Stoichiometric calculations are based on balanced chemical equations. They allow us to calculate the amount of product that would be formed from a given amount of reactant in the equation. In some instances, where amounts of two reactants are given, one reactant may be limiting. In some instances, only a percent of the theoretical yield is obtained. Chemical reactions obey the Law of Conservation of Mass/Energy. The change in enthalpy (ΔH) for a reaction measures the heat released or absorbed by a reaction carried out at constant pressure and determines whether the reaction is exothermic or endothermic. The amount of energy absorbed or released by a particular reaction can be calculated.

18. WHAT IS IMPORTANCE OF A SIMPLE CHEMICAL EQUATION?

The study of chemistry in fact is the study of chemical changes or, as they are more commonly called, chemical reactions. Some examples of chemical reactions are: the combination of elements to form compounds, the decomposition of compounds (such as sodium hydrogen carbonate or mercury (II) oxide) and reactions between compounds, such as the reaction of vinegar (a solution of acetic acid) with baking soda (sodium hydrogen carbonate). Reactions are usually described using chemical equations. Equations may be expressed in words, such as: mercury (II) oxide decomposes to mercury and oxygen. Using formulas, we state this reaction as:

$$2HgO \rightarrow 2Hg + O_2$$

A chemical equation has several parts: The reactants are those substances with which we start (here mercury (II) oxide, HgO, is the reactant). The arrow (→) means "reacts to form" or "yields." The products are the substances formed by the reaction (here mercury and oxygen are the products). The numbers preceding the formulas are called coefficients. Sometimes the physical state of the reaction component is shown; we use a lowercase, italic letter in parentheses following the substance to show its state. For example, if the equation for the decomposition of mercury (II) oxide were written as: $2HgO(s) \rightarrow 2Hg(l) + O2(g)$, we would know that the mercury (II) oxide was a solid, the mercury was a liquid and the oxygen was a gas when the reaction was carried out. The same equation is repeated below with all the parts labeled:

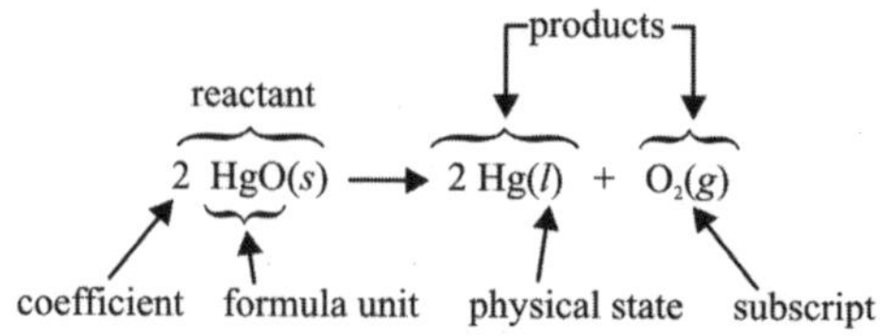

Table 6.1: Parts of an equation and the commonly used notations

Reactants	The starting substances, which combine in the reaction.
Products	The substances that are formed by the reaction.
Arrows	
$\rightarrow$	Found between reactants and products, means "reacts to form."
$\nrightarrow$	Used between reactants and products to show that the equation is not yet balanced.
$\uparrow$	Placed after the formula of a product that is a gas.
$\downarrow$	Placed after the formula of a product that is an insoluble solid, also called a precipitate.
Physical state	Indicates the physical state of the substance whose formula it follows.
(g)	Indicates that the substance is a gas
(l)	Indicates that the substance is a liquid
(s)	Indicates that the substance is a solid
(aq)	Means that the substance is in aqueous (water) solution
Coefficients	The numbers placed in front of the formulas to balance the equation.
Conditions	Words or symbols placed over or under the horizontal arrow to indicate conditions used to cause the reaction.
Δ	States that heat is supplied
RS	States that energy is supplied in the form of light
Elec	States that electrical energy is added

A Correctly Written Equation Obeys Certain Rules

The formulas of all reactants and products must be correct and correct formulas must be used in every case. An incorrect formula would represent a different substance and therefore completely change the meaning of the equation. For example, the equation: $2H_2O_2 \rightarrow 2H_2O + O_2$, describes the decomposition of hydrogen peroxide. This reaction is quite different from the decomposition of water, which is described by the equation:

$$2H_2O \rightarrow 2H_2 + O_2$$

An equation must be balanced by Mass

An equation is balanced by mass when the number of atoms of each element in the reactants equals the number of atoms of that element in the products. For example, the equation shown for the decomposition of water has four atoms of hydrogen in the two molecules of water on the reactant side and four atoms of hydrogen in the two molecules of hydrogen gas on the product side; therefore, hydrogen is balanced. It has two atoms of oxygen in the two reacting molecules of water and two atoms of oxygen in the single molecule of oxygen produced; therefore, oxygen is also balanced:

$$2H_2O \rightarrow 2H_2 + O_2$$

Four (2 × 2) H atoms on the left = four (2 × 2) H atoms on the right

Two (2 × 1) O atoms on the left = two (1 × 2) O atoms on the right

When the atoms are balanced, the mass is balanced and the equation obeys the Law of Conservation of Mass.

One can write and balanced equations in three steps:

1. Write the correct formulas of all the reactants. Use a plus sign (+) between the reactants and follow the final reactant with an arrow. After the arrow, write the correct formulas of the products, separating them with plus signs.
2. Count the number of atoms of each element on each side of the equation. Remember that all elements present must appear on both sides of the equation.
3. Change the coefficients as necessary so that the number of atoms of each element on the left side of the equation is the same as that on the right side. Only the coefficients may be changed to balance an equation; the subscripts in a formula must never be changed.

19. WHAT IS MEANT BY STOICHIOMETRY?

A balanced equation is a quantitative statement of a reaction. It relates amounts of reactants to amounts of products. Let us see what this statement means in terms of a particular reaction namely, burning of pentane, C_5H_{12}, which burns in oxygen to form carbon dioxide and water. The balanced equation for the combustion of pentane is:

$$C_5H_{12} + 8O_2 \rightarrow 5CO_2 + 6H_2O$$

While in qualitative terms, this equation shows that pentane reacts with oxygen to form carbon dioxide and water, in quantitative terms, the equation states that 1 molecule of pentane reacts with 8 molecules of oxygen to form 5 molecules of carbon dioxide and 6 molecules of water. For 15 molecules of pentane, (8 × 15) or 120 molecules of oxygen would be needed for complete reaction; (5 × 15) molecules of carbon dioxide and (6 × 15) molecules of water would be formed. Whatever number (n) of molecules of pentane was used in the reaction, 5n molecules of carbon dioxide and 6n molecules of water would be formed. If 6.02×10^{23} molecules (1 mole) of pentane are burned, $[8(6.02 \times 10^{23})]$ molecules (8 moles) of oxygen are needed. The reaction would form $[5(6.02 \times 10^{23})]$ molecules (5 moles) of carbon dioxide and $[6(6.02 \times 10^{23}]$ molecules (6 moles) of water. Any balanced equation gives the ratio between moles of reactants and moles of products. Given the number of moles of one component used or produced in a reaction, the number of moles or grams of any other component used or produced can be calculated. Such calculations are called stoichiometry. A stoichiometric problem can be stated in many ways, but it always contains these informations: 1. the reaction involved, 2. A stated amount of one component of the reaction and 3. A question asking "how much" of another substance is needed or formed in the reaction.

Stoichiometric problems can be solved by answering the same set of questions. The only difference is that some of the conversion factors are derived from the balanced chemical equation for the reaction involved. The steps to follow in solving a stoichiometric problem are the following:

1. Write the balanced equation for the reaction.
2. Decide which substance is wanted and in what units.
3. Decide which substance is given and in what units and what amount.
4. Determine the conversion factors required to convert:
   - the amount of given substance into moles
   - the moles of the given substance into moles of the wanted substance
   - the moles of wanted substance into the units wanted in the problem
5. Combine the amount of given substance and its units along with the appropriate conversion factors into an arithmetic equation in such a way that all factors except the wanted substance in the proper units will cancel.

Before we begin to do some stoichiometry problems, several points should be emphasized. First, to prevent confusion and errors, the name and units of each item in the equation are always shown. Second, no arithmetic is done until the whole equation is written out. Third, all units in the final equation must cancel except for those required in the answer.

20. WHY AND HOW DO WE CALCULATE PERCENT YIELD?

Very often, in fact, in most cases, the actual yield of a reaction does not equal the theoretical yield but is a lesser amount. The reasons for such discrepancies are varied. The reactant may have been impure, some of the product may have been lost, or small amounts of substances other than those shown in the equation may have been formed. In these cases the actual yield is less than the calculated theoretical yield. The percent yield is calculated by comparing the actual yield with the theoretical yield. The percent yield of a reaction depends on results obtained in the laboratory. If we know the amount of reactants used, we can calculate the theoretical yield. We then measure the amount of obtained product, the actual yield. The ratio between the two gives the percent yield.

$$\text{Percent yield} = (\text{actual yield} / \text{theoretical yield}) \times 100\%$$

As an example of this process, suppose that, for a reaction the actual yield is only 165 g CO_2 and the theoretical yield, the one calculated, is 175 g CO_2. The percent yield would be:

$$(165 \text{ g} / 187 \text{ g}) \times 100\% = 88.2\% \text{ yield}$$

As another example we may consider the reaction of ethane with chlorine and calculate the percent yield of ethyl chloride when 5.6 g ethane reacts with chlorine, 8.2 g ethyl chloride is obtained:

$$\underset{\text{Ethane}}{C_2H_6} + Cl_2 \longrightarrow \underset{\text{Ethyl cholirde}}{C_2H_5Cl} + HCl$$

Because this problem involves percent yield, the first step is to calculate the theoretical yield.

Equation:

$$C_2H_8 + Cl_2 \longrightarrow C_2H_5Cl + HCl$$

Wanted: Theoretical yield (----?-g C_2H_5Cl)

Given: 5.6 g C_2H_6 and 8.2 g C_2H_5Cl (actual yield)

Conversion factors: C_2H_6, mass to moles: 30.1 g C_2H_6 = 1 mol C_2H_6

mol C_2H_6 to mol C_2H_5Cl: 1 mol C_2H_6 yields 1 mol C_2H_5Cl

C_2H_5Cl, moles to mass: 1 mol C_2H_5Cl = 64.5 g C_2H_5Cl

Arithmetical equation

$$? \text{ g } C_2H_5Cl = 5.6\text{g } C_2H_6 \times \frac{1 \text{ mol } C_2H_6}{30.1\text{g } C_2H_6} \times \frac{1 \text{ mol } C_2H_5Cl}{1 \text{ mol } C_2H_6} \times \frac{64.5\text{g } C_2H_5Cl}{1 \text{ mol } C_2H_5Cl}$$

$$= 12\text{g } C_2H_5Cl$$

Answer

$$\text{Percent yield} = \frac{8.2\text{g } C_2H_5Cl}{12\text{g } C_2H_5Cl} \times 100\% = 68\%$$

The ratio of actual yield to theoretical yield can be used to measure the purity of a sample. Such a determination uses a reaction in which the active ingredient of the sample reacts quantitatively, that is, the reaction gives 100% yield. The percent yield of the reaction measures the percent of active ingredient in the sample. As an illustrative example we may consider that a sample of impure zinc weighting 7.45 g reacts with an excess of hydrochloric acid and 0.214 g hydrogen is released. Try to calculate the percent zinc in the sample.

The problem investigates the purity of a zinc sample and utilizes the reaction of zinc with hydrochloric acid, according to the equation:

$$Zn + 2HCl \rightarrow ZnCl_2 + H_2$$

This reaction is quantitative. A weighed sample of the impure zinc is immersed in excess hydrochloric acid. An excess of the acid assures that more than enough is present to react with all the zinc in the sample.

Analysis of the problem

1. Since the zinc sample is impure, the part that is zinc reacts to form hydrogen but the rest does not.

2. We can calculate the amount of zinc needed to form 0.214 g hydrogen. This amount is the actual amount of zinc in the sample; it must be less than the actual weight of the sample.
3. We can calculate the ratio between the amount of zinc needed to form 0.214 g H_2 and the actual weight of the sample. This ratio multiplied by 100% is the percent purity of the sample.

Solution

1. Calculate the amount of zinc that would yield 0.214 g hydrogen.

   Equation

$$Zn + 2HCl \rightarrow ZnCl_2 + H_2$$

   Wanted: ? g Zn

   Given: 0.214 g H_2, 7.45 g sample

   Conversion factors:

$$? \text{ g Zn} = 0.214\text{g } H_2 \times \frac{1 \text{ mol } H_2}{2.015\text{g } H_2} \times \frac{1 \text{ mol Zn}}{1 \text{ mol } H_2} \times \frac{64.5\text{g Zn}}{1 \text{ mol Zn}}$$

   H2, mass to moles: 2.016 g H_2 = 1 mol H_2

   mol H_2 to mol Zn: 1 mol H_2 requires 1 mol Zn (from the equation)

$$\% \text{ Zn} = \frac{6.94\text{g Zn}}{7.45\text{g sample}} \times 100\% = 93.2\%$$

   Zn, moles to mass: 1 mol Zn = 65.4 g Zn

   Arithmetic equation:

This amount is the actual amount of zinc in the sample.

2. Calculate the percent ratio, which will equal the percentage of zinc in the sample.

21. WHAT IS MEANT BY A LIMITING REACTANT?

All the stoichiometry problems generally encountered have two reactants, but the amount is given for only one of them. In calculating the solutions, we assume that enough of the second reactant was present to allow complete reaction of the first. However, this is not always the case. We may encounter many problems in which we know the amounts of all reactants and must calculate which reactant is present in excess before we can calculate a theoretical yield.

A problem of this type might be encountered if we owned a bicycle shop. Among other parts, each bicycle requires two wheels and one set of handlebars. Suppose that, after a busy season, our stockroom contains only 14 wheels and 6 sets of handlebars, although it contains huge quantities of all the other necessary parts. Now question arises as to how many bicycles can we build before a new shipment of wheels and handlebars arrives? We have enough wheels for seven bicycles (14 ÷ 2) and we have enough handlebars for six bicycles (6 ÷ 1). Clearly, we can put together only six bicycles and will have two unused wheels. No matter how we juggle the parts, we have only enough handlebars for six bicycles. Even with 100 wheels, we could make only six bicycles. The number of handlebars is the limiting factor in the number of bicycles we can make.

Similarly, in a chemical reaction in which the amount of each reactant is known, one reactant is usually present in excess and the amount of the other limits the amount of product that can be formed. The problem becomes how to decide which the limiting reactant is. Let us consider the reaction of hydrogen with chlorine to form hydrogen chloride. The balanced equation for this reaction is:

$$H_2 + Cl_2 \rightarrow 2HCl$$

According to the equation, each mole of hydrogen that reacts requires one mole of chlorine to form two moles of hydrogen chloride. If only 0.5 mol hydrogen is present, only 0.5 mole of chlorine is needed and 1.0 mol hydrogen chloride is formed. If 2.0 mole of hydrogen are to react, then 2.0 mole of chlorine are necessary and 4.0 mole of hydrogen chloride are formed. Suppose we have 2.0 moles of hydrogen but only 1.0 mole of chlorine. Only 1.0 mole of hydrogen can react, because we have only 1.0 mol chlorine. In this instance, chlorine is the limiting reactant; only 2.0 moles of hydrogen chloride are formed and the second mole of hydrogen remains unreacted. A limiting reactant problem, then, consists of two simpler problems. Suppose we have the reaction: $A + B \rightarrow C$ and we know now that much A and B are available. To determine how much C will be formed, we must calculate (1) how much C can be prepared from the given amount of A and (2) how much C can be prepared from the given amount of B. The smaller of these two amounts is the theoretical yield of C. In an illustrative example we may ask to ourselves as to what mass of lithium chloride can be formed by the reaction of 5.00 g lithium with 5.00 g chlorine.

Analysis of the problem: We know how much of each reactant is present. The theoretical yield of product might be calculated from either: The true or limiting theoretical yield is the smaller of the two.

Equation:

$$2Li + Cl_2 \rightarrow 2LiCl$$

Wanted: amount of LiCl

Given: Amounts of two reactants: 5.00 g Li and 5.00 g Cl2

Conversion factors: For lithium: 6.94 g Li = 1 mol Li

1 mol Li yields 1 mol LiCl

For chlorine: 70.9 g Cl2 = 1 mol Cl2

1 mol Cl2 yields 2 mol LiCl

$$? \text{ g LiCl} = 5.00\text{g Li} \times \frac{1 \text{ mol Li}}{6.94\text{g Li}} \times \frac{1 \text{ mol LiCl}}{1 \text{ mol Li}} \times \frac{42.4\text{g LiCl}}{1 \text{ mol LiCl}} = 30.6\text{g LiCl}$$

For both: 1 mol LiCl = 42.4 g LiCl

$$? \text{ g LiCl} = 5.00\text{g } Cl_2 \times \frac{1 \text{ mol } Cl_2}{70.9\text{g } Cl_2} \times \frac{1 \text{ mol LiCl}}{1 \text{ mol } Cl_2} \times \frac{42.4\text{g LiCl}}{1 \text{ mol LiCl}} = 5.98\text{g LiCl}$$

Arithmetic equation

There are two equations: Equation A is based on the amount of lithium present. Equation B is based on the amount of chlorine present.

Equation A is:

Equation B is:

We now ask ourselves "Will this reaction yield 30.6 g or 5.98 g of lithium chloride?" We must choose the smaller amount, 5.98 g, because only enough chlorine is available to prepare 5.98 g lithium chloride; we cannot get 30.6 g. Thus, chlorine is the limiting reagent and lithium is present in excess.

22. HOW CAN WE CALCULATE PERCENT COMPOSITION?

Percent means parts per hundred. The percent composition by mass is the ratio of the masses of one element in a compound to the mass of the entire compound expressed as a percent. Due to the law of constant composition, the relative amount of an element in a compound is independent of the source of the compound. That means that we can use either experimentally measured masses or atomic weights from the periodic table; either source of data would five the same result. The percent composition of a compound is the number of grams of each element or group of elements in 100 g of the compound, expressed as a percent.

Percent element = g element/g compound × 100%

For example, the percent composition of potassium chloride, KCl, can be calculated from the atomic weights of potassium and chlorine and the formula weight of KCl. Formula weight of KCl: 39.10 g + 35.45 g = 74.55 g

Percent potassium=39.10 g K/74.55 g KCl × 100% = 52.45% potassium

Percent chlorine= 35.45 g Cl/ 74.55 g KCl × 100% = 47.55% chlorine

From experimental results-to find the percent composition by mass from experimental data, divide the mass of the element by the mass of the compound and express the result as a percent.

From a formula-to find the percent composition by mass from atomic weights divide the atomic weight of the total amount of the element in the compound by the formula weight (or molecular weight) of the compound and express the result as a percent.

As an illustrative example we may try to calculate the percent silver in silver carbonate as under:

The formula of silver carbonate is Ag_2CO_3. Assume one mole.

$$\%Ag = 100\{2(107.8682\ g)/[2(107.8682\ g) + 12.0107\ g + 3(15.9994\ g)]\}$$
$$= 78.2376\%$$

The answer has seven significant figures. The small whole numbers and the 100 are exact numbers. The numerator has seven significant figures. The denominator has seven significant figures. Note that the number 12.0107 does not limit the significant figures because, by the addition rule, the result of the addition in the denominator is rounded to the fourth decimal place.

23. UNDER WHAT STOICHIOMETRY THE ENERGY CHANGES IN A REACTION TAKE PLACE?

While chemistry is the study of matter and the changes in matter, a chemical reaction is the process by which a substance changes into a different substance. During these transformations matter is neither created nor does it vanish. A fundamental law about the nature of matter is the Law of Conservation of Mass. This law states that during any chemical change, mass is neither lost nor gained; matter is merely transformed from one substance into another. We can understand why the Law of Conservation of Mass works if we think of Dalton's model of the atom. Atoms are neither created nor destroyed during chemical changes; they are just re-grouped into different compounds. The energy change associated with a reaction is a stoichiometric quantity and can be treated arithmetically, as happens with mass changes. For many reactions, enthalpy changes have been determined and tabulated in the chemical literature. The changes listed in such sources apply only to the form of the equation they accompany.

Chapter 6

CHARACTERISTICS AND HISTORICAL BACKGROUNDS BEHIND NAMING OF CHEMICAL ELEMENTS

QUESTION: WHAT ARE HISTORICAL BACKGROUNDS BEHIND NAMING OF CHEMICAL ELEMENTS AND WHAT ARE CHARACTER-ISTICS OF THE ELEMENTS?

Since Latin remained the language of communication for quite a long time between scientists from different nations, the use of Latin names such as aurum ("shining dawn") for gold, or ferrum, the Latin word for iron is understandable. Likewise, lead (Pb) and sodium (Na) are designated by their Latin names, plumbum and natrium, respectively. However some chemical elements are named for Greek or German words describing properties of the element, for example, bromine (Br), which comes from a Greek word meaning "stench." The name of cobalt comes from a German term meaning "underground gnome," because miners considered the metal a troublemaker. The names of several elements with high atomic numbers reflect the places where they were originally discovered or created: francium, germanium, americium and californium. Americium and californium, with atomic numbers of 95 and 98 respectively, are among those elements that do not occur naturally, but were created artificially. The same is true of several elements named after scientists, among them einsteinium, after Albert Einstein (1879–1955) and nobelium after Alfred Nobel (1833–1896), the Swedish inventor of dynamite who established the Nobel Prize. Let us take elements one by one in accordance with their group affiliation in the periodic table and delve into glances of some of their characteristics and naming background.

1. GROUP-1 (IA): HYDROGEN & ALKALI METALS (H, Li, Na, K, Rb, Cs and Fr)

The name hydrogen is taken from the Greek hydro + genes, or "water generator". The gas is colorless, odorless and tasteless and is the most abundant element in the known universe. It is also the lightest (in terms of atomic mass) and the simplest, having only one proton and one electron (and no neutrons in its most common isotope). First isolated and identified as an element by Cavendish in 1766, hydrogen was believed to have many different things. Cavendish himself thought that it was "inflammable air from metals", owing to its production by the action of acids on metals. Others thought it was pure phlogiston because of its flammability.

Hydrogen is among the ten most abundant elements on the planet, but very little is found in elemental form due to its low density and reactivity. Much of the terrestrial hydrogen is locked up in water molecules and organic compounds like hydrocarbons. The vast majority of hydrogen produced industrially today is made either from treatment of methane gas with steam or in the production of "water gas" from the reaction of coal with steam. Most of this hydrogen is used in the Haber process to manufacture ammonia. The element is also used for hydrogenating vegetable oils, turning them into margarine and shortening and some is used for liquid rocket fuel.

There are two important isotopes of hydrogen. Deuterium (^{2}H) has an abundance of 0.015% of terrestrial hydrogen and the nucleus of the isotope contains one neutron. Sometimes called heavy hydrogen, deuterium is separated by the fractional distillation of liquid hydrogen but it can also be produced by the prolonged electrolysis of ordinary water. Approximately 100,000 gallons of water will produce a single gallon of D_2O, "heavy water". This special kind of water is used as a moderator in some fission power reactors. Deuterium fuel is used in experimental fusion reactors. Tritium (^{3}H) contains two neutrons in its nucleus and is radioactive. It is very rare (about 1 in every 1018 atoms) and is formed in the environment by cosmic ray bombardment. Most tritium is manufactured by bombarding Li with neutrons. Tritium is used in thermonuclear weapons and experimental fusion reactors.

Lithium

Lithium is the lightest of all metals and is named from the Greek work for stone (lithos). It is the first member of the Alkali Metal family. It is less dense than water (with which it reacts) and forms a black oxide in contact with air. Lithium was discovered in 1817 by Johan August Arfwedson but not isolated until some time later by W.T. Brande and Sir Humphry Davy. In its mineral forms it accounts for only 0.0007% of the earth's crust. Its compounds are used in certain kinds of glass and porcelain products. More recently lithium has become important in dry-cell batteries and nuclear reactors. Some compounds of lithium have been used to treat manic depressives. In compounds lithium (like all the alkali metals) has a +1 charge. In its pure form it is soft and silvery white and has a relatively low melting point (181°C).

Sodium

Sodium is perhaps the most characteristic alkali metal, reacting violently with water and rapidly with the oxygen in air. Its symbol (Na) comes from its Latin name, Natria or Natrium, whereas its English name is taken from soda which contains it. Sodium was discovered and isolated in 1807 by Sir Humphry Davy. In its pure form it is silvery white and soft enough to cut with a knife. It is the sixth most abundant element in the earth's crust, occurring in large amounts in both sea water and soil in various mineral compounds, the most common of which is sodium chloride. The metal melts below the boiling point of water (97°C). In its elemental state it has been used as a molten coolant in nuclear reactors and is currently under research for sodium/sulfur batteries. Its most common compounds for industrial use include sodium chloride, sodium hydroxide (lye), sodium carbonate (washing soda) and sodium sulfate.

Potassium

Potassium reacts so violently with water that it bursts into flame. The silvery white metal is very soft and reacts rapidly with the oxygen in air. Its chemical symbol is derived from the Latin word kalium which means "alkali". Its English name is from potash which is the common name for a compound containing it. The seventh most abundant element, potassium was discovered and isolated in 1807 by Sir Humphry Davy. Important compounds of potassium include potassium hydroxide (used in some drain cleaners), potassium superoxide, KO_2, which is used in respiratory equipment and potassium nitrate, used in fertilizers and pyrotechnics. Potassium, like sodium, melts below the boiling point of water (63°C) and is less dense than water also. Like most of the alkali metals, potassium compounds impart a characteristic color to flames. In the case of this 19th element, the color is pale lavender. Like sodium ions, the presence of potassium ions in the body is essential for the correct function of many cells.

Rubidium

Rubidium (Latin: rubidius = red) is similar in physical and chemical characteristics to potassium, but much more reactive. It is the seventeenth most abundant element and was discovered by its red spectral emission in 1861 by Bunsen and Kirchhoff. Its melting point is so low that we can melt it in our hand if we had a fever (39°C). But that would not be a good idea because it would react violently with the moisture in our skin. Rubidium was once thought to be quite rare but recent discoveries of large deposits indicate that there is plenty to use. However at present it finds only limited application in the manufacture of cathode ray tubes.

Cesium

Cesium is a bright silvery metal which is a liquid in a warm room (28°C). Its name is from the Latin caesius which is a description of a sky blue spectral emission by which it was discovered in 1860 by Bunsen and Kirchhoff. Cesium is so reactive that it will even explode on contact with ice. It has been used as a "getter" in the manufacture of vacuum tubes (*i.e.*, it helps remove trace quantities of remaining gases). An isotope of cesium is used in the atomic clocks.

Francium

Francium is the last of the known alkali metals and does not occur to any significant extent in nature. Its all known isotopes are radioactive and have short half-lives (22 minutes is the longest). The existence of Francium was predicted by Dmitri Mendeleev in the 1870's and he presumed it would have chemical and physical properties similar to cesium. That may well be, but not enough francium has been isolated to test. Numerous historical claims to the discovery of element 87 were made resulting in the names russium, virginium and moldavium. However, the confirmed discovery is credited to Marguerite Perey who was an assistant to Marie Curie at the Radium Institute in Paris. She named the element after her native country, France.

2. GROUP-2 (IIA): THE ALKALINE EARTH METALS (Be, Mg, Ca, Sr, Ba and Ra)

Beryllium

The name Beryllium comes from the Greek 'beryllos' which is the name for the gemstone beryl. The element is a high-melting, silvery-white metal which is the first member of the alkaline earth metals. It is not abundant in the environment and occurs mainly in the mineral beryl with aluminum and silicon. Beryllium was first isolated in 1828 by Wöhler. It is used in specialty alloys such as spring metal in which it increases toughness. It was once known as glucinium because of the sweet taste of its compounds (which, alas, are toxic). The main mineral source of beryllium is also the same material of which emeralds are formed. High quality beryl crystals with chromium impurities have an exquisite green color which has been valued for thousands of years.

Magnesium

Magnesium takes its name from magnesite ore, named for the district Magnesia in Thessaly, Greece. It is a silvery-white metal which burns in air with a blinding white light. It reacts with hot water and acids. The metal is widely distributed in the environment in a number of minerals and a significant amount is found in sea water. Recognized as an element as back as 1775, it was first isolated in pure form by Davy in 1805. The metal and its alloys are widely used to manufacture light-weight mechanical parts. Magnesium is also an important nutritional element in trace amounts and is the central atom in the complex chlorophyll molecule.

Calcium

Calcium is a silvery-gray metal which takes its name from the Latin word 'calx', which means lime. It is the fifth most abundant element in the earth's crust and is widely distributed as limestone ($CaCO_3$), quicklime (CaO) and calcium fluoride. The pure metal was isolated in 1808 by Davy and its compounds give a characteristic brick-red color to flames. Calcium compounds are used in the manufacture of iron and steel, cements and plasters, as well as gypsum wall board. It is important biologically in the formation of bones and teeth. Calcium metal is fairly reactive and combines with water at room temperature to produce hydrogen gas and calcium hydroxide. It slowly oxidizes in air, becoming encrusted with white CaO and $CaCO_3$.

Strontium

The element strontium is named for a Scottish town, Strontian. It was isolated in 1808 by Davy and is a silvery and malleable metal that reacts vigorously with water to produce hydrogen gas. It has the same relative abundance as carbon and sulfur but does not occur in pure form. Strontium compounds are useful in pyrotechnic devices and signal flares because of the bright crimson coloring, they give to flames. Strontium-90, a radioactive isotope of the metal produced by fission reactions is a dangerous environmental menace because its chemistry is similar to calcium and it may take its place in bones. The strong radiation emitted by the isotope interferes with the production of new blood cells and can cause death.

Barium

Barium takes it name from the Greek word 'barys' for heavy. It was first isolated in 1808 by Davy and is a soft silvery-white metal. Like the lighter members of its family, barium reacts vigorously with water to produce hydrogen gas and so is commonly stored in oil. The metal does not occur free in nature but mainly as the sulfate and carbonate. The sulfate is used in X-ray diagnostics as a contrast medium (*i.e.*, in soft tissue like the digestive tract). Barium compounds (which are toxic) are also useful in pyrotechnic devices where they impart a characteristic green color.

Radium

Radium takes its name from the Latin word radius or ray. All isotopes of radium are radioactive and many exhibit luminescence, reacting readily with oxygen and water. The metal was discovered and isolated in 1911 by Marie Curie. When first discovered, compounds of the metal were used on watch dials for self-luminescence and in early cancer therapy. It has now been replaced by safer alternatives.

TRANSITION METALS

3. GROUP-3 (IIIB): Sc, Y, La and Ac

Scandium

Scandium was one of the elements predicted by Mendeleev in 1871 when he proposed his system for organizing the elements, the periodic table. At the time he called it ekaboron ("like boron") and suggested some of the physical and chemical properties it would have. The existence of scandium was confirmed only 8 years later by Lars Nilson. It is named for the Latin word for Scandinavia. The metal is lightweight and fairly corrosion-resistant. It also has a high melting point and therefore finds applications in the aerospace industry. Scandium is widely distributed on the earth and has abundance similar to cobalt and lithium. Most commercial scandium is obtained from the uranium refining process.

Yttrium

The current claim-to-fame for yttrium (named from a Swedish village, Ytterby) is its use in the so-called 1-2-3 oxide superconductors (along with barium and copper). These were the first superconducting materials to function at liquid nitrogen temperatures. The element was discovered in 1789 by Gadolin and finally isolated in 1828 by Wöhler. More than 15 tons of its oxides are now produced each year. In addition to its use in the research of superconductivity, it is also used in phosphors (red) for color television tubes. Yttrium metal is ductile and silvery. Powdered samples and turnings from machining can burst into flame. Most commercial yttrium is produced from monazite sands that are also the source for most of the rare earth elements.

Lanthanum

Discovered in 1893 by Mosander, lanthanum is named from the Greek lanthanein, "to lie hidden". That is an apt description since lanthanum generally occurs along with others the so-called rare earth elements and is very difficult to separate. The abundance of the metal is similar to that of zinc or nickel. Lanthanum is used in the electrodes for high-intensity carbon-arc lamps and also in the production of high-purity europium metal (element 63). The pure metal is silvery white and malleable. It is soft enough to cut with a knife and will oxidize readily in the air as well as react with moisture. For that reason the metal is usually stored under oil or kerosene. Most lanthanum is extracted from monazite sands.

Actinium

Actinium, named from the Greek aktinos (ray) is a rare, extremely radioactive metal that glows in the dark. It was discovered by Debierne in 1899 and again in 1902 by Giesel. Eventual priority was given to Debierne's work. Samples of actinium quickly decay to thorium and francium. The longest-lived isotope (Ac-227) has a half-life of 21.8 years. The element is obtained as an impurity in pitchblende, an ore mined for its uranium content. One-tenth of one gram of actinium can be recovered from 1 ton of pitchblende.

4. GROUP 4 (IVB): Ti, Zr, Hf and Rf

Titanium

Discovered independently by William Gregor and Martin Klaproth in 1795, titanium (named for the mythological Greek Titans) was first isolated in 1910. It is a highly corrosion-resistant metal with great tensile strength. It is ninth in abundance for elements in the earth's crust. Titanium is used for strengthening structural materials but is also useful in strengthening bones in the human body where it is used to make durable repairs since it does not react with tissue. Most titanium metal is refined from a mineral known as rutile, TiO_2 which is an intensely white powder that is also used as a high-hide paint pigment.

Zirconium

Named for the mineral zircon in which it is generally found, zirconium was discovered in 1789 by Klaproth and eventually isolated in 1824 by Berzelius. The metal reacts with oxygen and nitrogen in the atmosphere to form a protective coating that inhibits its further corrosion. It is resistant to weak acids and even forms a low-temperature superconductor when alloyed with niobium. Zirconium finds applications in industry which suit its high corrosion resistance and strength. It is very similar to the less plentiful Hafnium and the two are very difficult to separate. Most samples of either are contaminated with small amounts of the other element.

Hafnium

More abundant than better known metals such as silver and gold, hafnium (from the Latin Hafnia, a name for Copenhagen) was not discovered until 1923 by Coster and de Hevesy. The reason is the similarity of hafnium to zirconium. Mendeleev had predicted the existence of element 72 but had wrongly suggested that it might be found along with titanium ores. Instead it lay hidden with "pure" samples of zirconium. Later Niels Bohr predicted the arrangement of outer electrons for element 72 and it was using X-ray techniques to study this very thing that led to the identification of hafnium as a separate element (thus the connection to Copenhagen in the name for the metal: Bohr's hometown). Hafnium metal is used in the manufacture of control rods for nuclear reactors because of its ability to absorb neutrons.

Rutherfordium

In 1964 researchers in the Soviet Union at Dubna announced their discovery of element 104. A similar claim was made by researchers at the University of California at Berkeley. The Soviet scientists claimed to have bombarded a target of Pu-242 with Ne-22, resulting in a nucleus with 104 protons and a mass number of 260. The Berkeley team used a Cf-249 target and isotopes of carbon for projectiles, resulting in isotopes of 104 with mass numbers of 257 and 259. Several other isotopes were also prepared by the American team. The naming and priority of discovery controversy has raged ever since and Rutherfordium is the approved name selected by the IUPAC in August 1997.

5. GROUP 5 (VB): V, Nb, Ta and Db

Vanadium

Vanadium takes its name from the Scandinavian goddess Vanadis and was discovered in 1801 by Andrés Manuel del Rio. It was isolated in 1867 by Henry Roscoe as a silvery-white metal that is somewhat heavier than aluminum but lighter than iron. It has excellent corrosion resistance at room temperature. The history of its discovery is an interesting tale. del Rio sent his brown ore samples, containing what he thought was a new element to Paris for analysis and confirmation, along with a brief explanation that was ambiguous. The complete analysis and description of his work were lost in a shipwreck so the Paris lab saw nothing but brown powder and a brief confusing note. A second sample sent to Berlin was mislabeled lead chromate when it arrived. del Rio gave up, losing confidence in his discovery. The element was rediscovered in 1867 by Nils Sefstrôm. Vanadium has an unusually large number of stable oxidation states (+2, +3, +4, +5) each of which is characterized by a unique color in solution. The metal is used as an alloying agent for steel. It combines with nearly all non-metals in compounds.

Niobium

Discovered in a museum in 1801, niobium was named after Niobe, the mythological daughter of Tantalus (the metal is chemically related to Tantalum). Occurring in the earth's crust in proportions of about 20 parts-per-million, niobium is used as an alloying agent with iron and nickel. It has commercial uses in atomic reactors and has superconductive properties when alloyed with tin or aluminum. In many places niobium is known as columbium since it was originally "discovered" in a mineral named columbite. However, careful subsequent analysis indicated that the "new" element was actually Tantalum. The rediscovery of the element we know now as niobium was not even associated with columbium until after the official statistics were in. Pure niobium looks much like steel but resists corrosion better due to a thin coating of oxide that forms on all exposed surfaces. The only acid that attacks Nb at room temperature is hydrogen fluoride, HF. Above 200°C the metal becomes more reactive.

Tantalum

Named for the mythological character Tantalus by Anders Ekenberg who discovered it in 1802, tantalum is a heavy, gray metal that resembles the more expensive platinum in many respects and is sometimes used as an economical substitute for that element. The metal comprises only 0.0002% of the earth's crust. Interestingly, its most deposits lie outside the U.S. which is its chief consumer. Tantalum alloys are corrosion and wear resistant and find use in dental and surgical tools. Tantalum oxide is used in some electronic components and a composite of tantalum carbide (TaC) and graphite is one of the hardest materials ever produced.

Dubnium

The synthesis of element 105 was reported by Soviet scientists at the research station in Dubna as early as 1967 and by scientists at U.C. Berkeley in 1970. The Soviet scientists reported synthesis by bombarding Am-243 with Ne-22. At Berkeley, Cf-249 was bombarded with N-15. At the purported time of discovery the Soviets suggested the name Nielsbohrium (Ns) for the element while the American group suggested Hahnium (Ha). But the ensuing controversy over the discovery of this and other 6d-elements led the IUPAC to adopt its numerical system for naming elements beyond 103 until some agreements could be reached (hence Unp or Unnilpentium for 105). The longest lived isotope has a half-life of 34 seconds. Dubnium is the name approved by the IUPAC in August 1997.

6. GROUP 6 (VIB): Cr, Mo, W and Sg

Chromium

Discovered in 1797 by Louis-Nicholas Vauquelin and isolated a year later, chromium is a very hard steel-gray metal which takes its name from the Greek word chroma, for "color", suggesting the wide variety of colors which characterize its compounds, many of which are used as oil paint pigments. The pure metal is also a valuable alloying agent for spring steels and along with nickel, in stainless steels. Chromium(III) oxide (a popular green pigment) is the ninth most abundant chemical in the earth's crust. Similar in characteristics and practical uses to vanadium and titanium, pure chromium has a blue-white color. It is very brittle and corrosion resistant. It has three common oxidation states: +2, +3 and +6. The metal ore (principally chromite, $FeCr_2O_4$), is refined by heating in the presence of Si or Al. The resulting ferrosilicon or ferroaluminum compound is separated from the chromium by electrolysis.

Molybdenum

The first significant practical use of this metal (discovered in 1778 by Scheele) was in the alloy used to manufacture the German artillery gun known to the Allies as Big Bertha. The metal takes its name from the Greek word molybdos, meaning "lead" because some of its compounds were once confused with similar lead compounds before Scheele's discovery. The pure metal was first isolated in 1781 by Peter Jacob Hjelm. Chemically, molybdenum resembles a cross between Cr and W. It has a principal oxidation state of +6 although ionic compounds suggest lower states (*e.g.*, the common ore is MoS_2). Molybdenum contributes to the toughness and hardenability of quenched and tempered steels. It is also an essential plant nutrient. Some lands are actually barren for lack of this element in the soil.

Tungsten

The name for the metal tungsten is taken from the Swedish, Tung sten, for "heavy stone". The symbol, W, has its origins in the discovery of the element in the late 1700's linked to the mineral wolframite. Most German chemists still refer to the element as Wolfram (wolf's metal). Pure tungsten was eventually isolated in 1783 by brothers Juan and Fausto Elhuyar. Tungsten is known chiefly as the metal in incandescent light bulb filaments (tungsten has the highest melting point of any metal). But a lot of tungsten is also used in the manufacture of tungsten carbide, an extremely hard material used for making cutting tools and abrasives. The metal is also alloyed with steel for very high temperature applications such as rocket engine nozzles. Very pure tungsten is actually soft enough to be cut with an ordinary hacksaw but generally small amounts of impurities render the metal very hard and brittle. It is very similar to both Mo and Cr. It is straddled by elements having the second and third highest melting points, rhenium, Re and tantalum, Ta.

Seaborgium

The synthesis of element 106 was first reported by Soviet scientists in June of 1974. Three months later the Berkeley group claimed positive discovery of the same element. While, the Soviet claim was based on bombardment of three different isotopes of lead with high energy chromium-54 ions, the American group used a californium target and ions of oxygen-18. The longest lived isotope (which has no known practical uses) has a half-life of 800 milliseconds. In early1997 the first aqueous chemistry of Sg was reported out of Germany. The most stable oxidation state in aqueous solution appears to be +6 and formation of neutral or anionic oxo or oxohalide compounds occurs just as for Mo or W. The early experiments show, as expected, that Sg is chemically like other members of group 6 and not like uranium [this goes back to an early controversy about the so-called transuranium elements: would they exhibit chemical properties similar to uranium and thus need to be removed to a separate section of the periodic table, or would they, in spite of their radioactivity, exhibit expected chemical properties of their groups? Time and research have shown the latter to be correct]. Like elements 104 and beyond, the controversy over the discovery and therefore the naming of 106 led the IUPAC to adopt the numerical symbol Unh (unnilhexium) for this element 106 . But in 1993 the Berkeley group gave definitive proof of their priority in synthesis and gave the name Seaborgium (Sg) after Glenn T. Seaborg. While the choice of name was lauded in general, the IUPAC was awaiting a list of names from a committee for elements 104-109. The committee rejected Seaborgium because Seaborg was still living. The Berkeley group pointed out that Einsteinium and Fermium were both named after living scientists at the time (although both died before the names were officially accepted). The committee suggested Rutherfordium (Rf) but after many complaints about the choice of names, reverted to Seaborgium. Seaborgium is the name which the IUPAC finally approved in August 1997.

7. GROUP 7 (VIIB): Mn, Tc, Re and Bh

Manganese

The name of the 25th element comes from the Latin 'magnes', meaning magnet. It was discovered in 1774 by Scheele and isolated later that year by Johan Gahn. In pure form Manganese is a hard, brittle, gray-white metal. It is best known as an alloying agent in steel. It enhances the ability to hot-work steel and increases resistance to impact. In biological systems manganese is a crucial component of vitamin B1. The pure metal is produced from its most common compound (MnO_2, 10th most abundant compound in the earth's crust). It can be reduced chemically or refined electrolytically. The element has at least 5 stable oxidation states with distinctive colors (as is typical of transition metals). It is commonly encountered in the laboratory as the compound $KMnO_4$ which is a strong oxidizing agent. MnO_2 catalyzes the decomposition of H_2O_2 and is sometimes used for the small scale production of oxygen gas in the lab.

Technetium

Technetium has the distinction of being the first synthetically produced element. It was the last of the six elements predicted by Mendeleev to be discovered (in 1937 by Segrè and Perrier). Trace amounts were found in a sample of molybdenum that had been bombarded with deuterons in a cyclotron. All known isotopes of Tc are radioactive and none appears to occur naturally on earth. However, it is produced commercially in kg quantities because very tiny amounts (55 parts per million) can transform iron into a corrosion resistant alloy. All applications of the metal involve closed systems, however, because of the radioactivity. Pure Tc is a silvery-gray metal similar in appearance to platinum but it is normally produced in powdered form. Chemically it resembles a cross between Mn and Re. Although of little interest, compounds have been prepared in which Tc exhibits oxidation states of +4, +6 and +7.

Rhenium

A dense, silvery white metal, Rhenium takes its name from the Latin, Rhenus, for the Rhine River. It was discovered in 1925 by Ida and Walter Noddack along with Otto Berg. It is nearly twice as dense as lead (21.0 g/cm^3) and extremely rare (1 pound of rhenium per 1000 million pounds of earth). Nonetheless, the total annual U.S. production of Re is almost half a ton. It is used as a trace alloying element for metal components that are subjected to constant friction. Rhenium was another of the "missing" elements proposed by Mendeleev. The first sample was concentrated 100,000 fold from a gadolinium ore sample. Just enough was obtained for a spectroscopic study in which previously unseen lines were observed. The metal is acid resistant and is the one of the metals, having highest melting points. But its scarcity (and therefore expense) makes its practical use limited.

Bohrium

Element 107 (Uns) was first reported by Soviet scientists in 1976. This claim has since been substantiated by German researchers but no name was suggested and so it has been known by its numerical moniker, unnilseptium (107). Recently, however, a committee of the IUPAC suggested the name Bohrium (after Niels Bohr). This name was finally approved in August 1997. There are no known practical uses for Bohrium. All its known isotopes have extremely short half-lives.

8. GROUPS 8-10 (VIIIB): Fe, Ru, Os, Hs, Co, Rh, Ir, Mt, Ni, Pd, Pt and Ds

Iron

Iron, which takes its English name from the old Anglo-Saxon and its symbol from the Latin, ferrum, was identified and used in prehistoric times. It is a very common element, fourth most abundant in the earth's crust. In addition, two of the ten most common compounds in the earth's crust are the two common oxides of iron, FeO and Fe_2O_3. In its pure form, iron is a silvery-white metal, distinguished by its ability to take and retain a magnetic field and also dissolve small amounts of carbon when molten (thus yielding steel). Commercial refining of iron is based on the heating of Fe_2O_3 or Fe_3O_4 (magnetite) with a mixture of other substances in the high temperature environment of the blast furnace. The oxides are reduced to pure iron. In addition to hardening iron by adding small amounts of carbon and also some other metals to the molten iron, iron castings or forgings can be heat-treated to take advantage of the various physical properties of the different solid phases of iron. Pure iron reacts readily with oxygen and moisture in the environment and corrodes destructively. Even alloys such as steel need protection by painting or some other coating to prevent structural failure over time.

Ruthenium

Element 44 (named from the Latin, Ruthenia, for Russia) was originally discovered in 1807 by the Polish chemist Sniadecki, but the claim from a relatively unknown was not accepted by a Paris commission and he withdrew it. It was not until 1828 that Gottfried Osann claimed to have found three new elements in his platinum samples that interest in the metal increased. Although he could not substantiate his claim, he did not withdraw it. In 1844 Karl Klaus showed that two of Osann's metals were not new elements at all, but the third he was able to isolate and characterize. Ruthenium is a rare metal (less than 0.01 parts per million in the earth's crust). It is generally described as gray-white, hard and brittle. At room temperature it is resistant to virtually all acids. Its high melting point and brittle nature make casting difficult. Its chief commercial use is as a hardening agent in platinum jewelry. Recently interest has grown in some compounds of ruthenium which possess the ability to convert visible light into a suitable energy source for splitting water into hydrogen and oxygen (for fuel). Work in this field is going on.

Osmium

Discovered in 1803 by Smithson Tennant (most famous for his determination that diamond is just a form of carbon), osmium is a very dense, blue-white hard metal. Its name is taken from the Greek, osme, for "odor". The oxides of osmium emit highly toxic gases and form readily when the metal is exposed to air. Thus there are few commercial applications for osmium except as a minor alloying agent where it reduces frictional wear ("osmiroid" ball point pen tips, for example). Most osmium is recovered as a by-product of the refining of platinum and nickel ores.

Hassium

Like its immediate predecessors, element 108 (Hs), is of purely theoretical interest and has an extremely short half-life, decaying by spontaneous fission. It is so unstable, in fact, that it was not discovered until after element 109 had been positively confirmed. The search for elements beyond uranium has occupied the attention of various groups of scientists around the world since the late 1940's. The principals involved hail from the laboratories at Berkeley, California, what was previously the Soviet Joint Institute for Nuclear Research at Dubna and the Heavy-Ion Research Laboratory in Darmstadt, Germany. Originally a committee of the IUPAC recommended that element 108 be named Hahnium, after a German scientist, Otto Hahn. That name met with some opposition and the name approved in August 1997 is Hassium, named for the German state of Hesse.

Cobalt

Compounds of cobalt (name from the German Kobold or "goblin") were known in ancient times but the metal itself was isolated in 1739 by Georg Brandt. Cobalt is closely associated with cobalt glass and its characteristic deep blue color. But the metal is silvery-white and brittle, with a slight bluish color when polished. Cobalt is typically found with iron and nickel deposits. Its chemistry is similar to that of these elements but it has notable magnetic properties and one-fourth of the world's production of cobalt is used to make powerful "alnico" magnets (an alloy of Al, Ni and Co). Although the metal can be obtained from cobaltite and erythrite ores (the arsenide and arsenate, respectively) it is often less expensive to recover it from the waste material obtained from refining other metal ores. Cobalt(II) chloride is often used in humidity indicators because in dry form it is blue and when hydrated it is pink. The radioactive isotope of cobalt, Co-60, is used in portable "X-ray" devices. It actually produces gamma rays which can be used to detect internal flaws in metal in much the same way as X-rays by producing a shadowy image on film. Co-60 is also used in cancer treatment.

Rhodium

World production of rhodium (from the Greek rhodon, "rose") is about 10 tons. While the metal itself has few applications, it is an important alloying agent used as a hardener for platinum and palladium. As one of the so-called "platinum" metals, rhodium is often found in association with these other elements. It shares the properties of these metals: high corrosion resistance, hardness and ductility. It is the rarest of the group, occurring only to the extent of about 1 part per 200 million in the earth's crust. Rhodium was discovered in 1803 by William Hyde Wollaston who named it for the rose-red color of its salts.

Iridium

Iridium has the reputation of being the most corrosion resistant of all metals. It was discovered in 1803 by Smithson Tennant in 1803 and named for the Latin iris or "rainbow" because it forms a large number of very colorful compounds. The pure metal is very difficult to machine into useful shapes because of its hardness and its principal use is as a hardening agent for platinum. It will withstand attack by hot aqua regia (a mixture of concentrated hydrochloric and nitric acids which dissolves even gold) and can be used routinely at temperatures up to 2000°C.

Meitnerium

Element 109 was first synthesized by researchers at Darmstadt, West Germany in August 1983. For 10 days they hurled a beam of iron-58 ions into a bismuth-209 target. They detected the formation of one nucleus of Mt-267 which rapidly "boiled off" a neutron, reverting to Mt-266. This decayed within milliseconds to give (element 107). A committee of the IUPAC suggested the name Meitnerium (Mt) after the German physicist Lise Meitner. Final approval of the name and symbol was given in August 1997.

Nickel

Known since the early 1700's from a reddish-brown rock with splotches of green in it, nickel takes its name from the German kupfernickel, which can be variously translated as the name of a troll or "false copper". It is often found with copper and other similar metals. Before it was isolated in 1751 by Baron Axel Fredrik Cronstedt, it, like cobalt, was known for its ability to color glass (in this case, green). The U.S. five-cent piece is 25% nickel. The element is also an important alloying agent for stainless steels and in very powerful magnets. Nickel is found in the earth's crust to the extent of about 70 parts per million, about the same extent as copper and zinc. There is a good chance that a high proportion of the core of the earth is molten nickel. Unlike its near neighbor copper on the periodic table, nickel is only a fair electrical conductor. But like its other neighbor, cobalt, it is very useful in making strong permanent magnets. It is also highly resistant to attack by alkalis and is used to store and transport concentrated sodium and potassium hydroxide. Nickel reacts with most acids to produce hydrogen gas and the green Ni^{2+} ion.

Palladium

Discovered in 1803 by William Wollaston and named after the then recently discovered asteroid Pallas, palladium is a silvery-white, soft metal similar to platinum. It is a rare metal (only about 1 part per million in the earth) but occurs commonly along with copper, silver and gold. Palladium is used as an alloying agent with gold in jewelry ("white gold") and in some dental applications in place of silver or gold. Unlike the other so-called platinum metals, palladium is more susceptible to attack by acids, even hydrochloric acid. Palladium has the curious ability to absorb large quantities of hydrogen gas (up to 900 times its own volume) and this has generated some interest in its alloys as a storage system for hydrogen as a portable fuel for automobiles. It was also prominent in the "cold fusion" controversy some years ago when it was said (apparently falsely) that when it was made to absorb heavy hydrogen (deuterium), the atoms would undergo fusion and release more energy than was put into the process.

Platinum

Independently discovered in 1735 by Antonio de Ulloa and in 1741 by Charles Wood, platinum is named from the Spanish platina or "silver". The metal is classified as precious owing to its scarcity and commercial demand. It is very heavy and silvery white and is used in laboratory instruments, jewelry, medical and dental items and electrical contacts. Because it is immune to air oxidation, the metal is often found in native form in nature. It is sometimes found in a rare, naturally occurring alloy, platiniridium. Naturally occurring platinum is a mixture of six non-radioactive isotopes.

Darmstadtium

As early as 1991, scientists at Berkeley reported evidence of element 110, but definitive work seems to have emerged in November of 1994 from the GSI laboratory in Darmstadt, Germany. To produce a single atom of 110, the researchers used the UNILAC accelerator to bombard a target of lead over many days with more than a billion nickel atoms. Detectors searched each collision for element 110's distinctive (predicted) decay sequence. A team at Berkeley duplicated the synthesis, fusing nickel-64 and lead-208 nuclei in June 2003. The IUPAC approved the name "darmstadtium" (after the laboratory at which the discovery was made) in August of 2003.

9. GROUP-11 (IB): Cu, Ag, Au and Rg

Copper

The use of Copper goes quite far back into history. Copper beads have been found in what is now modern Iraq, dating back to 9000 BCE. The metal is relatively easy to mine and refine, contributing to its early and widespread use. Being soft, however, it is unsuitable for making reliable tools and weapons. Early metal smiths as far back as 3000 BCA learned to combine copper with other metals to produce more durable alloys. Brass (copper and zinc) and bronze (copper and tin) are two examples. The symbol and name for copper are from the Latin cuprum, which literally means "from the island of Cyprus", an early source of copper ore. Before 1982 U.S. pennies were pure copper. Now they are mostly zinc with a thin shell of copper. Most copper that is mined today is refined and drawn into wire for use in the electrical industries. A significant portion is also used in manufacturing water pipe. Copper, of course, has a characteristic color which most people recognize. It is one of the best electrical conductors and resists corrosion from most acids (except nitric and hot concentrated sulfuric). When exposed to the atmosphere for a period of time it develops a greenish coating or patina which is copper(II) carbonate, a protective coating that prevents further wear.

Silver

Silver (name from the Anglo-Saxon sioful and symbol from the Latin argentium) is considered a precious metal. It is found only to the extent of 0.05 parts per million in the earth (*i.e.*, we will have to dig up 20 million shovels full of dirt to get one shovel full of silver). Knowledge of the metal is of ancient origins. Refining methods are mentioned in the Hebrew Scriptures and ancient Egyptian writing. Silver was once considered more valuable than gold, probably because gold was easier to locate and refine at that time. Silver is ten times more abundant than gold. Pure silver metal is the best conductor of heat and electricity. Pure silver would be so soft for jewelry and utensils that it is usually alloyed with at least one other metal. Sterling silver, for example, is usually about 93% silver and 7% other metals, mostly copper. Silver can be found in high concentrations as argentite, Ag_2S, but most is recovered in the refining processes of other metals such as copper. Most compounds of silver are light sensitive and a lot of silver bromide and silver chloride are used by the photographic industry.

Gold

On an average, a million tons of earth contains just ten pounds of gold. This scarcity as well as its beauty and chemical properties, especially toxicity, account for its high value from ancient times. The name for the element is of Anglo-Saxon origin and the symbol comes from the Latin aurum, meaning "shining dawn". Gold metal has a distinctive yellow color and is incredibly malleable and ductile. A single ounce of pure gold can be beaten out to a sheet that is about 300 feet square. Pure gold is easily cut with a knife. Few elements react with gold under normal conditions and so most gold are recovered as small flakes of the pure element. Gold is a very good conductor and is often used to plate electrical contacts since it resists corrosion so well. It also is a good reflector of heat-carrying infra-red radiation. The world's oceans contain billions of tons of gold but it is too widely dispersed to be recovered (two-tenths of an ounce per million tons of water).

Roentgenium

Announcement of the discovery of element 111 was made in the popular press in January of 1995. Reports from the GSI labs in Germany indicate that nickel ions were used to bombard a bismuth target. The IUPAC has recently confirmed the discovery by S. Hoffmann and his coworkers in Darmstadt, Germany. The researchers proposed the name Roentgenium, with symbol Rg. Final approval was given in November 2004.

10. GROUP-12 (IIB): Zn, Cd, Hg and Uub

Zinc

The name for zinc is of German origins, Zink. It has been known since pre-historic times and compounds as well as the metal had been used for many years before anyone ever thought about elements at all: Pure zinc is a bluish-silver and ductile metal with a low melting and boiling point. Most zinc today is obtained from ZnS, extracted from zinc blende ore and roasted to remove the sulfur. Zinc can also be obtained by electrolysis of aqueous zinc sulfate, a common laboratory exercise. Zinc is reasonably resistant to corrosion and is used as a covering for baser metals like iron ("galvanizing"). Zinc can be readily cast or molded.

Cadmium

Discovered in 1817 by Strohmeyer (and independently in the same year by Hermann and Roloff), cadmium is named from the Greek kadmeia which is an ancient name for calamine (zinc oxide). Cadmium metal is soft and malleable when pure and is used mostly for electroplating steel (corrosion resistance) and in the manufacture of bearings. Cadmium compounds are also found in oil paints where they provide some brilliant colors (*e.g.*, "cadmium yellow"). Like zinc and tin, cadmium metal "cries" or emits a high pitched sound when bent rapidly. Most cadmium today is recovered as a by-product of the zinc refining process.

Mercury

Mercury or "quicksilver" has been known since ancient times, distinctive because of its liquid form at room temperature. The name and symbol are taken from the Latin hydrargyrum for "liquid silver". Of course, mercury is a good material for thermometers and barometers because it expands and contracts regularly with temperature and does not "wet" or stick to glass. Moreover it is very dense. It is also used in special motion-sensitive switches and in ordinary fluorescent lamps. Mercury alloys well with a number of metals. The alloys are known as amalgams. Dental alloy is an example of one such amalgam in common use. Virtually all mercury is derived from the mineral cinnabar which is mainly HgS.

Copernicium

Element 112 was synthesized in 1996 at the GSI in Darmstadt, Germany by Hofmann, et. al. Lead was bombarded with zinc atoms to achieve a mass total of 277. In July of 2009 the name "copernicium" and symbol "Cn" (in honor of Copernicus) were submitted to the IUPAC.

11. LANTHANIDES/LANTHANOIDS: Ce, Pr, Nd, Dm, Sm, Eu Gd, Tb, Dy, Ho, Er, Tm and Yb.

Cerium

Discovered by Berzelius and Hisinger in 1803, but not isolated as a metal until 1875, cerium (named for the asteroid Ceres) is the most abundant of the so-called rare-earth metals. It begins the series of lanthanides that runs from elements 58 to 71. In pure form the element is a malleable and ductile metal, similar in coloring to iron. It is much more reactive than iron, however, readily oxidizing in moist air and releasing hydrogen from boiling water. Friction from abrading a sample can cause it to ignite. Although the metal itself is too reactive for most uses, compounds of cerium are used in glass making and photography. It has limited use in some special alloys as well. Most commercial grade cerium is derived from monazite sand which is a mixture of phosphates of many of the rare earth metals along with calcium and thorium.

Praseodymium

Praseodymium, named from the Greek prasios + didymos (green twin), was isolated and identified by von Welsbach in 1885 from what was known at the time as didymium. von Welsbach's work revealed that this "substance" actually contained two new elements, one of which was praseodymium (neodymium was the other). Pure praseodymium is silvery-white and fairly soft. It oxidizes slowly in air and reacts vigorously with water to release hydrogen gas. It is used as an alloying agent along with magnesium for parts in aircraft engines. Misch metal is 5% praseodymium and is used for alloying steel and in flints used to create sparks in lighters. The glass in welder's goggles contains a mixture of praseodymium and neodymium.

Neodymium

Discovered in 1885 along with prascodymium, neodymium is named from the Greek neos + didymos (new twin). The silvery-white metal oxidizes easily in air and reacts with water, displacing hydrogen gas. Although another of the "rare" earth metals, neodymium is actually more abundant than many better known metals such as gold, silver, tin and lead. Misch metal, used in lighter flints, is about 18% neodymium. The element is also used in the manufacture of artificial rubies for laser applications.

Promethium

The existence of promethium (for the Greek god, Prometheus) was predicted in 1912 when Henry Moseley developed an X-ray method for determining integer atomic numbers of elements. An element was clearly missing between neodymium and samarium. Its existence was not confirmed until 1947 by Marinsky, Glendenin and Coryell. Historical claims for the discovery of element 61 create an interesting trail from around 1925 in Florence (suggested name: florentium) to America in 1926 (suggested name: illinium). None of the claims, however, could be substantiated and today we know they were not simply a result of fleetingly small samples but rather poor work. While spectral lines of promethium are evident in the light from some stars, it now seems apparent that no promethium is found in accessible areas of the earth, hence the difficulty in finding any. Initial attempts at synthesis of element 61 in a cyclotron at Ohio State University in 1941 led to the suggested name cyclonium. But the recognized synthesis and identification finally came at Oak Ridge in 1947. The longest-lived isotope of promethium is Pm-145 with a half-life of 17.7 years. There are no significant commercial uses of the metal and so very little have been produced except for theoretical studies.

Samarium

Named for the mineral samarskite from which it is extracted, samarium was isolated and identified by de Boisbaudran is 1879. The pure metal has a silver luster and tarnishes slowly at room conditions. It is readily magnetized and holds its magnetism extremely well. Rare earth magnets (samarium-cobalt, for example) exploit this property. Although it is present in samarskite, commercial production of samarium is from monazite sand which can contain as much as 2.8% Sm by weight.

Europium

Europium looks and seems a lot like lead, although it is not as dense. It was discovered in 1896 and isolated in 1901 by Demarcay, working with samples of supposedly "pure" samarium. Named for the continent of Europe, the element ranks thirteenth in abundance among the rare earth metals, but there is more of it than silver and gold combined. It is the most reactive of the rare earth metals, behaving with water in a manner similar to calcium. Generally refined from monazite sand, the pure metal has few applications, but we will find it less interesting to read this without some of its compounds which are used as activators and red phosphors in color CRT screens for television and computers.

Gadolinium

Gadolinium (from the mineral gadolinite and named for the Finnish chemist Gadolin) is a soft silvery-white metal that is used as an alloying agent in some steels and in the manufacture of some electronic components. Credit for its discovery is shared by de Marignac who did extensive spectroscopic studies on the mixture then known as didymia and by de Boisbaudran who finally isolated the metal in 1886. The metal has a very large capacity for absorbing thermal neutrons, making it an excellent material for control rods in fission power plants.

Terbium

Terbium is fourteenth in abundance among the 17 metals usually counted as "rare-earths", present in the earth's crust to the extent of only 0.9 ppm (about 1 teaspoon in every 63 tons of earth). Named for the Swedish village of Ytterby, the metal was discovered in 1843 by Mosander (along with erbium). Small amounts of terbium are used in special lasers and some solid state devices. The monazite sand from which terbium is generally extracted contains only about 0.03% by weight of Tb.

Dysprosium

The Greek word dysprositos (hard to get at) gives some indication of the scarcity of dysprosium, but only to a degree. It is about twice as abundant as uranium. The soft, silvery metal was discovered by de Boisbaudran in 1886 and it was finally isolated in 1906 by Urbain. A pure sample was not produced until the 1950s. The pure metal oxidizes readily in air.

Holmium

Holmium was discovered by Cleve in 1879 and named for the Latinized version of the name for Stockholm. Like most of the other rare-earth metals, it is silvery and soft and can be pounded or rolled into very thin sheets. At normal temperatures it is fairly inert but will oxidize at high temperatures and humidities. Like most of the rare-earth metals, holmium is generally obtained from monazite sand, where it is present to the extent of about 0.05%. Most holmium use is confined to research.

Erbium

Like the histories of the discoveries of many other rare-earths, the tale of erbium reads like a series of mistaken identities. These elements are generally found as oxides and most often together. Chemically, the oxides are very similar and at the time of their first examination were difficult to separate. Thus a sample of "lanthanum" might end up containing two additional elements that no one had bothered to look for. Many chemists thought that the oxides were elements themselves at one time. The oxide of yttrium (which along with scandium and lanthanum is generally included with the "rare-earths") known as yttria was eventually found to contain erbia and terbia as well, the oxides of, respectively, erbium and terbium. But the two are so similar that they were often confused in early work and what we now call erbium was originally terbium. In both cases, the credit for discovery goes to Mosander (1843 for erbium) and both elements were named for the Swedish town of Ytterby (which, by the way, also lends its name to Ytterbium and Yttrium, certainly some kind of record where naming elements is concerned). Like most of the rare-earth metals, erbium is silvery and soft, tarnishing slightly in air.

Thulium

The rarest of the naturally-occurring rare-earth metals, thulium was discovered in 1879 by Cleve, working from samples of erbia, an oxide of erbium. The metal is named for the ancient name for Scandinavia, Thule. Like others in the lanthanide series, thulium is silvery in color but it is also very soft, soft enough to cut with a knife.

Ytterbium

The first of the so-called "rare-earths" to be discovered, ytterbium takes its name from the Swedish village Ytterby (also the source of the names for terbium, erbium and yttrium). Discovery is credited to de Marignac in 1878. Initial identification was tediously made from the same mixture that most chemists of the time worked from: oxides of the lanthanides which gave rise to the term "rare-earth" due to its powdery consistency and often brownish color. But with the chemical separation techniques available at the time, it was very difficult to distinguish these similar elements. Even ytterbium itself turned out to hide another element. Lutetium was separated from it in 1907. Pure ytterbium is like most of the lanthanides: silvery and ductile, reacting slowly with air to form an oxide. Mostly obtained from monazite sand, ytterbium makes up about 0.03% of that mixture.

Lutetium

Lutetium ranks among the rare-earths in abundance only above thulium and promethium (and there's none of that anyway). Its official name comes from an ancient name for Paris, Lutecia, but it has had many names, most recently lutecium (only a change in official spelling). It was discovered independently by von Welsbach and Urbain in 1907-08. The refinement of ion exchange methods and their application to the separation of the rare-earths made the separation of lutetium from ytterbium possible. von Welsbach decided to rename ytterbium aldebaranium and picked cassiopium for element 71. Urbain preferred neoytterbium and lutecium. Urbain's choices eventually were accepted, although the prefix was dropped from ytterbium and the spelling of lutecium was eventually changed. The metal is the hardest and densest of the rare-earths and is the last of the lanthanides.

12. ACTINIDES/ACTINOIDS: Th, Pa, U, Np, Pu, Am, Cm, Bk, Cf, Es, Fm, Md and No

Thorium

Thorium was discovered by Berzelius in 1828 and named for the Norse god of thunder, Thor. It is a gray, radioactive metal which is fairly abundant in the earth's crust (more than twice as much as tin) and is the first of the so-called "actinide" series which ends with lawrencium (element 103). The long half-life of the principal isotope, Th-232, (about 10^{10} years) insures that there will be plenty for quite some time to come. The metal is fairly soft and malleable but darkens slowly in air due to oxidation. It reacts slowly with water at room temperature. Applications of thorium include some special magnesium alloys and photo sensors. The oxide is used in high-quality lenses. An isotope of thorium can be "bred" into uranium-234 by bombardment with slow neutrons. The U-234 is a fissile form of uranium and can be used in power plants.

Protactinium

Discovered in 1913 by Fajans and Göhring and then isolated in 1934 by Grosse, protactinium is named from the Greek proto + actinium (parent of actinium). The silvery-white metal is extremely rare, very radioactive and highly poisonous. Originally called brevium by its discoverers because of the short life-time of the transition between Th-234 and U-234, a longer-lived isotope was eventually isolated and called protoactinium by Grosse. The name was shortened to its present form in 1949. About 60 tons of pitchblende ore (which contains uranium, radium and a host of other radioactive elements) yields about 125 grams of protactinium.

Uranium

Uranium was discovered in 1789 by Klaproth and named for the planet Uranus. Isolation came in 1841 by Péligot and the radioactivity of the element was noted by Becquerel in 1896. The pure metal is heavy, silvery and lustrous. It tarnishes slowly in air and reacts with boiling water. Most of the naturally occurring uranium is the isotope U-238. This form of uranium is not fissionable, i.e., it cannot be used in atomic weapons or power plants. A much smaller percentage of naturally occurring uranium is the isotope U-235, which is fissionable. The process of "enriching" uranium to increase the proportion of U-235 in a sample is expensive and tedious but necessary to produce fuel that is usable in power plants and material for weapons. The U-238 is not completely useless, however, as it can be "bred" into Pu-239 by bombardment with slow neutrons. This is how weapons-grade plutonium is produced. The use of Pu-239 in power plants is a controversial subject due to its toxicity and the fear of diversion by terrorist groups.

Neptunium

Neptunium (named for the planet Neptune) was the first of the transuranium elements to be synthesized. The synthesis took place at Berkeley, California, after initial examination of the decay products of U-235 suggested the possibility of a new element. Credit for its discovery goes to McMillan and Abelson in 1940. Although the neptunium on which the characterization work was done was synthesized in a cyclotron, we now know that minute amounts of the element exist in the environment (the longest-lived isotope has a half-life of about 2 million years). All isotopes of the metal are radioactive.

Plutonium

Discovered by Seaborg in 1941, plutonium was named for the 9th planet, completing the series of elements named for planets which begins with uranium. Evidence for the existence of plutonium was obtained from work with samples of neptunium and the eventual synthesis in fact results from the beta decay of neptunium-239. The longest-lived isotope of plutonium is Pu-244 with a half-life of 82 million years. However the isotope of chief interest is Pu-239 which, like U-235, is fissionable. Most of the nuclear weapons built by the "great powers" today are based on Pu-239 which is derived from U-238 in special "breeder" reactors. Pu-239 is also a by-product of normal fission power reactors and accounts for a good deal of the concern over nuclear waste by-products since it is both highly radioactive and exceptionally toxic. Also, since the critical mass of plutonium is only about one third that of U-235, the possibility for terrorist diversion of the material is considered a serious matter. A piece of plutonium about the size of a soft ball would feel hot to the touch because of the high level of alpha particle radiation given off. A somewhat larger piece of the metal would boil water within minutes. Plutonium is occasionally used in deep-space probes as a source of energy (too far from the sun for effective solar power), the heat being directly converted into electricity by a special device.

Americium

Element 95, americium, was discovered after element 96. Seaborg, James, Morgan and Ghiorso share credit for its synthesis in 1944. Since the element is similar to europium (and below it in the periodic table) it seemed sensible to name element 105 for the American continent. Americium-241 is the isotope first produced and is manufactured today in kilogram quantities for use in high-precision measuring devices and some home smoke detectors. All isotopes of the metal are radioactive, the longest half-life being about 7400 years.

Curium

"Simply" by bombarding the newly characterized Pu-239 with alpha particles, the Berkeley team of Seaborg, James and Ghiorso managed to produce element 96 even before any element 95 had been made. Named in honor of Pierre and Marie Curie in 1944, curium is another actinide metal.

Berkelium

Berkelium was synthesized by Thompson, Ghiorso and Seaborg in 1949 and named for the site of its discovery: Berkeley, California. The initial synthesis was achieved by bombarding Am-241 with alpha particles. Two neutrons are ejected and a nucleus of Bk-243 results. The longest-lived isotope of berkelium is Bk-247 with a half-life of 1400 years.

Californium

The fifth element in succession to emerge from the Berkeley, California cyclotron was element 98, californium (named after the State of origin). Credit for the synthesis is given to the team of Thompson, Street and Ghiorso and, of course, Seaborg. Synthesis was achieved by bombarding Cm-242 with alpha particles. A neutron is ejected and a nucleus of Cf-245 results. The longest-lived isotope of californium, Cf-251, has a half-life of 890 years.

Einsteinium

Credit for the synthesis of element 99 was given to the Berkeley team of Ghiorso et. al. in 1952 and the radioactive metal was named for Albert Einstein. Evidence for the feasibility of its synthesis came from the debris of the first hydrogen bomb explosion in which U-238 was apparently transformed into U-253 which becomes Es-253 by the loss of seven beta particles. The laboratory synthesis of Es-253 begins with Pu-239 and goes through a five step process. The longest-lived isotope is Es-254 with a half-life of 276 days.

Fermium

From evidence detected in the debris of the first hydrogen bomb test, the Berkeley team of Ghiorso et. al. succeeded in the synthesis of element 100 in 1952 and named it fermium after the Italian physicist Enrico Fermi. Fm-254 was produced by neutron bombardment of Es-253. The longest-lived isotope of fermium is Fm-257 with a half-life of 80 days.

Mendelevium

Mendelevium (after Dmitri Mendeleev) was first synthesized in 1955 by Ghiorso, Harvey, Choppin, Thompson and Seaborg. Es-253 was bombarded with alpha particles, ejecting a neutron and yielding a nucleus of Md-256. The most stable isotope is Md-258 with a half-life of 56 days.

Nobelium

Discovered by Ghiorso, Sikkeland, Walton and Seaborg and named for Alfred Nobel, element 102 was originally named by a Swedish group working at the Nobel Institute of Physics in Stockholm. Details of their work, however, did not yield the expected results and the credit eventually shifted to the American team which decided to retain the original given name. The most stable isotope is No-259 with a half-life of just about 1 minute.

Lawrencium

The last of the actinide metals and only the second of the transuranium elements that Glenn Seaborg was not involved with, element 103 was synthesized in 1961 by Ghiorso,

Sikkeland, Larsh and Latimer at the Lawrence Radiation Laboratory (and named for Ernest Lawrence, the inventor of the cyclotron). Lr-256 is the most stable isotope with a half-life of 28 seconds.

13. GROUP-13 (IIIA): THE BORON-ALUMINUM FAMILY (B, Al, Ga, Ln and Tl)

Boron

The name Boron comes from the Arabic and Persian words for borax, its principal ore. Although compounds of boron were known in ancient times, it was first isolated in 1808 by Gay-Lussac and Thénard and independently by Sir Humphry Davy (who has a lot of elements to his credit). Boron exists in the earth's crust to the extent of only about 10 ppm (about the same abundance as lead). The pure element is shiny and black. It is very hard and in extremely pure form is nearly as hard as diamond, but much too brittle for practical use. At high temperatures it is a good conductor but at room temperature and below is an insulator. This behavior as well as many of its other properties earns it the classification of a metalloid. In addition to the crystalline form of boron there is also an amorphous dark brown powder (as shown above). The element can be prepared by the reduction of borax ($Na_2B_4O_7$) with carbon. High-purity boron can be produced by electrolysis of molten potassium fluoroborate. Common compounds of boron include borax and boric acid (H_3BO_3).

Aluminum

Aluminum ranks third on the list of the ten most abundant elements in the earth's crust, while its oxide is fourth among the ten most common compounds in the crust. It is the most abundant metal on the planet. Its name is taken from the Latin alumen for alum. Soft, lightweight and silvery, its existence was proposed by Lavoisier in 1787; it was named by Davy in 1807 and finally isolated by Orsted in 1825. In its purest form the metal is bluish-white and very ductile. It is an excellent conductor of heat and electricity and finds use in some wiring. When pure it is too soft for construction purposes but addition of small amounts of silicon and iron hardens it significantly. Although it does not seem to be particularly reactive, aluminum is considered an active metal. Its behavior is deceptive because it reacts rapidly with the oxygen in the air to form aluminum oxide (Al_2O_3), or alumina, which is tightly bound to the metal and exists as a dense coating (unlike the oxides of iron). This coating protects it from further reaction. Clearly, however, this coating is not entirely foolproof since aluminum does not exist in native form. Most of the aluminum today is produced by the Hall process which uses significant amounts of energy in the form of electricity to electrolyze aluminum metal from a molten salt mixture. The large initial outlay of energy is one important reason why recycling aluminum is such a good and cost-effective idea.

Gallium

Gallium is one of the elements originally predicted by Mendeleev in 1871 when he published the first form of the periodic table. He dubbed it ekaaluminum, indicating that it should have chemical properties similar to aluminum. The actual metal was isolated and named (from the Latin Gallia, for France) by Paul-Emile Lecoq de Boisbaudran in 1875. The detective work behind the isolation of gallium depended on the recognition of unexpected lines in the emission spectrum of a zinc mineral, sphalerite. Eventual extraction and characterization followed. Today, most gallium is still extracted from this zinc mineral. Although once considered fairly obscure, gallium became an important commercial item in the '70s with the advent of gallium arsenide LEDs and laser diodes. At room temperature gallium is as soft as lead and can be cut with a knife. Its melting point is abnormally low and it will begin to melt in the palm of a warm hand. Gallium is one of a small number of metals that expands when freezing.

Indium

The element indium (named from the Latin indicum, for the color indigo) was discovered in 1863 by Reich and Richter. It is a rare metal, with abundance similar to that of silver. It is generally found in deposits with zinc and refineries which produce this more common metal often sell indium as well. The pure metal is so soft that we can "wipe" it onto other materials in much the same way as lead (or even pencil graphite). It is corrosion resistant. As with gallium, identification of indium involved the recognition of new emission spectrum lines (its name was chosen because of indigo lines in its spectrum). Curiously enough, Reich who did the initial chemical isolation work was color blind and had to turn over his experiment to an assistant (Richter) who was the first to observe the characteristic lines. Like pure tin, pure indium emits a squealing sound when bent.

Thallium

Sir William Crookes discovered thallium in 1861, positively identifying it by a bright green line in its spectrum (hence the name, which is from the Greek, thallos, for "green twig"). Although in appearance thallium resembles lead, it does not have the corrosion resistance of lead and so has few commercial applications. Thallium compounds are quite toxic and some have been used as rat poisons. A few compounds are used in glasses for special infra-red lenses.

Ununpentium

In studies announced jointly by the Joint Institute for Nuclear Research in Dubna, Russia and the Lawrence Livermore National Laboratory in the U.S., four atoms of element 113 were produced in 2004 via decay of element 115 after the fusion of Ca-48 and Am-243.

14. GROUP-14 (IVA): THE CARBON FAMILY (C, Si, Ge, Sn and Pb)

Carbon

Carbon is the fourth most abundant element in the known universe but not nearly as common on the earth, despite the fact that living organisms contain significant amounts of the element. Common carbon compounds in the environment include the gases carbon dioxide (CO_2) and methane (CH_4). Carbon exists in several forms called allotropes. Diamond is one form with a very strong crystal lattice, known as a precious gem from the most ancient records. Graphite is another allotrope in which the carbon atoms are arranged in planes which are loosely attracted to one another (hence it is used as a lubricant). The recently discovered fullerenes are yet another form of carbon. Most elemental carbon is taken from the ground in the form of coal but certainly diamonds should not be ignored. Carbon has a very high melting and boiling point and rapidly combines with oxygen at elevated temperatures. In small amounts it is an excellent hardener for iron, yielding the various steel alloys upon which so much of modern construction depends. An important (but rare) radioactive isotope of carbon, C-14, is used to date ancient objects of organic origin. It has a half-life of 5730 years but there is only 1 atom of C-14 for every 10^{12} atoms of C-12 (the usual isotope of carbon).

Silicon

The name for silicon is taken from the Latin silex which means "flint". The element is second only to oxygen in abundance in the earth's crust and was discovered by Berzelius in 1824. The most common compound of silicon, SiO_2, is the most abundant chemical compound in the earth's crust. We know it better as common beach sand. Silicon is a crystalline semi-metal or metalloid. One of its forms is shiny, grey and very brittle (it will shatter when struck with a hammer). In another allotropic form silicon is a brown amorphous powder most familiar in "dirty" beach sand. The crystalline form of silicon is the foundation of the semiconductor age.

Germanium

Like silicon, germanium is used in the manufacture of semi-conductor devices. Unlike silicon, it is rather rare (only about 1 part in 10 million parts in the earth's crust). The metalloid was one of the elements predicted by Mendeleev in 1871 (ekasilicon) to fill out his periodic table and was discovered in 1886 by Winkler. The physical and chemical properties of germanium closely parallel those of silicon. Although it forms a compound, germanium dioxide, just like silicon, it is generally extracted from the by-products of zinc refining.

Tin

Mentioned in the Hebrew Scriptures, tin is of ancient origins. Early metal smiths were quick to learn that mixing copper with tin created a more durable metal (bronze) and it is principally for its alloys that tin is valued today. Named after the Etruscan god Tinia, the chemical symbol for tin is taken from the Latin stannum. The metal is silvery white and very soft when pure. It has the look of freshly cut aluminum but the feel of lead. Polished tin is slightly bluish. It has been used for many years in the coating of steel cans for food because it is more resistant to corrosion than iron. It forms a number of useful low-melting alloys (solders) which are used to connect electrical circuits. Bending a bar of tin produces a characteristic squealing sound called "tin cry". Perhaps one of the most familiar of tin compounds, SnF_2, tin(II) fluoride, goes by the trade name of fluoristan and is found in some fluoride toothpastes.

Lead

Known to the ancients, lead takes its name from the Anglo-Saxon word for the metal and its symbol comes from the Latin plumbum (from which we get the modern word "plumber" since old plumbing was done with lead pipes). Although lead is not very common in the earth's crust, what is there is readily available and easy to refine. Its chief use today is in lead-acid storage batteries such as those used in automobiles. In pure form it is too soft to be used for much else. Lead has a blue-white color when first cut but quickly dulls on exposure to air, forming Pb_2O, one of the few lead(I) compounds. Most stable lead compounds contain lead in oxidation states of +2 or +4. Various isotopes of lead come at the end of the natural decay series of elements like uranium, thorium and actinium. These are Pb-206, Pb-207 and Pb-208

Ununquadium

The synthesis of element 114 was reported in January of 1999 by scientists from the Joint Institute for Nuclear Research in Dubna (near Moscow) and Lawrence Livermore National Laboratory (in California). In an experiment lasting more than 40 days Russian scientists bombarded a film of Pu-244 supplied by Livermore scientists with a beam of Ca-48. One atom of element 114 was detected with a half-life of more than 30 seconds. This is about 100,000 times as long as the previously longest-lived isotope of 112 produced (element 113 has yet to be made) and bolsters the arguments of theorists who envision an "island of stability" in the super-heavy elements. Confirmation of the work has not been done and no name or symbol has been suggested.

15. GROUP-15 (VA): THE NITROGEN FAMILY (N, P, As, Sb and Bi)

Nitrogen

Nitrogen, which makes up about 78% of our atmosphere, is a colorless, odorless, tasteless and chemically unreactive gas at room temperature. It is named from the Greek nitron + genes for soda forming. Discovered in 1772 by Daniel Rutherford (and independently by others such as Priestly and Cavendish), it is more abundant in the known universe than carbon or silicon. Most commercially produced nitrogen gas is recovered from liquefied air. Of that amount, the majority is used to manufacture ammonia (NH_3) via the Haber process. Much is also converted to nitric acid (HNO_3). Small amounts of nitrogen (for laboratory study) can be prepared by heating solid ammonium chloride (NH_4Cl) with solid ammonium nitrite (NH_4NO_2). Sodium chloride and water are by-products. Nitrogen forms a surprising number of compounds with oxygen, exhibiting a wide range of oxidation states. Some of the more familiar ones include nitrogen dioxide (NO_2), a component of photochemical smogs and a contributor to acid rain in some areas and dinitrogen monoxide (N_2O) commonly called "laughing gas".

Phosphorus

Named from the Greek word phosphoros ("bringer of light"), phosphorus was first isolated in 1669 by Henning Brand (a painting commemorating this achievement is in the rotating gallery). The original extraction was made from about 60 pails of urine. Industrial processes for isolating phosphorus today involve heating calcium phosphate, carbon and silicon dioxide. Phosphorus occurs in at least 10 allotropic forms, the most common (and reactive) of which is so-called white (or yellow) phosphorus which looks like a waxy solid or plastic. It is very reactive and will spontaneously inflame in air so it is stored under water. The other common form of phosphorus is red phosphorus which is much less reactive and is one of the components on the striking surface of a match book. Red phosphorus can be converted to white phosphorus by careful heating. Commercially, phosphorus compounds are used in the manufacture of phosphoric acid (H_3PO_4) (found in soft drinks and used in fertilizer compounding). Other compounds find applications in fireworks and, of course, phosphorescent compounds which glow in the dark.

Arsenic

Arsenic has been known for a very long time and the person who may have first isolated it is not known but credit generally is given to Albertus Magnus in about the year 1250. The element, which is classified as a metalloid, is named from the Latin arsenicum and Greek arsenikon which are both names for a pigment, yellow orpiment. Tiny amounts of arsenic are used today in the semiconductor industry to create LEDs. Some compounds are used in the manufacture of everything from wallpaper to ceramics. Of course, arsenic is most often thought of in terms of its toxicity and it is used as a weed killer and rat poison. Intentional arsenic poisoning is now pretty much the stuff of old murder mysteries. Originally arsenious oxide was administered in small doses over a period of time, precipitating death accompanied by symptoms reminiscent of pneumonia. Today the arsenic remaining in the body can be detected during autopsy so this is not (happily) a very effective way to dispatch people undetected. Commercially, arsenic is produced by heating the pigment orpiment (As_2S_3) in air and then reacting the oxide product with carbon.

Antimony

Antimony and its compounds have been known for centuries. Scientific study of the element began during the early 17th century, much of the important work being done by Nicolas Lemery. The name of the element comes from the Greek anti + monos for "not alone", while the modern symbol is rooted in the Latin-derived name of the common ore, stibnite. Antimony is a hard, brittle metalloid which is alloyed with other metals to increase hardness. It is also used in some semi-conductor devices. The recovery of elemental antimony parallels that of arsenic: the sulfide ore (stibnite) is roasted in air and then heated with carbon.

Bismuth

Bismuth, the heaviest non-radioactive naturally occurring element, was isolated by Basil Valentine in 1450. It is a hard, brittle metal with an unusually low melting point (271°C). Alloys of bismuth with other low-melting metals such as tin and lead have even lower melting points and are used in electrical solders, fuse elements and automatic fire sprinkler heads. The metal can be found in nature, often combined with copper or lead ores, but can also be extracted from bismuth (III) oxide by roasting with carbon. Compounds of bismuth are used in pigments for oil painting and one is in a popular pink preparation for the treatment of common stomach upset.

Ununtritium

In studies announced jointly by the Joint Institute for Nuclear Research in Dubna, Russia and the Lawrence Livermore National Laboratory in the U.S., four atoms of element 113 were produced in 2004 via decay of element 115 after the fusion of Ca-48 and Am-243.

16. GROUP-16 (VIA): THE CHALCOGENS (O, S, Se, Te and Po)

Oxygen

Named from the Greek oxys + genes, "acid-former", oxygen was discovered in 1772 by Scheele and independently by Priestly in 1774. Oxygen is the most common element in the earth's crust and makes up about 20% of the air we breathe. Historically the discovery of oxygen as an element essential for combustion stands at the heart of the phlogiston controversy. Oxygen is a gas at room temperature and is colorless, odorless and tasteless. Liquid oxygen has a slight blue color. Virtually every element known forms a compound with oxygen except for some of the noble gases. Two allotropes of oxygen are known: the ordinary oxygen (O_2) and the ordinary ozone (O_3) in air and above it. Ozone occurs naturally in the upper atmosphere where it acts as a shield protecting the earth's biosphere from ultraviolet radiation. Closer to the surface ozone can be a nuisance as it is very reactive. But in a controlled environment it is also handy: it makes a good bleach and disinfectant.

Sulfur

Known from ancient times (mentioned in the Hebrew Scriptures as brimstone) sulfur was classified as an element in 1777 by Lavoisier. Pure sulfur is tasteless and odorless with a light yellow color. Samples of sulfur often encountered in the lab have a noticeable odor. Sulfur is the tenth most abundant element in the known universe. There are three allotropic forms of sulfur. Two are crystalline and one is amorphous (sometimes called "plastic" sulfur). This last form gradually reverts to one of the more stable crystalline forms. Most sulfur is recovered directly as the element from underground deposits by injecting super-heated water and piping out molten sulfur (sulfur melts at 112°C). Perhaps the most significant compound of sulfur used in modern industrialized societies is sulfuric acid (H_2SO_4). Sulfur dioxide (SO_2) finds practical applications in bleaching and refrigeration but it is also a nuisance gas resulting from the burning of sulfurous coals. Sulfur dioxide gas then reacts with the water vapor in the air to produce a weak acid, sulfurous acid (H_2SO_3) which contributes to the acid rain problem. Other than sulfuric acid, perhaps the most familiar compound of sulfur in the chemistry lab is the foul-smelling hydrogen sulfide gas, H_2S, which smells like rotten eggs.

Selenium

Selenium was discovered by Berzelius in 1818. It is named for the Greek for "moon", Selene. Most selenium is recovered from the electrolytic copper refining process. This is usually in the form of the red allotrope. There are at least two other allotropes of the element, including a semi-metallic state. Selenium is an important semi-conductor which is particularly sensitive to light. Thus it finds applications that range from light sensors to xerography. In large amounts the element is toxic, but it is a trace element in living organisms.

Tellurium

Discovered by von Reichenstein in 1782, tellurium is a brittle metalloid that is relatively rare. It is named from the Latin tellus for "earth". Tellurium can be alloyed with some metals to increase their machinability and is a basic ingredient in the manufacture of blasting caps. Elemental tellurium is occasionally found in nature but is more often recovered from various gold ores, all containing $AuTe_2$.

Polonium

Polonium was discovered in 1898 by Marie Curie and named for her native country of Poland. The discovery was made by extraction of the remaining radioactive components of pitchblende following the removal of uranium. There is only about $10^{-6}{}_{g}$ per ton of ore. Current production for research purposes involves the synthesis of the element in the lab rather than its recovery from minerals. This is accomplished by producing Bi-210 from the abundant Bi-209. The new isotope of bismuth is then allowed to decay naturally into Po-210. The sample pictured above is actually a thin film of polonium on stainless steel. Although radioactive, polonium has a few commercial uses. We can buy our own sample of polonium at a photography store. It is part of the special anti-static brushes for dusting off negatives and prints.

Ununhexium

In studies reported by the Joint Institute for Nuclear Research in Dubna, Russia in 2000/2001, atoms of element 116 were observed to decay following the complete fusion of Ca-48 and Cm-248. No independent confirmation of the experiments has been claimed (an earlier claim to the discovery of elements 116 and 118 at Berkeley was withdrawn).

17. GROUP-17 (VIIA): THE HALOGENS (F, Cl, Br, I and At)

Fluorine

Fluorine (from the Latin fluere, for "flow") was isolated by Henri Moissan in 1886. It is a highly toxic and reactive greenish-yellow gas at room temperature. Because of its reactivity, elemental fluorine is never found in nature and no other chemical element can displace fluorine from its compounds. In the late 1600's minerals which we now know contain fluorine, were used in etching glass. The discovery of the element was prompted by the search for the chemical substance which was able to attack glass (it is HF, a weak acid). The early history of the isolation and work with fluorine and hydrogen fluoride is filled with accidents since both are extremely dangerous. Eventually, electrolysis of a mixture of KF and HF after carefully ensuring that the resulting hydrogen and fluorine would not come in contact, in a platinum apparatus, yielded the element. Compounds of fluorine are present in fluoridated toothpaste and in many municipal water systems where they help to prevent tooth decay. And, of course, fluorocarbons such as Teflon have made a major impact on life in the 20th century.

Chlorine

Chlorine, which is similar to fluorine but not as reactive, was prepared by Sheele in the late 1700's and shown to be an element by Davy in 1810. It is a greenish-yellow gas with a disagreeable odor (we can detect it near poorly balanced swimming pools). Its name comes from the Greek word Chloros, meaning greenish-yellow. In high concentration it is quite toxic and was used in World War I as a poison gas. Like fluorine and the other members of the halogen family, chlorine is diatomic in nature, occurring as Cl_2 rather than Cl. It forms -1 ions in ionic compounds with most metals. Perhaps the best known compound of that type is sodium chloride, common table salt (NaCl). Small amounts of chlorine can be produced in the lab by oxidizing HCl with MnO_2. On an industrial scale, chlorine is produced by electrolysis of brines or even sea water. Sodium hydroxide (also in high demand) is a by-product of the process. In addition to the ionic compounds that chlorine forms with metals, it also forms molecular compounds with non-metals such as sulfur and oxygen. There are four different oxides of the element. Hydrogen chloride gas (from which we get hydrochloric acid) is an important industrial product.

Bromine

Bromine is a reddish-brown fuming liquid at room temperature with a very disagreeable chlorine-like smell. In fact its name is derived from the Greek bromos or "stench". It was first isolated in pure form by Balard in 1826. It is the only non-metal that is a liquid at normal room conditions. Bromine on the skin causes painful burns that heal very slowly. It is an element to be treated with the utmost respect in the laboratory. Most bromine is produced by displacement from ordinary sea water. Chlorine (which is more active) is generally used to dislodge the bromine from various compounds in the water. Before leaded gasolines were removed from the market, bromine was used in an additive to help prevent engine "knocking". Production now is chiefly devoted to dyes, disinfectants and photographic chemicals.

Iodine

Elemental iodine is a dark grey solid with a faint metallic luster. When heated at ordinary air pressures it sublimes to a violet gas. The name iodine is taken from the Greek ioeides which means "violet colored". It was discovered in 1811 by Courtois. Commercially iodine is recovered from seaweed and brines. It is an important trace element in the human diet, required for proper function of the thyroid gland. Thus iodine is added to table salt ("iodized") to insure against iodine deficiencies. Radioactive isotopes of iodine are used in medical tracer work involving the thyroid and also to treat diseases of that gland.

Astatine

Astatine is the last of the known halogens and was synthesized in 1940 by Corson and others at the University of California. It is radioactive and its name, from the Greek astatos, means "unstable". The element can be produced by bombarding targets made of bismuth-209 with high energy alpha particles (helium nuclei). Astatine 211 is the product and has a half-life of 7.2 hours. The most stable isotope of astatine is 210 which has a half-life of 8.1 hours. Not much is known about the chemical properties of astatine but it is expected to react like the other halogens, although much less vigorously and it should be more metallic than iodine. There should be tiny quantities of astatine in the earth's crust as products of other radioactive decays, but their existence would be short-lived.

18. GROUP-18 (VIIIA/0): THE NOBLE GASES (He, Ne, Ar, Kr, Xe and Rn)

Helium

With a name taken from the Greek helios for "sun", helium is the second lightest and second most abundant gas in the known universe (hydrogen being number one). Because of its scarcity in our atmosphere its existence was not suspected until spectroscopic measurements revealed an unknown element present in the sun. The discovery of helium is generally credited to Janssen and Lockyear in 1868. As no helium compounds are known, this family of gases was once thought to be inert. In 1962 the first noble gas compound was prepared with xenon. Still, helium only occurs in uncombined form and must either be extracted from the atmosphere by liquefaction of air or separated from deposits of natural gas. It is thought that some of the terrestrial helium is the product of the alpha decay of radioactive isotopes beneath the crust. Helium is the only known element which cannot be converted to a solid simply by cooling. It has 98% the lifting power of hydrogen with none of the Hindenburg-type drawbacks and is still used today in airships.

Neon

The fourth most abundant element in the known universe, neon was discovered in 1898 by Ramsay and Travers. Its name is taken from the Greek neos or "new". This is perhaps the best known noble gas because of its use in so-called "neon" lights (many of which actually contain other gases) and is relatively plentiful in the earth's atmosphere (fifth in abundance, following carbon dioxide). No stable compounds of neon are known to date. The gas is extracted from air by liquefaction.

Argon

Argon is third in abundance in the earth's atmosphere (about 1% by volume). Rayleigh and Ramsay isolated and identified the gas in 1894. Like the rest of the noble gases, it is colorless, tasteless and odorless. The name for argon is taken from the Greek Argos for "inactive". For years argon has been used in ordinary incandescent light bulbs to replace the oxygen that would otherwise shorten the lifetime of the filament. It is used in some types of welding where active atmospheric gases would interfere with the process. Argon is also used in various types of "black lights" or UV lamps since excitation of the gas produces a significant amount of ultraviolet radiation. A few curious compounds have been made with argon but they are not very stable.

Krypton

Named from the Greek kryptos or "hidden", krypton is neither green nor a solid material that can defeat Superman. Rather it is another noble gas discovered in 1898 by Ramsay and Travers. It ranks sixth in abundance in the atmosphere. Krypton gas is used in various kinds of lights, from small bright flashlight bulbs to special strobe lights for airport runways. As with the other noble gases, krypton is isolated from the air by liquefaction. One of the naturally occurring non-radioactive isotopes of krypton, Kr-86 (17.3%) is used as the basis for the current international definition of the metre. One metre is 1,650,762.73 wavelengths of the red-orange spectral line of krypton-86.

Xenon

Also discovered by Ramsay and Travers in 1898, xenon (from the Greek xenos for "strange") is the rarest of the stable noble gases in the air. It is still recovered by liquefaction techniques and is widely used in strobe lamps. In 1962 the first noble gas compound was produced by Neil Bartlett, combining xenon, platinum and fluorine. It is now possible to produce xenon compounds in which the oxidation states range from +2 to +8.

Radon

Discovered in 1900 by Friedrich Dorn, radon is a radioactive noble gas now regarded as a potential health hazard in some homes. It also has medical applications for cancer treatment. Its original name was to be niton for "shining" but it was eventually named as a derivative of radium. Radon is found in underground deposits where is it produced by uranium and radium decay. Radon fluoride (RnF) has been produced and the element glows with a yellow light in the solid state.

Chapter 7

THE CHEMICAL BOND

1. WHAT ARE CHEMICAL BONDS AND HOW DO THEY RESULT?

When two or more atoms are brought closer and closer, an attractive force between the electrons of individual atoms and the nuclei of one or more of the atoms can result. If this force is large enough to keep the atoms together, a chemical bond is said to be formed. All chemical bonds result from the simultaneous attraction of one or more electrons by more than one nucleus. In fact a chemical bond is any force of attraction that holds two atoms or ions together. In most cases, the force of attraction is between one or more electrons held by one of the atoms and the positively charged nucleus of the second atom. Chemical bonds vary widely in their stability, ranging from relatively strong covalent bonds to very weak hydrogen bonds. According to the octet rule atoms bond together to form molecules in such a way that each atom participating in a chemical bond acquires an electron configuration resembling that of the noble gas nearest to it in the periodic table. Thus the outer shell of each bonded atom will contain eight electrons (or two electrons for hydrogen and lithium). The simplest chemical bond is that formed between two hydrogen atoms. Each hydrogen atom has one electron. As the two atoms approach each other, the nucleus of one atom attracts the electron of the other. Eventually the two orbitals overlap, becoming a single orbital containing two electrons as is evident in the following figure (Fig. 7.1).

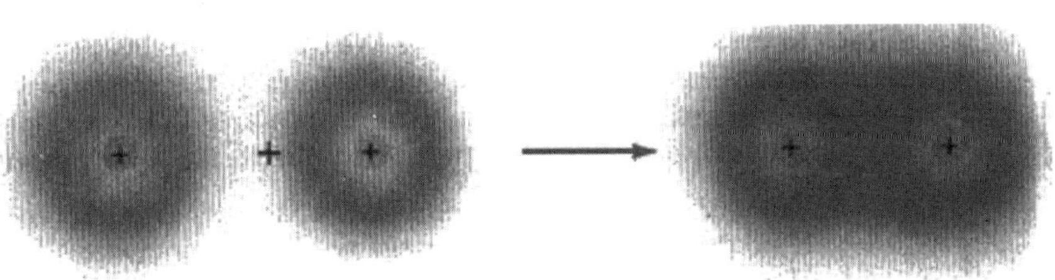

H(g) + H(g) ⟶ $H_2(g)$

Fig 7.1: Two hydrogen atoms, each with one electron, combine to form a hydrogen molecule, in which the two electrons are shared between the atoms and serve to give each atom a filled valence shell.

This orbital encompasses space around both nuclei. Although the electrons may be in any part of this orbital, we can predict that they are most likely to be in the space between the nuclei, shielding one nucleus from the other and being attracted by both. In the resulting molecule, both atoms have two electrons and a filled outer (valence) shell. These shared electrons form a bond between the two atoms. This chemical bond is a covalent bond, a pair of electrons shared between two atoms. When this bond forms, energy is released. This release of energy shows that the molecule of hydrogen is more stable than the separate atoms.

2. WHAT BONDS ARE DUBBED AS 'COVALENT, POLAR COVALENT AND IONIC BONDS'?

Because the hydrogen molecule contains two identical atoms, it can be assumed that the bonding electrons in this covalent bond are shared equally by these atoms. But most chemical bonds are not between like atoms rather they are between atoms of different elements. These bonds are slightly different from that in a hydrogen molecule. We may consider the bond between a hydrogen atom and a chlorine atom: again both atoms require one more electron to satisfy the octet rule. As the atoms come together, their orbitals overlap and the two atoms share a pair of electrons. However, the hydrogen-chlorine bond differs from the hydrogen-hydrogen bond because the electrons are not shared equally between hydrogen and chlorine but are more strongly attracted to the chlorine. They are more apt to be found close to the chlorine in comparison to the hydrogen. Because of this unequal sharing, the chlorine atom assumes a slightly negative character and the hydrogen atom a slightly positive character.

We say that the bond is polar covalent, meaning that the bond consists of electrons shared between two atoms (therefore covalent) but shared unequally, thus giving the bond a positive and a negative end, a condition described by the term polar. We can also say that the bond is a dipole or has a dipole moment, meaning that the bond has a positive end (the hydrogen) and a negative end (the chlorine). The more negative atom in a bond is often shown with the δ^- symbol and the more positive atom is shown with the δ^+symbol. The bond between hydrogen atoms is nonpolar (has no positive and negative ends) covalent (electrons are shared).

An ionic bond is the extreme case of a polar covalent bond. In an ionic bond, the bonding atoms differ so markedly in their attraction for electrons that one or more electrons are essentially transferred from one atom to the other. The sodium-chlorine bond which is in fact a bond between sodium and chloride ions is an example of an ionic bond. The attraction of the chlorine atom for electrons is so much greater than that of a sodium atom that the 3s electron of sodium is said to be completely transferred from sodium to chlorine.

In summary, then, the three types of bonds are: (1) a covalent bond, in which the electrons are shared equally; (2) a polar covalent bond, in which the electrons are shared unequally; and (3) an ionic bond, in which electrons are transferred from one atom to the other. These bonds may be illustrated as under (Fig. 7.2.)

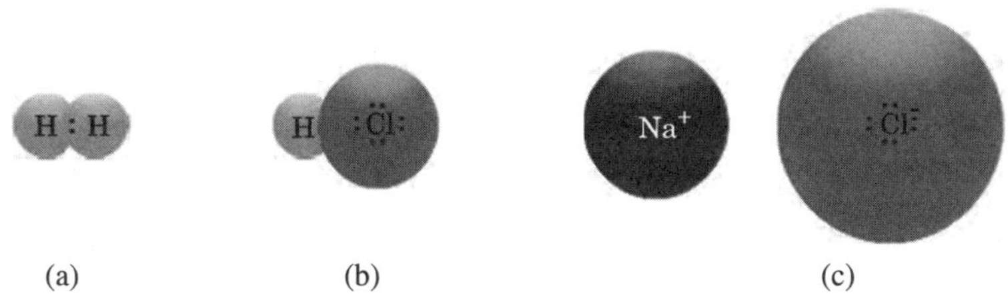

Fig. 7.2: Electrons in nonpolar covalent, polar covalent and ionic bonds: (a) the electrons are shared equally; (b) the electrons are held closer to the more-negative chlorine atom; (c) one electron has been transferred from sodium to chlorine.

3. HOW CAN WE PREDICT BOND TYPES ON THE BASIS OF ELECTRONEGATIVITY CONSIDERATION?

The electronegativity (EN) of an element measures of its atoms attraction for the electrons in a chemical bond. One scale of electronegativity was developed by the American chemist Linus Pauling. On this scale, fluorine, the most electronegative element, has been assigned an electronegativity value of 4.0. Carbon has an electronegativity of 2.5, hydrogen, 2.1 and sodium 0.9. The electronegativity of most metals is close to 1.0 and that the electronegativity of a nonmetal, although dependent on its location in the table, is always greater than 1.0. In general, electronegativity increases from bottom to top in a column and from left to right across a period. We further note that the noble gases, Group 18 elements, do not appear in this table. Since electronegativity measures the relative attraction of atoms for electrons in chemical bonds and the noble gases react differently from the halogens and other nonmetals, the concepts of electronegativity do not apply to them.

It is possible to predict the type of bond that will result between atoms of two elements. The farther apart (left to right) the two elements are in the periodic table, the more ionic and the less covalent will be the bond between their atoms. Thus, metals react with nonmetals to form ions joined predominantly by ionic bonds. Bonds with the highest degree of ionic character are formed by the reaction of alkali or alkaline earth metals with the halogens, particularly with fluorine or chlorine. Nonmetals react together to form covalent bonds. If the bond is between two neighbors in the table, the bond will be less polar one than if the nonmetals are separated by other element. For example, carbon and nitrogen are in neighboring columns and carbon and fluorine are in Groups 14 and 17, respectively. A carbon-nitrogen bond will be less polar than a carbon-fluorine bond. Finally, if the two atoms are of the same element, as is the case in a hydrogen molecule or a chlorine molecule, the bond will be essentially nonpolar.

The quantitative concepts floated in the previous paragraph can be quantified by the concept of electronegativity which is tendency for an atom to attract a pair of electrons that it shares with another atom. For example, the molecule hydrogen chloride, HCl, consists of a hydrogen atom, H and a chlorine atom, Cl, sharing a pair of electrons.

If the pair of electrons are not shared equally, *i.e.*, if they spend more time with one atom than with the other, the favored atom is said to be more electronegative. In the case of HCl, measurements indicate that the molecule has a dipole moment, that is, the chlorine end is relatively negative and the hydrogen end is relatively positive. This means that the electron pair spends more time with the chlorine atom than with the hydrogen atom and thus chlorine is more electronegative than hydrogen. Nonmetals have much higher electronegativities than metals; of the nonmetals, fluorine is the most electronegative, followed by oxygen, nitrogen and chlorine. The larger is the difference in electronegativity between two atoms, the more polar will be the bond between them. In the extreme case of a bond between a metal and a nonmetal, a complete transfer of electrons takes place.

When two atoms combine, the nature of the bond between them is determined by the difference between their electronegativities (denoted by EN). If the atoms forming the bond differ in electronegativity by more than 1.7 units, the bond will be at least 50% ionic (referred to as percent ionic character); we treat such a bond as wholly ionic. If the values differ by less than 0.4 units, we consider the bond to be wholly nonpolar. If the difference is between 0.4 and 1.7 electronegativity units, the bond is considered to be polar covalent. Remember that electronegativities have been calculated from fairly imprecise data for particular bonding situations. Though electronegativity is useful in predicting the nature of a bond and for comparing bond types, the prediction is only an approximation. It is worth remembrance too that no sharp distinction exists between ionic, polar covalent and nonpolar bonds; rather, they form a continuum. Even the most ionic bond (between cesium and fluorine) has some covalent character and only bonds between atoms of the same element have no ionic character. In these bonds, the atom with the higher electronegativity will be the negative end of the bond and, in extreme situations, will become the negative ion. To show these partial charges on a polar covalent bond, we mark the positive end of the bond with a δ^+ and the negative end of the bond with a δ^-. As an illustrative example we may predict the nature of the bond between the following pairs of atoms as predominantly nonpolar covalent, polar covalent, or ionic. For each polar covalent bond, we may use a small Greek letter δ to show which atoms bears a partial positive charge (δ^+) and which a partial negative charge (δ^-).

(a). S-O (b). C-O (c). Al-F

The electronegativity of oxygen is 3.5 and that of sulfur is 2.5. The difference is 1.0 unit; we predict the S-O bond to be polar covalent. The oxygen is partially negative and the sulfur is partially positive, so we write:

$$\overset{\delta^+}{\text{S}}\text{—}\overset{\delta^-}{\text{O}}$$

The electronegativity difference between oxygen and carbon is 1.0 unit (3.5–2.5). Therefore, we predict the C - O bond to be polar covalent. Because oxygen is the more electronegative of the two, it carries the negative charge.

$$\overset{\delta^+}{\mathrm{C}}\text{—}\overset{\delta^-}{\mathrm{O}}$$

The electronegativity difference between fluorine and aluminum is 2.5 units (4.0 - 1.5). Therefore, we predict the Al - F bond to be largely ionic. The aluminum forms a cation, the fluoride an anion.

4. WHAT ARE SINGLE, DOUBLE AND TRIPLE BONDS?

A covalent bond represents the sharing of electrons between two atoms. Single bonds result from the sharing of a single pair of electrons. Usually, as in the hydrogen molecule, each atom forming the bond contributes one electron to the bond. Sometimes, as in the reaction of ammonia, NH_3, with a hydrogen ion, H^+, to form the ammonium ion, NH_4^+, both electrons come from the same atom:

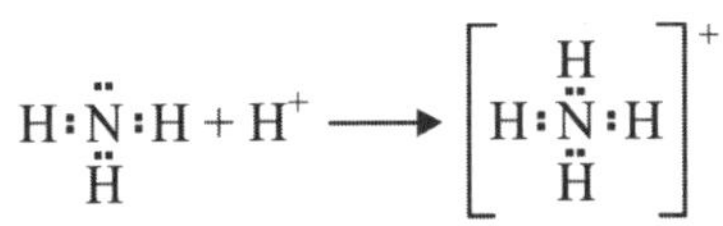

It is a common practice to use a dash to represent a pair of electrons. In this text we will use dashes for shared electrons and dots for unshared (lone-pair) electrons. With this notation, the above equation is written:

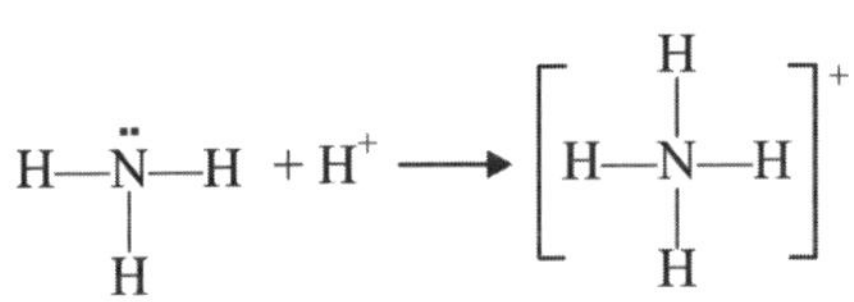

In the ammonia molecule, the nitrogen atom shares a pair of electrons with each of the three hydrogen atoms. In each bond, one electron comes from nitrogen and one from hydrogen. The nitrogen still has an unshared pair of electrons. A hydrogen ion has no electrons; the single hydrogen electron was lost when the atom became an ion and gained a positive charge. When the hydrogen ion is bonded to the ammonia molecule both the electrons of the bond come from nitrogen only. A bond in which one atom has donated both electrons is often referred to as a coordinate covalent bond. It is most important to realize that the different name refers only to the method of formation. Once the ammonium ion is formed, all hydrogen-nitrogen bonds in the ion are equivalent. Notice, too, that the entire ammonium ion now carries a positive charge, denoted by placing brackets around the ion and writing a superscript +.

In addition to single bonds, there are double bonds and triple bonds. A double bond represents the sharing of four electrons by two atoms. The bond between carbon and oxygen is often a double bond, as in formaldehyde, CH_2O.

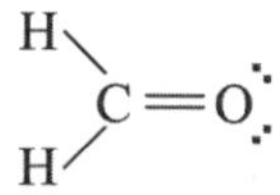

Here carbon is singly bonded to each of the hydrogen atoms and doubly bonded to oxygen. Of this double bond, two electrons have come from carbon and two from oxygen. The single carbon-hydrogen bonds are nonpolar (EN = 0.4); the double carbon-oxygen bond is polar covalent (EN = 1.0). Note that each atom in the diagram of formaldehyde now follows the octet rule. Each hydrogen atom has two electrons; the carbon and the oxygen have eight electrons each. Notice too that the oxygen has two pairs of unshared electrons. Such an unshared pair is sometimes known as a lone pair. We will see that the negative end of a polar bond often holds unshared electron pairs. A triple bond is formed when two atoms share six electrons (three pairs). The nitrogen molecule contains a triple bond. Its structure is shown below wherein every nitrogen atom donates three electrons to the bond and retains a lone pair.

$$:N\equiv N:$$

5. WHAT STRUCTURES ARE DUBBED AS 'LEWIS STRUCTURES'?

A compound contains two or more atoms of different elements joined by chemical bonds. The properties of the compound depend on the arrangement of atoms in the compound and the types of bonds between them. To help in gaining this information for covalently bonded molecules, we draw structures for the molecules such as generally shown for hydrogen, ammonia, formaldehyde and nitrogen. These structures are called Lewis structures. A Lewis structure shows each atom in the molecule or ion and its relationship to the other atoms. It also shows all bonding electrons as well as those valence electrons that are nonbonding. Lewis structures (also called electron-dot structures) are formed when Lewis symbols (also called electron-dot symbols) are combined. Lewis symbols are a simple way of visualizing the valence electrons in an atom. In a Lewis symbol, the symbol for the element is used to represent the atom and its core electrons. Dots placed around the atom are used to indicate the valence electrons. The drawing of Lewis dot symbols is a useful way to show the position and number of valence electrons in an atom. Remember, the valence electrons are the electrons in the outermost energy level of an atom and are instrumental in the bonding process. But an even better use for electron dot symbols is to join them together to form the Lewis structures of molecules or polyatomic ions. However, the purpose of Lewis structures is not to show the three dimensional shape of a molecule or polyatomic ion.

The purpose of Lewis structures is to show the number and kinds of bonds and the order in which the atoms or ions are connected in the molecule or polyatomic ion. Many of these structures are simple and can be determined by inspection. While, others are a bit more complicated and require some thinking. Many covalent compounds can be drawn by inspection using the valence electrons and the knowledge that covalent bonds are shared bonds. Just determine the Lewis dot structures of the atoms and the number of bonds for each atom. Then pair up the available electrons for the covalent bonds and draw the molecule. Although, many molecules can be drawn by inspection others require the use of a few rules to help put them together.

6. HOW ARE ATOMS ARRANGED IN COMPOUNDS?

Most often the formula of a compound is written in a way that predicts the arrangement of its atoms. Each covalently bonded molecule or ion has one central atom or a chain of central atoms. The central atom or chain of atoms will be the least electronegative element in the structure. This chain is often of carbon atoms, although it may be of nitrogen or some other nonmetal. Methane, CH_4, has one central carbon atom. Butane, C_4H_{10}, has a chain of four carbon atoms. The element symbols that directly follow the central atom in the formula indicate that these atoms are bonded to the central atom(s). Thus, methane, CH_4, has four hydrogen atoms bonded to its central carbon atom. Butane, C_4H_{10}, has ten hydrogen atoms bonded to the four carbon atoms. Formaldehyde, whose formula is written as CH_2O or HCHO, has two hydrogen atoms and one oxygen atom bonded to the central carbon atom. Methyl amine, CH_3NH_2, has three hydrogen atoms bonded to the carbon atom. The carbon atom is also bonded to a nitrogen atom and two more hydrogen atoms are also bonded to the nitrogen atom.

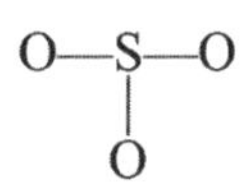 rather than S–O–O–O

If more than one arrangement of atoms seems possible, we choose the one with the most symmetry. In carbon dioxide, CO_2, the atoms are arranged O-C-O, a more symmetrical arrangement than C-O-O. Similarly, sulfur trioxide, SO_3 is written as has been shown above. In showing the arrangement of atoms, one should keep in mind two things. Hydrogen rarely bonds to more than one atom. Halogens are usually bonded to only one atom. Only in polyatomic ions or molecules, such as bromate ion, BrO_3^-, or chloric acid, $HClO_3$, are halogens the central atom and thus bonded to more than one other atom.

Oxyacids like H_2SO_4, H_3PO_4, $HClO_3$ and acetic acid are a little different. The acidic hydrogen atoms, that are those hydrogen atoms which are written at the beginning of the formula, are bonded to oxygen and not to the central atom. Therefore, these compounds have the atomic arrangements as under:

As illustrative examples we may show the arrangement of atoms in the following compounds.

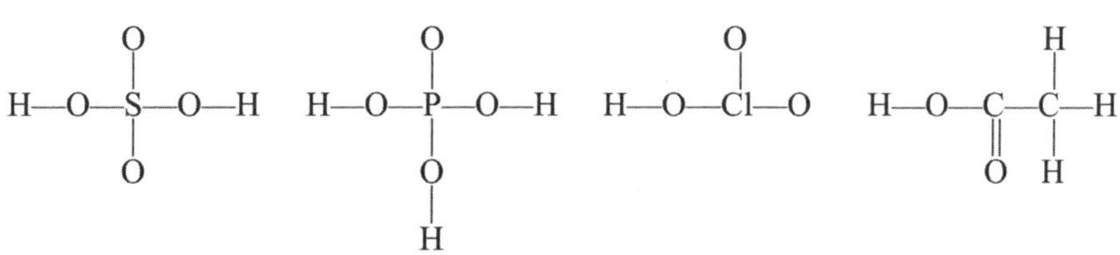

(a). SO_2 (b). C_2H_5Cl and (c). C_2H_2

***SO_2*:** From its formula we can predict that the two oxygen atoms are bonded to the sulfur; or we can choose the most symmetrical arrangement with the least electronegative atom, the sulfur atom, at the center. Both arrangements are the same: O- S-O. however, the molecule is a bent one for obvious reasons to follow soon.

***C_2H_5Cl*:** The two carbon atoms will be bonded together, forming a chain. Three hydrogen atoms are bonded to one carbon atom; two hydrogen atoms and the chlorine atom are bonded to the other carbon. The arrangement is:

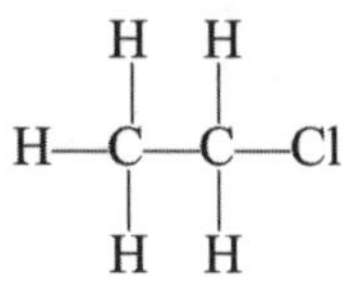

***C_2H_2*:** Again the two carbon atoms will be bonded together. The most symmetrical arrangement for the two hydrogen atoms is one on each carbon, giving the arrangement: H-C-C-H in which there is a triple bond between carbon atoms.

7. HOW ARE ELECTRONS NUMBERED AND PLACED IN LEWIS STRUCTURES?

The Lewis structure of a molecule or ion shows the arrangement of atoms and the distribution of electrons in that molecule or ion. While it has already been shown how to predict the arrangement of atoms, in this section we may show how to predict the number and distribution of electrons. Some of these electrons will be shared; some will be unshared. The number in each category can be determined using the following steps. We may illustrate each step of this process using a molecule of formaldehyde, CH_2O. We should note that in these examples all the atoms follow the octet rule.

1. Each atom in the molecule except hydrogen is assumed to require eight electrons; hydrogen will require only two. Formaldehyde, CH_2O, will require electrons to fill 20 spaces.

   $2(2) + (8 + 8) = 20$ spaces to fill Hydrogen carbon oxygen

2. Each atom contributes its valence electrons toward filling these spaces. Formaldehyde has 12 available valence electrons:

   $2(1) + 4 + 6 = 12$ valence electrons availabl Hydrogen carbon oxygen

3. The difference between these numbers is the number of bonding electrons shared by two atoms. Formaldehyde will contain: $20 - 12 = 8$ shared electrons

4. The number of unshared electrons is the difference between the number of valence electrons available and the number of shared or bonding electrons. Formaldehyde will have $12 - 8 = 4$ unshared electrons. Now we draw the Lewis

structure. Having determined the arrangement of atoms, we place a pair of electrons between each set of neighboring atoms. For formaldehyde, we write

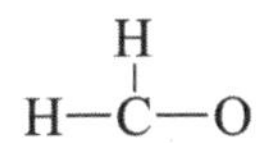

which uses six electrons and leaves two more electrons to be shared? These electrons will group with another pair to form a double bond. Carbon-hydrogen bonds are always single; therefore, the double bond will be between carbon and oxygen, giving the structure

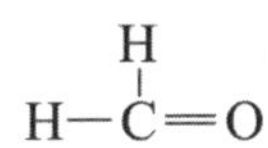

5. The unshared electrons are then added where needed to give an octet. In the structure for formaldehyde, we have four unshared electrons. Oxygen still does not have an octet, so the unshared electrons are added to the oxygen, giving

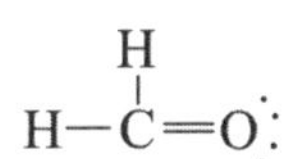

This structure is the complete Lewis structure of formaldehyde.

8. UNDER WHAT RULES WE CAN DRAW LEWIS STRUCTURES OF MOLECULES?

For doing so first we draw skeleton of the molecule concerned keeping in mind that the least electronegative atom will be at center and hydrogen and halogens will be at perimeter. Then we calculate the following:

a. Number of spaces to fill if each atom obeys octet rule (A)

b. Total number of valence electrons available (V)

c. Number of bonding electrons (B): B = A – V

d. Number of nonbonding (unshared) electrons (N): N = V – B

1. Assign two bonding electrons to each bond. Assign any bonding electrons remaining to double or triple bonds between appropriate atoms (not hydrogen or halogen).
2. Assign nonbonding electrons (N) where needed to complete octets.

As illustrative examples we may draw the Lewis structure of the following:

(a). Ethylene, C_2H_4 (b). Phosphorus trichloride, PCl_3 (c). Methyl alcohol, CH_3OH

Ethylene, C_2H_4: The most symmetrical arrangement for ethylene, C_2H_4, is:

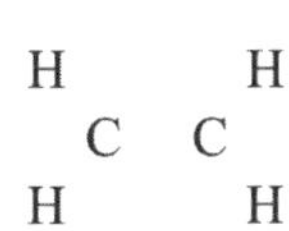

The number of electron spaces is

4(2)	+	2(8)	= 24
Hydrogen		carbon	

The number of valence electrons is

4(1)	+	2(4)	= 12
Hydrogen		carbon	

The number of bonding electrons is $24 - 12 = 12$
The number of unshared electrons is $12 - 12 = 0$

Using a pair of electrons for each bond predicted in the most symmetrical arrangement gives:

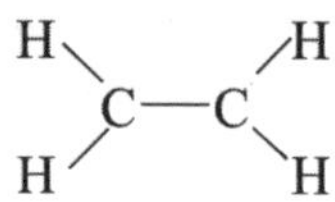

Two bonding electrons remain unused and each carbon atom still needs two more electrons. Two electrons can fill this need by forming a double bond between the carbons, which gives the structure:

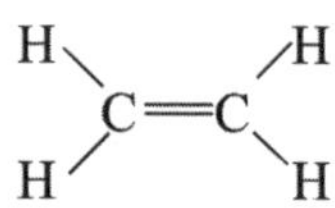

All atoms in the structures as shown satisfy the octet rule; therefore no nonbonding (unshared) electrons are needed.

Phosphorus trichloride, PCl_3: The symmetrical atomic arrangement for phosphorus trichloride, PCl_3 is

Cl

Cl P Cl

The number of electron spaces is

8	+	3(8)	= 32
Phosphorus		chlorine	

The number of valence electrons is

5	+	3(7)	= 26
Phosphorus		chlorine	

The number of bonding electrons is $32 - 26 = 6$
The number of unshared electrons is $26 - 6 = 20$
Adding six bonding electrons to the predicted arrangement gives:

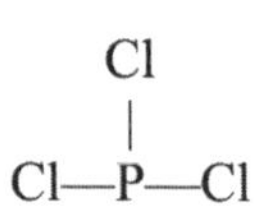

Adding the unshared electrons completes the octet of each atom and gives:

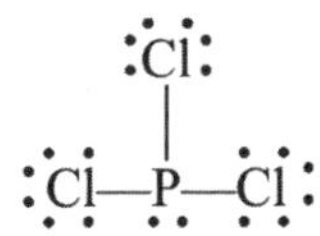

Methyl alcohol, CH_3OH: Reading the formula for methyl alcohol, CH_3OH, we predict the atomic arrangement to be three hydrogen atoms and the oxygen atom bonded to the carbon atom, with another hydrogen bonded to the oxygen:

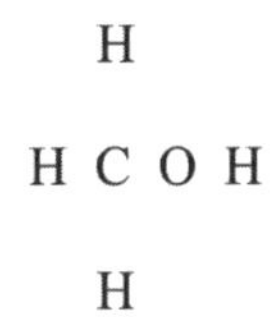

The number of electron spaces available is:

8	+	3(2)	+	8	+	2	= 24
Carbon		hydrogen		oxygen		hydrogen	

The number of valence electrons is $4 + 3(1) + 6 + 1 = 14$
The number of bonding electrons is $24 - 14 = 10$

Using these bonding electrons for single bonds in the predicted arrangement gives:

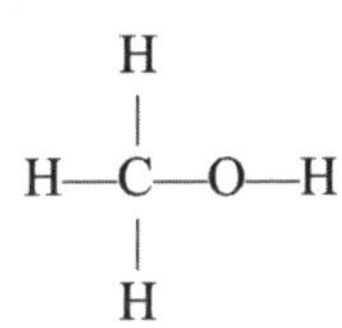

The number of unshared electrons is $14 - 10 = 4$

The only atom still missing an octet of electrons is oxygen, which needs four more electrons. Putting the unshared electrons in the structure around oxygen, we get a complete Lewis structure of methyl alcohol as under:

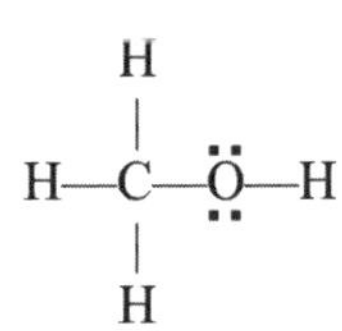

9. HOW CAN WE WRITE LEWIS STRUCTURES OF IONS?

To draw the Lewis structure of an ion, we follow the same steps as for drawing the Lewis structure of a molecule with one exception: In calculating the number of valence electrons available, one additional electron is added for each negative charge on the ion or one electron is subtracted for each positive charge on the ion. The entire structure is enclosed in brackets and the charge is shown as a superscript outside the brackets.

As illustrative examples we may draw the Lewis structure of the following ions:

(a). Hydronium ion, H_3O^+ (b). Ammonium ion, NH_4^+ (c). Chlorate ion, ClO_3^-

Hydronium ion, H_3O^+: The atomic arrangement for the hydronium ion, H_3O^+ is:

H

H O H

The number of spaces is 3(2) + 8 = 14

The number of valence electrons is

3(1) +	6	−1 = 8
Hydrogen	oxygen	charge

The number of shared electrons is 14 – 8 = 6

The number of unshared electrons is 8 – 6 = 2

The Lewis structure is:

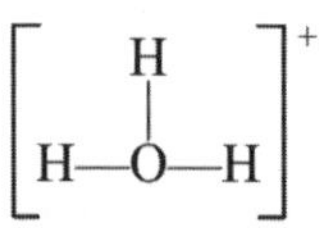

H

H N H

H

Ammonium ion, NH_4^+: The atomic arrangement for the ammonium ion, NH_4^+ is:

The number of spaces is

4(2) +	8	= 16
Hydrogen	nitrogen	

The number of valence electrons is

4(1) +	5	−1	= 8
Hydrogen	nitrogen	charge	

The number of shared electrons is 16 – 8 = 8

The number of unshared electrons is 8 – 8 = 0

The Lewis structure is:

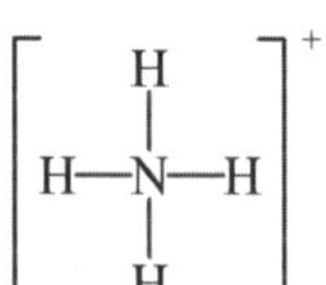

Chlorate ion, ClO_3^-: The most symmetrical arrangement of atoms for the chlorate ion, ClO_3^- is:

O Cl O

O

The number of spaces is

3(8)	+	8	= 32
Oxygen		chlorine	

The number of valence electrons is

3(6)	+	7	+	1	= 26
Oxygen		chlorine	charge		

The number of shared electrons is 32 – 26 = 6
The number of nonbonding electrons is 26 – 6 = 20
The Lewis structure of this ion is:

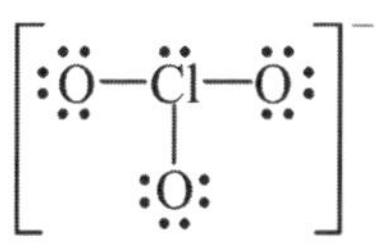

10. WHAT IS MEANT BY RESONANCE AND CONTRIBUTING STRUCTURES?

As chemists began to work with Lewis structures, it became more and more obvious that, for a great many molecules and ions, no single Lewis structure provided a truly accurate representation. For example, a Lewis structure for the carbonate ion, CO_3^{2-}, shows carbon bonded to three oxygen atoms by a combination of one double bond and two single bonds. Three possible Lewis structures for CO_3^{2-} are shown as under:

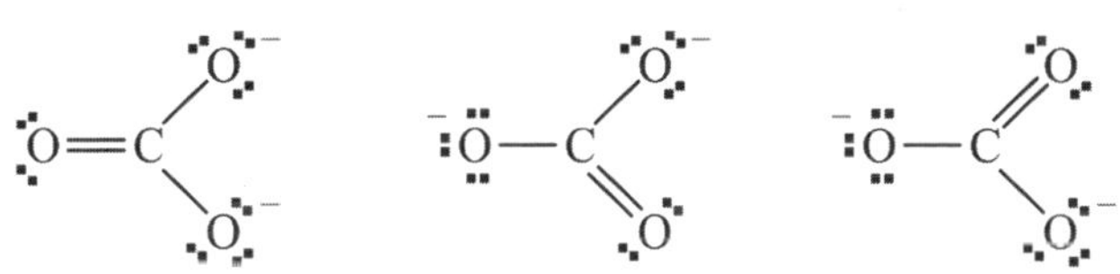

Fig. 7.3(a): Three possible Lewis structures for the carbonate ion, CO_3^{-2}

Each implies that onc carbon oxygen bond is different from the other two. However, this difference in bonding is not the case; rather, it has been shown that all three bonds are identical. To describe molecules and ions, like the carbonate ion, for which no single Lewis structure is adequate, the theory of resonance was developed by Linus Pauling in the 1930s. According to resonance theory, many molecules and ions are best described by drawing two or more Lewis structures and considering the real molecule or ion as a hybrid (composite) of these structures. The individual Lewis structures are called contributing structures. We show that the real structure is a hybrid of the various contributing structures by connecting them with double-headed arrows as has been shown in the following figure.

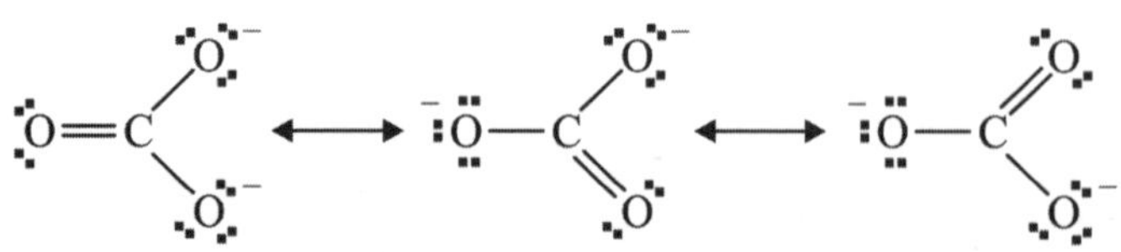

Fig. 7.3(b): The carbonate ion can be represented as its three possible contributing structures connected with double-headed arrows to imply resonance

Student should remember that the carbonate ion, or any other compound we describe in this way, has one and only one real structure. The problem is that our systems of representation are not adequate to describe the real structures of molecules and ions. The resonance method is a particularly useful way to describe the structure of these compounds, for it retains the use of Lewis structures with electron-pair bonds. We fully realize that the carbonate ion is not accurately represented by any single contributing structure.

For an elaborative account let us show that sulfur trioxide can be represented by a resonance hybrid of three contributing structures. The most symmetrical arrangement for sulfur trioxide is:

O S O

O

The number of electron spaces is $8 + 3(8) = 32$

The number of valence electrons is $6 + 3(6) = 24$

The number of shared electrons is $32 - 24 = 8$

The number of unshared electrons is $24 - 8 = 16$

The Lewis structure of sulfur trioxide is either

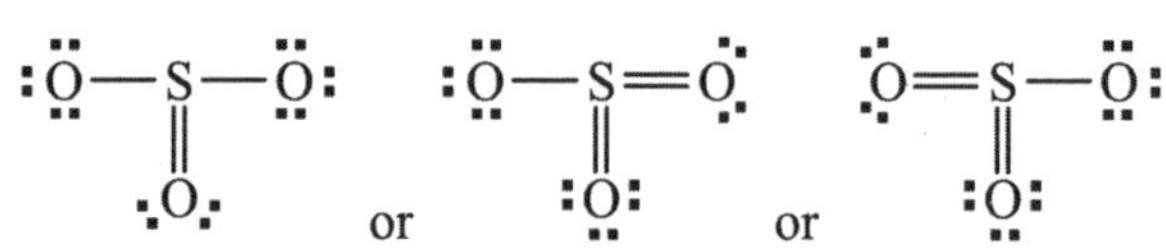

All of these structures are equivalent, therefore the molecule exhibits resonance and might be represented by:

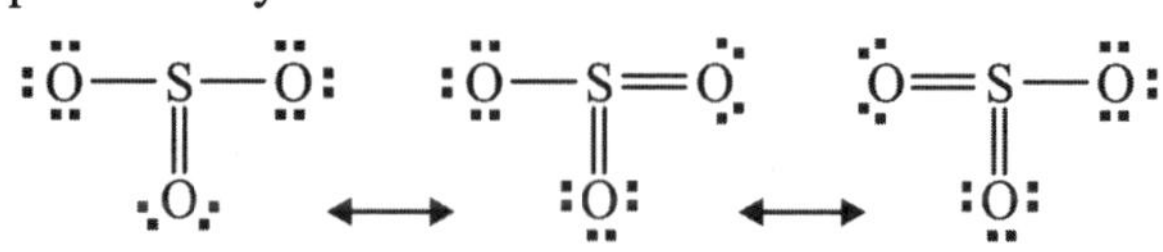

Fig. 7.3(c): The sulfurtrioxide molecule can be represented as its three possible contributing structures connected with double-heated arrows to imply resonance

11. WHAT CORRELATIONS HAVE BEEN NOTED BETWEEN BOND ANGLES AND THE SHAPES OF LINEAR MOLECULES?

We have already considered shared pair of electrons as the fundamental unit of the covalent bond and have drew Lewis structures for several small molecules and ions containing various combinations of single, double and triple bonds. In this section, let us use the valence-shell electron-pair repulsion (VSEPR) model to predict the geometry of these and other covalently bonded molecules and ions. The VSEPR model can be explained in the following way.

We know that an atom has an outer shell of valence electrons. These valence electrons may be involved in the formation of single, double, or triple bonds or they may be unshared. Each set of electrons, whether unshared or in a bond, creates a negatively charged region of space. We have already learned that like charges repel each other. The VSEPR model states that the various regions containing electrons or electron clouds around an atom spread out so that each region is as far from the others as possible.

If a molecule contains only two atoms, those two atoms are in a straight line and thus form a linear molecule. Some three-atom molecules also have straight-line geometry. For example:

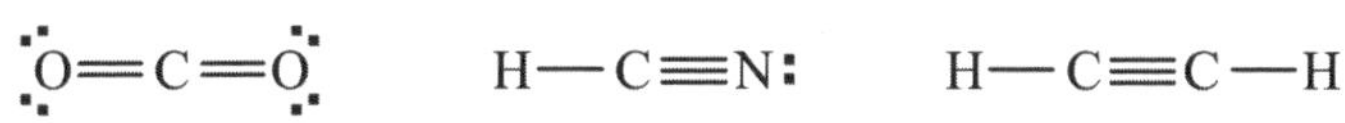

carbon dioxide   hydrogen cyanide   acetylene

Notice that, in the Lewis structure of these molecules, the central atom(s) bonds with only two other atoms and has no unshared electrons. Only two electron clouds emerge from that central atom. For these two clouds to be as far away from each other as possible, they must be on opposite sides of the central atom, forming a bond angle of 180° with each other. An angle of 180° gives a straight line. The VSEPR theory says, then, that the geometry around an atom that has only two bonds and no unshared electrons is a straight line. Figure 7.4 shows the linear nature of these molecules.

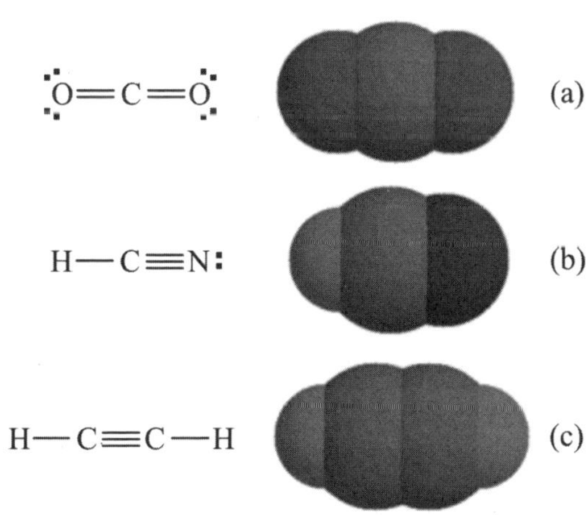

Fig. 7.4: Linear molecules: (a) Carbon dioxide, CO_2; (b) Hydrogen cyanide, HCN; and (c) Acetylene, C_2H_2.

12. WHAT CORRELATIONS HAVE BEEN NOTED BETWEEN BOND ANGLES AND THE SHAPES IN CASES OF STRUCTURES WITH THREE REGIONS OF HIGH ELECTRON DENSITY AROUND THE CENTRAL ATOM?

In case of structures with three regions of high electron density around the central atom, let us look at the following Lewis structures:

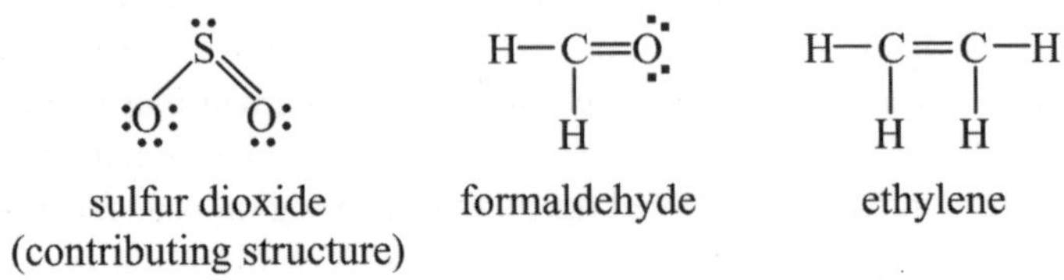

In these molecules, each central atom has three electron clouds emanating from it. In sulfur dioxide, the sulfur atom is bonded to two oxygen atoms and has one unshared pair of electrons. In formaldehyde and ethylene, each carbon atom has two single bonds to hydrogen, a double bond to another atom and no unshared pair. The sulfur atom in sulfur dioxide and the carbon atom in ethylene and formaldehyde are surrounded by three clouds of high electron density. For these clouds to be as far as possible from one another, they will form a plane containing the central atom and will emanate from the central atom at angles of 120° to each other. The structure will be trigonal planar. The central atom will be in the center of the triangle and the ends of the electron clouds at the corners of the triangle. If we experiment with a marshmallow as the central atom

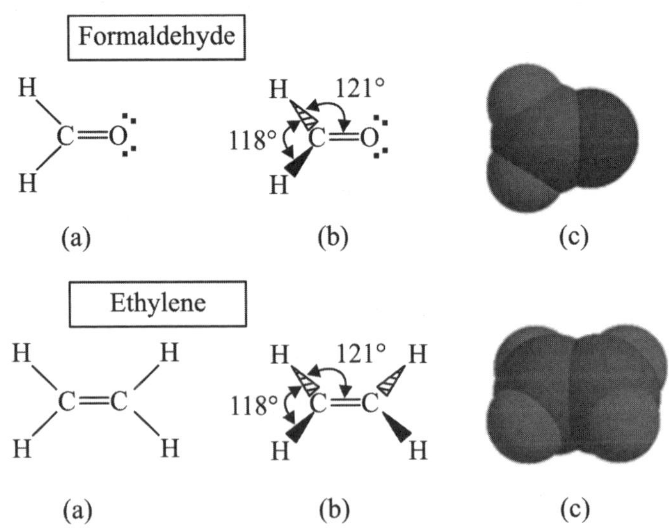

Fig. 7.5: Shapes of the formaldehyde and ethylene molecules: (a) Lewis structure;(b) trigonal planar arrangement of the three areas of electron density around the carbon atom; (c) space-filling models. In these figures, [wedge] represents a bond projecting in front of the plane of the screen and [hashed wedge] represents a bond projecting behind the plane of the screen.

and three toothpicks as electron clouds, we can prove to ourselves that the toothpicks are farthest apart when using trigonal planar structures. illustrated below. Note that the angles are not exactly 120° but are remarkably close to that predicted value.

Although the electron clouds of these molecules give a trigonal planar shape around each carbon atom, one describes the geometry of a molecule only on the basis of the relationships between its atoms. A formaldehyde molecule is trigonal planar because it has an atom at the end of each electron cloud. The ethylene molecule has trigonal planar geometry around each of its carbon atoms. The whole molecule is planar and its shape resembles two triangles joined point to point. In sulfur dioxide, there are three electron clouds around the sulfur. Only two of these connect two atoms. In the molecule, the oxygen-sulfur-oxygen atoms make a 120° angle. The molecule is bent. A central atom surrounded by three clouds of high electron density will have trigonal planar geometry if it is bonded to three atoms. Its geometry will be called bent if it is bonded to two atoms and also has an unshared pair of electrons.

13. WHAT CORRELATIONS HAVE BEEN NOTED BETWEEN BOND ANGLES AND THE SHAPES IN CASES OF STRUCTURES WITH FOUR REGIONS OF HIGH ELECTRON DENSITY AROUND THE CENTRAL ATOM?

The following Lewis structures show three molecules containing central atom surrounded by four clouds of high electron density:

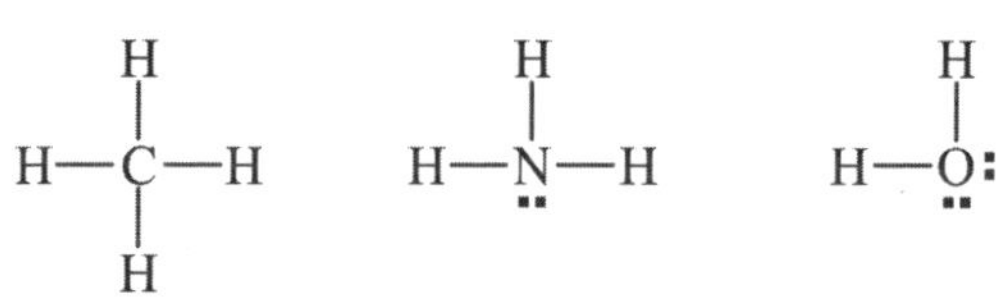

methane ammonia water

These molecules are alike in that in each central atom is surrounded by four pairs of electrons, but they differ in the number of unshared electron pairs on the central atom. We must note that, although we have drawn them in a plane, the molecules are three-dimensional and atoms may be in front of or behind the plane of the paper. Now question arises what geometry does the VSEPR theory predict for these molecules?

Let us predict the shape of methane, CH_4 as the prototype. The Lewis structure of methane shows a central atom surrounded by four separate regions of high electron density. Each region consists of a pair of electrons bonding the carbon atom to a hydrogen atom. According to the VSEPR model, these regions of high electron density spread out from the central carbon atom in such a way that they are as far away from one another as possible.

We can predict the resulting shape using a Styrofoam ball or marshmallow and four toothpicks. Poke the toothpicks into the ball, making sure that the free ends of the toothpicks are as far from one another as possible. If we have positioned them correctly, the angle between any two toothpicks will be 109.5°. If we now cover this model with four triangular pieces of paper, we will have built a four-sided figure called a regular tetrahedron. The following figure shows: (a) the Lewis structure for methane, (b) the tetrahedral arrangement of the four regions of high electron density around the central carbon atom and (c) a space-filling model of methane.

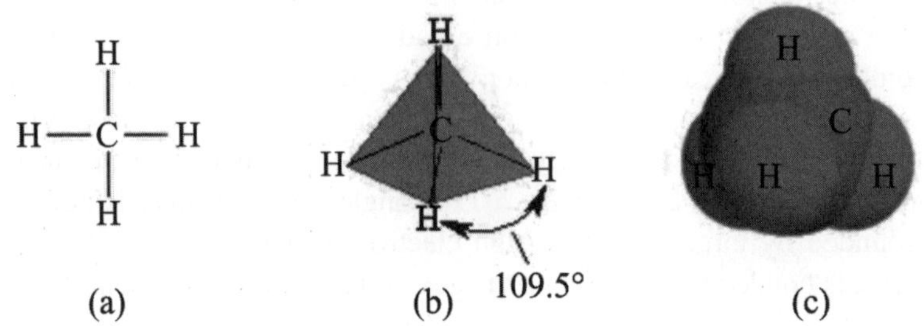

Fig. 7.6: The shape of a methane molecule, CH_4: (a) Its Lewis structure; (b) Its tetrahedral shape; (c) A space-filling model.

According to the VSEPR model, the H - C - H bond angle in methane should be 109.5°. This angle has been measured experimentally and found to be 109.5°. Thus, the bond angle predicted by the VSEPR model is identical to that observed. We say that methane is a tetrahedral molecule in which the carbon atom is at the center of a tetrahedron and each hydrogen atom is at one of the corners of the tetrahedron.

14. WHAT STRUCTURES WILL BE PREDICTED FOR AMMONIA AND WATER MOLECULES ON THE BASIS OF VSEPR MODEL?

We can predict the shape of the ammonia and water molecules in exactly the same manner as has been adopted for methane molecule. The Lewis structure of NH_3 shows a central nitrogen atom surrounded by four separate regions of high electron density. Three of these regions consist of a single pair of electrons forming a covalent bond with a hydrogen atom; the fourth region contains an unshared pair of electrons. According to the VSEPR model, the four regions of high electron density around the nitrogen are arranged in a tetrahedral manner, so we predict that each H - N - H bond angle should be 109.5°. The observed bond angle is 107.3°. This small difference between the predicted angle and the observed angle can be explained by proposing that the unshared pair of electrons on nitrogen repels the adjacent bonding pairs more strongly than the bonding pairs repel each other.

Ammonia is not a tetrahedral molecule. The atoms of ammonia form a pyramidal molecule with nitrogen at the peak and the hydrogen atoms at the corners of a triangular base. Just as the unshared pair of electrons in sulfur dioxide contributes to the geometry of the molecule but is not included in the description of its geometry, the unshared pair of electrons in ammonia gives it a tetrahedral shape but its geometry is based only on

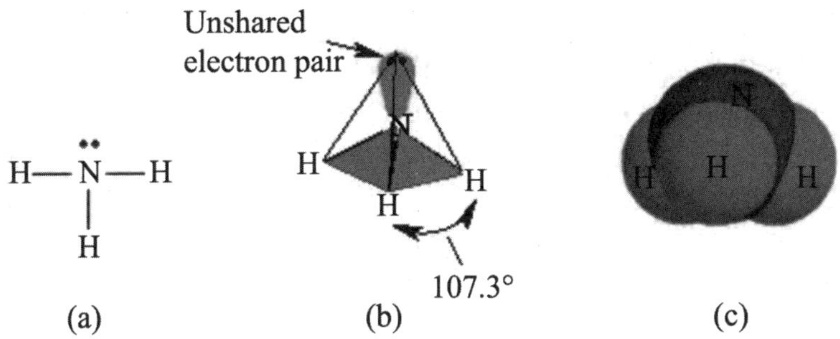

Fig. 7.7: The shape of an ammonia molecule, NH_3: (a) its Lewis structure; (b) its geometry; (c) a space filling model. We notice as to how the unshared electrons serve to create its shape.

the arrangement of atoms, which is pyramidal. The following figure shows the Lewis structure of the water molecule. In H_2O, a central oxygen atom is surrounded by four separate regions of high electron density. Two of these regions contain a pair of electrons forming a covalent bond between oxygen and hydrogen; the other two regions contain an unshared electron pair.

The four regions of high electron density in water are arranged in a tetrahedral manner around oxygen. Based on the VSEPR model, we predict an H - O - H bond angle of 109.5°. Experimental measurements show that the actual bond angle is 104.5°. The difference between the predicted and observed bond angles can be explained by proposing, as we did for NH_3, that unshared pairs of electrons repel adjacent bonding pairs more strongly than the bonding pairs repel each other. Note that the variation from 109.5° is greatest in H_2O, which has two unshared pairs of electrons; it is smaller in NH_3, which has one unshared pair; and there is no variation in CH_4. To describe the geometry of the water molecule, remember that the geometry of a molecule describes only the geometric relationships between its atoms. The three atoms of a water molecule are in a bent line like those of sulfur dioxide. We say the water molecule is bent.

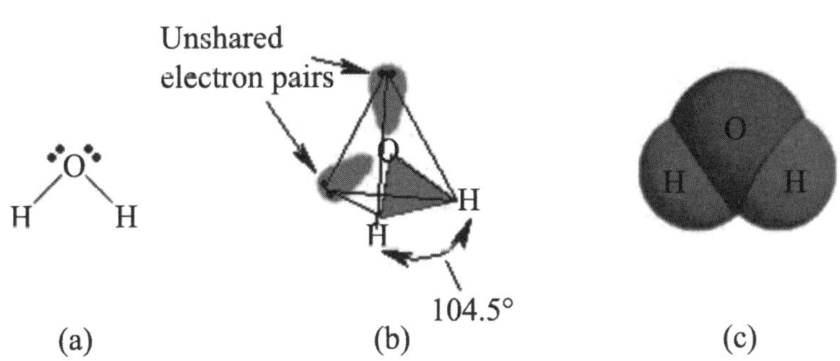

Fig. 7.8: The shape of a water molecule, H_2O: (a) its Lewis structure; (b) its geometry; (c) a space-filling model. Notice how the unshared pairs of electrons affect the tetrahedral geometry

15. WHAT GENERALIZATIONS EMERGE FROM OUR DISCUSSION OF THE SHAPES OF METHANE, AMMONIA AND WATER MOLECULES?

A general prediction emerges from our discussions of the shapes of methane, ammonia and water molecules. Whenever four separate regions of high electron density surround a central atom, we can accurately predict a tetrahedral distribution of electron clouds and bond angles of approximately 109.5°. The geometries of molecules can be predicted. The shape of a molecule whose central atom is bonded to four other atoms is tetrahedral. One in which the central atom has one unshared pair of electrons and bonds to three other atoms will be pyramidal and one in which the central atom has two unshared pairs of electrons and bonds to two other atoms will be bent. Molecular shapes and bond angles may be summarized in a tabular form as under:

Table 7.1: Molecular shapes and bond angles

Number of regions of high electron density around central atom	Arrangement of regions of high electron density in space	Predicted bond angles	Example	Geometry of molecule
4	Tetrahedral	109.5°	CH_4, methane	Tetrahedral
			NH_3, ammonia	Pyramidal
			H_2O, water	Bent
3	Trigonal planar	120°	H_2CO, formaldehyde	Trigonal planar
			C_2H_4, ethylene	Planar
			SO_2, sulfur dioxide	Bent
2	Linear	180°	CO_2, carbon dioxide	Linear
			C_2H_2, acetylene	Linear

As further illustrative examples we may try to predict all bond angles in the following molecules.

a. CH_3Cl b. CH_3CN c. CH_3COOH

a. CH_3Cl: The Lewis structure of methyl chloride is:

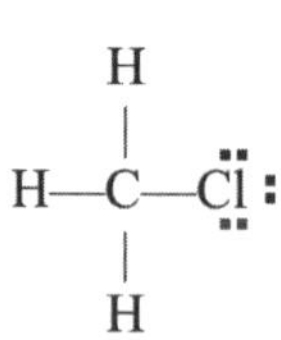

In the Lewis structure of CH_3Cl carbon is surrounded by four regions of high electron density, each of which forms a single bond. Based on the VSEPR model, we predict a tetrahedral distribution of electron clouds around carbon, H - C - H and H - C - Cl bond angles of 109.5° and a tetrahedral shape for the molecule. Note the use of doted lines to represent a bond projecting behind the plane of the paper and a solid wedge to represent a bond projecting forward from the plane of the paper.

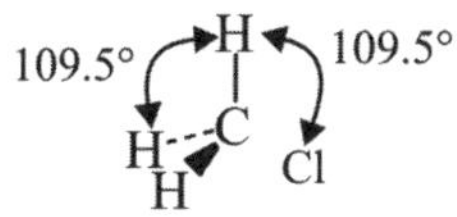

b. CH_3CN: The Lewis structure of acetonitrile, CH_3CN is:

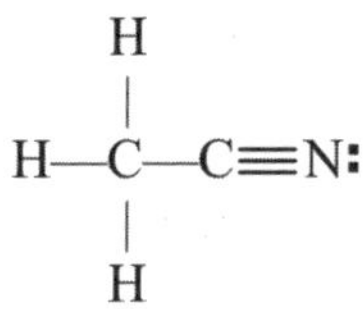

The methyl group, CH_3^-, is tetrahedral. The carbon of the -CN group is in the middle of a straight line stretching from the carbon of the methyl group through the nitrogen.

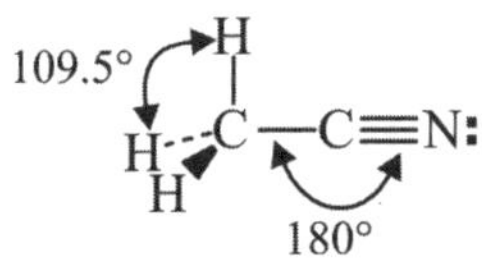

c. CH_3COOH: The Lewis structure of acetic acid is:

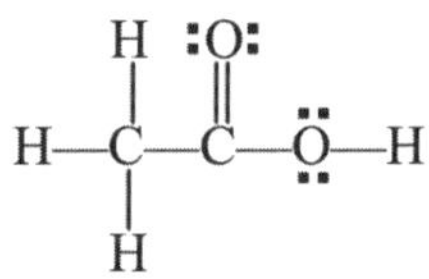

Both the carbon bonded to three hydrogen atoms and the oxygen bonded to carbon and hydrogen atoms are centers of tetrahedral structures. The central carbon will have 120° deg bond angles.

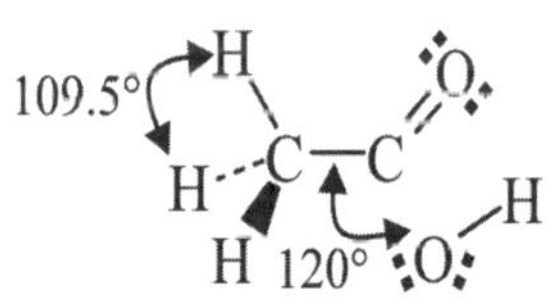

The geometry around the first carbon is tetrahedral, around the second carbon atom is trigonal planar and around the oxygen is bent.

16. HOW DOES POLARITY DEVELOP IN MOLECULES?

Some molecules such as methane, CH_4 contain only nonpolar bonds. Such molecules are nonpolar molecules. Other molecules contain polar bonds, that is, bonds between atoms whose electronegativities differ by more than 0.4 units. Whether these latter molecules are polar or nonpolar depends on the arrangement in space of these bonds and the resulting geometry of the molecules. If we picture the geometry of a molecule and show its polar bonds with an arrow (⟶ with a crossed tail) aimed at the more electronegative atom, we can usually obtain a picture of the molecule that indicates whether or not it is polar. The following figure illustrates this method for several molecules.

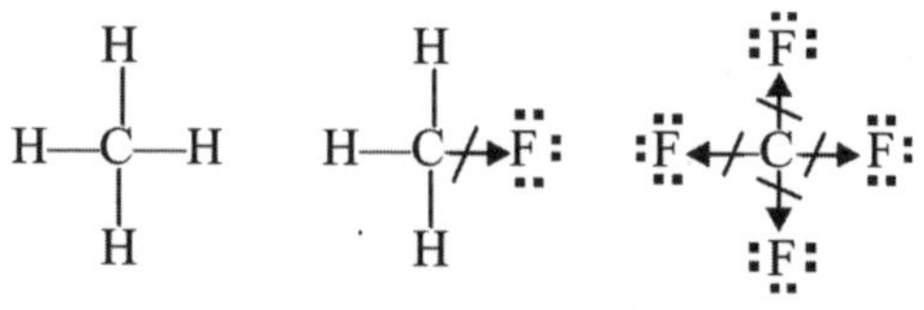

(a) (b) (c)

Fig. 7.9: Polarity in molecules: (a) methane has no polar bonds and is a nonpolar molecule; (b) methyl fluoride has one polar bond (denoted by) and is therefore a polar molecule; (c) carbon tetra fluoride has four counter-balanced polar bonds and thus is a nonpolar molecule.

One may ask which of the molecules in the above figure is polar? Methane, which contains no polar bonds, is clearly nonpolar. Methyl fluoride contains one polar bond between carbon (EN = 2.5) and fluorine (EN = 4.0). Methyl fluoride is a polar molecule; the negative end of the dipole is at the fluoride atom. Now look at carbon tetra fluoride. It contains four polar carbon-fluorine bonds but they counteract one another, so the molecule itself is nonpolar. The principle is like that of erecting an antenna tower (shown below). If the guy wires are correctly balanced against one another, the tower stays erect. (Compare the balanced pull of the carbon-fluorine bonds in carbon tetra fluoride.) If the guy wires are not balanced, the tower topples. (Compare the unbalanced pull of the carbon-fluorine bond in methyl fluoride).

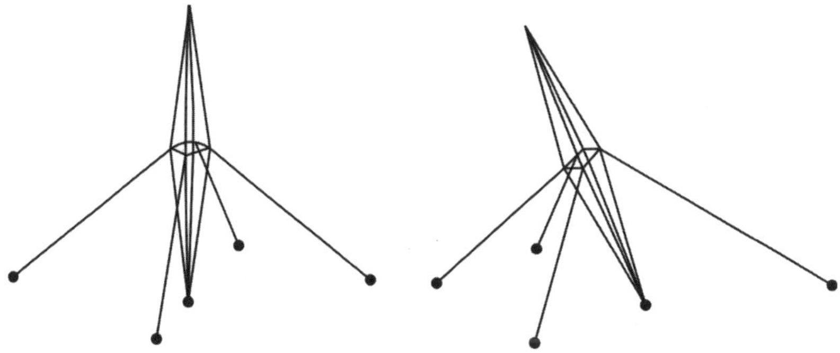

(a) (b)

Fig. 7.10: Erecting an antenna tower: (a) antenna tower with balanced guy wires (compare with carbon tetra fluoride); (b) antenna tower with unbalanced guy wires (compare with methyl fluoride)

The following figure shows three other molecules and their polarities. Notice that both ammonia and water have unshared electrons on the atom at the negative end of the dipole. These unshared electrons enhance the polarity of the bond to make these molecules very polar.

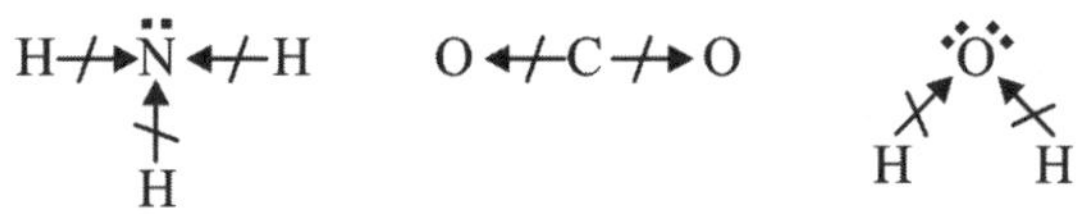

(a) (b) (c)

Fig. 7.11: Predicting the polarity of molecules: (a) ammonia, NH_3, has unbalanced polar bonds and is a polar molecule; (b) carbon dioxide, CO_2, has balanced polar bonds and is a nonpolar molecules; (c) water, H_2O, has unbalanced polar bonds and is a polar molecule.

17. HOW CAN WE PREDICT THE POLARITY IN THE FOLLOWING MOLECULES?

(a). CH_2O (b). $SiBr_4$ (c). SO_2

Solution

a. CH_2O: The Lewis structure of CH_2O is:

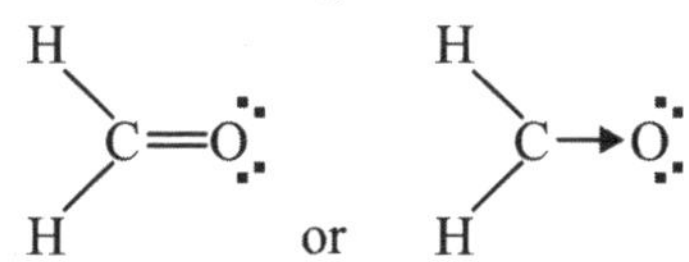

The electronegativity difference between carbon and oxygen is 1.0. Thus, the C = O bond will be polar, with a partial positive charge on the carbon and a partial negative charge on the oxygen. The C - H bond is not polar. By showing the polar bond as an arrow, the diagram shows clearly that the molecule is polar (the bond is not balanced). The unshared electrons on the oxygen should enhance this polarity.

b. $SiBr_4$: The Lewis structure of silicon tetra bromide is:

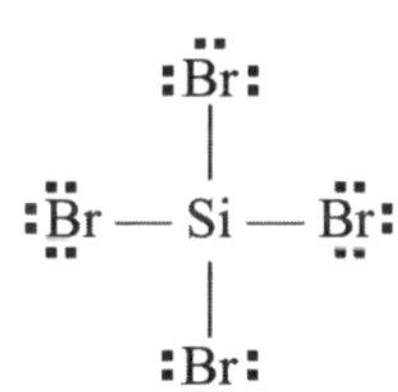

The electronegativity of silicon is 1.8 and that of bromine is 2.8. The silicon - bromine bond is polar. Showing these bonds as arrows in a tetrahedral structure clarifies that silicon tetra bromide is a nonpolar molecule. The polar bonds balance each other as in carbon tetra fluoride and the molecule is nonpolar.

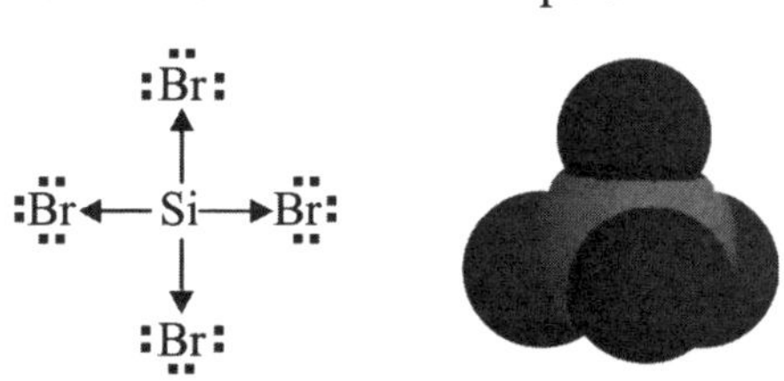

c. SO_2: The Lewis structure of sulfur dioxide is:

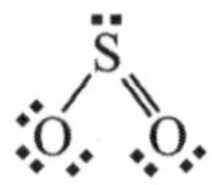

The electronegativity of sulfur is 2.5 and that of oxygen is 3.5; thus the sulfur-oxygen bonds are polar. By drawing these polar bonds as arrows in the bent molecule of sulfur dioxide, we show its polar nature:

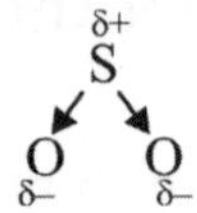

The molecule is a polar one. This molecule is a resonance hybrid, but this fact does not affect its polarity.

18. HOW DO POLAR MOLECULES INTERACT?

Polar molecules interact with one another. This interaction is governed by the fact that like charges repels and opposite charges attract. The following figure (7.12) shows a collection of polar molecules aligned so that the partially positive end of one molecule is near the partially negative end of another.

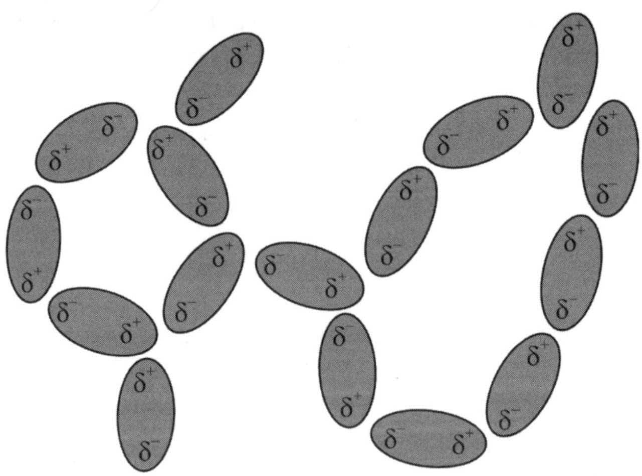

Fig. 7.12: Collection of polar molecules alligned at attractive ends.

Polar molecules aligned by charge need not all be of the same compound. We know that water is a polar molecule. When another polar molecule such as ammonia is added to water, it dissolves because of the interaction between the two types of

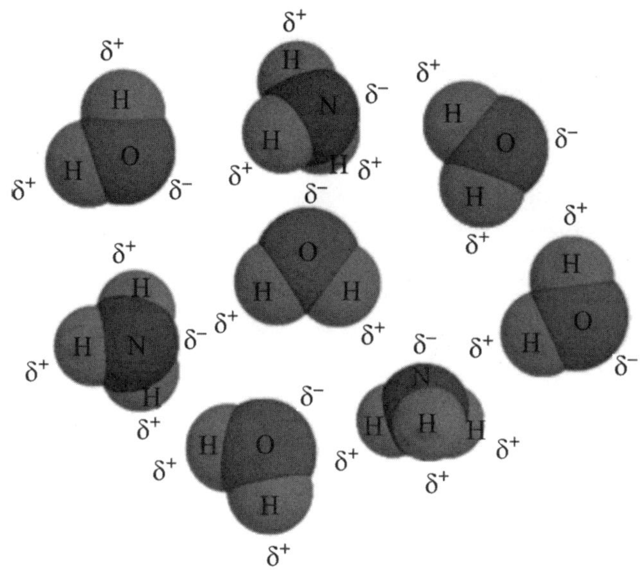

Fig. 7.13: Illustration of interaction between two types of polar molecule

molecules. The following figure shows the polarity of water molecules and ammonia molecules and the interactions between them.

19. WHAT SPECIES EXIST IN AQUEOUS SOLUTIONS OF IONIC COMPOUNDS?

When an ionic compound dissolves in water, the resulting solution contains ions rather than neutral particles. For example, when sodium chloride, NaCl, dissolves in water, the solution contains sodium ions (Na^+) and chloride ions (Cl^-), rather than neutral units of NaCl. A polyatomic ion does not break up into separate atoms in solution. Thus, a solution of sodium nitrate, $NaNO_3$, contains sodium ions and nitrate ions; no nitrogen ions or oxygen ions are present. A solution of ammonium sulfate, $(NH_4)_2SO_4$, contains ammonium ions and sulfate ions.

This dissolution of ionic compounds takes place because of interactions like that between ammonia and water. The following figure shows at the molecular level what happens when sodium chloride, an ionic compound, dissolves in water. Each ion of the solid crystal becomes surrounded by water molecules, with the negative end of the water molecules approaching closest to the positive sodium ions and the positive end of the water molecules surrounding the negative chloride ions. The water molecules pull these ions, one by one, away from the rest of the crystal. Other ionic compounds that are virtually insoluble in water have such strong interactions between their ions that the pull of the polar water molecules is not strong enough to break the ions apart.

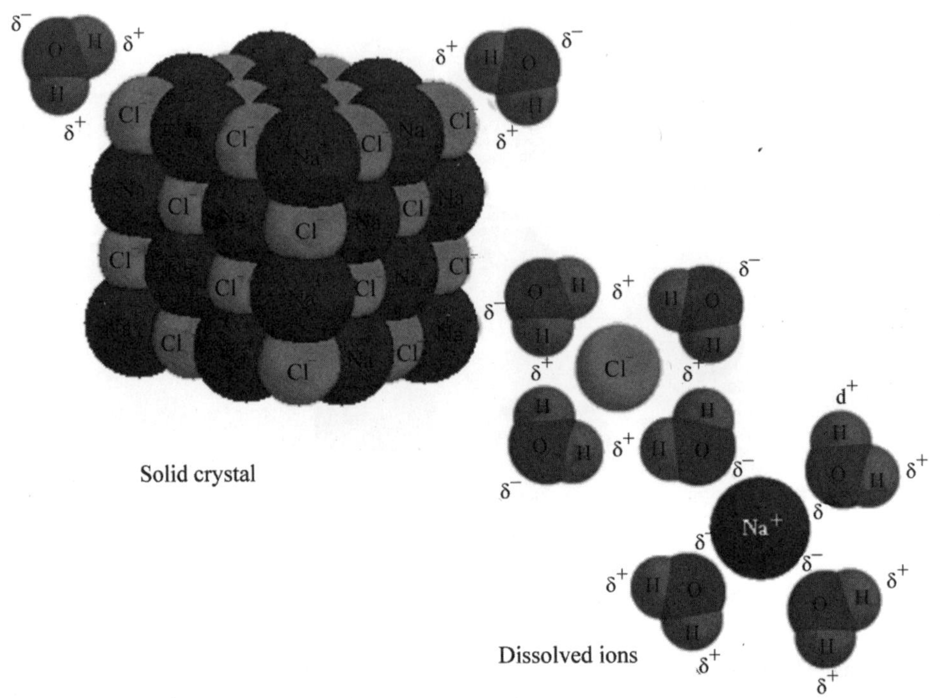

Fig. 7.14: The dissolution of sodium chloride in water. Notice how the polar water molecules are oriented in one way around the positive sodium ions and in another way around the oppositely charged chloride ions

Frequently, when a very polar molecule such as hydrogen chloride dissolves in water, there is sufficient interaction between the two molecules to cause one of them to break up into ions. With hydrogen chloride, the equation for this reaction is

$$HC1 + H_2O \longrightarrow H_3O^+ + C1^-$$

$$NH_3 + H_2O \longrightarrow NH_4^+ + OH^-$$

To some extent the same process occurs when ammonia is dissolved in water. The equation for that reaction is

20. WHAT THE TERMS "ELECTROLYTES, NONELECTROLYTES AND WEAK ELECTROLYTES" STAND FOR?

The presence of ions in a solution of an ionic compound can be demonstrated with an apparatus like the one shown in the following figure. The apparatus consists of two electrodes--one connected to one pole of a power source and the other connected to a light bulb that is, in turn, connected to the other pole of the power source. If the two electrodes touch, an electric current flows through the completed circuit and the bulb lights. If the two electrodes are separated and then immersed in water, no current flows. Pure water cannot carry an electric current. However, if the pure water is replaced by an aqueous (water) solution of an ionic compound, the bulb lights, indicating a flow of electricity through the solution. The electric current is carried through the solution by the dissolved ions.

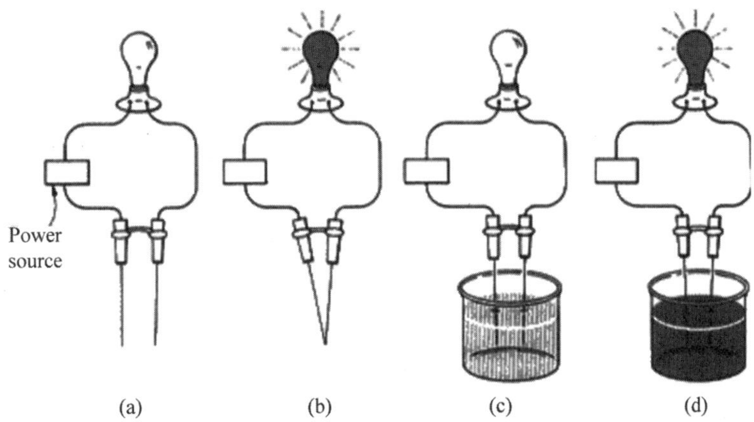

Fig. 7.15: Conductivity apparatus for showing the presence of ions in an aqueous solution: (a) electrodes apart, circuit broken; (b) electrodes touching, circuit complete; (c) electrodes in pure water, circuit broken; (d) electrodes in an ionic solution, circuit complete.

Compounds whose aqueous solutions conduct electricity are called electrolytes. This group of compounds includes hydroxides, salts (metal-nonmetal and metal-polyatomic ion compounds) and some acids (nitric, sulfuric, hydrochloric). Compounds that dissolve in water without forming a solution that carries an electric current are called nonelectrolytes. These compounds dissolve as molecules. Many compounds containing only nonmetals are nonelectrolytes. If a compound containing only nonmetals ionizes, its molecule will usually contain hydrogen bonded to a very electronegative atom like oxygen (as in acetic acid) or chlorine (as in hydrochloric acid). Compounds that react somewhat but not completely with water to form ions are said to be partially ionized; their solutions are poor conductors of electricity. They are called weak electrolytes. Acetic acid is a weak electrolyte. In the above apparatus, a solution of acetic acid causes the bulb to glow only faintly, showing that not many ions are present in solution.

There are some other differences between ionic and covalent compounds too. Ionic compounds are usually solids at room temperature. They have high melting points. They dissolve only in very polar liquids like water. Some are only sparingly soluble in water. Covalent compounds are found in all three states: solids, liquids and gases, at room temperature. Those of low molecular weight and those that are polar may dissolve in water, but a covalent solid is much more apt to be soluble in a nonpolar (carbon tetrachloride) or slightly polar (ethyl alcohol) liquid.

Chapter 8

CHARACTERISTICS OF THE GASEOUS STATE

1. WHAT IS RELATIONSHIP BETWEEN PHYSICAL STATE AND NORMAL MELTING AND BOILING POINTS OF SUBSTANCES?

It is well known that the physical properties of a particular substance determine its actual state at room temperature. If both its normal melting point and normal boiling point are below room temperature (approximately 20°C), the substance is a gas under normal conditions. For example, the normal melting point and boiling points of oxygen are –218°C and –189°C respectively and oxygen is a gas at room temperature. If the normal melting point of a substance is below room temperature, the substance is a liquid at room temperature. For example, benzene melts at 6°C and boils at 80°C and it is a liquid at room temperature. If both the normal melting point and the normal boiling point are above room temperature, the substance is a solid. For example, sodium chloride melts at 801°C and boils at 1413°C and is a solid under normal conditions. We may note that while the solids melt and boil above room temperature and the liquids melt below room temperature and boil above room temperature, the gases melt and boil below room temperature.

2. WHAT IS RELATIONSHIP BETWEEN SHAPE AND VOLUME OF A SUBSTANCE?

A solid has a fixed shape and volume that do not change with the shape of its container. We may consider a rock and the fact that its size and shape stay the same, regardless of where we put it. A liquid has a constant volume, but its shape conforms to the shape of its container. If we consider a sample of milk, its volume stays the same, whether we put it in a saucer a or in a glass; clearly its shape changes to match the shape of the container. A gas changes both its shape and volume to conform to the shape and volume of its container. Let us consider a sample of air. It will fill an empty room, a balloon, a tire, or a rubber raft. Its shape and volume conform to the shape and volume of the container in which it is placed.

3. WHY DO WE USUALLY SPECIFY THE TEMPERATURE AND PRESSURE AT WHICH THE DENSITY OF A SUBSTANCE IS MEASURED?

Density is ratio of the mass of a substance to its volume, expressed, for example, in units of grams per cubic centimeter or pounds per cubic foot. The density of a pure substance varies little from sample to sample and is often considered a characteristic property of the substance. Most substances undergo expansion when heated and therefore have lower densities at higher temperatures. Many substances, especially gases, can be compressed into a smaller volume by increasing the pressure acting on them. For these reasons, the temperature and pressure at which the density of a substance is measured are usually specified. The density of a gas is often converted mathematically to what it would be at a standard temperature and pressure (STP). Water is unusual in that it expands and thus decreases in density, as it is cooled below 3.98°C (its temperature of maximum density). For every substance at a certain temp and pressure there is only one unique density value. This value is the same no matter how much or how little of the substance is used. Density can be used to determine what certain substances are. For example, if there is a substance which has a density of 1.00 g/mL, a chart of densities can be used to show that the unknown substance is pure water. The densities of liquids and solids are measured in grams per milliliter and grams per cubic centimeter, respectively and change very little as the temperature of the sample changes. Gases have much lower densities, so much lower that gas densities are measured in grams per liter instead of grams per milliliter. The density of a gas varies considerably as the temperature of the gas changes.

4. WHAT INFERENCES CAN BE DRAWN FROM INTERMOLECULAR STRUCTURES OF SUBSTANCES?

The volume of a solid or a liquid does not change very much with pressure. We can neither change the volume of a brick by squeezing it, nor can we squeeze one liter of liquid into a 0.5-L bottle. The volume of a gas does change a great deal with pressure; we can squeeze a 1.0-L balloon into a 0.5-L space. The constant shape and volume of a solid suggest that its particles (atoms, ions, or molecules) are held together by fairly rigid bonds. The variable shape and constant volume of a liquid suggest that there is some bonding between its particles but that these bonds are not rigid and probably are less strong than those in a solid. The fact that a gas has neither constant shape nor constant volume suggests that there are no bonds and only very slight interactive forces between the particles of a gas. The variety in compressibility suggests other hypotheses. If solids and liquids cannot be compressed, the particles of which they are composed must be very close together. The high compressibility of a gas implies that the particles of a gas are very far apart with a great deal of space between them. This last hypothesis is supported by the difference between the densities of solids and liquids and the densities of gases. One mL of a solid or liquid always has much more mass than does one mL of a gas.

5. HOW IS KINETIC ENERGY DISTRIBUTED?

Any consideration of the properties of a collection of particles such as molecules requires knowledge of their energy. Part of this energy is kinetic energy, the energy of motion. The kinetic energy (KE) of an object is determined by the equation:

$$KE = 1/2mv^2 \text{ where } m = \text{mass; } v = \text{velocity}$$

This equation states that the kinetic energy of an object is dependent on both its mass and its velocity. A semi trailer truck and a subcompact car traveling at the same velocity have different kinetic energies. We would be aware of this difference if they crashed, for the truck would demolish the subcompact. For the two vehicles to have the same kinetic energy, the subcompact would have to be traveling at a much higher velocity than the truck.

In a collection of molecules, each molecule has a kinetic energy that can be calculated by the equation given above. Even if the molecules have a constant mass, they differ in velocity, so that a collection of molecules will have a wide range of kinetic energies, from very low to very high. Every molecule may change its kinetic energy often, but the overall distribution will remain the same. The following figure (Fig. 8.1) shows a typical distribution of kinetic energies in a collection of molecules. In the graph, kinetic energy is plotted along the horizontal axis and the percent of molecules having a particular kinetic energy is shown by the height of the curve at that point. Several observations can be made by studying the graph:

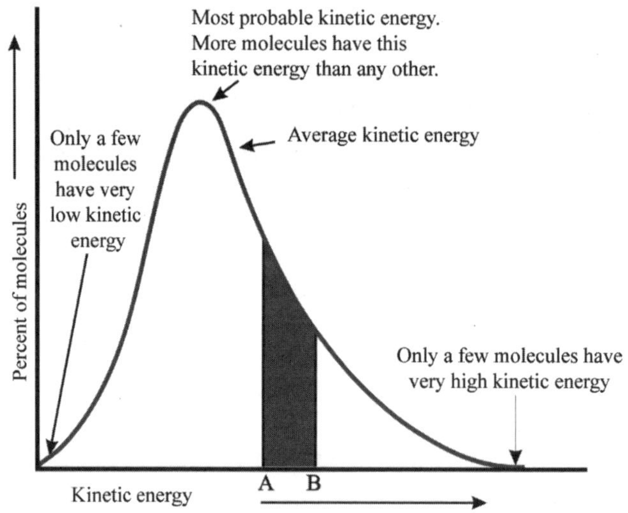

Fig. 8.1: Distribution of kinetic energy in a collection of molecule

The area under the curve represents the total number of molecules in the sample. Between any two points on the horizontal axis, the area under the curve represents the number of molecules that have kinetic energies in that range. For example, the shaded area between A and B represents the number of molecules that have kinetic energies between A and B. The following points need consideration:

1. The peak of the curve shows the most probable kinetic energy. More molecules have this energy than any other.

2. The average kinetic energy is slightly greater than the most probable kinetic energy.
3. Some molecules have a kinetic energy much higher than the average value.
4. Some molecules have a kinetic energy much lower than the average value.
5. Here we notice that the distribution of energies is much like the distribution of grades among students in a test.

6. WHAT IS CORRELATION BETWEEN KINETIC ENERGY AND TEMPERATURE?

The average kinetic energy of a collection of molecules is directly proportional to its temperature. At absolute zero (–273°C), the molecules have a minimum kinetic energy. As the temperature of the sample increases, so does its average kinetic energy. As the temperature rises, the distribution of kinetic energies among the molecules in the sample also changes. The following figure (Fig. 8.2) shows the distribution of kinetic energies in a sample at two different temperatures. Curve A is at the lower temperature while curve B is at the higher temperature. We may note the following differences between the two curves:

1. The peak of curve B is lower and broader than the peak of curve A. This difference in the curves means that, at the higher temperature, fewer molecules have the average kinetic energy and the distribution of energies is more spread out.
2. The peak of curve B is at a higher kinetic energy than the peak of curve A. This difference means that, at the higher temperature, the average kinetic energy of the molecules is higher.

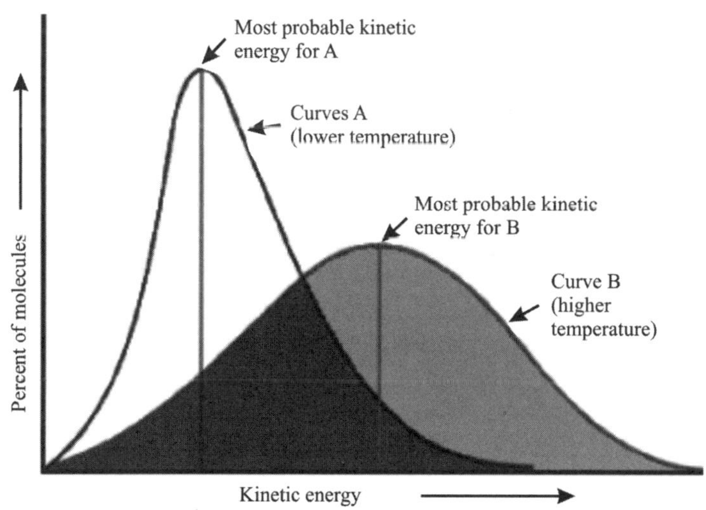

Fig. 8.2: Plot of distribtion of kinetic energies at two different temperatures

We can conclude that, as the temperature of a sample increases, not only does the average kinetic energy increase but also fewer molecules have the average energy and the distribution of energies among the molecules is more uniform.

7. WHAT ARE POSTULATES OF KINETIC MOLECULAR THEORY IN RESPECT OF GASES?

The kinetic molecular theory describes the properties of molecules in terms of motion (kinetic energy) and temperature. Though the theory is most often applied to gases, it is helpful in explaining molecular behavior in all states of matter. As applied to gases, the kinetic molecular theory has the following postulates:

1. Gases are composed of very tiny particles (molecules). The actual volume of these molecules is so small as to be negligible compared with the total volume of the gas sample. A gas sample is, then, mostly empty space. This fact explains the compressibility of gases.
2. There are no appreciable attractive forces between the molecules of a gas. This postulate explains why, over a period of time, the molecules of a gas do not cluster together at the bottom of its container.
3. The molecules of a gas are in constant, rapid, random and straight-line motion. This postulate explains why a gas spreads so rapidly through the available space, for example, why the smell of hot coffee can spread quickly from the kitchen throughout the house.
4. During their motion, the gas molecules constantly collide with one another and with the walls of the container. (The collision with the walls provides the pressure exerted by a gas.) None of these collisions is accompanied by any loss of energy; instead, they are what are known as elastic collisions. We know that a "new" tennis ball collides more elastically than a "dead" tennis ball.
5. The average kinetic energy of the molecules in a gas sample is proportional to its temperature (Kelvin) and is independent of the composition of the gas. In

Table 8.1: Postulates and the experimental evidence of the kinetic molecular theory

Postulate	***Evidence***
1. Gases are tiny molecules mostly in empty space.	The compressibility of gases.
2. There are no attractive forces between molecules.	Gases do not clump.
3. The molecules move in constant, rapid, random and straight-line motion.	Gases mix rapidly.
4. The molecules collide elastically with container walls and with one another.	Gases exert pressure that does not diminish over time.
5. The average kinetic energy of the molecules is proportional to the Kelvin temperature of the sample.	Charles' Law

other words, at the same temperature, all gases have the same average kinetic energy. It also follows from this postulate that at zero Kelvin all molecular motion ceases. These postulates and the experimental evidence for them may be summarized in Tabular form as under:

Clearly, the actual properties of individual gases vary somewhat from these postulates, as their molecules do have a real volume and there is some attraction between the molecules. However, our discussions generally ignore these variations and concentrate on an ideal gas, one that behaves according to this model.

8. HOW CAN WE MEASURE BOTH THE GAS SAMPLES AND THEIR PROPERTIES?

A gas sample obeys a number of laws that relate its volume to its pressure, temperature and mass. Mass and volume are familiar concepts and can be measured with familiar apparatus. Temperature can be measured on any scale: Celsius, Fahrenheit, or Kelvin; however, if the temperature is to be used in a calculation involving gases, the Kelvin scale must be used. Standard temperature for gases, the temperature at which the properties of different gases are compared, is 273 K (0°C).

Pressure is defined as force per unit area and is measured in units that have dimensions of force per unit area. For example, the air pressure in tires is measured in pounds per square inch. The pressure of the atmosphere is frequently measured with a mercury barometer. Pressure can be more easily understood if we consider how a barometer measures pressure. In preparing a barometer, a glass tube at least 760 mm long and closed at one end is filled with mercury and then carefully inverted into a pool of mercury. The level of the mercury in the column will fall slightly and then become steady.

The height of the column of mercury measures the pressure of the atmosphere. To understand this concept, consider the pressure on the surface of the mercury pool at the base of the column. Above this surface rises the "sea" of air (the atmosphere) that surrounds the Earth. On each square centimeter of the surface, we can visualize a 20-km column of air pressing down. On the surface under the mercury column, the mercury is pressing down. The two pressures must be equal. If they were not, mercury would be flowing into or out of the column and the height of the column would not be steady. The atmosphere must be exerting a pressure equal to that exerted by the mercury column. We know that pressure is force per unit area. The total area under the atmosphere or under the column of mercury is not critical, because the force that is measured is the force on each unit of area under the column or the atmosphere, not the total force.

When this experiment is performed in dry air at sea level and at 0°C, the column of mercury is 760 mm high; therefore, we say that the atmosphere is exerting a pressure equal to that of 760 mm of mercury. This amount of pressure has been defined as one atmosphere (1 atm) of pressure and designated as standard pressure. Thus STP is used to mean standard temperature and pressure or standard conditions (0°C, 1 atm). The values of standard pressure measured in units other than atmospheres are shown:

1 atmosphere = 1.01325×10^5 Pascals (the Pascal, Pa, is the SI unit)

= 76 cm, or 760 mm, mercury

= 760 torr (1 torr = the pressure exerted by 1 mm mercury)

= 29.92 in. mercury (used to report atmospheric pressure in weather reports)

= 1.013 bar (used in meteorology) (1 cm mercury = 13.3 millibars)

Each of these relationships can be used as a conversion factor.

9. WHY COMMERCIAL JET-AIRCRAFT CABINS ARE PRESSURIZED?

We know that the average air pressure of dry air at sea level is 1 atm and atmospheric pressure decreases as altitude increases, because the sea of air above it becomes less dense. Our bodies become adjusted to the normal pressure of the altitude at which we live. Minor problems of adjustment can occur when we move from sea level to the mountains and vice versa. Major problems develop at higher altitudes. Commercial jet-aircraft cabins must therefore be pressurized, because humans cannot survive the low pressure of the atmosphere at the altitudes at which jet aircraft fly. For the same reason, travelers in space must wear pressurized suits.

10. HOW DO MANOMETERS MEASURE PRESSURE?

While barometers measure the pressure of the atmosphere, manometers measure the pressure of isolated gas samples. Some manometers measure pressure with a column

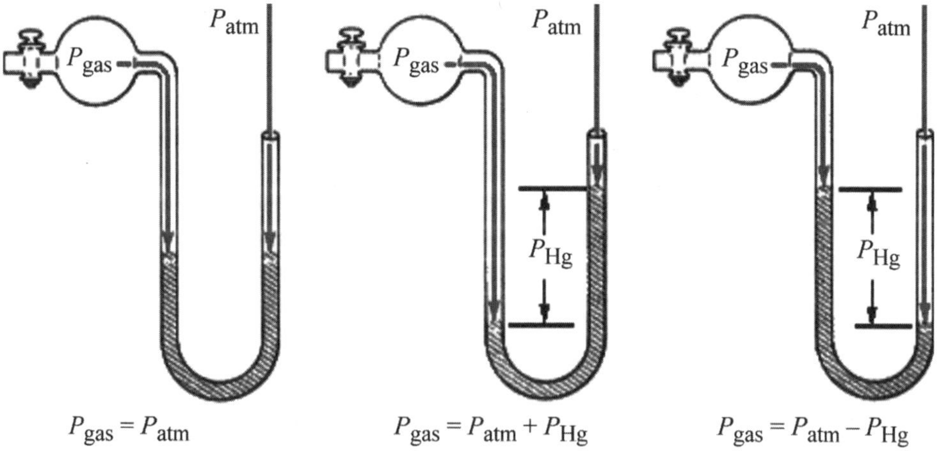

Fig. 8.3: A manometer. The height difference between the mercury levels in the two sides of the tube measures the pressure difference between the gas sample and the atmosphere.

of mercury, like a mercury barometer. This type of manometer has a U-shaped tube partially filled with mercury. One end of the tube is open to a chamber holding a gas sample and other end is open to the atmosphere. If the mercury level on the side of the tube open to the gas sample is lower than that on the side open to the atmosphere, the pressure of the gas is greater than that of the atmosphere by an amount equal to the difference in height between the two mercury columns. If the mercury level on the side of the gas sample is higher than that on the side open to the atmosphere, the pressure of the gas is less than the atmospheric pressure by the difference in the heights of the two columns.

11. WHAT IS BOYLE'S LAW?

Boyle's Law states that if the temperature of a gas sample is kept constant, the volume of the sample will vary inversely as the pressure varies. This statement means that, if the pressure increases, the volume will decrease. If the pressure decreases, the volume will increase. This law can be expressed as an equation that relates the initial volume (V_1) and the initial pressure (P_1) to the final volume (V_2) and the final pressure (P_2). At constant temperature: $V_1 / V_2 = P_2 / P_1$

Rearranging this equation gives: $V_1P_1 = V_2P_2$ or $V_2 = V_1 X P_1 / P_2$

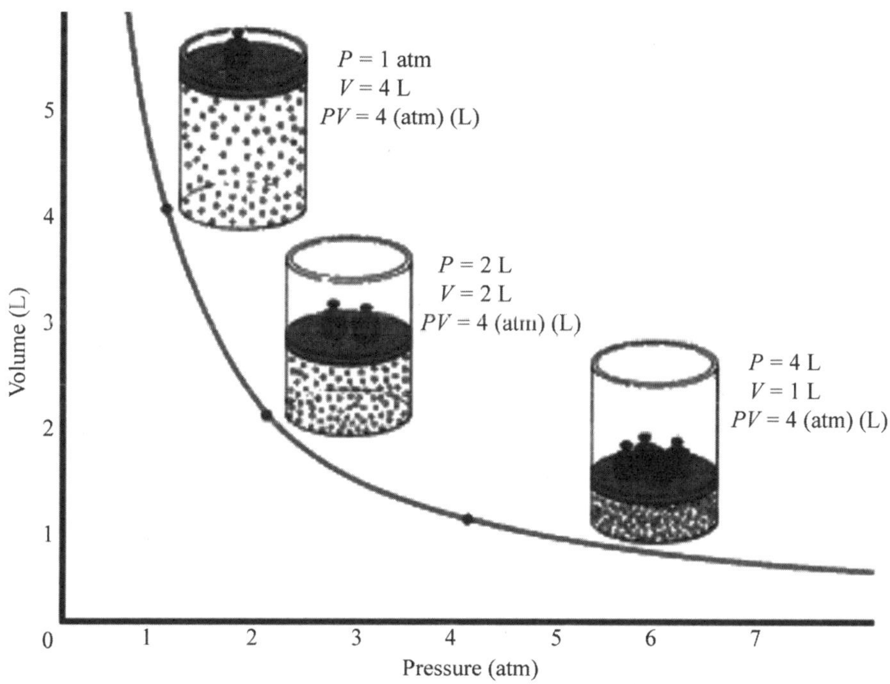

Fig. 8.4: Boyle's law: At constant temperature, the volume of a gas sample is inversely proportional to the pressure. The curve is a graph based on the data listed in the figure.

Boyle's Law may be illustrated diagrammatically as under showing a sample of gas enclosed in a container with a movable piston. The container is kept at a constant temperature and subjected to a regularly increasing amount of pressure. When the piston is stationary, the pressure it exerts on the gas sample is equal to the pressure the gas exerts on it. When the pressure on the piston is doubled, it moves downward until the pressure exerted by the gas equals the pressure exerted by the piston. At this point the volume of the gas is halved. If the pressure on the piston is again doubled, the volume of gas decreases to one-fourth its original volume.

12. HOW DOES BOYLE'S LAW RELATE TO THE KINETIC MOLECULAR THEORY?

At the molecular level, the pressure of a gas depends on the number of collisions its molecules have with the walls of the container. If the pressure on the piston is doubled, the volume of the gas decreases by one-half. The gas molecules, now confined in a smaller volume, collide with the walls of the container twice as often and their pressure once again equals that of the piston. The first postulate of the theory states that a gas sample occupies a relatively enormous empty space containing molecules of negligible volume. Changing the pressure on the sample, changes only the volume of that empty space, not the volume of the molecules.

13. HOW DOES CHARLES' LAW RELATE TO THE POSTULATES OF THE KINETIC MOLECULAR THEORY?

Charles' Law states that if the pressure of a gas sample is kept constant, the volume of the sample will vary directly with the temperature in Kelvin. As the temperature increases, so will the volume; if the temperature decreases, the volume will decrease. This relationship can be expressed by an equation relating the initial volume (V_1) and initial temperature (T_1 measured in K) to the final volume (V_2) and final temperature (T_2 measured in K). At constant pressure, $V_1/V_2 = T_1/T_2$

Rearranging this equation we get: $V_2 = V_1 X T_2/T_1$ or $V_2/T_2 = V_1/T_1$

The kinetic molecular theory states that the molecules in a gas sample are in constant, rapid and random motion. This motion allows the tiny molecules to effectively occupy the relatively large volume filled by the entire gas sample. For understanding the notion of "effectively occupy", we may consider a basketball game, with thirteen persons on the court during a game (ten players and three officials). Standing still, they occupy only a small fraction of the floor. During play they are in constant, rapid motion effectively occupying the entire court. We could not cross the floor without danger of collision. The behavior of the molecules in a gas sample is similar. Although the actual volume of the molecules is only a tiny fraction of the volume of the sample, the constant motion of the molecules allows them to effectively fill that space. As the temperature increases, so does the kinetic energy of the molecules. As they are all of the same mass, an increased kinetic energy must mean an increased velocity. This increased velocity allows the molecules to occupy or fill an increased volume, as do the basketball players in fast action. Similarly, with decreased temperature, the molecules move less rapidly and fill a smaller space.

14. WHAT IS THE FREQUENTLY USED FORM OF THE COMBINED GAS LAW?

Frequently, a gas sample is subjected to changes in both temperature and pressure. In such cases, the Boyle's Law and Charles' Law equations can be combined into a single equation, representing the Combined Gas Law, which states: "The volume of a gas sample changes inversely with its pressure and directly with its Kelvin temperature." $V_2 = V_1 X T_2 / T_1 X P_{1/} P_{2.}$ Suppose V_1, P_1 and T_1 are the initial conditions and V_2, P_2 and T_2 are the final conditions. The Combined Gas Law equation can be rearranged to another frequently used form: $P_1V_1/T_1 = P_2V_2/T_2$.

15. WHAT IS CORRELATION BETWEEN AVOGADRO'S HYPOTHESIS AND MOLAR VOLUME?

Avogadro's Hypothesis states that at the same temperature and pressure, equal volumes of gases contain equal numbers of molecules. This statement means that, if one liter of nitrogen at a particular temperature and pressure contains 1.0×10^{22} molecules, then one liter of any other gas at the same temperature and pressure will also contain 1.0×10^{22} molecules.

The reasoning behind Avogadro's Hypothesis is not always immediately apparent. But consider that the properties of a gas that relate its volume to its temperature and pressure have been described using the postulates of the kinetic molecular theory without mentioning the composition of the gas. One of the conclusions we drew from those postulates was that, at any pressure, the volume a gas sample occupies depends on the kinetic energy of its molecules and the average of those kinetic energies is dependent only on the temperature of the sample. Stated slightly differently, at a given temperature, all gas molecules, regardless of their chemical composition, have the same average kinetic energy and therefore occupy the same effective volume.

16. WHAT COROLLARIES HAVE BEEN EXTRACTED FROM AVOGADRO'S HYPOTHESIS?

One corollary of Avogadro's Hypothesis is the concept of molar volume. The molar volume (the volume occupied by one mole) of a gas under 1.0 atm pressure and at 0°C (273.15 K), *i.e.* at STP is 22.4 L (to three significant figures). Molar volume can be used to calculate gas densities, d_{gas}, under standard conditions. The equation for this calculation is: At STP, d_{gas} = formula or molecular weight in grams/22.4 liters per mole. A second corollary of Avogadro's Hypothesis is that, at constant temperature and pressure, the volume of a gas sample depends on the number of molecules (or moles) the sample contains. Stated a little differently, if the pressure and temperature are constant, the ratio between the volume of a gas sample and the number of molecules the sample contains is a constant. Stating this ratio as an equation, we have: Volume of sample 1/Volume of sample 2= Number of molecules in sample 1/Number of molecules in sample 2.

17. WHAT IS SIGNIFICANCE OF THE IDEAL GAS EQUATION?

The various statements relating the pressure, volume, temperature and number of moles of a gas sample can be combined into one statement: The volume (V) occupied by a gas is directly proportional to its Kelvin temperature (T) and the number of moles (n) of gas in the sample and it is inversely proportional to its pressure (P). In mathematical form, this statement becomes: V= nRT/P where V = volume, n = moles of sample, P = pressure, T = temperature in K and R = a proportionality constant known as the gas constant. This equation, called the ideal gas equation, is often seen in the form:

PV= nRT. The term ideal gas means a gas that obeys exactly the gas laws. Real gases, those gases whose molecules do not follow exactly the postulates of the kinetic molecular theory, exhibit minor variations in behavior from those predicted by the gas laws. The value of the gas constant R can be determined by substituting into the equation the known values for one mole of gas at standard conditions.

18. WHAT IS DALTON'S LAW OF PARTIAL PRESSURES?

We know that the composition of a gas does not affect the validity of the gas laws. It follows then that mixtures of gases must follow those laws in the same way that a single gas does. They do and, when Boyle's Law is applied to a mixture of gases, there is a relationship between the composition of a gas sample and its total pressure. This relationship is known as Dalton's Law of Partial Pressures (the same Dalton who proposed the atomic theory). Dalton's Law of Partial Pressures states: (1) Each gas in a mixture of gases exerts a pressure, known as its partial pressure that is equal to the pressure the gas would exert if it were the only gas present; (2) the total pressure of the mixture is the sum of the partial pressures of all the gases present. This law is based on the postulate of the kinetic molecular theory which states that a gas sample is mostly empty space. The gas molecules are so far apart from one another that each acts independently. A mathematical expression of the Law of Partial Pressures is:

$\mathbf{P_{Total} = P_1 + P_2 + P_3 + \cdots}$ where P_{Total} equals the total pressure of the mixture and P_1, P_2, P_3, etc are the partial pressures of the gases present in the mixture. Suppose we have 1 L oxygen at 1 atm pressure in one container, 1 L nitrogen at 0.5 atm pressure in a second container and 1 L hydrogen at 3 atm pressure in a third container. If we combine the samples in a single 1-L container, the total pressure is 4.5 atm (1 atm + 0.5 atm + 3 atm). A corollary of this law is that, in a mixture of gases, the percent of each gas in the total volume is the same as the percent of each partial pressure in the total pressure.

From the total pressure of a mixture of gases and its percent composition, we can calculate the partial pressure of the individual gases.

$$V_{gas}/V_{Total} = P_{gas}/P_{Total}$$

19. WHAT WEAK INTERMOLECULAR FORCES COME INTO PLAY IN GASEOUS MOLECULES?

While discussing laws pertaining to gases, we assume that all gases are ideal and behave in accordance with the postulates of the kinetic molecular theory and the ideal gas equation. Under standard conditions of temperature and pressure and also at higher temperatures and lower pressures, the behavior of most real gases such as oxygen, nitrogen and carbon dioxide is that predicted by the gas laws and the kinetic molecular theory. It is for this reason that we study the properties of ideal gases. However, as the temperature of a gas is decreased, the kinetic energy of the molecules decreases, their movement becomes more sluggish and the attractive forces that exist between real molecules play a larger role in determining the behavior of the sample. Likewise, if the pressure is increased and the volume decreased until the volume of the space between the molecules approximates the volume of the molecules themselves, the molecules can no longer act as the wholly independent particles postulated by the kinetic molecular theory.

Under these conditions of low temperature and high pressure, any attractive forces that exist between the molecules of the gas come into play. These attractive forces are dipole-dipole interactions and dispersion forces. Dispersion forces also called London or Van der Waal's forces are weak forces of attraction that exist between all molecules without regard to the polarity of the molecules. To understand the nature of these forces, we need to remember that, even though a molecule may have no permanent dipole, it does have a cloud of rapidly moving electrons. If this cloud is distorted, no matter how briefly, the molecule will then have a temporary negative charge at one end and a temporary positive charge at the other end. In other words, the molecule has a temporary dipole. This temporary dipole can distort the electron clouds of nearby molecules so that they, too, have temporarily induced dipoles. The forces of attraction between the temporary partial positive charges on some molecules and the temporary partial negative charges on neighboring molecules are the dispersion forces.

20. HOW DO REAL GASES DIFFER FROM IDEAL GASES?

Under standard conditions of temperature and pressure, molecules move freely without intermolecular attraction. As the molecules moving more slowly are closer together they interact. The dipole in one molecule, regardless of whether it is real or temporary, interacts with the dipole of its neighbors. The lower the temperature and the closer the molecules are together (a result of higher pressure), the more effective are these dipole-dipole interactions in preventing the free movement of molecules required by the kinetic

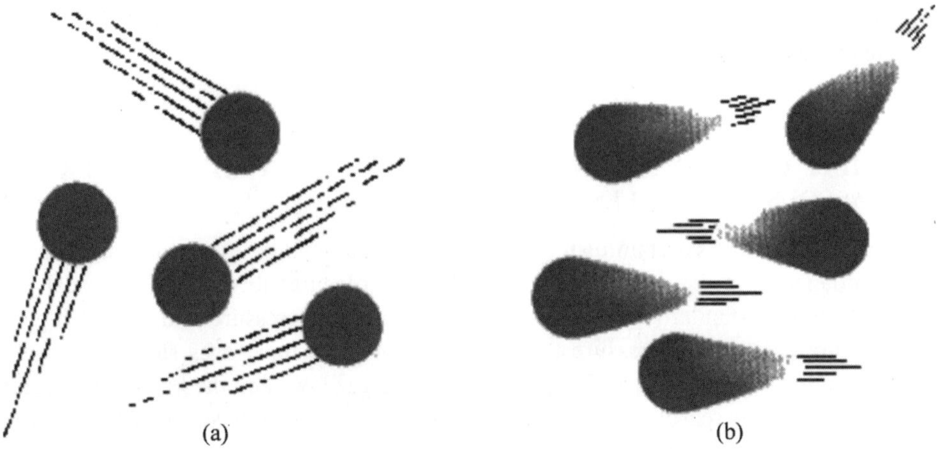

(a) (b)

Fig. 8.5: The interaction of polar molecules: (a) gas molecules move freely at STP, without interaction; (b) interaction occurs when the gas is at low temperature or high pressure, causing temporary dipoles

molecular theory. Gases that show these tendencies are said to be real gases, as opposed to ideal gases (those whose behavior is close to that predicted by the gas laws). Gases of low molecular weight and no polarity are the most ideal for example, hydrogen and helium. We expect the behavior of real gases to deviate more and more from the ideal as the polarity (either real or induced) and the molecular weight of the molecules increase. Molecular weight is a factor because the size and mass of a molecule increase as its molecular weight increases.

Chapter 9

SOLUTIONS

1. WHAT ARE CHARACTERISTICS OF SOLUTIONS?

A solution is a homogeneous mixture that results as a result of dissolution of one substance, the solute, in another substance, the solvent. Although the term solution usually calls to one's mind a liquid solvent, both solute and solvent can be in any of the three states of matter. For example, a solid can be dissolved in a liquid (salt-water), a liquid can be dissolved in a liquid (rubbing alcohol), a gas can be dissolved in a liquid (carbonated beverages), or a solid can be dissolved in another solid (alloys). As for as its terminology is concerned, it is too obvious. If the solution is liquid and the solute is solid or gas, the liquid component is usually called the solvent. If both solute and solvent are liquids, the component present in the larger amount is usually considered to be the solvent. We should keep in mind that a solution is homogeneous and only one phase or physical state is visible. A cloudy liquid is not a solution, for the cloudiness means that a substance insoluble in the solvent has been suspended in the solvent. For example, milk is a suspension, not a solution, while tea is a solution.

Though solutions in one form or another have been used by people throughout history, the scientific nature of solutions has been explored and established quite recently. A solution is distinct from a colloid or a suspension. In most common solutions, the solvent is a liquid, often water and the solute may be a solid, gas, or liquid. For example, syrups are solutions of sugar, a solid, in water, a liquid; household ammonia is a solution of ammonia gas in water; and vinegar is a solution of acetic acid, a liquid, in water. When two liquids, *e.g.*, water and ethanol, can be mixed in any proportions, the solvent is commonly considered to be the one present in greater proportion. Some alloys are solutions of one solid in another, as are many rocks. A mixture of gases, such as air, is usually not thought of as a solution.

The solute particles in a solution are generally of molecular size or even smaller, much smaller than those in a colloid or a suspension. The solute particles cannot be observed even with an ultramicroscope. They do not settle out from the solvent on standing and they cannot be separated from the solvent by physical means, such as filtration or centrifugation. A solution differs from a compound in that its components can occur in continuously varying proportions, within certain limits (although within a given solution they are present in the same proportions throughout the solution), while the components of a compound can occur only in certain fixed proportions.

The addition of a solute affects the boiling point, freezing point and vapor pressure of the solution concerned, in general raising the boiling point, depressing the freezing point and lowering the vapor pressure. A number of substances such as acids, bases and salts exhibit characteristic behavior in aqueous solution. These substances dissociate in water to form charged species *i.e.* positive and negative ions, which enable the solution to conduct electricity. Such solutions are called electrolytic ones. The proportion of the solute to the solvent in a given solution is expressed in terms of the concentration of the solution. Concentrations may be stated in a number of ways, such as giving the amount of solute contained in a given volume of solution or the amount of the solute dissolved in a given mass of the solvent. A solution having a relatively high concentration is said to be concentrated and a solution having a low concentration is said to be dilute. In many solutions the concentration has a maximum limit that depends on various factors, such as temperature, pressure and the nature of the solvent. This maximum concentration is called the solubility of the solute under specified conditions. When a solution contains the maximum amount of the solute, it is said to be saturated and if it contains less than that amount, it is said to be unsaturated.

2. WHAT ARE CHARACTERISTICS OF A SOLVENT?

The solvent is the constituent of a solution that acts as the dissolving agent. In solutions of solids or gases in a liquid, the liquid concerned is the solvent. In all other solutions (*i.e.*, liquids in liquids or solids in solids) the constituent that is present in larger quantity is considered the solvent. The most familiar and widely used solvent is water. Other compounds are valuable as solvents because they dissolve materials that are insoluble or nearly insoluble in water. Some of such solvents are: acetone, alcohol, benzene (or benzol), carbon disulfide, carbon tetrachloride, chloroform, ether, ethyl acetate, furfural, gasoline, toluene, turpentine, xylene (or xylol) etc. these solvents are often organic compounds which may be divided into polar and nonpolar types. Polar solvents, of which water is an example, have molecules having electric charges unequally distributed, making one end of each molecule more positive than the other end. Nonpolar solvents, of which carbon tetrachloride is an example, have molecules whose electric charges are equally distributed.

3. WHAT THE TERM 'SOLUBILITY' STAND FOR?

Solubility in the general sense refers to the property of being soluble, that is being able to dissolve, usually in a liquid. Chemists, however, use the word solubility to also mean the maximum amount of a chemical substance that dissolves in a given amount of solvent at a specific temperature. The most obvious factor affecting solubility is the nature of the solvent. Ordinary table salt (sodium chloride) is soluble in water, but only slightly soluble in ethanol and insoluble in diethyl ether. Temperature is also important in determining solubility. Solids are usually more soluble at higher temperatures; more salt will dissolve in warm water than in an equal amount of cold water. Graphs showing the solubility of different solids as a function of temperature are called solubility curves and are very useful in chemical analysis. Solubility also depends on pressure, especially in the case of gases as they are more soluble at higher pressures. Under certain conditions a solution may be made to contain more solute than a saturated solution at the same temperature and pressure; such a solution is called supersaturated. If even a single crystal of undissolved solute is added to a supersaturated solution, all the excess solute above the normal solubility concentration will immediately crystallize out of the solution.

Solubility is a quantitative term. Solubilities vary enormously. The terms soluble and insoluble are relative. A substance is said to be soluble if more than 0.1 g of that substance dissolves in 100 mL solvent. If less than 0.1 g dissolves in 100 mL solvent, the substance is said to be insoluble or, more exactly, sparingly soluble. The terms miscible and immiscible may be encountered when considering the solubility of one liquid in another. Miscible means soluble without limits; for example, alcohol is miscible with water. Immiscible and insoluble mean the same; oil is immiscible with water, as in oil and vinegar salad dressing.

Fig. 9.1: Soluble and insoluble substances. Alcohol is soluble in water; when added to water, it forms a clear solution. Oil is insoluble in water; when added to water, the two liquids form separate layers.

4. HOW CAN WE DETERMINE SOLUBILITY OF A SUBSTANCE IN A SOLVENT?

In order to determine the solubility of a substance in a solvent, a known amount of the solvent, for example, 100 mL is put in a container. Then the substance whose solubility is to be determined is added until, even after vigorous and prolonged stirring, some of that substance does not dissolve. Such a solution is said to be saturated because it contains as much solute as possible at that temperature. In this saturated solution, the amount of solute is the solubility of that substance at that temperature in that solvent. Doing this experiment with water as the solvent and sodium chloride which is a moderately soluble salt, as the solute, we find that, at 20°C, 35.7 g of the salt dissolve in 100 mL water. The solubility of sodium chloride is, then, 35.7 g/100 mL water at 20°C. The solubility of sodium nitrate which is a very soluble salt is 92.1 g/100 mL water at 20°C. At the opposite end of the scale is barium sulfate, which has a solubility of 2.3×10^{-4} g/100 mL water at 20°C. Barium sulfate is an insoluble salt.

5. WHAT ARE CHARACTERISTICS OF SATURATED SOLUTIONS?

A saturated solution is one in which the dissolved solute is in equilibrium with the undissolved solute. The container in the following figure (Fig. 9.2) holds a saturated solution of sucrose (cane sugar); at the bottom of the container is some undissolved sucrose. If we could see the individual molecules in this solution, we would see that some of the molecules of sucrose are moving away from the solid crystals at the bottom of the container as they dissolve. The same number of molecules is coming out of solution to become part of the undissolved sucrose.

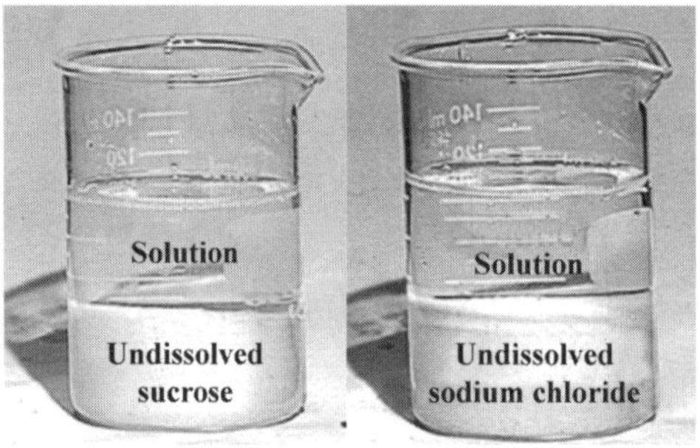

Fig. 9.2: Equilibrium in solutions: (a) the equilibrium in a sucrose solution between dissolved and undissolved molecules and (b) the equilibrium in an ionic solution between dissolved ions and undissolved sodium chloride.

Dissolution (dissolving) and precipitation are occurring at the same rate, thereby satisfying the requirement for a dynamic equilibrium. This requirement is set forth when we discuss the equilibrium between a liquid and its vapor. We can express the equilibrium in the sucrose solution with the equation:

Sucrose ⇄ sucrose (aq)

Two statements can be made about this solution: (1) the two processes dissolution and precipitation are going on at the same time and (2) the number of molecules in the solution remains constant. A saturated solution of an ionic compound is slightly different from that of a covalent compound like sucrose while the ionic compound dissolves and exists in solutions as ions; the covalent compound dissolves and exists in solutions as molecules. The equilibrium of sodium chloride with its ions in a saturated solution would be shown by the equation:

$$NaCl(s) \rightleftarrows Na^{+}(aq) + Cl^{-}(aq)$$

Since an unsaturated solution contains less solute than does a saturated solution, no equilibrium is present in it. When additional solute is added to an unsaturated solution, it dissolves but when additional solute is added to a saturated solution, the amount of dissolved solute does not increase, because the limit of solubility has already been reached. Adding more solute to a saturated solution simply increases the amount of undissolved solute.

One should remember in discussing solutions that solubility changes with temperature. A solution that is saturated at one temperature may be unsaturated at a different temperature. At this spot it seems appropriate to mention a few comments on preparation of solutions. Dissolution requires interaction between the molecules (or ions) of the solute and the molecules of the solvent. A finely divided solute will dissolve more rapidly than one that is in large chunks, because more contact is made between solute and solvent in the first case. Constant stirring increases the rate of dissolution, because stirring changes the particular solvent molecules that are in contact with undissolved solute. Because the solubility of solids and liquids generally increases with temperature, solids and liquids are often dissolved in warm solvents.

Table 9.1: Solubility-Temperature correlation (solubility (g/100 mL water)

Compound	*Type of bonding*	*At 20°C*	*At 100°C*
Sodium chloride	Ionic	35.7	39.1
Barium sulfate	Ionic	2.3×10^{-4}	4.1×10^{-4}
Ammonia	Polar covalent	89.9	7.4
Hydrogen chloride	Polar covalent	82.3	56.1
Oxygen	Nonpolar covalent	4.5×10^{-3}	3.3×10^{-3}

6. HOW AND TO WHAT EXTENT INTERMOLECULAR FORCES AFFECT SOLUBILITY OF ONE SUBSTANCE IN ANOTHER?

Many factors affect the solubility of one substance in another. One factor that affects solubility is the nature of the intermolecular forces or interionic forces in both the solute and the solvent. When one substance dissolves in another, the attractive forces in both must be overcome. The dissolving solute must be able to break up the aggregation of molecules in the solvent (that is, the hydrogen bonds between molecules or the dispersion forces between molecules in a nonpolar solvent) and the molecules of the solvent must have sufficient attraction for the solute particles to remove them one by one from their neighbors in the undissolved solute. If the solute is ionic, only a very polar solvent like water provides enough interaction to affect dissolution. In the ionic compounds like barium sulfate that are called insoluble, the interaction between the ions is greater than can be overcome by interaction with the polar water molecules. If the solute particles are polar molecules, polar solvents such as alcohols can usually affect dissolution. If the solute is nonpolar, it may dissolve only in nonpolar solvents, not because polar solvent molecules are unable to overcome the weak dispersion forces between the solute molecules, but because these dispersion forces are too weak to overcome the dipole-dipole interaction between the solvent molecules.

A general rule is that like dissolves like. Ionic and polar compounds are most apt to be soluble in polar solvents like water or liquid ammonia. Nonpolar compounds are most apt to be soluble in nonpolar solvents, such as carbon tetrachloride and hydrocarbon solvents like gasoline. The solubility of gases in water depends greatly on the polarity of the gas molecules. Those gases whose molecules are polar ones are much more soluble in water than are nonpolar gases. Ammonia, a strongly polar molecule, is very soluble in water (89.9 g/100 g H_2O); so is hydrogen chloride (82.3 g/100 g H_2O). Since helium and nitrogen are nonpolar molecules, both are only slightly soluble; helium is soluble to the tune of 1.8×10^{-4} g/100 g H_2O and nitrogen is soluble to the tune of 2.9×10^{-3} g/100 g H_2O.

7. WHY DOES SURVIVAL OF FISH BECOME QUESTIONABLE IN THERMALLY POLLUTED WATER BODIES?

We know that the solubilities of solids and liquids in liquid solvents usually increase as the temperature increases, but the solubilities of gases decrease with increasing temperature. This property of gases causes our concern for the fish population of lakes, oceans and rivers threatened with thermal (heat) pollution as fish require dissolved oxygen to survive but when the temperature of water increases, the concentration of dissolved oxygen decreases and the survival of the fish becomes questionable.

8. WHY PRESSURE CONSIDERATION IS THE MAIN CONCERN FOR CARBONATED BEVERAGES AND SCUBA DIVERS?

Though the pressure on the surface of a solution has very little effect on the solubilities of solids and liquids, it does have an enormous effect on the solubility of gases. As the partial pressure of a gas in the atmosphere above the surface of a solution increases, the solubility of that gas increases. A carbonated beverage is bottled and capped under a high partial pressure of carbon dioxide so that a great quantity of carbon dioxide dissolves. When the bottle is uncapped, the partial pressure of carbon dioxide above the liquid drops to that in the atmosphere, the solubility of carbon dioxide decreases and the gaseous CO_2 comes out of solution as fine bubbles.

This same phenomenon is of concern in scuba diving. As divers descend, they are under ever-increasing pressure due to the increasing mass of water column above them. Each 33 ft of water exert a pressure of 1 atm. As the divers descend the pressure on the air in their lungs increases rapidly and the blood dissolves more than normal amounts of both nitrogen and oxygen. As the diver comes up, the pressure decreases and the gases come out of solution (blood). If the change in pressure is too swift, the bubbles of gas are released into the blood stream and tissues, instead of into the lungs. The bubbles cause sharp pain wherever they occur, most notably in the joints and ligaments. This dangerous and sometimes fatal condition is called bends, the name deriving from the temporary deformities caused by the affected diver's being unable to straighten his or her joints.

9. WHAT ARE BENEFITS OF HYPERBARIC MEDICAL TREATMENTS?

Because helium is less soluble than nitrogen in human blood, at all pressures, divers are advised to breathe a mixture of helium and oxygen instead of air which is a nitrogen rich, nitrogen-oxygen mixture. In hyperbaric medical treatments, patients are subjected to a pressure greater than the atmospheric pressure. The term hyperbaric means greater-than-normal atmospheric pressure. This pressure increases the amount of oxygen dissolved in the blood. Treatment in hyperbaric, or high-pressure, chambers is of particular value for patients who are suffering from severe anemia, hemoglobin abnormalities, or carbon monoxide exposure or undergoing skin grafts, because this treatment increases the amount of oxygen transported by the blood to the tissues.

10. WHAT IS CORRELATION BETWEEN TEMPERATURE AND THE HEAT OF A SOLUTION?

While some solutes raise the temperature of a solution and some lower the temperature, there are many which have no noticeable effect. This behavior of a solute depends on the heat of solution of the solute in the given solvent. The heat of solution, *i.e.*, the amount of heat given off or absorbed during the process of dissolution, is equal to the difference between the energy that must be supplied to break up the crystals of the solute and the energy that is released when the solute particles are taken into solution by the solvent. If the heat of solution is negative (*i.e.*, more energy is required to break up the crystal than is released in forming the solution), then the temperature will decrease while if the heat of solution is positive, the temperature will increase.

11. WHAT ARE THE VARIOUS CONCENTRATION UNITS?

There are five units of concentration that are particularly useful to chemists. The first three: molality, molarity and normality are dependant upon the mole unit. The last two: percent by volume and percent by weight have nothing to do with mole, rather they are concerned with only weight or volume of the solute or substance to be diluted, versus the weight or volume of the solvent or substance in which the solute is diluted. Percentages can also be determined for solids within solids.

Molality

The molal unit is not used nearly as frequently as the molar unit. A molality is the number of moles of solute dissolved in one kilogram of solvent. Students should be careful not to confuse molality and molarity. One liter of water has a specific gravity of 1.0 and weighs one kilogram; so one can measure out one liter of water and add the solute to it. Most other solvents have a specific gravity greater than or less than one.

Therefore, one liter of anything other than water is not likely to occupy a liter of space. To make a one molal aqueous (water) solution of sodium chloride (NaCl), measure out one kilogram of water and add one mole of the solute, NaCl to it. The atomic weight of sodium is 23 and the atomic weight of chlorine is 35. Therefore the formula weight for NaCl is 58 and 58 grams of NaCl dissolved in 1kg water would result in a 1 molal solution of NaCl.

Molarity

The molar unit is probably the most commonly used chemical unit of measurement. Molarity is the number of moles of a solute dissolved in a liter of solution. A molar solution of sodium chloride is made by placing 1 mole of a solute into a 1-liter volumetric flask. (Taking data from the example above, we may use 58 grams of sodium chloride). Water is then added to the volumetric flask up to the one liter line. The result is a one molar solution of sodium chloride. Molality is represented by a small "m," whereas molarity is represented by an upper case "M".

Normality

There is a relationship between normality and molarity. Normality can only be calculated when we deal with reactions, because normality is a function of equivalents. The example below uses potassium hydroxide (KOH) to neutralize arsenic acid. By studying the reaction it is possible to determine the proton exchange number to determine the normality of the arsenic acid.

Look at the equation $H_3AsO_4 + 2KOH \rightarrow K_2HAsO_4 + 2H_2O$

Equivalent weight = molar mass/(number of H^+ per mole)

Equivalent = mass of compound/Equivalent weight

And Normality = (equivalents of X)/Liter

And the part that is of interest to us is that Normality = molarity/n (where n = the number of protons exchanged in a reaction). We know that when a hydrogen atom is ionized and loses its electron, we are left with only a proton. So a hydrogen ion is basically a proton. Let's assume that we have a 0.25 M solution of H_3AsO_4 and want to determine the normality of it if it participates in the reaction

$$H_3AsO_4 + 2KOH \rightarrow K_2HAsO_4 + 2H_2O$$

When H_3AsO_4 is neutralized by KOH, H_3AsO_4 provides two protons to form $2H_2O$. Note that H_3AsO_4 has three hydrogens, but K_2HAsO_4 has only one hydrogen atom. That means that 2 protons were exchanged,

$$\text{Again normality} = \text{molarity} \times n$$

Remember that normality of the solution is 0.25 mol H_3AsO_4 and there were two protons exchanged (2 equivalents/mole)

$$\frac{0.25 \text{ moles } H_3AsO_4}{1 \text{ L } H_3AsO_4} \times \frac{2 \text{ equivalents } H_3AsO_4}{1 \text{ L } H_3AsO_4} =$$

$$\frac{0.0 \text{ equivalents } H_3AsO_4}{1 \text{ L } H_3AsO_4} = 0.5 \text{ Normal } H_3AsO_4$$

So, in short, while there is a relationship between the normality of a solution and the molarity of a solution, the normality can only be determined by examining reaction, determining the proton exchange and multiplying molarity by that number.

Normality is particularly useful in titrations calculations.

Where N = normality, V = volume, a = the substance on the left of the equation involved in proton exchange and b = substance on the right of the equation involved in proton exchange:

$$Na \times Va = Nb \times Vb$$

Percentages

Percentages are easy to calculate because they do not require information about the chemical nature of the substance. Percentages can be determined as percent by weight or percent by volume. Percentages are used more in the technological fields of chemistry, such as environmental technologies, than they are in pure chemistry.

Percent by Weight

To make up a solution based on percentage by weight, one would simply determine what percentage was desired (for example, a 20% by weight aqueous solution of sodium chloride) and the total quantity to be prepared. If the total quantity needed is 1 kg, then it would simply be a matter of calculating 20% of 1 kg which, of course is: 0.20 NaCl * 1000 g/kg = 200 g NaCl/kg. In order to bring the total quantity to 1 kg, it would be necessary to add 800g water.

Percent By Volume

Solutions based on percent by volume are calculated the same as for percent by weight, except that calculations are based on volume. Thus one would simply determine what percentage was desired (for example, a 20% by volume aqueous solution of sodium chloride) and the total quantity to be prepared. If the total quantity needed is 1 liter, then it would simply be a matter of calculating 20% of 1 liter which, of course is: 0.20 NaCl × 1000 ml/l = 200 ml NaCl/l of solution.

12. HOW ARE QUANTITATIVE RELATIONSHIPS BETWEEN SOLUTES AND SOLVENTS EXPRESSED?

A complete description of a solution, besides divulging the identity of solute and solvent, also states as to how much solute is dissolved in a given amount of the solvent or the solution. The quantitative relationship between solute and solvent is known as the concentration of the solution. This concentration may be expressed using several different methods, as outlined below.

Concentration By Mass

The concentration of a solution may be given as the mass of solute in a given amount of solution, as is evident in the statements: 'The northern part of the Pacific Ocean contains 35.9 g salt in each 1000 g of seawater and the North Atlantic Ocean has a higher salt concentration, 37.9 g salt/1000 g seawater.

Concentration By Percent

The concentration of a solution is often expressed as percent concentration by mass or percent by volume of solute in solution. Percent by mass is calculated from the mass of solute in a given mass of solution. For example, a 5%-by-mass aqueous solution of sodium chloride contains 5 g sodium chloride and 95 g water in each 100 g solution. Percent by mass = (mass of solute/ mass of solution) X 100%

As an illustrative example we may try to know how many grams of glucose and water are present in 500 g of a 5.3% by-mass glucose solution.

We know that 5.3% of the solution is glucose:

$$\frac{5.3 \text{ g glucose}}{100 \text{ g solution}} \times 500 \text{ g solution} = 26.5 \text{ g glucose}$$

The remainder of the 500 g is water:

$$500 \text{ g} - 26.5 \text{ g} = 437.5 \text{ g water}$$

If both solute and solvent are liquids, the concentration may be expressed as percent by volume. Both ethyl alcohol and water are liquids; the concentration of alcohol-water solutions is often given as percent by volume. For example, a 95% solution of ethyl alcohol contains 95 mL ethyl alcohol in each 100 mL solution.

Percent by volume = (volume of solute/volume of solution) X 100%

Because the density of liquids changes slightly as the temperature changes, a concentration given in percent by mass is accurate over a wider range of temperatures than is a concentration given in percent by volume. Sometimes a combination of mass and volume is used to express the concentration, the mass of solute dissolved in each 100 mL solution. Using this method, a 5% (wt/vol) solution of sodium chloride contains 5 g sodium chloride in each 100 mL solution.

Concentration in Parts Per Million (ppm) and Parts Per Billion (ppb)

The terms: Parts per million (ppm) and parts per billion (ppb) are encountered more and more frequently as we become aware of the effects of substances present in trace amounts in water and air and as we develop instruments sensitive enough to detect substances present in such low concentrations. In discussing mass, by parts per million, we mean concentration in grams per 10^6 grams, or micrograms per gram. In discussing volume, by parts per million, we may mean milliliters per cubic meter, or the mixed designation of milligrams per cubic meter. For parts per billion, the general trend is toward the use of micrograms per liter when discussing water contaminants, micrograms per cubic meter for air and micrograms per kilogram for soil concentrations.

Concentration in Terms of Moles

The concentration of a solution may be stated in terms of molarity (M), which is the number of moles of solute per liter of solution or the number of millimoles (mmol) (1 millimole = 10^{-3} mole) per milliliter of solution.

Molarity (M) = moles solute/volume (liter) solution
Millimoles solute = solute/volume milliliter solution

A 6 M (six molar) solution of hydrochloric acid contains 6 mol of hydrochloric acid in 1 L solution. The molarity of a solution gives a ratio between moles of solute and volume of solution. It can be used as a conversion factor between these two units in calculations involving solutions. As a conversion factor, it can be used in two ways:

1. Moles/volume (L) states the number of moles in one liter of solution. This conversion factor is used in calculating the number of moles of solute in a given volume of solution.
2. Volume (L)/moles, states that one liter contains some number of moles of solution. This conversion factor is used to calculate the volume of a solution that contains a given quantity of solute.

Every time the volume of a solution is stated, the concentration of the solution is given. This form may look confusing, but without this marking it is easy to forget which solution we are referring to.

13. HOW ARE CONCENTRATION-DEPENDENT CALCULATIONS PERFORMED?

The stoichiometric calculations involve pure substances and samples whose mass could be determined by weighing. In case of a gaseous reactant the mass of the sample could be determined if we knew its pressure, volume and temperature by applying the ideal gas equation and the concept of molar volume. Now, by using the concentration of a solution as a conversion factor, we can extend stoichiometric calculations to include solutions.

As an illustrative example we may try to know what weight of barium sulfate is precipitated by the addition of an excess of sulfuric acid to 55.6 mL of 0.54 M barium chloride. We may proceed and perform calculations as under:

Equation: $BaCl_2 + H_2SO_4 \rightarrow BaSO_4 + 2HCl$
Given: 55.6 mL of 0.54 M $BaCl_2$

***Conversion factors*:**

Since 0.54 mol $BaCl_2$ in 1 L of 0.54 M $BaCl_2$
Hence 0.54 mmol $BaCl_2$ in 1 mL of 0.54 M $BaCl_2$
1 mol $BaCl_2$ forms 1 mol $BaSO_4$
1 mol $BaSO_4$ weighs 233.4 g

Arithmetic equation

$$? \text{ g BaSO}_4 = 55.6 \text{ ml of } 0.54 \text{ M BaCl}_2 \times \frac{0.54 \text{ mmol BaCl}_2}{1 \text{ mL of } 0.54 \text{ M BaCl}_2}$$

$$\times \frac{1 \text{ mmol BaSO}_4}{1 \text{ mmol BaCl}_2} \times \frac{10^{-3} \text{ mol BaSO}_4}{1 \text{ mmol BaCl}_2} \times \frac{233.4 \text{ g BaSO}_4}{1 \text{ mol BaSO}_4}$$

Answer: 7.0 g $BaSO_4$

As another example we may consider two solutions together. Let us try to find out what volume of 0.154 M sodium hydroxide will completely react with 25.0 mL of 0.0952 M hydrochloric acid. The concentration and volume of the first are known; only the concentration of the second is known. Its volume can be calculated as under.

Equation: $NaOH + HCl \rightarrow NaCl + H_2O$
Wanted: ? ml of 0.154 M NaOH
Given: 25.0 mL of 0.0952 M HCl

Conversion factors:

1 mL of 0.0952 M HCl contains 0.0952 mmol HCl
1 mol NaOH reacts with 1 mol HCl
1 mL of 0.154 M NaOH contains 0.154 mmol NaOH

Arithmetic equation:

$$? \text{ mL of } 0.154 \text{ M NaOH} = 25.0 \text{ mL of } 0.0952 \text{ M Hcl}$$

$$\times \frac{0.0952 \text{ mmol Hcl}}{1 \text{ mL of } 0.0952 \text{ M Hcl}} \times \frac{1 \text{ mmol NaOH}}{1 \text{ mmol HCl}} \times \frac{1 \text{ mL of } 0.0154 \text{ M NaOH}}{0.154 \text{ mmol NaOH}}$$

Answer: 15.5 mL of 0.154 M NaOH

As yet another example we may make use of molarity in stoichiometric problems involving gases. For example, what is the concentration of an aqueous solution of hydrochloric acid if 25.0 mL of the acid reacts with an excess of solid calcium carbonate to yield 0.307 L carbon dioxide, measured at 0.95 atm and 25°C?

Analysis of the Problem

1. The product of the reaction is a gas whose volume was not measured at standard conditions of temperature and pressure. We can calculate the number of moles of gas produce by using the ideal gas equation: PV = nRT
2. Using the number of moles of carbon dioxide produced, we can calculate the number of moles of hydrochloric acid used. The moles of HCl are in 25.0 mL solution. Molarity is a ratio between moles of solute and volume of solution. By dividing the number of moles of HCl by the volume (L) of solution in which it was dissolved, we will obtain the molarity of the acid solution.

Calculations

$$PV = nRT$$

$$0.95 \text{ atm} \times 0.307 \text{ L} = n \times 0.0821 \frac{\text{L-atm}}{\text{mol-K}} \times (25 + 273)\text{K}$$

$$n = \frac{0.95 \text{ atm} \times 0.307 \text{ L}}{0.0821 \frac{\text{L-atm}}{\text{mol-K}} \times 298\text{K}} = 0.012 \text{ mol } CO_2$$

Carry out the stoichiometric calculation

Equation: $2\ HCl + CaCO_3 \rightarrow CO_2 + CaCl_2 + H_2O$
Wanted: The molarity of the acid solution or mol HCl/L solution
Given: 0.012 mol CO_2
Conversion factor: 2 mol HCl yield 1 mol CO_2

Arithmetic equation:

$$? \text{ M HCl} = \frac{? \text{ mol HCl}}{1 \text{ L solution}}$$

$$= 0.012 \text{ mol } CO_2 \times \frac{2 \text{ mol HCl}}{1 \text{ mol } CO_2} \times \frac{1}{25 \text{ mL}} \times \frac{1000 \text{ mL}}{1 \text{ L}}$$

Answer: 0.96 M HCl

14. WHAT IONS ARE DUBBED AS SPECTATOR IONS?

Ionic compounds are different from those containing only covalent bonds. The latter ones exist as molecules whether in the pure state (crystals, liquids) or in solution. The former are a combination of two different kinds of particles. One kind (cations) carries a positive charge and the second kind (anions) carries a negative charge. Sodium chloride is a typical ionic compound. It exists not as sodium chloride units but as sodium ions and chloride ions. In the solid state, these ions are arranged in a three-dimensional lattice, one kind of ion alternating with the other. In solution, the sodium ions and the chloride ions behave independently of each other.

Let us consider a particular example of this independent behavior. Suppose we have two solutions. One contains silver ions and an anion whose silver salt is soluble. The second contains chloride ions and a cation whose chloride is soluble. A white precipitate of silver chloride forms when the two solutions are combined. We can write an equation for the reaction:

$$Ag^+ + Cl^- \rightarrow AgCl(\downarrow)$$

The other ions are still present in the solution surrounding the precipitate (the supernatant liquid), but they do not participate in the reaction. Earlier we would have written the equation differently. Assuming that nitrate was the anion with silver and sodium the cation with chloride, the equation would have been

$$AgNO_3 + NaCl \rightarrow AgCl(\downarrow) + NaNO_3$$

Another way to write the equation shows the ionic compounds in solution as separate ions:

$$Ag^+ + NO_3^- + Na^+ + Cl^- \rightarrow AgCl(\downarrow) + Na^+ + NO_3^-$$

This last equation shows clearly that only the silver and the chloride ions take part in the reaction; the sodium and nitrate ions do nothing. These nonparticipating ions are called spectator ions, meaning that they do not participate in the reaction, just as people in the stands at a football game are spectators, meaning that they do not participate in the game. The best equation for a reaction is one that shows only the participating ions; such an equation is called the net ionic equation. The net ionic equation for the reaction of silver ion with chloride is

$$Ag^+ + Cl^- \rightarrow AgCl\ (\downarrow)$$

Let us consider another situation. When an acid reacts with a hydroxide, water and a salt are formed. All acids release hydrogen ions in solution. Hydroxides furnish hydroxide ions. Hydrogen ions and hydroxide ions react to form water, a covalent un-ionized molecule. The equation for this reaction is

$$H^+ + OH^- \rightleftarrows H_2O$$

Previously we would have written this neutralization reaction using complete formulas. For example, the equation for the reaction of hydrochloric acid with sodium hydroxide would be

$$HCl + NaOH \rightarrow NaCl + H_2O$$

Rewriting this equation to show the acid, the hydroxide and the salt as ions - which is the way these compounds exist in solution - would give the equation

$$H^+ + Cl^- + Na^+ + OH^- \rightarrow Na^+ + Cl^- + H_2O$$

This equation shows clearly that the hydrogen and the hydroxide ions react and that the sodium and chloride ions are spectator ions. They would not appear in the net ionic equation.

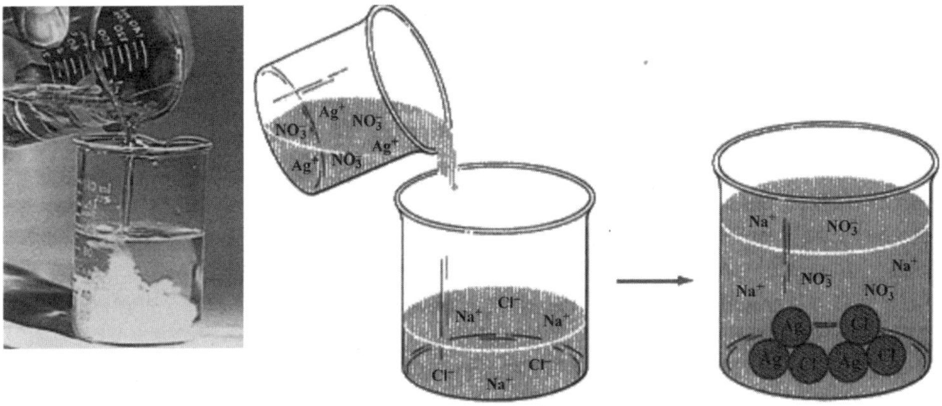

(a) $Ag^+ + Cl^- \rightarrow AgCl(\downarrow)$

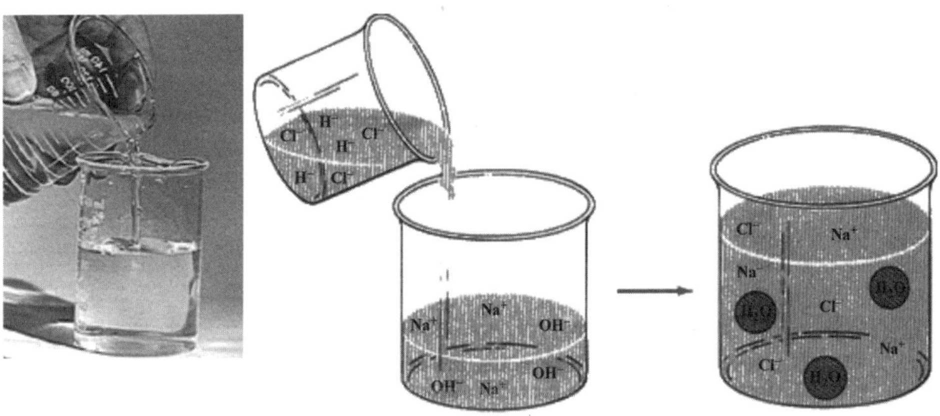

(b) $H^+ + OH^- \rightarrow H_2O$

Fig. 9.3: (a) When a solution of silver nitrate is added to a solution of sodium chloride, the silver ions combine with the chloride ions to form a precipitate of silver chloride. The sodium and the nitrate ions are nonparticipating spectator ions. (b) When hydrochloric acid is added to a solution of potassium nitrate, the hydrogen ions of the acid combine with the hydroxide ions of the potassium hydroxide to form molecules of water. The chloride and potassium ions are nonparticipating spectator ions.

Net ionic equations are also useful in showing displacement reactions. When magnesium is added to an acid, hydrogen gas and magnesium ions are formed. This statement gives us enough information to write the equation:

$$Mg + 2H^+ \rightarrow Mg^{2+} + H_2$$

Again, the identity of the anion of the acid is not important. It is present as a spectator ion and need not be shown in the equation. Like all equations, net ionic equations must be balanced. Previously we have been concerned only with balancing equations by numbers and kinds of atoms. Now we must also be certain that equations are balanced by charge; that is, the total charge on the reactants must equal the total charge on the products.

Net ionic equations can be derived from complete equations. The complete equation is written first. It is examined to determine whether there has been a change in oxidation numbers. If there has been such a change, those ions or molecules that contain the atoms changing oxidation number are isolated for the net ionic equation. If no oxidation numbers change, the products are examined to find whether a covalent compound was formed. If so, the ions that combined to form that molecule are isolated for the net ionic equation. If all the products are ionic and no oxidation number change occurred, one of the products must be a precipitate. The solubility rules are used to determine which product is the precipitate and the ions it contains are isolated for the net ionic equation.

15. WHAT COLLIGATIVE PROPERTIES OF SOLUTIONS ARE IMPORTANT?

The physical properties of a solution are different from those of the pure solvent. Many differences in physical properties are predictable if the solute in the pure state is nonvolatile, that is, if it has a very low vapor pressure. Sugar, sodium chloride and potassium nitrate are examples of nonvolatile solutes. Colligative properties are those physical properties of solutions of nonvolatile solutes that depend only on the number of particles present in a given amount of solution, not on the nature of those particles. Generally we consider four colligative properties: vapor pressure lowering, boiling point elevation, freezing point depression and osmotic pressure.

Vapor Pressure Lowering

At any given temperature, the vapor pressure of a solution containing a nonvolatile solute is less than that of the pure solvent. This effect is called vapor pressure lowering. The solid line in Fig. 9.4 is a plot of the vapor pressure of pure water against temperature.

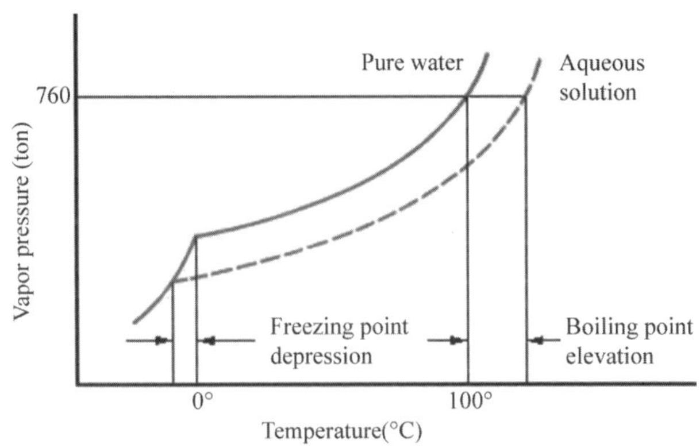

Fig. 9.4: The vapor pressure of pure water is shown as a solid line and the vapor pressure of an aqueous solution is shown as a dashed line. Note the differences between the solution and the pure substance in melting point and boiling point.

The break in the curve at 0°C is the intersection of the curve of the vapor pressure of the solid with the curve of the vapor pressure of the liquid. The dashed line in this Fig. 9.6 is a plot of the vapor pressure of an aqueous solution of sugar against temperature. Notice that the vapor pressure of the solution is always less than that of the pure solvent. What causes this difference?

The surface of a pure solvent (Fig. 9.5a) is populated only by solvent molecules. Some of these molecules are escaping from the surface and others are returning to the liquid state. The surface of a solution is populated by two kinds of molecules; some are solvent molecules,

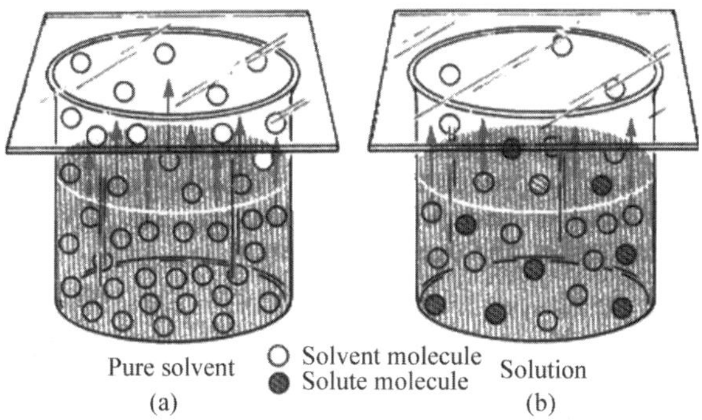

Fig. 9.5: Vapor pressure lowering: (a) the vapor pressure of a pure liquid; (b) the vapor pressure of a solution. In (b), the number of solvent molecules on the surface of the liquid has been decreased by the presence of the solute molecules. Fewer solvent molecules can vaporize and the vapor pressure is lower.

Others are solute molecules. Only the solvent molecules are volatile. They alone can escape to build up the vapor pressure of the solution. There are fewer solvent molecules on the surface of the solution than on the surface of the pure liquid. Fewer will vaporize and, as a consequence, the vapor pressure of the solution will be less than that of the pure liquid at the same temperature.

16. WHAT ARE CAUSES OF BOILING POINT ELEVATION AND FREEZING POINT DEPRESSION?

The boiling point of a substance is the temperature at which the vapor pressure of the substance equals the atmospheric pressure. A solution containing a nonvolatile solute, having a lower vapor pressure than the pure solvent, must be at a higher temperature before its vapor pressure equals atmospheric pressure. Thus, the boiling point of a solution containing a nonvolatile solute is higher than that of the pure solvent. This effect is called boiling point elevation.

We may recall that freezing and melting point are two terms that describe the same temperature, the temperature at which the vapor pressure of the solid equals the vapor pressure of the liquid and at which the solid and the liquid are in equilibrium.

We should remember, too, that vapor pressure decreases as the temperature decreases. The vapor pressure of a solution is lower than that of the solvent, so the vapor pressure of a solution will equal that of the solid at a lower temperature than in the case of the pure solvent. Thus, the freezing point will be lower for a solution than for the pure solvent. This effect is called freezing point depression. Remember that, just as it is the solvent that vaporizes when a solution boils, it is the solvent, not the solution that becomes solid when a solution freezes. When a salt solution freezes, the ice is pure water (solid); the remaining solution contains all the salt. Application of this principle leads us to add antifreeze (a nonvolatile solute) to the water in the radiators in our cars. We thus lower the freezing point of the solvent (water) and the solution remains a liquid even at subfreezing temperatures.

17. WHAT IS CORRELATION BETWEEN OSMOSIS, DIALYSIS AND DIFFUSION?

Osmosis and osmotic pressure depend on the ability of small molecules to pass through semi permeable membranes like a thin piece of rubber, a cell membrane, or a thin piece of plastic wrap. We may think of the membrane as a sieve with very tiny holes. Solvent particles are small and can very easily pass through these holes; solute particles are larger and cannot pass through.

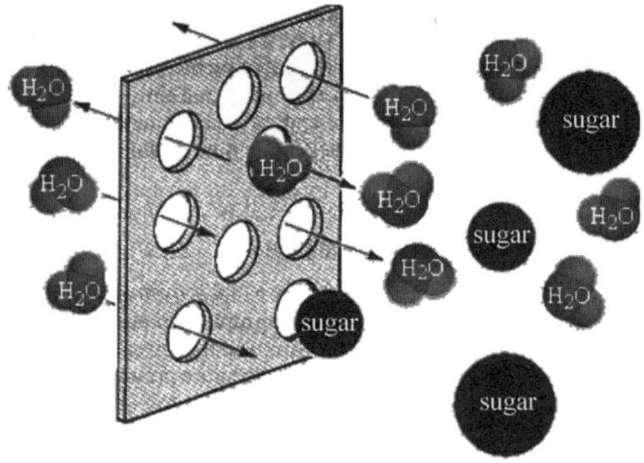

Fig. 9.6: A semipermeable membrane allows small solvent molecules to pass through but prevents the passage of larger particles like those of a nonvolatile solute

When a semipermeable membrane separates a solution from pure solvent, solvent molecules move back and forth through the membrane, but not in equal numbers. More molecules move from the pure solvent into the solution than from the solution into the solvent. The movement of solvent molecules will continue to be uneven until the number of solvent molecules is the same on both sides of the membrane. The process is called osmosis. The above Fig. 9.6 illustrates these points. In Fig. 9.7a, different amounts of pure solvent (water) are separated by a semipermeable membrane. Water molecules from both sides move through the membrane until the pressure of solvent on both sides of the membrane is equal. This equality is indicated by the equal heights of the columns (Fig. 9.7b).

In Figure 9.7c, solute molecules have been added on one side of the membrane, creating a solution. Now there are fewer solvent molecules next to this side of the membrane than there are on the side of the pure solvent. To overcome this difference, solvent molecules move more rapidly from the solvent side than from the solution side in an effort to equate these numbers (Notice the bigger arrow meaning migration from the solvent side). In Fig. 9.7d, the height of the solution is greater than that of the solvent, but now the rate at which the solvent molecules pass through the membrane is the same from both sides. The difference in heights of the columns is proportional to the osmotic pressure of the original solution.

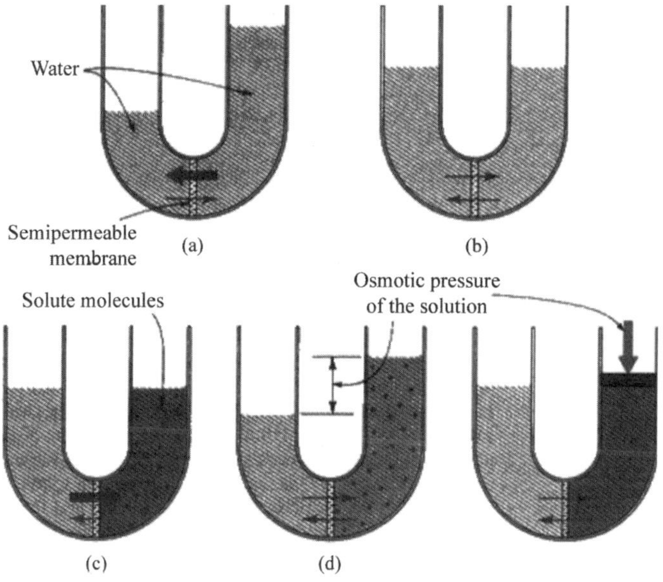

Fig. 9.7: Osmosis and osmotic pressure. Differing amounts of pure solvent on either side of a semipermeable membrane (a) will, through osmosis, become equally divided on either side of the membrane (b). However, if solute molecules are added to one side (c), some of the solvent will migrate into the solution side, causing a difference in osmotic pressure (d). The difference in pressure can be counteracted by increased surface pressure on the solution side.

Osmotic pressure is also being measured in last part of Fig. 9.7. Here, pressure is being applied to the surface of the solution. The osmotic pressure of the solution is the pressure that must be applied to the solution to prevent migration of solvent molecules from the more dilute (or pure solvent) side into the solution.

Osmosis is transfer of a liquid solvent through a semipermeable membrane that does not allow dissolved solids (solutes) to pass. Osmosis refers only to transfer of solvent; transfer of solute is called dialysis. In either case the direction of transfer is from the area of higher concentration of the material transferred to the area of lower concentration. This spontaneous migration of a material from a region of higher concentration to a region of lower concentration is called diffusion. Osmosis is the movement of water across a membrane which is selectively permeable. In osmosis, water moves across a membrane from a region with low solute concentration to a region with high solute concentration. Thus, osmosis tends to equalize the solute concentrations in two separate membrane-enclosed regions. In living cells, water moves by osmosis across membranes between cells.

18. WHAT ARE ISOTONIC SOLUTIONS?

Osmosis will occur if a vessel is separated into two compartments by a semipermeable membrane and both the compartments are filled to the same level with a solvent and solute is added to one side. If an external pressure is exerted on the side containing the solute, the transfer of solvent can be stopped and even reversed (reverse osmosis). Two solutions separated by a semipermeable membrane are said to be isotonic if no osmosis occurs. If osmosis occurs, transfer of solvent is from the hypotonic solution to the hypertonic solution, which has the higher osmotic pressure.

19. WHAT IS IMPORTANCE OF CELL MEMBRANE?

Properties such as osmotic pressure, freezing point and boiling point, which depend on the number of particles present rather than on their size or chemical nature, are called colligative properties. A number of theories explaining osmotic pressure by analogy to gases have been devised, but most have been discarded in favor of thermodynamic interpretations using such concepts as the entropy of dilution. Osmosis and dialysis are of prime importance in living organisms, where they influence the distribution of nutrients and the release of metabolic waste products. Living cells of both plants and animals are enclosed by a semipermeable membrane called the cell membrane, which regulates the flow of liquids and of dissolved solids and gases into and out of the cell. The membrane forms a selective barrier between the cell and its environment; not all substances can pass through the membrane with equal facility. Without this selectivity, the substances necessary to the life of the cell would diffuse uniformly into the cell's surroundings and toxic materials from the surroundings would enter the cell.

20. WHY DO COLLIGATIVE PROPERTIES OF A SOLUTION INCLUDE FREEZING POINT, BOILING POINT, OSMOTIC PRESSURE AND SOLVENT VAPOR PRESSURE?

Colligative properties are properties of a solution that depend on the number of solute particles present but not on the chemical properties of the solute. Colligative properties of a solution include freezing point, boiling point, osmotic pressure and solvent vapor pressure. By measuring these properties and comparing them with the corresponding properties of the pure solvent, it is possible to determine the number of particles of solute present in the solution. If the mass of solute present is also known, the number-average molecular weight can be calculated by dividing the mass of solute by the number of particles present to obtain the average mass per particle.

For any solution, the amount that the vapor pressure is lowered, the freezing point depressed, or the boiling point elevated with respect to the properties of the pure solvent depends on the number of solute particles in solution, not on the nature of those particles. Similarly, the osmotic pressure of a solution is dependent only on the number of solute particles, not on their nature. The number of moles of glucose (therefore, the number of glucose molecules in a given amount of water) differs among these solutions. The greater is the number of molecules of solute, the greater is the difference between the properties of the pure solvent and those of the solution.

One mole of glucose, a molecular compound, dissolves to yield one mole of particles; however, one mole of sodium chloride, an ionic compound, dissolves to yield two moles of particles, one mole of sodium ions and one mole of chloride ions. One mole of sodium sulfate (Na_2SO_4) dissolves to yield three moles of particles, two of sodium ions and one of sulfate ions. Because the number of particles determines the colligative properties of a solution, one mole of sodium chloride dissolves in a given amount of water causes approximately twice the change in colligative properties than does one mole of glucose dissolved in the same volume of water. One mole of sodium sulfate dissolved in the same amount of water causes approximately three times the change.

Chapter 10

ACIDS AND BASES

1. WHAT ARE GENERAL CHARACTERISTICS OF ACIDS AND BASES?

Acids and bases are two related classes of chemicals, the members of each class have a number of common properties when dissolved in a solvent, usually water. Acids in water solutions exhibit some common properties: they taste sour; turn litmus paper red; and react with certain metals, such as zinc, to yield hydrogen gas. Similarly bases in aqueous solutions exhibit some common properties: they taste bitter; turn litmus paper blue; and feel slippery. When an aqueous (water) solution of an acid is mixed with a water solution of a base, neutralization reaction takes place and water and a salt are formed. This process of neutralization is complete only if the resulting solution has neither acidic nor basic properties.

1. Generally an acid is taken as a chemical that provides hydrogen ions, H^+ and conversely a base is taken as a chemical that consumes hydrogen ions, or generates hydroxyl, OH^- ions. Chemically acids and bases may be considered opposites of each other.
2. The concept of acids and bases is so important in chemistry that several useful definitions of "acid" and "base" have been given which pertain to different chemical environments.
3. The neutralization reaction is characteristically very rapid. For example, sulfuric acid and sodium hydroxide, NaOH, react instantaneously to yield water and sodium sulfate according to the following equation:

$$H_2SO_4 + 2NaOH \rightarrow 2H_2O + Na_2SO_4$$

Acids and bases can either be classified as inorganic and organic or as strong and weak. While hydrogen sulfide, phosphoric acid, hydrogen chloride and sulfuric acid are some of the common inorganic acids, some common inorganic bases are: sodium hydroxide, sodium carbonate, sodium bicarbonate, calcium hydroxide and calcium carbonate. Some of the more common organic acids are: citric acid, carbonic acid, hydrogen cyanide, salicylic acid, lactic acid and tartaric acid. Pyridine and ethylamine are two important examples of organic bases. Acids, such as hydrochloric acid and bases, such as potassium hydroxide, that have a great tendency to dissociate in water are completely ionized in solution. They are called strong acids or strong bases. Acids, such as acetic acid and bases, such as aqueous ammonia, that are reluctant to dissociate in water are only partially ionized in solution. They are called weak acids or weak bases. Strong acids in solution produce a high concentration of hydrogen ions and strong bases in solution produce a high concentration of hydroxide ions and a correspondingly low concentration of hydrogen ions. The hydrogen ion concentration is often expressed in terms of its negative logarithm, or pH. Strong acids and strong bases make very good electrolytes, *i.e.*, their solutions readily conduct electricity. Weak acids and weak bases make poor electrolytes.

2. HOW IMPORTANT ARE ACID-BASE CONCEPTS?

Modern understanding of acids and bases began with the discovery in 1834 by the English physicist Michael Faraday that acids, bases and salts are electrolytes; that is, when dissolved in water, they produce a solution that contains charged particles, *i.e.* ions and can conduct an electric current. Generally people are acquainted with three concepts/theories that identify a singular characteristic which defines an acid and a base: the Arrhenius theory, for which the Swedish chemist Svante Arrhenius was awarded the 1903 Nobel Prize in chemistry; the Bronsted-Lowry, or proton donor, theory, advanced in 1923; and the Lewis, or electron-pair, theory, which was also presented in 1923. Each of the three theories has its own advantages and disadvantages and each is useful under certain conditions. However, there are many other definitions of acids and bases which are correct and useful as for as they go.

It is obvious that an acid-base reaction is a chemical reaction that occurs between an acid and a base. Several concepts that provide alternative definitions for the reaction mechanisms involved and their application in solving related problems exist. Despite several differences in definitions, their importance becomes apparent as different methods of analysis when applied to acid-base reactions for gaseous or liquid species, or when acid or base character may be somewhat less apparent.

3. WHAT ARE HISTORIC ACID-BASE CONCEPTS?

The first scientific concept of acids and bases was provided by Antoine Lavoisier circa 1776. Since Lavoisier's knowledge of strong acids was mainly restricted to oxoacids, such as HNO_3 (nitric acid) and H_2SO_4 (sulfuric acid), which tend to contain central atoms in high oxidation states surrounded by oxygen and since he was not aware of the true composition of the hydrohalic acids (HF, HCl, HBr and HI), he defined acids in terms of their oxygen containing nature, which in fact he named from Greek words meaning "acid-former" "acid" or "sharp" and "engender". Interestingly, in spite of its narrowness, the Lavoisier definition was held as absolute truth for over 30 years, until the 1810 article and subsequent lectures by Sir Humphry Davy in which he proved the lack of oxygen in acidic species such as H_2S, H_2Te and the hydrohalic acids. However, Davy failed to develop a new theory, concluding that "acidity does not depend upon any particular elementary substance, but upon peculiar arrangement of various substances". One notable modification of oxygen theory was provided by Berzelius, who stated that acids are oxides of nonmetals while bases are oxides of metals.

Based on his extensive works on the chemical composition of organic acids, another definition of acids and bases was proposed by Justus von Liebig circa 1838. This concept put an end to the doctrinal shift from oxygen-based acids to hydrogen-based acids, started by Davy. According to Liebig, an acid is a hydrogen-containing substance in which the hydrogen could be replaced by a metal. Liebig's definition, while completely empirical, remained in use for almost 50 years until the adoption of the Arrhenius definition.

4. WHAT ARE COMMON ACID-BASE CONCEPTS?

Arrhenius Concept

In 1884 Svante Arrhenius, a Swedish chemist and later Wilhelm Ostwald, a German chemist proposed that an acid be defined as a hydrogen-containing compound which, when dissolved in water, increases the concentration of hydrogen ions, greater than that of pure water. Similarly, Arrhenius also proposed that a base be defined as a substance which when dissolved in water, produces an excess of hydroxyl ions, OH^-. The neutralization reaction then simply becomes a reaction between hydrogen and hydroxyl ions: $H^+ + OH^- \rightarrow H_2O$

When an acid or base dissolves in water, a certain percentage of the acid or base particles will break up, or dissociate, into oppositely charged ions. The Arrhenius theory defines an acid as a compound that can dissociate in water to yield hydrogen ions, H^+ and a base as a compound that can dissociate in water to yield hydroxide ions, OH^-. For example, hydrochloric acid, HCl, dissociates in water to yield the required hydrogen ions, H^+ and also chloride ions, Cl^-. The base sodium hydroxide, NaOH, dissociates in water to yield the required hydroxide ions, OH^- and also sodium ions, Na^+. A number of criticisms of the Arrhenius-Ostwald theory have been made, the two prominent ones being: 1. Acids are restricted to hydrogen-containing species and bases to hydroxyl-containing species only and 2. The theory applies to aqueous solutions exclusively, whereas many acid-base reactions are known to take place in the absence of water.

Solvent System Concept

One of the limitations of Arrhenius definition has been its reliance on aqueous/water solutions. E.C. Franklin studied the acid-base reactions in liquid ammonia in 1905 and pointed out the similarities to water-based Arrhenius theory and Albert F.O. Germann, working with liquid $COCl_2$, generalized Arrhenius definition to cover aprotic solvents and formulated the solvent system theory in 1925. Germann pointed out that in many solvents there is a certain concentration of a positive species, solvonium (earlier lyonium) cations and negative species, solvate (earlier lyate) anions, in equilibrium with the neutral solvent molecules. For example, water and ammonia undergo such dissociation into hydronium and hydroxide and ammonium and amide ions, respectively as under:

$$2\,H_2O \leftrightarrow H_3O^+ + OH^-$$

$$2\,NH_3 \leftrightarrow NH_4^+ + NH_2^-$$

Some aprotic systems also undergo similar dissociation, as do dinitrogen tetroxide into nitrosonium and nitrate; antimony trichloride into dichloroantimonium and tetrachloroantimonate and phosgene into chlorocarboxonium and chloride.

$$N_2O_4 \leftrightarrow NO^+ + NO_3^-$$
$$2\ SbCl_3 \leftrightarrow SbCl_2^+ + SbCl_4^-$$
$$COCl_2 \leftrightarrow COCl^+ + Cl^-$$

A solute causing an increase in the concentration of the solvonium ions and a decrease in the concentration of solvate ions is defined as an acid and one causing the reverse is defined as a base. Thus, in liquid ammonia, KNH_2 (supplying NH_2^-) is a strong base and NH_4NO_3 (supplying NH_4^+) is a strong acid. In liquid sulfur dioxide (SO_2), thionyl compounds (supplying SO^{2+}) behave as acids and sulfites (supplying SO_3^{2-}) behave as bases. The non-aqueous acid-base reactions in liquid ammonia are similar to the reactions in water. The unique strength of this definition lies in describing the acid-base reactions in aprotic solvents, for example, the following reaction in liquid N_2O_4:

$$AgNO_3\ (\text{base}) + NOCl\ (\text{acid}) \rightarrow N_2O_4\ (\text{solvent}) + AgCl\ (\text{salt})$$

Since solvent-system definition depends on the solvent as well as on the compound concerned, the same compound can change its role depending on the choice of the solvent. Thus, $HClO_4$ is a strong acid in water, a weak acid in acetic acid and a weak base in fluorosulfonic acid. This was seen as both strength and a weakness of the solvent system concept, since some substances, such as SO_3 and NH_3 were felt to be acidic or basic on their own right. Moreover the solvent system theory was criticized as too general to be useful; it was felt that there was something intrinsically acidic about hydrogen compounds, not shared by non-hydrogenic solvonium salts.

Bronsted-lowry Concept

We now know that whereas the Arrhenius definitions of acids and bases describe the characteristics of aqueous solutions of acids and bases, some substances act as acids or bases when they are dissolved in solvents other than water, such as liquid ammonia. A more satisfactory acid-base theory was proposed in 1923 by the Danish chemist Johannes Bronsted and a British chemist, Thomas Lowry, not only independently but also without knowing each other. The Bronsted-Lowry theory provides a more general definition of acids and bases that can be used to deal with both aqueous and non-aqueous solutions. It defines an acid as a proton donor and a base as a proton acceptor. In the Bronsted-Lowry theory, water, H_2O, can be considered as both an acid and a base since it can lose a proton to form a hydroxide ion, OH^-, or accept a proton to form a hydronium ion, H_3O^+.

When an acid loses a proton, the remaining species can be a proton acceptor and is called the conjugate base of the acid. Similarly when a base accepts a proton, the resulting species can be a proton donor and is called the conjugate acid of that base. For example, when a water molecule loses a proton to form a hydroxide ion, the hydroxide ion can be considered the conjugate base of the acid, water. When a water molecule accepts a proton to form a hydronium ion, the hydronium ion can be considered the conjugate acid of the base, water.

In the Bronsted-Lowry system, an acid reacts by donating a proton to a base and in doing so, the left over part of acid becomes its own conjugate base. The formula of the conjugate base is the formula of the acid with one hydrogen atom less. The resulting species of the base becomes its own conjugate acid. The formula of the conjugate acid is the formula of the base plus one hydrogen ion. Let us illustrate this system using the neutralization of hydrochloric acid with sodium hydroxide. When hydrochloric acid reacts with hydroxide ion, water and chloride ion are formed. In the equation for the reaction each acid-base pair has the same subscript. $Acid_1$ is HCl, its conjugate base is $base_1$; hydroxide ion is $base_2$ and its conjugate acid (water) is $acid_2$.

$$\underset{acid_1}{HCl} + Na^+ + \underset{base_2}{OH^-} \longrightarrow \underset{acid_2}{H_2O} + Na^+ + \underset{base_1}{Cl^-}$$

Chloride ion is the conjugate base of hydrochloric acid and water is the conjugate acid of the hydroxide ion. In this equation the sodium ion is a spectator ion. The equation for the reaction of hydrochloric acid with ammonia may be written as:

$$\underset{acid_1}{HCl} + \underset{base_2}{NH_3} \longrightarrow \underset{acid_2}{NH_4} + \underset{base_1}{Cl^-}$$

When water reacts with ammonia, it acts as an acid:

$$\underset{acid_1}{H_2O} + \underset{base_2}{NH_3} \longrightarrow \underset{acid_2}{NH_4} + \underset{base_1}{OH^-}$$

Hydroxide ion is the conjugate base of water and when water reacts with an acid, it acts as a base:

$$\underset{base_1}{H_2O} + \underset{acid_2}{HCl} \longrightarrow \underset{acid_1}{H_3O^+} + \underset{base_2}{Cl^-}$$

The conjugate acid of water is the hydronium ion, H_3O^+, an ion formed by the association of a hydrogen ion with a water molecule. All of the reactions in these examples fit the general equation for a Bronsted-Lowry acid-base reaction:

$$\underset{acid_1}{HA} + \underset{base_2}{B^-} \longrightarrow \underset{base_1}{A^-} + \underset{acid_2}{HB}$$

The Bronsted-Lowry acid-base definitions greatly expand the classes of substances called acids and bases. Members of the acid group need not be compounds only but can also be ions and a base need not contain in its formula the hydroxide ion. The usefulness of these definitions is emphasized by considering the Lewis structures of typical members of the groups. Following are Lewis structures for some bases.

Each of these bases has at least one unshared pair of electrons on a very electronegative atom. When a proton, hydrogen ion, is added to these structures, it forms a covalent bond with one of these unshared electron pairs. Because a hydrogen ion consists of a nucleus containing a single proton, the terms hydrogen ion and proton are synonymous.

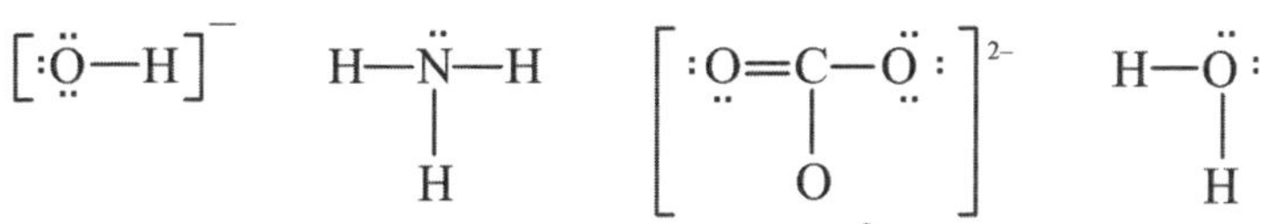

OH^- NH_3 CO_3^{2-} H_2O

These definitions somewhat broaden the category of substances that are acids or bases. The category of acids now includes those listed above as well as ions such as ammonium ion, NH_4^+ and bicarbonate ion, HCO_3^-. Among Bronsted-Lowry bases are the hydroxide ion, OH^-, the anion of any acid; and ammonia, NH_3. Many species such as water, bicarbonate ion and ammonia can act as either an acid or a base. This theory states that an acid is a proton (hydrogen ion, H^+) donor and a base a proton acceptor. Although the acid must still contain hydrogen, the Brønsted-Lowry theory does not require an aqueous medium. For example, liquid ammonia, which acts as a base in aqueous solution, can act as an acid in the absence of water by transferring a proton to a base and forming the amide anion, NH_2^-.

$$NH_3 + \text{base} \rightarrow NH_2^- + (\text{base} + H^+)$$

The Bronsted-Lowry definition of acids and bases also explains as to why a strong acid displaces a weak acid from its compounds and likewise it takes care of the behavior of strong and weak bases too. Here acid-base reactions are viewed as a competition for protons. In terms of a general chemical equation, the reaction of Acid (1) with Base (2), shown as:

$$\text{Acid (1)} + \text{Base (2)} \rightarrow \text{Acid (2)} + \text{Base (1)}$$

results in the transfer of a proton from Acid (1) to Base (2). In losing the proton, Acid (1) becomes its conjugate base, Base (1) and in gaining a proton, Base (2) becomes its conjugate acid, Acid (2). The equilibrium represented by the equation above may be displaced either to the left or to the right and the actual reaction will take place in the direction that produces the weaker acid-base pair. For example, hydrogen chloride (HCl) is a strong acid in water because it readily transfers a proton to water to form a hydronium ion:

$$HCl + H_2O \rightarrow H_3O^+ + Cl^-$$

The equilibrium lies mostly to the right because the conjugate base of HCl, Cl^-, is a weak base and H_3O^+, the conjugate acid of H_2O, is a weak acid. In contrast, hydrogen fluoride, HF, is a weak acid in water because it does not readily transfer a proton to water:

$$HF + H_2O \rightarrow H_3O^+ + F^-$$

This equilibrium lies mostly to the left because H_2O is a weaker base than F^- and because HF is a weaker acid (in water) than H_3O^+. The Brønsted-Lowry theory also explains as to why water can be amphoteric, that is, why it can serve as either an acid or a base. Water serves as a base in the presence of an acid that is stronger than water (such as HCl), in other words, an acid that has a greater tendency to dissociate than does water:

$$HCl + H_2O \rightarrow H_3O^+ + Cl^-$$

Water can also serve as an acid in the presence of a base that is stronger than water (such as ammonia):

$$NH_3 + H_2O \rightarrow NH_4^+ + OH^-$$

The strength of an acid can be measured by the extent to which it transfers a proton to water to produce the hydronium ion, H_3O^+. Conversely, the strength of a base is indicated by the extent to which it removes a proton from water. A convenient acid-base scale is calculated from the amount of H_3O^+ that is formed in water solutions of acids or of OH^- formed in water solutions of bases. The former is known as the pH scale and the latter as the pOH scale. The value for pH is equal to the negative logarithm of the hydronium ion concentration and for pOH; it is the negative logarithm of the hydroxyl ion concentration in an aqueous solution:

$$pH = -\log [H_3O^+] \text{ and } pOH = -\log [OH^-]$$

Pure water has a pH of 7.0. When an acid is added, the hydronium ion concentration $[H_3O^+]$ becomes larger than that in pure water and the pH becomes less than 7.0, depending on the strength of the acid concerned. The pOH of pure water is also 7.0 and in the presence of a base, the pOH drops to values lower than 7.0.

Lewis Concept

Another regular acid-base theory that provides a very broad definition of acids and bases was put forth by the American chemist Gilbert Lewis, who is also known for his electron pair theory. Lewis theory has the further advantage of not requiring the acid to contain hydrogen. This theory states that acids are electron-pair acceptors and bases are electron-pair donors. Boron trifluoride, BF_3, can be considered a Lewis acid and ethyl alcohol can be considered a Lewis base. This theory also has the advantages that it works when solvents other than water are involved and it does not require the formation of a salt or of acid-base conjugate pairs. Thus, ammonia is viewed as a base because it can donate an electron pair to the acid boron trifluoride, for example to form an acid-base association pair as under.

$$H_3N{:} + BF_3 \rightarrow H_3N\text{-}BF_3$$

5. WHAT ACID-BASE CONCEPTS ARE KNOWN AS 'OTHER ACID-BASE CONCEPTS'?

Though there are some more acid-base theories pertaining to different aspects of chemical phenomena, they are either too general or too specialized to be of any benefit to the beginners and learners. However, instead of allowing these concepts to hibernate till the learners come to know about them and feel the need to haunt for them, it is better to introduce these concepts for awakening them.

Usanovich Concept

Simultaneously with Lewis, a Soviet chemist Mikhail Usanovich from Tashkent, developed a general theory that does not restrict acidity to hydrogen-containing compounds, but his approach, published in 1938, was even more general than Lewis theory. Usanovich's theory can be summarized as defining an acid as anything that accepts negative species or donates positive ones and a base as the reverse. This pushed the concept of acid-base reactions to its logical limits and even redefined the concept of redox (oxidation-reduction) as a special case of acid-base reactions and so did not become wide spread, despite being easier to understand than Lewis theory, which required detailed familiarity with atomic structure.

Lux-Flood Concept

Proposed by German chemist Hermann Lux in 1939 and further improved by Håkon Flood circa 1947, this acid-base theory was a revival of oxygen theory of acids and bases and is still used in modern geochemistry and electrochemistry of molten salts. This definition describes an acid as an oxide ion (O^{2-}) acceptor and a base as an oxide ion donor. For example:

$$MgO \text{ (base)} + CO_2 \text{ (acid)} \rightarrow MgCO_3 \text{ (salt)}$$
$$CaO \text{ (base)} + SiO_2 \text{ (acid)} \rightarrow CaSiO_3 \text{ (salt)}$$

Pearson Concept (HsaB Concept)

In 1963 Ralph Pearson proposed an advanced qualitative concept known as Hard Soft Acid Base principle, later made quantitative with help of Robert Parr in 1984. The term 'Hard' applies to species that are small, have high charge states and are weakly polarizable. The term 'Soft' applies to species that are large, have low charge states and are strongly polarizable. Acids and bases interact and the most stable interactions are hard-hard and soft-soft. This theory has found use in both organic and inorganic chemistry.

In fact the HSAB concept is an acronym for 'hard and soft (Lewis) acids and bases. Also known as the Pearson acid base concept, HSAB is widely used in chemistry for explaining stability of compounds, reaction mechanisms and pathways. It assigns the terms 'hard' or 'soft' and 'acid' or 'base' to chemical species. 'Hard' applies to species which are small, have high charge states (the charge criterion applies mainly to acids and to a lesser extent to bases) and are weakly polarizable. 'Soft' applies to species which are big, have low charge states and are strongly polarizable.

The theory is used in contexts where a qualitative, rather than quantitative description would help in understanding the predominant factors which drive chemical properties and reactions. This is especially so in transition metal chemistry, where numerous experiments have been done to determine the relative ordering of ligands and transition metal ions in terms of their hardness and softness. HSAB theory is also useful in predicting the products of metathesis reactions. Quite recently it has been shown that even the sensitivity and performance of explosive materials can be explained on basis of HSAB theory. In fact Ralph Pearson introduced the HSAB principle in the early 1960s as an attempt to unify inorganic and organic reaction chemistry.

The gist of this theory is that soft acids react faster and form stronger bonds with soft bases, whereas hard acids react faster and form stronger bonds with hard bases, all other factors being equal. The classification in the original work was mostly based on equilibrium constants for reaction of two Lewis bases competing for a Lewis acid. Hard acids and hard bases tend to have: small atomic/ionic radius, high oxidation state, low polarizability and high electronegativity. Hard bases have highest-occupied molecular orbitals (HOMO) of low energy and hard acids have lowest-unoccupied molecular orbitals (LUMO) of high energy. Examples of hard acids are: H^+, alkali ions, Ti^{4+}, Cr^{3+}, Cr^{6+}, BF_3, etc. Examples of hard bases are: OH^-, F^-, Cl^-, NH_3, CH_3COO^- and CO_3^{2-}. The affinity of hard acids and hard bases for each other is mainly ionic in nature. On the other hand soft acids and soft bases tend to have: large atomic/ionic radius, low or zero oxidation state, high polarizability and low electronegativity. Soft bases have HOMO of higher energy than hard bases and soft acids have LUMO of lower energy than hard acids. (However the soft-base HOMO energies are still lower than the soft-acid LUMO energies.) Examples of soft acids are: CH_3Hg^+, Pt^{2+}, Pd^{2+}, Ag^+, Au^+, Hg^{2+}, Hg_2^{2+}, Cd^{2+} and BH_3. Examples of soft bases are: H^-, R_3P, SCN^- and I^-. The affinity of soft acids and bases for each other is mainly covalent in nature.

6. HOW ARE ACIDS NAMED?

Commonly acids are substances that react with water to form an aqueous hydronium ion, $H^{+1}(aq)$. These compounds have characteristics in common with both ionic and molecular compounds. They are studied extensively in General Chemistry, in part, because they are useful compounds and, in part, because they are very good examples of equilibria. There is more than one way to define an acid. Unfortunately, compounds that fit one definition might not be considered acids by another definition. In the broadest and most general sense, acids are compounds similar to ionic compounds but with H^{+1} as their positive ions. This will be written first in the formula. Acid names end with the word 'acid' making them easy to identify from the name.

There are two categories of acids: binary acids and oxyacids.

Binary Acids

The binary compound formed when a halogen or any element, except oxygen, from Group 16 of the periodic table combines with hydrogen can be named as were the binary nonmetallic compounds. However, when these compounds are dissolved in water, the solution contains hydrogen ions. Because this property identifies an acid, these compounds must also be named as acids. Therefore, these compounds have two sets of names; one for the pure state and one for the compound dissolved in water.

Here two points should be noted:

1. The acid name has the prefix hydro and the suffix -ic;
2. These formulas are always written with hydrogen first. Other nonmetals form compounds with hydrogen, but they are not acids; their formulas are written with hydrogen last. Methane, CH_4, ammonia, NH_3 and arsine, AsH_3, are examples.

Table 10.1: Nomenclature for Binary Acids

Formula	*Name in pure state*	*Name in water solution*
HCl	Hydrogen chloride	Hydrochloric acid
H_2S	Hydrogen sulfide	Hydrosulfuric acid
HBr	Hydrogen bromide	Hydrobromic acid

Oxyacids/Oxoacids

These acidic compounds contain hydrogen, oxygen and at least one other element. The non-hydrogen part of the compound will be an anion, usually from the list in the ionic nomenclature section. To name an oxyacid, identify the name of the anion. If the ending of that ion is –ate, change it to –ic. If the ending of that ion is ite, change it to –ous. End the name with the word 'acid.' For example, H_2SO_4 is sulfuric acid and HNO_2 is nitrous acid. Note that these names do not include the prefix 'hydro' like the binary acids.

7. HOW DO WE WRITE FORMULAE OF ACIDS?

When given the name of an acid, we must first identify the negative ion. If the acid name ends in '-ic' the negative ion is a simple, monatomic ion and we can determine its charge from its location on the periodic table. If the acid name ends in –ic, it contains a negative ion whose name ends in –ate. If the acid name ends in –ous, it contains a negative ion whose name ends in –ite. Write H^{+1} as the positive ion followed by the negative ion. Use subscripts to make the compound neutral. For example, hydrochloric acid is HCl, chromic acid is H_2CrO_4 and hypochlorous acid is HClO. Note that this last example contains the hypochlorite ion rather than having a misspelled binary acid prefix.

8. HOW ARE TERNARY ACIDS AND THEIR ANIONS NAMED?

When a ternary compound contains hydrogen and a polyatomic anion (for example, HNO_3), its name in the pure state is hydrogen followed by the name of the anion. Pure HNO_3 has the name hydrogen nitrate. When this compound is dissolved in water, it is an acid and is named as such. HNO_3 in water solution is named nitric acid. Table 10.2 lists the formulas of some of these compounds, the names they carry when in water solution, the oxidation number of the nonmetal other than oxygen that they contain and the name and formula of their anions. The rules for naming these compounds as acids follow the table. One should be sure to study the table as he/she reads the rules and notice the pattern shown in the names and formulas. The rules for naming ternary acids are as follows:

1. The name of the most common oxyacid for a particular nonmetal is the root of the element's name plus the suffix ic. It has no prefix. The name of the anion of this acid is the root of the element's name plus the suffix ate. The oxidation number of the nonmetal in this acid is high but not necessarily the highest possible. These acids are sometimes referred to as ic-ate acids. Of the oxyacids: nitric, sulfuric, phosphoric and chloric are in the category of "most common acids."
2. The name of the acid in which the nonmetal has the next lower oxidation number is the element's root plus the suffix ous. The name of its anion is the root plus ite. These acids may be called ous-ite acids. The acids: nitrous, sulfurous and chlorous fall into this group. Their formulas can be predicted if we have learned the formulas in the first group. The anion of an ous-ite acid contains one fewer oxygen atom than that of the ic-ate acid.
3. As with the halogens, if there is an oxyacid in which the nonmetal has an even lower oxidation number, that acid is named using the prefix-suffix hypo-ous

Table 10.2: Some acidic compounds and their anions in aqueous systems

Formula of the acid	*Oxidation number of the nonmetal*	*Name of the acid in aqueous solution*	*Name and formula of the anion*
HNO_3	+5	Nitric acid	Nitrate, NO_3^-
HNO_2	+3	Nitrous acid	Nitrite, NO_2^-
H_2SO_4	+6	Sulfuric acid	Sulfate, SO_4^{2-}
H_2SO_3	+4	Sulfurous acid	Sulfite, SO_3^{2-}
H_3PO_4	+5	Phosphoric acid	Phosphate, PO_4^{3-}
H_2CO_3	+4	Carbonic acid	Carbonate, CO_3^{2-}
$HClO_4$	+7	Perchloric acid	Perchlorate, ClO_4^-
$HClO_3$	+5	Chloric acid	Chlorate, ClO_3^-
$HClO_2$	+3	Chlorous acid	Chlorite, ClO_2^-
$HClO$	+1	Hypochlorous acid	Hypochlorite, ClO^-
HCl	−1	Hydrochloric acid	Chloride, Cl^-

and its anion using hypo-ite. Hypochlorous is in this category. Its formula can be predicted if we know the formula of chloric acid. The anions of these acids contain two fewer oxygen atoms than the anions of the ic-ate acids.

4. Again as with the halogens, if there is an oxyacid in which the nonmetal has a higher oxidation number than in the most common acid, that acid is named per-ic acid and its anion per-ate. Of the acids in the proceeding table, only perchloric is in this group. Its formula can be predicted from the formula of chloric acid. The anion will contain one more oxygen atom than the anion of the ic-ate acid. Some of the compounds that behave as acid in aqueous systems and their anions may be mentioned here in a tabular form.

The formulas of the salts of these acids are neutral combinations of ions. To name them, first name the cation according to the rules given previously.

9. HOW DO WE NAME THE TERNARY ACIDS CONTAINING CARBON?

Many acids contain only carbon, hydrogen and oxygen. Acetic acid is an example. Its formula can be written as $HC_2H_3O_2$ or HCH_3CO_2 or CH_3COOH or CH_3CO_2H. Regardless of how it is written, it is clear that there is one hydrogen atom only in acetic acid that can exhibit acidic nature; the other three hydrogens do not separate as hydrogen ions in aqueous solution. Notice how the acidic hydrogen is placed by itself in each of the formulas to signify this difference. Many acids, like acetic acid, contain a group of atoms bonded to a -COOH group. Only the hydrogen of the -COOH group is an acidic hydrogen. For example, C_6H_5COOH benzoic acid, C_2H_3COOH acrylic acid. These acids are called carboxylic acids. In naming the anions of these acids, the ic of the acid is replaced by ate. Thus, acetic acid $HC_2H_3O_2$, yields acetate ion, $C_2H_3O_2^-$, benzoic acid C_6H_5COOH, yields benzoate ion, $C_2H_5COO^-$ and acrylic acid C_2H_3COOH, yields acrylate ion, $C_2H_3COO^-$.

10. WHAT SHOULD BE NAME OF HBr(aq)?

HBr(aq) stands for hydrobromic acid. Since it does not contain a metal or the NH_4^+ ion, it is not ionic. The formula does begin with H, so it is an acid. The anion is Br^{-1}, bromide. This ends in -ide and the compound is aqueous, so it is named as an acid (The format is hydro -ic acid and the root is -brom-. The name is hydrobromic acid

11. WHAT ARE NEUTRALIZATION REACTIONS?

In chemistry, the process in which an acid and a base react with each other to form a salt and water is known as neutralization. Just as a neutral color contains no vivid colors and a neutral person has no strong opinions, a neutralization reaction between an acidic and a basic (alkaline) solution lessens the acidic and basic properties of both solutions. According to the Arrhenius theory of acids and bases, neutralization is a chemical reaction, in which a water solution of acid is mixed with a water solution of base to form a salt and water; this reaction is complete only if the resulting solution has neither acidic nor basic properties. Such a solution is called a neutral solution.

Complete neutralization can take place when a strong acid, such as hydrochloric acid, HCl, is mixed with a strong base, such as sodium hydroxide, NaOH. Strong acids and strong bases completely break up, or dissociate, into their constituent ions when they dissolve in water. In the case of hydrochloric acid, hydrogen ions, H^+ and chloride ions, Cl^-, are formed and in case of sodium hydroxide, sodium ions, Na^+ and hydroxide ions, OH^-, are formed. The hydrogen and hydroxide ions readily unite to form water. If the number of hydrogen ions in the hydrochloric acid solution is equal to the number of hydroxide ions in the sodium hydroxide solution, complete neutralization occurs when the two solutions are mixed. The resulting solution contains sodium ions and chloride ions that congregate, when water evaporates, to form sodium chloride, common table salt.

In a neutralization reaction in which either a weak acid or a weak base is used only partial neutralization occurs. In a neutralization reaction in which both a weak acid and a weak base are used, complete neutralization can occur if the acid and the base are equally weak. The heat produced in the reaction between an acid and a base is called the heat of neutralization. When any strong acid is mixed with any strong base, the heat of neutralization is always about 13,700 calories for each equivalent weight of acid and base neutralized. The neutralization reaction of hydrochloric acid with sodium hydroxide is the reaction of an acid with a base to form water and a salt as under:

$$\underset{\text{hydrochloric acid}}{HCl} + \underset{\text{sodium hydroxide}}{NaOH} \longrightarrow \underset{\text{sodium chloride}}{NaCl} + \underset{\text{water}}{H_2O}$$

The net ionic equation is: $H^+ + OH^- \rightleftarrows H_2O$

We have seen acid-base reaction within the Bronsted-Lowry system as the reaction of a proton donor with a proton acceptor, as in the reaction of hydrochloric acid with the acetate ion:

$$\underset{\text{acid}_1}{HCl} + \underset{\text{base}_2}{C_2H_3O_2^-} \longrightarrow \underset{\text{acid}_2}{HC_2H_3O_2} + \underset{\text{base}_1}{Cl^-}$$

or with ammonia:

$$\underset{\text{acid}_1}{HCl} + \underset{\text{base}_2}{NH_3} \longrightarrow \underset{\text{acid}_2}{NH_4^+} + \underset{\text{base}_1}{Cl^-}$$

12. HOW DO WE SYMBOLIZE DISSOCIATION OF WEAK AND STRONG ACIDS?

Acids differ enormously in the extent to which they dissociate into ions in aqueous solution. Some acids, such as hydrochloric and nitric acids, are strong electrolytes and completely dissociated into ions; these acids are known as strong acids. Others, such as acetic and nitrous acids, are only partially dissociated in solution. These acids are weak electrolytes and are known as weak acids. While the solution of a strong acid contains no acid molecules, the solution of a weak acid contains both undissociated molecules and ions.

The ions of a weak acid tend to recombine to form molecules of the parent acid. We can show this recombination in the ionization equation with double arrows, which indicate that the reaction is reversible and goes both ways. Some molecules are dissociating into ions and some ions are recombining to form molecules. Acetic acid is a weak acid; we show its ionization with double arrows to indicate that a solution of acetic acid contains both molecules and ions. When using double arrows, it is customary to show a longer or heavier arrow in the direction of the predominant reaction. Thus, in the ionization of a weak acid, in which there are fewer ions than molecules, the longer arrow points toward the molecules. Because the ions of a strong acid do not recombine, we show its ionization with a single arrow, meaning that only the ions of the acid are present in its aqueous solution. For the strong acid nitric acid, the ionization would be shown as:

$$HNO_3 \rightarrow H^+ + NO_3^-$$

In the Bronsted-Lowry system, a strong acid is one that easily donates a proton. The conjugate base of a strong Bronsted-Lowry acid is a weak base, one that has little affinity for a proton. Thus, for hydrochloric acid,

$$\underset{\text{base}_1}{H_2O} + \underset{\text{strong acid}_2}{HCl} \longrightarrow \underset{\text{weak conjugate base}_2}{Cl^-} + \underset{\text{acid}_1}{H_3O^+}$$

A weak acid does not readily donate a proton. Its conjugate base (or anion) is a relatively strong base, one with considerable attraction for a proton. Thus, for acetic acid,

$$\underset{\text{base}_1}{H_2O} + \underset{\text{weak acid}_2}{HC_2H_3O_2} \rightleftarrows \underset{\text{strong conjugate base}_1}{C_2H_3O_2^-} + \underset{\text{acid}_2}{H_3O^+}$$

13. HOW DO WE REPRESENT IONIZATION OF POLYPROTIC ACIDS?

Molecules of some acids, such as sulfuric acid, H_2SO_4 and phosphoric acid, H_3PO_4, have more than one ionizable hydrogen atoms. These acids are called polyprotic acids. The ionization of these acids occurs in steps, with the molecule losing one proton at a time. The ionization of sulfuric acid, a diprotic acid, occurs as follows:

$$H_2SO_4 \rightarrow H^+ + HSO_4^-$$
$$HSO_4^- \rightleftarrows H^+ + SO_4^{2-}$$

We notice that, although the first ionization is that of a strong acid, the second is that of a weak acid. The ionization of a phosphoric acid, a triprotic acid, occurs as follows:

$$H_3PO_4 \rightleftarrows H^+ + H_2PO_4^-$$
$$H_2PO_4^- \rightleftarrows H^+ + HPO_4^{2-}$$
$$HPO_4^{-2} \rightleftarrows H^+ + PO_4^{3-}$$

For sulfuric acid and indeed for all polyprotic acids, the first ionization is much more complete than the subsequent ionizations.

14. HOW DO WE EXPRESS EQUILIBRIA AND REPRESENT EQUILIBRIUM CONSTANTS IN SOLUTIONS OF WEAK ACIDS?

We know that equilibrium exists when two opposing reactions occur at the same rate. For example, in case of evaporation and condensation the equilibrium situation will be

Represented as: $Rate_{evaporation} = Rate_{condensation}$

We also know that there is equilibrium between dissolved and undissolved solute in saturated solutions, in which the opposing reactions are precipitation and dissolution: $Rate_{dissolution} = Rate_{precipitation}$

The ionization of a weak acid is also an equilibrium situation. Here the equilibrium is between molecules and ions. We have implied the existence of this equilibrium with double arrows in the ionization equations for weak acids. In a solution of acetic acid, a weak acid, the two ongoing reactions are dissociation of molecules into ions and recombination of ions into molecules. This equilibrium is shown by the equation

$$HC_2H_3O_2 \rightleftarrows H^+ + C_2H_3O_2^-$$

The opposing reaction rates are: $Rate_{dissociation} = Rate_{recombination\ of\ ions}$

There is another criterion for equilibrium. For a system in equilibrium, a mathematical relationship exists between the concentrations of the components of the equilibrium. This relationship is known as the equilibrium constant, K_{eq}. If the ionization is that of a weak acid, the equilibrium constant is known as the acid dissociation constant and has the symbol K_a. In general terms, for a reaction: $aA + bB \rightleftarrows cC + dD$, the equilibrium constant expression is:

$K_{eq} = [C]^c[D]^d / [A]^a[B]^b$, where the square brackets mean molar concentration. We may note that the numerator of the equilibrium constant expression is the product of the concentrations of the products, each raised to a power equal to its coefficient in the balanced equation for the equilibrium reaction. The denominator is the product of the concentrations of the reactants, each raised to a power equal to its coefficient in the balanced equation for the equilibrium reaction. A weak acid with the formula HA would show in solution the following equilibrium

$$HA \rightleftarrows H^+ + A^-$$

The ionization constant expression for this equilibrium is

$$K_a = [H^+][A^-]/[HA]$$

By convention, these ionization equilibria are always written with the ions as products. Thus, in an ionization constant expression, the concentrations of the ions are always in the numerator, that of the un-ionized molecules in the denominator. Therefore, for acetic acid, with the ionization equilibrium

$$HC_2H_3O_2 \rightleftarrows H^+ + C_2H_3O_2^-$$

the equilibrium constant expression is

$$K_a = [H^+][C_2H_3O_2^-]/[HC_2H_3O_2]$$

We have already pointed out that a weak acid is only slightly ionized and that its solution contains mostly molecules and a few ions. Therefore, the concentration value of the un-ionized molecules in the denominator of the equilibrium constant is much larger than those of the ions in the numerator. Consequently, values of the ionization constants for weak acids are always much less than 1.

15. HOW DO WE REPRESENT THE ACID DISSOCIATION CONSTANTS OF POLYPROTIC ACIDS?

In regard to the acid dissociation constants of polyprotic acids, we have already noted that polyprotic acids ionize stepwise and that each anion is less ionized than its conjugate acid. Each of the ionizations has an equilibrium equation and equilibrium constant. Phosphoric acid has three ionizable hydrogens. The acid molecule ionizes to yield a hydrogen ion and a dihydrogen phosphate ion:

$$H_3PO_4 \rightleftarrows H^+ + H_2PO_4^- \quad K_a = 7.5 \times 10^{-3}$$

The dihydrogen phosphate ion ionizes to yield another hydrogen ion and the monohydrogen phosphate ion:

$$H_2PO_4^- \rightleftarrows H^+ + HPO_4^{2-} \quad K_a = 6.2 \times 10^{-8}$$

Finally, the monohydrogen phosphate ion ionizes to yield another hydrogen ion and the phosphate anion:

$$HPO_4^{2-} \rightleftarrows H^+ + PO_4^{3-} \quad K_a = 2.2 \times 10^{-12}$$

The acid dissociation constant becomes smaller with every ionization step. For acids with more than one ionizable hydrogen, the first dissociation constant is always the largest; the value for each successive dissociation constant decreases. This progression goes along with the prediction that, in each successive ionization process, the anion formed is a stronger base.

Carbonic acid (H_2CO_3) is a solution of carbon dioxide in water. Molecules of carbonic acid are not stable. The mixture of carbon dioxide and water ionizes in stepwise manner like phosphoric acid:

$$CO_2 + H_2O \rightleftarrows H^+ + HCO_3^-$$
$$HCO_3^- \rightleftarrows CO_3^{2-} + H^+$$

Ammonium ion acts as a weak acid, ionizing to ammonia and a hydrogen ion:

$$NH_4^+ \rightleftarrows H^+ + NH_3$$

16. HOW CAN WE CALCULATE THE HYDROGEN ION CONCENTRATION IN ACID SOLUTIONS?

Acid dissociation constants are used to calculate the hydrogen ion concentration in the solution of an acid.

Hydrogen Ion Concentration in Solutions of Strong Acids

Strong acids with one ionizable hydrogen atom are completely ionized in aqueous solution; therefore, the hydrogen ion concentration of these solutions is equal to the molar concentration of the acid.

Hydrogen Ion Concentration in Solutions of Weak Acids

The hydrogen ion concentration of an aqueous solution of a weak acid depends on the value of its acid dissociation constant and is always less than the concentration of the weak acid. The hydrogen ion concentration can be calculated using the value of K_a and the molar concentration of the weak acid. Acetic acid is a weak acid that ionizes according to the equation

$$HC_2H_3O_2 \rightleftarrows H^+ + C_2H_3O_2^-$$

Its acid dissociation constant is

$$K_a = [H^+][C_2H_3O_2^-] / [HC_2H_3O_2] = 1.8 \times 10^{-5}$$

The hydrogen ion concentration of a 1.0 M acetic acid solution can be calculated as follows:

The solution contains 1.0 mol acetic acid in 1.0 L solution. Because acetic acid is a weak electrolyte, only a small fraction of the molecules ionize to hydrogen and acetate ions; most remain as un-ionized acetic acid molecules. Let 'x' stand for the number of moles of acetic acid that ionizes. If x moles ionize, then (1.0 – x) moles of acetic acid remain un-ionized. For x moles of acetic acid that ionize, x moles of H^+ and x moles of $C_2H_3O_2^-$ are formed. The resulting concentrations of acetic acid, hydrogen ion and acetate ion at equilibrium are

$HC_2H_3O_2$	$\rightleftarrows$	H^+	$C_2H_3O_2^-$
1.0 – x		x	x

Substituting these values into the expression for the acid dissociation constant gives:

$$K_a = \frac{[H^+]\,[C_2H_3O_2^-]}{[HC_2H_3O_2]} = \frac{(x)(x)}{1.0 - x} = 1.8 \times 10^{-5}$$

The small value of the acid dissociation constant suggests that the amount of acid dissociated is very small (less than 0.01 M). Using the rules for significant figures in addition and subtraction), we know that the quantity 1.0 – 0.01 expressed to two significant figures is 1.0. If x has a value less than 0.01, it is appropriate to disregard x in the expression (1.0 – x) and change the K_a expression to:

$$\boxed{K_a = x^2/1.0 = 1.8 \times 10^{-5}}$$

Solving this equation we get:

$$x^2 = 1.8 \times 10^{-5} = 18 \times 10^{-6}$$
$$x = 4.2 \times 10^{-3} = [H^+] = [C_2H_3O_2^-]$$

The above information may be given in a tabular form as under:

Table 10.3: Concentrations of species in 1.0 M acetic acid solution

	In calculation	*Calculated value*
Acetic acid molecules	1.0 – x	1.0 M
Hydrogen ions	X	4.2×10^{-3} M
Acetate ions	X	4.2×10^{-3} M

The hydrogen ion concentration in 1.0 M acetic solution is, then, 4.2×10^{-3} M, or 0.0042 M. The number 0.0042 is not significant when subtracted from 1.0, so our simplification of the original equation was justified. If the acid is very dilute, for example 10^{-3} M, or if it is one with a large dissociation constant, such as 10^{-2}, this simplification would not be valid. These calculations emphasize the difference between strong and weak acids. Though the hydrogen ion concentration of a 1.0 M solution of a strong acid is 1.0 M, the hydrogen ion concentration of a 1.0 M solution of a weak acid can be calculated from the acid dissociation constant of the weak acid and is much less than 1.0 M.

17. WHAT IS MEANT BY 'COMMON-ION EFFECT'?

Common-ion effect is decrease in solubility of an ionic salt, i.e., one that dissociates in solution into its ions, caused by the presence in solution of another solute that contains one of the same ions as the salt. In fact the common-ion effect is an example of chemical equilibrium. For example, silver chloride, AgCl, is a slightly soluble salt that in solution dissociates into the ions Ag^+ and Cl^-, the equilibrium state being represented by the equation: AgCl (solid) ~ $Ag^+ + Cl^-$. According to Le Châtelier's principle, when a stress is placed on a system in equilibrium, the system responds by tending to reduce that stress. In the system taken as an example, if another solute containing one of those ions is added, *e.g.*, sodium chloride, NaCl, which supplies Cl^- ions, the solubility equilibrium of the solution will be shifted to remove more Cl^- from the solution, so that at the new equilibrium point there will be less Ag^+ ions in solution and more AgCl will be precipitated out as a solid. We know that the solution of a weak acid contains the equilibrium; HA ⇄ H^+ + A^- and, as long as the equilibrium exists, everything is in balance so that

$$K_a = [H^+][A^-] / [HA]$$

If a substance is added to the solution that changes concentration of one of the species, the system goes out of equilibrium. Then either more molecules dissociate or more ions combine until concentrations again fit the equilibrium constant expression. When this set of concentrations is established, equilibrium is again present, the opposing rates are equal and the concentrations become constant.

It is possible to calculate the new equilibrium concentrations after a change in one of the concentrations. Suppose we have one liter of a solution containing one mole of acetic acid and one mole of sodium acetate. The acetic acid is present as an equilibrium mixture of acetic acid molecules, hydrogen ions and acetate ions. The sodium acetate is present only as ions (that is salts are completely ionized in solution). The acetate ions from both the acetic acid and the sodium acetate participate in the acetic acid equilibrium:

$$HC_2H_3O_2 \rightleftarrows H^+ + C_2H_3O_2^- \quad K_a = 1.8 \times 10^{-5}$$

Although sodium ions are also present in solution, they do not participate in the equilibrium, they do not appear in the equation for the equilibrium and they play no role in determining the concentrations of those substances whose formulas do appear in the equilibrium expression. To calculate the hydrogen ion concentration in this solution, we assume that x equals the concentration of ionized acetic acid molecules and the concentrations at equilibrium are: (1.0 – x) = the concentration of un-ionized acetic acid molecules

x = the concentration of hydrogen ions

1.0 + x = the concentration of acetate ions (the concentration of sodium acetate in the solution plus the acetate ions from the ionization of acetic acid), in the equilibrium equation, the concentrations are:

$HC_2H_3O_2$	⇄	H^+	+	$C_2H_3O_2^-$
(1.0 – x)		x		(1.0 + x)

As in the case of a solution containing only acetic acid, we predict that the concentration of ionized acetic acid and therefore of hydrogen ions, is very small and not significant when added to or subtracted from 1.0 M. Given this assumption, (1.0 +

x) is approximately equal to 1.0 and (1.0 – x) is also approximately equal to 1.0. Thus, the concentrations can be expressed as

$HC_2H_3O_2$	⇄	H^+	+	$C_2H_3O_2^-$
(1.0 – x)		x		1.0

Substituting these values into the expression for the acid dissociation constant for acetic acid and solving gives:

K_a =	$\frac{[H^+][C_2H_3O_2^-]}{[C_2H_3O_2]}$	=	$\frac{(x)(1.0)}{(1.0)}$	= 1.8 X 10^{-5}
x =	1.8 X 10^{-5}			
$[H^+]$ =	1.8 X 10^{-5} M			

The addition of sodium acetate has decreased the hydrogen ion concentration from 4.2×10^{-3} M in 0.1 M acetic acid solution to 1.8×10^{-5} M in a 1.0 M acetic acid solution that is also 1.0 M in acetate ion. This decrease is tremendous. These calculations have shown that an ionic equilibrium such as the ionization of a weak acid can be shifted by the addition of another ionic substance (such as salt) that contains one of the ions present in the equilibrium. This effect is known as the common-ion effect because it is caused by the addition of an ion common to both substances.

18. WHAT IS DIFFERENCE BETWEEN ANHYDRIDES OF INORGANIC AND ORGANIC ACIDS?

An acid anhydride is a chemical compound that reacts with water to form an acid. Anhydrides of inorganic acids are usually oxides of nonmetallic elements. For example, carbon dioxide, CO_2, is the anhydride of carbonic acid, H_2CO_3, nitrogen pentoxide, N_2O_5, is the anhydride of nitric acid, HNO_3, phosphorus pentoxide, P_2O_5, is the anhydride of phosphoric acid, H_3PO_4, sulfur dioxide, SO_2, is the anhydride of sulfurous acid, H_2SO_3 and sulfur trioxide, SO_3, is the anhydride of sulfuric acid, H_2SO_4. Anhydrides of organic acids, like the acids themselves, contain the carbonyl group, CO. Organic anhydrides include acetic anhydride or ethanoic anhydride, $(CH_3C{=}O)_2O$ and benzoic anhydride, $(C_6H_5C{=}O)_2O$.

19. HOW ARE TITRATIONS CONDUCTED AND DESIRED INFORMATION EXTRACTED FROM THEM?

Laboratories, whether medical, industrial, or institutional, are frequently asked to determine the exact concentration of a particular species in a solution. For example, very frequently it is desired to know the concentration of acetic acid in a sample of vinegar and the concentrations of iron, calcium and magnesium ions in a hard-water sample. Such determinations may be made using a quantitative technique known as titration. In a titration, a known volume of a solution of unknown concentration is reacted with, or titrated by, an unknown volume of a solution of known concentration. By knowing the titration volumes and the mole ratio in which the solutes react, the concentration of the first solution can be calculated. The solution of unknown concentration may contain an acid (such as stomach acid), a base (such as ammonia), an ion (such as iodide ion), or any other substance whose concentration must be determined.

In fact titration is gradual addition of an acidic solution to a basic solution or vice versa. Titrations are used to determine the concentration of acids or bases in solution. For example, a given volume of a solution of unknown acidity may be titrated with a base of known concentration until complete neutralization has occurred. This point is called the equivalence point and is generally determined by observing a color change in an added indicator such as phenolphthalein. From the volume and concentration of added base and the volume of acid solution, the unknown concentration of the solution before titration can be determined. Titrations can also be used to determine the number of acidic or basic groups in an unknown compound. A specific weight of the compound is titrated with a known concentration of acid or base until the equivalence point has been reached. From the volume and concentration of added acid or base and the initial weight of the compound, the equivalent weight and thus the number of acidic or basic groups, can be computed.

Instead of adding an indicator to observe the equivalence point, one can construct a graph on which the pH at regular intervals is plotted along one axis and the number of moles of added acid or base at these intervals along the other axis; such a plot is called a titration curve and is usually sigmoid (S-shaped), with the inflection point, where the curve changes direction, corresponding to the equivalence point. From the pH at the equivalence point, the dissociation constant of the acidic or basic group can be determined. If a compound contains several different acidic or basic groups, the titration curve will show several sigmoid-shaped curves like steps and the dissociation constant of each group can be obtained from the pH at its corresponding equivalence point.

20. WHAT ARE REQUIREMENTS FOR ANALYTICAL TITRATIONS?

There are several requirements for analytical titrations; some of them may be mentioned as under:

- The equation for the reaction must be known, so that the stoichiometric ratio can be used in calculations.
- The reaction must be fast and complete.

- When the reactants have combined exactly, there must be a clear-cut change in some measurable property of the reaction mixture. The occurrence of this change is called the endpoint of the reaction.
- There must be a way of measuring accurately the amount of each reactant, whether that reactant is initially in solution or is a solid to be dissolved.

We may discuss these requirements as they apply to a particular titration, that of a solution of sulfuric acid of known concentration with a sodium hydroxide solution of unknown concentration. The balanced equation for this acid-base reaction is:

$$2\,NaOH + H_2SO_4 \rightarrow Na_2SO_4 + 2H_2O$$

Sodium hydroxide, NaOH, dissociates in water to sodium ions, Na^+ and hydroxide ions, OH^-; sulfuric acid, H_2SO_4, first ionizes to form hydrogen ions, H^+ and bisulfate ions, HSO_4^- and then to hydrogen, H^+ and sulfate ions, SO_4^{2-}. The reaction between hydroxide ions and hydrogen ions is rapid and complete; thus, the second requirement for an analytical titration is met. Here we may desire to know 'what clear-cut change in property will occur when the reaction is complete'. Suppose the sodium hydroxide solution is slowly added to the acid solution. As each hydroxide ion is added, it reacts with a hydrogen ion to form a water molecule. As long as unreacted hydrogen ions remain in solution, the solution is acidic. When the number of added hydroxide ions exactly equals the original number of hydrogen ions, the solution becomes neutral. If any extra hydroxide ions are added, the solution becomes basic. Here a pertinent question arises as to how will the experimenter know when the solution becomes basic? We have learned about organic compounds called indicators that have one color in acidic solutions and another in basic solutions. If such an indicator is present in acid-base titration, it changes color when the solution changes from acidic to basic.

21. WHAT ARE ACID-BASE INDICATORS AND HOW DO THEY WORK?

Indicators are species obtained from plants and animals that, by their presence, abundance, lack of abundance, or chemical composition, demonstrate some distinctive aspect of the character or quality of an environment. For example, in places where metal-rich minerals occur at the soil surface, indicator species of plants can be examined to understand the patterns of naturally occurring pollution. Acid-base indicators are organic compounds that, in aqueous solution, exhibit color changes indicative of the acidity or basicity of the solution. Common indicators include p-nitrophenol, which is colorless in the pH range:1 to 5 and yellow in the pH range: 5 to 9; methyl orange, yellow in basic and neutral solutions and reddish below pH 3.7; phenolphthalein, colorless in acid and neutral solutions, pink at about pH 8.5 and purplish at pH 10; and litmus. Most indicators are also used in large amounts for dyeing; small quantities are nonetheless invaluable for use as indicators in chemical laboratories.

An acid-base indicator is not always a synthetic chemical. It is often a complex organic dye that undergoes a change in color when the pH of a solution changes over a specific pH range. Many plant pigments and some natural products are good indicators and synthetic species like phenolphthalein and methyl red are also available and widely used.

Phenolphthalein is an acid-base indicator that is colorless in acidic solutions and pink in basic solutions. If phenolphthalein is added to the original sample of sulfuric acid, the solution will be colorless and will remain so as long as hydrogen ions are in excess. After enough sodium hydroxide solution has been added to react with all of the hydrogen ions, the next drop of base will provide a slight excess of hydroxide ions and the solution will turn pink. Thus, there will be a visible and clear-cut indication of the occurrence of the endpoint. The following three indicators are generally used in acid-base titrations.

Indicator	*Color in Acid*	*Color in Base*
Phenolphthalein	Colorless	Pink
Methyl orange	Red	Yellow
Bromothymol blue	Yellow	Blue

The requirement to measure accurately the volumes of solutions used is met by the use of volumetric glassware, in particular, burets, to measure the volumes of the solutions. The precision of a buret is one part per thousand. The following figure (Fig. 10.6) shows a typical titration setup.

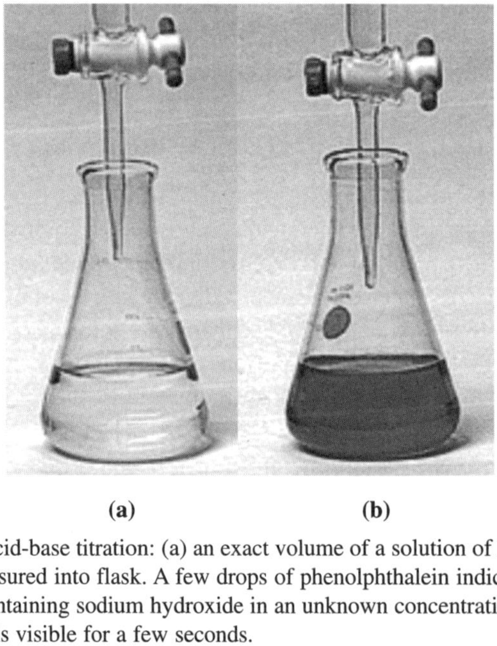

(a) **(b)**

Fig. 10.1: A typical acid-base titration: (a) an exact volume of a solution of known concentration of acid is measured into flask. A few drops of phenolphthalein indicator are added. (b) The solution containing sodium hydroxide in an unknown concentration is added until a faint pink color is visible for a few seconds.

Data and calculations for a typical acid-base titration are shown in Tabular (Table 10.3) form wherein we notice that three trials were run (it is a standard procedure to check the precision of titration). Titration reactions are not always acid-base reactions. They may be oxidation-reduction reactions, precipitation reactions, or combination reactions. The endpoint may be determined by a change in color of an added indicator or by a change in color of the solution when the reaction is complete. In other titrations, the endpoint may be marked by a change in electrical conductivity of the reaction mixture, by the formation of a precipitate, or by a variety of other means.

Table 10.4: Data of the titration of 0.108 M sulfuric acid with a solution of sodium hydroxide of unknown concentration

***Equation*:** 2 NaOH + $H_2SO_4 \rightarrow Na_2SO_4 + 2\ H_2O$
***Molarity of H_2SO_4*:** 0.108 M = 0.108 mmol/mL

	Trial I	*Trial II*	*Trial III*
Volume of 0.108 M H_2SO_4:	25.0 mL	25.0 mL	25.0 mL
Volume of NaOH solution:			
Buret readings:			
At end	34.12 mL	0.64 mL	39.61 mL
At beginning	6.15 mL	35.84 mL	2.34 mL
Volume of NaOH at endpoint:	33.48 mL	33.46 mL	33.50 mL
Arithmetic equation*			
0.180 mmol H_2SO_4	2 mmol NaOH		1
? M NaOH = 25.0 mL H_2SO_4	X	X	X
	1 mmol H_2SO_4	33.48 mL NaOH	1 mL H_2SO_4

Answer: 0.161 M NaOH

* The first three factors in the equation give the millimoles of H_2SO_4 used. Dividing by the volume of acid used gives the molarity (mol/L). The 1 in the last factor is a dimensionless number.

22. WHAT IS LITMUS?

Litmus is a water-soluble mixture of different dyes extracted from lichens, especially Roccella tinctoria. It is a vegetable dye obtained from lichens, usually of the genus Variolaria and used in chemistry to determine the presence of acids and bases in a solution. Strips of paper impregnated with a blue or red litmus solution, or small quantities of the solution itself are used to indicate the presence of an acid or a base. It is often absorbed on to filter paper. The resulting piece of paper or solution with water becomes a pH indicator (one of the oldest), used to test materials for acidity. Blue litmus paper turns red under acidic conditions and red litmus paper turns blue under basic (*i.e.* alkaline) conditions, the color change occurring over the pH range 4.5–8.3 (at 25°C). Neutral litmus paper is purple in color. The mixture contains 10 to 15 different dyes.

Its main use is to test whether a solution is acidic or alkaline. Wet litmus paper can also be used to test water-soluble gases; the gas dissolves in the water and the resulting solution colors the litmus paper. Other reactions can cause a color-change to litmus paper; for instance, chlorine gas turns blue litmus paper white, the litmus paper is bleached. This reaction is irreversible and therefore here the litmus is not acting as an indicator. To find out if a substance is neutral, a blue and red sheet of litmus paper is needed. When the substance is placed on it, the color should remain the same. Universal indicator is a pH indicator that transitions through numbers 3-12 (on the pH chart) to indicate the acidity or alkalinity of solutions. Although there are a number of commercially available universal pH indicators, most are a variation of a formula patented by Yamada in 1923.

23. WHAT IS INDIGO?

Indigo [from Latin =Indian] is an important blue dyestuff used in printing inks and for vat dyeing of cotton. It was anciently produced in India and was known in Egypt, probably c.1600 B.C.; mummies of the XVIII dynasty have been discovered wrapped in indigo-dyed cloth. Indigo is obtained from leguminous plants of the genus Indigofera, chiefly from the Asian species Indigofera tinctoria, but also from several other species. The plants contain a colorless, soluble glucoside called indican. When the macerated plants are allowed to ferment in vats of water, the colorless form of indigo is liberated and stirring of the liquid causes oxidation of the colorless material to form blue sediment. The natural indigo gives a strong blue color of great permanence. Use of the natural dye greatly decreased after the synthesis of indigo was accomplished. Adolf von Baeyer was the first to synthesize it, but others developed the methods used for its commercial production from aniline and chloroacetic acid.

24. WHAT IS MEANING AND SIGNIFICANCE OF HYDROGEN ION CONCENTRATION (pH)?

The term pH is used to indicate the hydrogen ion (positively charged nucleus of hydrogen atom) concentration of a solution, a measure of the solution's acidity. Hydrogen ions are usually represented by the symbol H^+. The term pH (from French pouvoir hydrogène, "hydrogen power") is defined as the negative logarithm of the concentration of H^+ ions: $pH = -\log_{10} [H^+]$, where $[H^+]$ is the concentration of H^+ ions in moles per liter. Because H^+ ions associate with water molecules to form hydronium (H_3O^+) ions, pH also is often expressed in terms of the concentration of hydronium ions. In pure water at 22°C (72° F), H_3O^+ and hydroxyl (OH^-) ions exist in equal quantities; the concentration of each is 1×10^{-7} moles/liter, creating a neutral solution.

Consequently, the pH of pure water is $-\log (1 \times 10^{-7})$, which equals $\log (1 \times 10^{7})$, or 7. If acid is added to water, however, an excess of H_3O^+ ions is formed: H^+ (acid) plus H_2O (water) yields H_3O^+ (hydronium ions). When the concentration of H_3O^+ exceeds the concentration of OH^-, the solution becomes acidic. In an acidic solution, the concentration of hydronium (H_3O^+) ions can range from 1 to 1×10^{-7} moles/liter (but not including 1×10^{-7}), depending on the strength and amount of the acid. Therefore, acid solutions have a pH ranging from 0 up to, but not including, 7. Acids with lower pH values are stronger. Inversely, when the concentration of OH^- exceeds the concentration of H_3O^+, the solution becomes basic/alkaline. In a basic solution, the concentration of hydroxyl (OH^-) ions can range from 1 to 1×10^{-7} moles/liter (but not including 1×10^{-7}). This corresponds to a concentration of hydronium ions that ranges from 1×10^{-14} to (but not including) 1×10^{-7} moles/liter. Therefore, basic solutions can have a pH ranging from 14 down to, but not including, 7. Bases with higher pH values are stronger. The pH of a solution can be measured by titration, which consists of the neutralization of the acid (or base) by a measured quantity of base (or acid) of known concentration, in the presence of an indicator (a compound the color of which depends on the pH). The pH of a solution can also be determined directly by measuring the electric potential arising at special electrodes immersed in the solution.

The term pH is used because the hydrogen ion concentration in solutions of weak acids and in many other fluids is frequently much less than 1. Therefore, when the concentration is expressed exponentially, it contains a negative exponent. Many students find numbers with negative exponents to be confusing and they answer with some hesitation such questions as: Is 1.8×10^{-4} larger or smaller than 3.6×10^{-5}? (To answer the question, we may state both numbers with the same exponent of 10. This restatement changes 3.6×10^{-5} to 0.36×10^{-4}, a value that is clearly less than 1.8×10^{-4}.) To avoid confusion when dealing with small numbers, dissociation constants and ion concentrations are stated not in exponential form, but as the negative logarithms of the actual values. The letter p has been chosen to mean "negative logarithm of." Thus, pH means the negative log of the hydrogen ion concentration and pOH means the negative log of the hydroxide ion concentration:

$$pH = -\log [H^+] \quad pOH = -\log [OH^-]$$

25. WHAT IS IMPORTANCE OF A BUFFER IN BIOLOGICAL SYSTEMS?

In chemistry the term Buffer is used for a solution containing chemical species that inhibit the solution's change in pH, or hydrogen-ion concentration. Such substances consist either of the pairing of a weak acid and a related salt of the acid, or of the similar pairing of a weak base and a salt of the base. Fluids in living organisms are strongly buffered. Sea water and soil substances are other examples of buffers found in nature. Buffers are used in chemistry and serve as references in the measurement of pH.

In fact in chemistry and biochemistry a buffer is a system, usually an aqueous (water) solution that resists change in its pH when an acid or a base is added to it. Normally, the addition of acid to a solution will lower its pH and the addition of a base will raise its pH. If the solution is a buffered one, however, its pH will be changed to a much lesser extent than would be expected from the amounts of added acid or base. In actuality a buffer is a solution that can keep its relative acidity or alkalinity constant, *i.e.*, keep its pH constant, despite the addition of strong acids or strong bases. Buffer solutions are frequently solutions that contain either a weak acid or one of its salts or a weak base and one of its salts. Many acid-base reactions take place in living organisms. However, for organisms to perform certain vital functions, the body fluids associated with these functions must maintain a constant pH. For example, blood must maintain a pH close to 7.4 in order to carry oxygen from the lungs to cells; blood is therefore a powerful buffer.

26. WHAT IS CORRELATION BETWEEN pH AND pK_a?

When the hydrogen ion concentration is stated in exponential notation, the smaller is the exponent, the greater will be the acidity of the solution. Consequently, with pH values, the lower is the pH; the more acidic will be the solution. The pK_a of an acid is the negative logarithm of its acid dissociation constant. Just as pH can be used to describe the hydrogen ion concentration of a solution, pK_a can be used to describe the dissociation constant of a weak acid. The higher the pK_a of an acid, the weaker is the acid. For example, formic acid ($K_a = 1.8 \times 10^{-4}$) is a stronger acid than acetic acid ($K_a = 1.8 \times 10^{-5}$). The pK_a of formic acid is 3.74, a smaller number than 4.74, the pK_a of acetic acid. Notice too that, for the polyprotic acids, the pK_a increases with each ionization. For example, the pK_a for the first ionization of phosphoric acid:

$$H_3PO_4 \rightleftarrows H^+ + H_2PO_4^- \quad pK_a = 2.12$$

is much smaller than that of the second ionization:

$$H_2PO_4^- \rightleftarrows H^+ + HPO_4^{2-} \quad pK_a = 7.21$$

Phosphoric acid is a much stronger acid and therefore much more completely ionized in solution than the dihydrogen phosphate ion.

27. WHAT THE TERM 'Kw' STANDS FOR?

To a very small but very important extent, water is a weak acid that ionizes to hydrogen and hydroxide ions. This equilibrium reaction has the equation:

$$H_2O \rightleftarrows H^+ + OH^-$$

As with all other weak acids, this reaction has an acid dissociation constant expression:

$$K_a = [H^+]\,[OH^+]/[H_2O]$$

In any amount of water, the concentration of water is so high (55.5 mol water molecules in 1000 mL water) and the number of ionized water molecules is so low (1.0×10^{-7} mol in 1000 mL water) that the concentration of the water molecules is a constant. Therefore we use a different expression when considering the ionization of water. This expression is an ion product called K_w, which has the value 1.0×10^{-14}.

$$K_w = K_a[H_2O] = [H^+][OH^-] = 1 \times 10^{-14}$$

The pK_w of water is the negative logarithm of this constant

Therefore pKw = 14

In pure water, the hydrogen ion concentration, $[H^+]$, equals the hydroxide ion concentration, $[OH^-]$. These concentrations can be calculated from the equation for the ionization of water:

$$H_2O \rightleftarrows H^+ + OH^-$$

Let us suppose that x stands for the hydrogen ion concentration, $[H^+]$. Then x also equals the hydroxide ion concentration, $[OH^-]$. Substituting this into the equilibrium expression we obtain

$[H^+][OH^-]$	=	$(x)(x) = 1 \times 10^{-14}$
x^2	=	1×10^{-14}
x	=	1×10^{-7}
$[H^+]$	=	$[OH^-] = 1 \times 10^{-7}$ M

The pH of pure water is 7, the negative logarithm of 1×10^{-7}. A neutral solution is one that is neither acidic nor basic. The hydrogen ion concentration equals the hydroxide ion concentration and both equal 1×10^{-7} M. In a neutral solution, then, pH = pOH = 7.

An acidic solution is one in which the hydrogen ion concentration is greater than the hydroxide ion concentration; in other words, the hydrogen ion concentration is greater than 1×10^{-7} M and the hydroxide ion concentration is less than 1×10^{-7} M. In terms of pH, an acidic solution has a pH less than 7. In order to find the hydroxide ion concentration in an acid solution, we should note that the following relationships exist whenever water is present:

$$[H^+][OH^-] = 1 \times 10^{-14} \text{ and } pH + pOH = 14$$

If the hydrogen ion concentration is known, the first relationship can be used to calculate the hydroxide ion concentration. Suppose the hydrogen ion concentration is 0.10 M. substituting into and rearranging the first relationship gives

$$[OH^-] = 1.0 \times 10^{-14} /0.10 = 1 \times 10^{-13} \text{ M}$$

Suppose the pH of an acidic solution is 1.0. The second relationship can be used to calculate the pOH.

Since 1.0 + pOH = 14, pOH = 14 – 1 = 13

An alkaline or basic solution is one in which the hydrogen ion concentration is less than 1×10^{-7} M. In terms of pH, an alkaline solution is one in which pH is greater than 7.0.

Chapter 11

OXIDATION-REDUCTION (REDOX) REACTIONS

1. WHY REDOX REACTIONS ARE CHARACTERIZED AS COMPLEMENTARY CHEMICAL REACTIONS?

Oxidation-reduction reactions, also known as Redox Reactions, are chemical processes in which electrons are transferred from one atom, ion, or molecule to another. Explosions, fires, batteries and even our own bodies are powered by oxidation-reduction reactions. When iron rusts or colored paper bleaches in the sun, oxidation-reduction process is involved. Originally the term oxidation was used to refer to a reaction in which oxygen combined with an element or compound, *e.g.*, the reaction of magnesium with oxygen to form magnesium oxide or the combination of carbon monoxide with oxygen to form carbon dioxide. Similarly, reduction referred to a decrease in the amount of oxygen in a substance or its complete removal, *e.g.*, the reaction of cupric oxide and hydrogen to form copper and water.

When an atom or molecule combines with oxygen, it tends to give up electrons to the oxygen in forming a chemical bond. Similarly, when it loses oxygen, it tends to gain electrons from oxygen. Such changes are now described in terms of changes in the oxidation number, or oxidation state, of the atoms concerned. In fact oxidation and reduction are complementary chemical reactions characterized by the loss or gain, respectively, of one or more electrons by an atom, ion or molecule. Thus oxidation has come to be defined as a loss of electrons or an increase in oxidation number, while reduction is defined as a gain of electrons or a decrease in oxidation number, whether or not oxygen itself is actually involved in the reaction.

2. HOW ARE OXIDATION-REDUCTION REACTIONS BALANCED?

Oxidation-reduction reactions are most simply balanced in the form of chemical equations by arranging the quantities of the substances involved so that the number of electrons lost by one substance is equal to the number gained by another substance. In such reactions, the substance losing electrons (undergoing oxidation) is said to be an electron donor, or reductant, since its lost electrons are given to and reduce the other substance. Conversely, the substance that is gaining electrons (undergoing reduction) is said to be an electron acceptor, or oxidant. The corrosion of metals is a naturally occurring redox reaction. Industrially, many redox reactions are of great importance:

combustion of fuels; electrolysis in which oxidation occurs at the anode and reduction at the cathode and metallurgical processes in which free metals are obtained from their ores. it may not be out of context to add that in plants osmosis is at least partially responsible for the absorption of soil water by root hairs and for the elevation of the liquid to the leaves of the plant. However, plants wilt when watered with saltwater or treated with too much fertilizer, since the soil around their roots then becomes hypertonic. Each type of atom has a certain number of electrons (which varies from atom to atom) in its elemental form. When an atom forms a bond or otherwise interacts with another atom, it is possible that it will lose or gain some electrons. The oxidation state of an atom is a description of how many electrons it has lost or gained from its original state.

When an iron nail is immersed in a solution of copper(II) ion, a thin coating of copper is deposited on the nail and iron(II) ions are found in the solution. The equation for this reaction may be written as:

$$\underset{0}{Fe(s)} + \underset{+2}{Cu^{2+}(aq)} \longrightarrow \underset{0}{Cu(s)} + \underset{+2}{Fe^{2+}(aq)} \quad \text{.....................} (1)$$

Here we have used a net ionic equation to describe the overall reaction as the essential components of an oxidation-reduction reaction are much more easily identified in a net ionic equation than in a formula equation. Hence we use almost exclusively net ionic equations in discussion of redox reactions. Below each component of the equation its oxidation number has been written. This has been done to stress that a reaction is classified as oxidation-reduction if in its course two elements change oxidation number; one must increase in oxidation number and the other must decrease in oxidation number. we may review the rules for assigning oxidation numbers as under.

Table 11.1: Rules for assigning oxidation numbers to elements

Substance	*Oxidation number*
Any uncombined element	0
A monatomic ion	The charge on the ion
Oxygen (except in peroxides)	–2
A compound	The sum of charges equals zero
A polyatomic ion	The sum of charges equals the charge on the ion

In the reaction whose equation is shown above under equation number (1), iron has been oxidized: its oxidation number has increased from 0 to +2. If an element's oxidation number increases, the element must have lost electrons, which is the criterion for oxidation. Therefore, iron was oxidized in the reaction. The following table (Table 11.2) lists the criteria of oxidation and reduction and states that the substance oxidized is the reducing agent in the reaction. Therefore, iron is the reducing agent in this reaction. Copper(II) ion is reduced in this reaction. It meets the requirements of having decreased in oxidation number and gained electrons. Copper(II) ion is the oxidizing agent in this reaction.

Table 11.2: Criteria for oxidation and reduction

Substance oxidized	*Substance reduced*
Loses electrons	Gains electrons
Attains a more positive oxidation number	Attains a more negative oxidation number
is the reducing agent	is the oxidizing agent

In the balanced equation for this or any other oxidation-reduction reaction, the number of electrons lost by the substance oxidized equals the number of electrons gained by the substance reduced. Electrons can neither be plucked from the air to use in an equation nor left over after a reaction.

3. WHAT ARE HALF-REACTIONS?

Although oxidation and reduction processes proceed simultaneously and an oxidation-reduction reaction can be shown in a single equation, the processes of oxidation and reduction are often shown as separate equations known as half-reactions. The half-reactions for the oxidation of sodium and magnesium are:

$$Na \rightarrow Na^+ + e^-$$
$$Mg \rightarrow Mg^{2+} + 2e^-$$

In these oxidation half-reactions, electrons are found as products. Similarly, we may consider the reduction half-reactions for chlorine and oxygen:

$$Cl_2 + 2e^- \rightarrow 2Cl^-$$
$$O_2 + 4e^- \rightarrow 2O^{2-}$$

It is obvious that in a reduction half-reaction, the electrons are reactants. An oxidation-reduction reaction results from the combination of an oxidation half-reaction with a reduction half-reaction. The reaction of iron with copper(II) ion combines the two half-reactions:

$$\text{Oxidation: } Fe(s) \rightarrow Fe^{2+}(aq) + 2e^-$$
$$\text{Reduction: } Cu^{2+}(aq) + 2e^- \rightarrow Cu(s)$$

The equations have been labeled oxidation and reduction. It is to be noted that electrons are a product (have been lost) in the oxidation half-reaction and are a reactant (have been gained) in the reduction half-reaction. By adding together an oxidation half-reaction and a reduction half-reaction, the net ionic equation for an oxidation-reduction reaction is obtained. This equation will be balanced if the number of electrons in the two half-reactions is equal. Both half-reactions shown above involve two electrons. Adding them together gives the balanced net ionic equation for the overall reaction.

$$Fe(s) + Cu^{2+}(aq) + \cancel{2e} \longrightarrow Fe^{2+}(aq) + \cancel{2e} + Cu(s)$$

If the number of electrons in the two half-reactions is not the same, as, for example, in the reaction of aluminum with hydrogen ion, each equation must be multiplied by an appropriate factor. The unbalanced equation for this reaction is:

$$Al(s) + H^+ \rightarrow Al^{3+} + H_2(g)$$

The half-reactions are:

$$\text{Oxidation: } Al(s) \rightarrow Al^{3+}(aq) + 3e^-$$
$$\text{Reduction: } 2H^+(aq) + 2e^- \rightarrow H_2(g)$$

Three electrons are lost in the oxidation half-reaction, but only two are gained in the reduction half-reaction. Before we can add them to obtain the net ionic equation for the overall reaction, we must multiply the oxidation half-reaction by 2 and the reduction half-reaction by 3. We then obtain

$$\text{Oxidation: } 2Al(s) \rightarrow 2Al^{3+}(aq) + 6e^-$$
$$\text{Reduction: } 6H^+(aq) + 6e^- \rightarrow 3H_2(g)$$

This can be added to give the balanced equation:

$$2Al(s) + 6H^+(aq) + \cancel{6e} \longrightarrow 2Al^{3+}(aq) + \cancel{6e} + 3H_2(g)$$

As an illustrative example we may write the equation for the reaction of zinc with hydrochloric acid by identifying the half-reactions involved and show that the balanced net ionic equation can be obtained from these half-reactions. The equation for the reaction of zinc with hydrochloric acid is: $Zn + 2HCl \rightarrow H_2 + ZnCl_2$

Since zinc's oxidation number changes from 0 to +2, it is oxidized.

$$Zn \rightarrow Zn^{2+} + 2e^-$$

As hydrogen's oxidation number changes from +1 to 0, it is reduced. Because the product hydrogen is a diatomic gas, we must use two hydrogen ions as the reactant.

$$2H^+ + 2e^- \rightarrow H_2$$

Both half-reactions use two electrons; therefore, they can be added as under to give:

$$Zn + 2H^+ + \cancel{2e} \longrightarrow H_2 + Zn^{2+} + \cancel{2e}$$

4. HOW ARE COMPLEX REDOX EQUATIONS BALANCED?

Though the equations for some redox reactions can be balanced by inspection, many redox reactions have equations much too complex to be balanced by inspection without a great deal of frustrating trial and error. Such complex equations can more easily be balanced in a series of steps that utilize the changes in oxidation numbers and the half-reactions involved. In both methods we use ionic equations rather than complete formulas.

Balancing Redox Equations Using Oxidation Numbers

This method of writing balanced equations for redox reactions depends on knowing how oxidation numbers change during the reaction. Each step has been illustrated by developing the equation for the reaction of copper metal with nitric acid to form copper(II) ion and nitrogen dioxide.

***Step-1*:** a skeletal equation is written for the reaction using the formulas of all the ions, atoms and molecules participating in the reaction. For example:

$$Cu + H^+ + NO_3^- \nrightarrow Cu^{2+} + NO_2$$

***Step-2*:** If the reaction occurs in an acid or in a base, hydrogen ion and water molecules are added as reactants or products. For example: the above reaction takes place in an acid. We will leave H^+ as a reactant and add water as a product, knowing we might need H^+ as a product or water as a reactant.

$$Cu + H^+ + NO_3^- \nrightarrow Cu^{2+} + NO_2 + H_2O$$

***Step-3*:** an oxidation number is assigned to each element in the skeletal equation. For example:

$$\underset{0}{Cu} + \underset{+1}{H^+} + \underset{+5,\,-2}{NO_3^-} \nrightarrow \underset{+2}{Cu^{2+}} + \underset{+4,\,-2}{NO_2} + \underset{+1,\,-2}{H_2O}$$

***Step-4*:** For atoms of those elements that change oxidation number, a line is used above or below the equation to connect the reactant form with the product form. Above or below the line the loss or gain of electrons that accompanies the change is added. For example:

$$\overset{-2\,e^-\ (Cu \to Cu^{2+})}{Cu + H^+ + NO_3^- \nrightarrow Cu^{2+} + NO_2 + H_2O}$$
$$+1\,e^-\ (NO_3^- \to NO_2)$$

***Step-5*:** Using multipliers, equate the loss and gain of electrons. For example:

$$\overset{-2\,e^-\ (Cu \to Cu^{2+})}{Cu + H^+ + NO_3^- \nrightarrow Cu^{2+} + NO_2 + H_2O}$$
$$2(+1\,e^-)\ (NO_3^- \to NO_2)$$

***Step-6*:** the coefficient of the substance oxidized or reduced is multiplied by the multiplier used to equate the electron changes. For example,

$$\overset{-2\,e^-\ (Cu \to Cu^{2+})}{Cu + H^+ + 2NO_3^- \nrightarrow Cu^{2+} + 2NO_2 + H_2O}$$
$$2(+1\,e^-)\ (NO_3^- \to NO_2)$$

The oxygen is balanced by changing the number of water molecules present and the hydrogen ions are balanced to match the water molecules. For example, each nitrate ion yields one NO_2 and one O^{2-} so two nitrate ions yield two O^{2-}. These require four hydrogens to form two water molecules, so the coefficient of H^+ is changed to 4. The equation should now be balanced.

$$Cu + 4H^+ + 2NO_3^- \nrightarrow Cu^{2+} + 2NO_2 + 2H_2O$$

***Step-7*:** it is checked that the charge is balanced. For example, in our case the charge on the left is +2; the charge on the right is also +2. The equation is balanced. As an illustrative example let us consider that Potassium permanganate reacts with hydrochloric acid to produce manganese(II) ion and free chlorine. We may write the balanced equation for the reaction using oxidation numbers as under:

***Step-1*:** The skeletal equation for the reaction is written below.

$$MnO_4^- + 2Cl^- \nrightarrow Mn^{2+} + Cl_2$$

Because the product chlorine is diatomic, a multiple of two chlorine atoms must be used in the equation. We prepare for this requirement by using $2Cl^-$ as a reactant.

***Step-2*:** The reaction takes place in an acid; add H^+ as a reactant and water as a product.

$$MnO_4^- + H^+ + 2Cl^- \nrightarrow Mn^{2+} + Cl_2 + H_2O$$

***Step-3*:** The oxidation numbers are:

$$MnO_4^- + H^+ + 2Cl^-\ 0 \nrightarrow Mn^{2+} + Cl_2 + H_2O$$
$$+7, -2\ +1\ -1\ +2\ 0\ +1, -2$$

***Step-4*:** Connect the two forms of the substances that change oxidation number and show electron change.

$$+5\ e^-$$
$$MnO_4^- + H^+ + 2Cl^- \nrightarrow Mn^{2+} + Cl_2 + H_2O$$
$$-2\ e^-$$

***Step-5*:** Equate the loss and gain of electrons.

$$2(+5\ e^-)$$
$$MnO_4 + H^+ + 2Cl^- \nrightarrow Mn^{2+} + Cl_2 + H_2O$$
$$5(+2\ e^-)$$

***Step-6*:** Multiply the coefficients of the substances oxidized or reduced by the multipliers used to equate the electron change.

$$2MnO_4^- + H^+ + 10Cl^- \nrightarrow 2Mn^{2+} + 5Cl_2 + H_2O$$

Balance the oxygen and hydrogen using water and H^+. We have 8 oxygens on the left; we need 8 water molecules on the right. This change will require $16H^+$ on the left. Inserting these numbers gives the balanced equation.

$$2MnO_4^- + 16H^+ + 10Cl^- \rightarrow 2Mn^{2+} + 5Cl_2 + 8H_2O$$

***Step-7*:** Check that the charges are balanced. If they are, the equation is probably balanced too. On the left, –2 + 16 –10 = 4 and on the right it is + 4. The equation is balanced.

As yet another illustrative example, we may write the balanced ionic equation for the reaction of metallic zinc with acidified potassium dichromate solution to yield zinc(II) and chromium(III) ions, using oxidation numbers as under:

***Step-1*:** Write the skeletal equation for the reaction.

$$Zn + Cr_2O_7^{2-} + H^+ \not\rightarrow Zn^{2+} + Cr^{3+}$$

***Step-2*:** The reaction takes place in acid, so we should add H^+ and H_2O as needed. At the same time, we note that the dichromate ion has two chromium atoms, so we will need a multiple of two chromium atoms in the product. To meet this requirement, put the coefficient 2 in front of the chromium(III) ion.

$$Zn + Cr_2O_7^{2-} + H^+ \not\rightarrow Zn^{2+} + 2Cr^{3+} + H_2O$$

***Step-3*:** Assign oxidation numbers

$$\underset{0}{Zn} + \underset{+6,\ -2}{Cr_2O_7^{2-}} + \underset{+1}{H^+} \not\rightarrow \underset{+2}{Zn^{2+}} + \underset{+3}{2Cr^{3+}} + \underset{+1,\ -2}{H_2O}$$

***Step-4*:** Connect the two forms of the substances that change oxidation number and show electron change.

$$\overbrace{Zn + Cr_2O_7^{2-} + H^+ \not\rightarrow Zn^{2+}}^{-2\ e^-}$$
$$Zn + \underbrace{Cr_2O_7^{2-} + H^+ \not\rightarrow Zn^{2+} + 2Cr^{3+}}_{+6\ e^-} + H_2O$$

***Step-5*:** Using multipliers, equate the loss and gain of electrons.

$$\overbrace{Zn + Cr_2O_7^{2-} + H^+ \not\rightarrow Zn^{2+}}^{3(-2\ e^-)}$$
$$Zn + \underbrace{Cr_2O_7^{2-} + H^+ \not\rightarrow Zn^{2+} + 2Cr^{3+}}_{+6\ e^-} + H_2O$$

***Step-6*:** Multiply the coefficients of the oxidized and reduced substances in the equation by the same number used to balance the electrons. Balance the oxygen and hydrogen using water and hydrogen ions. There are 7 oxygen atoms on the left; we will need 7 water molecules on the right, which will require 14 hydrogen ions on the left.

$$3Zn + Cr_2O_7^{2-} + 14H^+ \rightarrow 3Zn^{2+} + 2Cr^{3+} + 7H_2O$$

***Step-7*:** Check that the charges are balanced. On the left, 0 -2 -14 =12; on the right, 3(+2) + 2(+3) = 12. The charges are balanced; the equation is balanced.

5. HOW ARE REDOX EQUATIONS BALANCED BY USE OF HALF-REACTIONS?

The equations for oxidation-reduction reactions can also be balanced using half-reactions. We may illustrate this method using the reaction of nitric acid with copper metal to form nitrogen dioxide and copper(II) ion. The steps using this method are as follows:

***Step-1*:** the skeletal equation for the reaction is written using the ions, atoms and molecules that participate in the reaction. For example:

$$Cu + H^+ + NO_3^- \rightarrow Cu^{2+} + NO_2$$

***Step-2*:** the product and reactant form of each substance is isolated except H^+ that occurs in the skeletal equation. For example,

$$Cu \rightarrow Cu^{2+} \text{ and } NO_3^- \rightarrow NO_2$$

***Step-3*:** these half-reactions are balanced by mass using hydrogen ions and water in acid solutions and hydroxide ions and base in basic solutions. For example,

$$Cu \rightarrow Cu^{2+} \text{ and } 2H^+ + NO_3^- \rightarrow NO_2 + H_2O$$

***Step-4*:** Balance the charges in these half-reactions using electrons as reactants or products. Remember that electrons are negative; they must be added to the more positive side of the equation. Example:

$$Cu \rightarrow Cu^{2+} + 2\,e^-$$

$$2H^+ + NO_3^- + e^- \rightarrow NO_2 + H_2O$$

***Step-5*:** Using multipliers, equate the numbers of electrons in the half-reactions. Example:

$$Cu \rightarrow Cu^{2+} + 2\,e^-$$

$$2(2H^+ + NO_3^- + e^- \rightarrow NO_2 + H_2O)$$

***Step-6*:** Add the multiplied half-reactions. Delete the electrons from both sides of the equation. Example:

$$Cu + 4H^+ + 2NO_3^- + \cancel{2e} \longrightarrow Cu^{2+} + \cancel{2e} + 2NO_2 + 2H_2O$$

***Step-7*:** Check that the equation is balanced by mass and by charge. This equation has on the left 4(+1) + 2(-1) = +2; on the right, +2. The equation is balanced. Notice that, in this method, it is not necessary to determine the oxidation numbers of the various elements present.

As an illustrative example let us write the balanced equation for the reaction of sulfuric acid with iodide ion to form free iodine and hydrogen sulfide gas using half-reactions.

***Step-1*:** Write the skeletal equation for the reaction.

$$H^+ + SO_4^{2-} + I^- \not\rightarrow H_2S + I_2$$

It may be noted that hydrogen sulfide is in the gaseous state; it is therefore not ionic and is written as a molecule.

***Step-2*:** Isolate the half-reactions.

$$H^+ + SO_4^{2-} \nrightarrow H_2S$$
$$I\text{-} \nrightarrow I_2$$

***Step-3*:** Balance these half-reactions by mass using hydrogen ions and water.

$$10H^+ + SO_4^{2-} \nrightarrow H_2S + 4H_2O$$

We have 4 oxygen atoms on the left, so we add $4H_2O$ on the right. They will require 8 hydrogen atoms. The hydrogen sulfide requires 2 hydrogen atoms. Together we need 10 hydrogen ions on the left. The second equation requires only a change of coefficient on the left.

$$2I^- \nrightarrow I_2$$

***Step-4*:** Balance these half-reactions by charge by adding electrons to the more positive side.

$$10H^+ + SO_4^{2-} + 8e^- \nrightarrow H_2S + 4H_2O$$
$$2I\text{-} \nrightarrow I_2 + 2e^-$$

***Step-5*:** Equate the electrons in these half-reactions using multipliers.

$$10H^+ + SO_4^{2-} + 8e^- \nrightarrow H_2S + 4H_2O$$
$$4(2\,I^- \nrightarrow I_2 + 2e\text{-})$$

***Step-6*:** Add the multiplied half-reactions and delete the electrons.

$$10H^+ + SO_4^{2-} + 8e^- + 8I^- \rightarrow H_2S + 4H_2O + 4I_2 + 8e^-$$

***Step-7*:** Check that the charges are balanced. On the left this equation has 10 + (-2) + 8(-1) = 0; on the right, 0. The charges are balanced; the equation is balanced.

As yet another illustrative example let us consider that the reaction of permanganate ion, MnO_4^-, with oxalate ion, $C_2O_4^{2-}$, in acid solution forms manganese(II) ion and carbon dioxide. Using half-reactions, we can write the balanced equation for this reaction.

***Step-1*:** Write the skeletal equation for the reaction.

$$MnO_4^- + H^+ + C_2O_4^{2-} \nrightarrow Mn^{2+} + CO_2$$

***Step-2*:** Isolate the half-reactions.

$$MnO_4^- \nrightarrow Mn^{2+}$$
$$C_2O_4^{2-} \nrightarrow CO_2$$

***Step-3*:** Balance these half-reactions by mass.

$$MnO_4^- + 8H^+ \rightarrow Mn^{2+} + 4H_2O$$
$$C_2O_4^{2-} \nrightarrow 2CO_2$$

***Step-4*:** Balance these half-reactions by charge by adding electrons to the more positive side.

$$MnO_4^- + 8H^+ + 5e^- \nrightarrow Mn^{2+} + 4H_2O$$
$$C_2O_4^{2-} \nrightarrow 2CO_2 + 2e^-$$

***Step-5*:** Equate the electrons in these half-reactions using multipliers.

$$2(MnO_4^- + 8H^+ + 5e^- \rightarrow Mn^{2+} + 4H_2O)$$
$$5(C_2O_4^{2-} \nrightarrow 2CO_2 + 2e^-)$$

***Step-6*:** Add the multiplied half-reactions and delete the electrons.

$$2MnO_4^- + 16H^+ + 10e^- + 5C_2O_4^{2-} \rightarrow 3Mn^{2+} + 8H_2O + 10CO_2 + 10e^-$$

***Step-7*:** Check that the equation is balanced by checking the total charge on each side of the equation: on the left, 2(–1) + 16 + 5(–2) = +4; on the right, 2(+2) = +4.

6. WHAT INFORMATION IS IMPARTED BY REDUCTION POTENTIALS?

Both elements and ions differ in the ease with which they are reduced or oxidized. Hence after each equation in a redox reaction, a number called the reduction potential (symbol, E^o) is used, that compares the tendency for the listed reaction to occur with the tendency of aqueous hydrogen ion to be reduced:

$$2H^+ + 2e^- \rightarrow H_2 \quad E^0 = 0.0 \text{ V.}$$

The reduction potential for the reduction of hydrogen ion to hydrogen has been arbitrarily assigned a value of 0.0 volts (V). A substance whose reduction potential is less than that of hydrogen ion (has a negative value) and is reduced less easily than is hydrogen ion. A substance whose reduction potential is positive (is greater than that of a hydrogen ion) is more easily reduced than is hydrogen. The reduction potentials listed in tables were measured at 25°C and 1 atm pressure, using solutions in which one mole of reactant was dissolved in 1L solution. At different concentrations, temperatures, or pressures, the values of the reduction potentials are slightly different.

Polyatomic ions such as, MnO_4^-, $Cr_2O_7^{2-}$, NO_3^-, etc are often involved in electron-transfer reactions. Reduction potentials for these ions are generally shown in the tabular form. Here one may wonder as to why we show the whole ion instead of just the element concerned? The element is not alone in solution but is part of the polyatomic ion. For example, in the permanganate ion, MnO_4^-, manganese has an oxidation number of +7. Manganese in this state is not an ion but part of a covalently bonded polyatomic ion. The half-reactions show only substances that actually exist in solution.

7. WHY REDUCTION POTENTIAL TABLES ARE USED IN WRITING REDOX EQUATIONS?

We know that an oxidation-reduction equation is the sum of an oxidation half-reaction and a reduction half-reaction. Reduction half-reactions can be obtained from a table of reduction potentials, but what about oxidation half-reactions? These equations can be obtained in a balanced form by reversing the reduction equations of the table. Let us see how equations can be used in balancing redox equations. As an illustrative example we may consider the situation when copper reacts with dilute nitric acid and copper(II) ion and nitrogen oxide (NO) are formed:

$$Cu + HNO_3 \not\rightarrow Cu^{2+} + NO$$

We may write the balanced ionic equation for this reaction as under:

Copper has changed oxidation number from 0 to +2; it has been oxidized. We can find the equation for the reduction of copper ion in the table and reverse it to represent oxidation.

$$Cu \rightarrow Cu^{2+} + 2e^-$$

The nitrogen of nitric acid was reduced in this reaction. We can find the equation for this reduction of nitric acid to nitrogen oxide.

$$NO_3^- + 4H^+ + 3e^- \rightarrow NO + 2H_2O$$

We can balance the loss and gain of electrons by multiplying the oxidation half-reaction by 3 and the reduction half-reaction by 2. Adding the resulting equations gives us the equation:

$$3Cu + 2NO_3^- + 8H^+ + 6e^- \rightarrow 4H_2O + 2NO + 3Cu^{2+} + 6e^-$$

After canceling electrons, we have the balanced equation for this reaction.

8. WHAT REDOX REACTIONS ARE SPONTANEOUS?

A table of reduction potentials allows prediction of the spontaneity of a redox reaction. We know that a spontaneous reaction has a negative free energy change and in such reactions, the sum of the potentials of the half-reactions is positive. The potential of the reduction half-reaction is obtained from a table of reduction potentials. The potential of the oxidation half-reaction is equal in magnitude to the reduction potential of the reversed reduction half-reaction but is opposite in sign. For example, we can use the reduction half-reaction

$$Mg^{2+} + 2e^- \rightarrow MgO \quad E^0 = -2.38 \text{ V}$$

To write the oxidation half-reaction

$$MgO \rightarrow Mg^{2+} + 2e^- \quad E^0 = +2.38 \text{ V}$$

The sum of the potentials of a redox reaction can be calculated as follows. For the redox reaction

$$Fe + Cu^{2+} \rightarrow Fe^{2+} + Cu$$

The half-reactions are:

$$\text{Oxidation: } Fe \rightarrow Fe^{2+} + 2e^-$$
$$\text{Reduction: } Cu^{2+} + 2e^- \rightarrow Cu$$

The E^0 of the reduction half-reaction is obtained from Tables:

$$Cu^{2+} + 2e^- \rightarrow Cu \quad E^0 = +0.34 \text{ V}$$

These tables also give the potential for the reaction

$$Fe^{2+} + 2e^- \rightarrow Fe \quad E^0 = -0.44 \text{ V}$$

If the equation is reversed, we must also reverse the sign of the E^0 for the oxidation half-reaction

$$Fe \rightarrow Fe^{2+} + 2e^- \quad E^0 = +0.44 \text{ V}$$

The sum of the potentials for the overall reaction is:

$$E^0 = -0.34 \text{ V} -0.44 \text{ V} = +0.78 \text{ V}$$

The reaction is spontaneous and will occur without the net input of energy.

We may consider another situation. Suppose a student proposed preparation of hydrogen gas by adding copper metal to hydrochloric acid. Is this proposal sensible? The equation for the proposed reaction is:

$$Cu + 2HCl \rightarrow Cu^{2+} + 2Cl^- + H_2(g)$$

The half-reactions for this equation and their potentials are:

$$\text{Oxidation: } Cu \rightarrow Cu^{2+} + 2e^- \quad E^0 = -0.34 \text{ V}$$
$$\text{Reduction: } 2H^+ + 2e^- \rightarrow H_2(g) \quad E^0 = 0.0 \text{ V}$$

The potential for the reduction half-reaction was obtained from a table. The potential for the oxidation half-reaction was obtained by reversing the equation and sign of the corresponding reduction half-reaction in the table. The sum of the potentials is negative.

$$E^0 = -0.34 \text{ V} + 0.0 \text{ V} = -0.34 \text{ V}$$

The reaction is not spontaneous. The students will not be able to prepare hydrogen gas by this reaction. In summary, a redox reaction is spontaneous if the sum of the potentials of its half-reactions is positive. If the sum of the potentials is negative, the reaction will not occur without the net input of energy.

As another illustrative example we may consider whether or not acidified potassium dichromate solution oxidize chloride ion? The equation for the proposed reaction is:

$$2Cl^- + 14H^+ + Cr_2O_7^{2-} \rightarrow 2Cr^{3+} + Cl_2 + 7H_2O$$

***Solution*:** The half reactions for the equation are:

$$\text{Oxidation: } 2Cl^- \rightarrow Cl_2 + 2e^- \quad E^o = -1.36 \text{ V}$$
$$\text{Reduction: } Cr_2O_7^{2-} + 14H^+ \rightarrow 2Cr^{3+} + 7H_2O \quad E^o = +1.33 \text{ V}$$

The potential of the reduction half-reaction was obtained from the table. The potential for the oxidation of chloride ion was obtained by reversing the sign of the potential for the half-reaction

$$Cl_2 + 2e^- \rightarrow 2Cl^-$$

The sum of these potentials is negative.

$$Eo = -1.36 \text{ V} + 1.33 \text{ V V} = -0.03 \text{ V}$$

Acidified potassium dichromate will not, under standard conditions, oxidize chloride ion.

9. WHAT IS SIGNIFICANCE OF ACTIVITY SERIES?

The activity of an element determines the ease with which the free element becomes an ion. For metals to become ions, they must be oxidized as is evident in case of lithium here: $Li \rightarrow Li^+ + e^-$

This equation is the reverse of the equation shown in case of reduction. A metal is said to be very active (easily oxidized) if the reduction potential of its ion is near the top of the table (that is, has a very large negative value). For example, lithium has a reduction potential of –3.05 V. Lithium is a very active metal and is very easily oxidized to lithium ion. An activity series of the metals lists the metals in order of decreasing ease of oxidation. In terms of reduction potentials, this listing would be in order of increasing reduction potentials. Activity series for metals are shown in tables. This activity series can be used to predict displacement reactions. A metal will displace any metal below it in the series. For example, copper is above silver in the activity series, copper will displace silver, as shown in the equation

$$Cu + 2Ag^+ \rightarrow Cu^{2+} + 2Ag$$

The activity of a nonmetal also depends on the relative ease with which it forms ions. For a nonmetal, the formation of ions involves reduction. The reduction potential of chlorine measures the tendency of the following reaction to occur:

$$Cl_2 + 2e^- \rightarrow 2Cl^-$$

The most active nonmetal has the lowest reduction potential.

10. WHAT ARE ELECTROCHEMICAL CELLS?

We assume that the oxidation-reduction reactions take place in a single container and that electrons are transferred directly from the substance oxidized to the substance reduced. Although a redox reaction may take place in this manner, it is also possible to transfer the electrons through a wire or electrical circuit. Such an arrangement is called an electrochemical cell. An electrochemical cell that utilizes a spontaneous redox reaction to produce a flow of electrons is called either a galvanic cell, after the Italian chemist Luigi Galvani (1737–1798), or a voltaic cell, after Galvani's friend Alessandro Volta (1745–1827). Galvani and Volta were pioneers in the study of the production of electricity by chemical means. In a voltaic cell, the half-reactions of the spontaneous redox reaction are kept separate and the electrons produced flow through an external wire. The following figure shows one such cell.

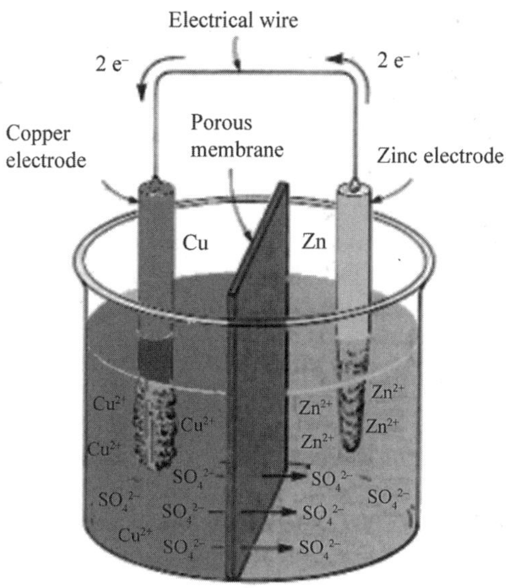

Fig. 11.1: A galvanic cell

The overall reaction of the cell shown is:

$$Zn(s) + CuSO_4(aq) \rightarrow Cu(s) + ZnSO_4(aq)$$

In this apparatus, the two half-reactions are separated from each other by a porous membrane. On one side of the membrane is a solution of zinc sulfate and a piece of copper. A wire connects the zinc electrode and the copper electrode. At the zinc electrode the reaction is:

$$Zn \rightarrow Zn^{2+} + 2\ e^-$$

The electrons produced in this half-reaction move through the wire to the copper-containing solution, where they combine with the copper ions to form copper atoms:

$$Cu^{2+} + 2\ e^- \rightarrow Cu$$

The excess sulfate ions left behind as the copper ions are reduced, pass through the porous membrane to balance the charge of the newly formed zinc ions. The electrons flowing through the wire are an electric current and the arrangement or cell we have just described is a battery. The voltage of the cell can be calculated from the reduction potentials of the two half-reactions. In the cell of the above figure, the half reactions are:

$$Cu^{2+} + 2e^- \rightarrow Cu\ E^0 = 0.34\ V$$
$$Zn \rightarrow Zn^{2+} + 2e^-\ E^0 = 0.76\ V$$

The voltage of the cell (if the concentrations are 1 M) is 1.10 V. Theoretically, any oxidation-reduction reaction can be used in a cell that produces an electrical current. The batteries that we use in flashlights, calculators, radios and so on, produce electricity by various oxidation-reduction reactions.

11. HOW REDOX REACTIONS CAN BE SUMMARIZED?

Reactions in which electrons are transferred from one species to other are oxidation-reduction or redox reactions. In a redox reaction, the substance oxidized increases in oxidation number and thus loses electrons and the substance reduced decreases in oxidation number and thus gains electrons. The substance oxidized is the reducing agent and the substance reduced is the oxidizing agent. Simple redox reactions can be balanced by inspection but, for more complex reactions, either oxidation numbers or half-reactions can be used to aid in balancing. The reduction potential of an element measures its tendency to gain electrons in comparison with the tendency of hydrogen ion to gain electrons. The potential of an oxidation half-reaction is equal in magnitude but is opposite in sign to the potential of the reduction half-reaction found in tables. A redox reaction is spontaneous if the sum of its potentials is positive. An active metal is easily oxidized and an active nonmetal is easily reduced. A spontaneous redox reaction can be used to produce an electric current by physically separating the half-reactions and transferring the electrons through a wire. Non-spontaneous redox reactions are often used in industry to produce chemicals. These processes require the input of large amounts of electrical energy.

Chapter 12

CHEMICAL REACTIONS

1. WHAT ENERGY CHANGES ARE ASSOCIATED WITH REACTIONS?

Energy is the ability or capacity to do work or to produce changes. Forms of energy include heat, light, sound, electricity and chemical energy. Energy and work are measured in the same units: foot-pounds, joules, ergs, etc, depending on the system of measurement being used. When a force acts on a body, the work performed (and the energy expended) is the product of the force and the distance over which it is exerted. Energy is a state function commonly defined as the capacity to do work. Since work is defined as the movement of an object through a distance, energy can also be described as the ability to move an object through a distance. In fact energy is the ability of a system to do work. Important forms of energy include kinetic (motion), potential (location), thermal (heat), gravitational and electrical. In chemistry, energy is required to break and form bonds of molecules. Common units of energy are the calorie and joule. According to the Law of Conservation of Energy, energy can neither be created nor destroyed, though it can be transformed from one form to another. Every reaction has an energy change associated with it. This energy change can be shown either as the enthalpy change (ΔH) or as the free energy change (ΔG). Free energy, G is the energy available to do useful work. These two energy changes are related by the equation: $\Delta G = \Delta H - T\Delta S$, in which T is the temperature (K) at which the reaction is taking place and ΔS is the associated entropy change.

2. WHAT DEDUCTIONS CAN BE MADE FROM ENTROPY CHANGES?

Since entropy measures disorderness, the entropy change accompanying a reaction measures how the reaction affects the orderliness of the system. Other factors being equal, it is the nature of matter to move toward a state of maximum disorderness. We know that when rocks or bricks are dumped from a truck, they do not spontaneously arrange themselves in a neat wall; rather, they land in an untidy pile. If we emptied a bag of red, white and blue jelly beans onto a flat surface, they would not spontaneously settle into a replica of the flag but would become even more disordered than they were in the bag, for they would now be spread over a larger surface. In either case the entropy will increase and energy must be added to decrease the entropy, *i.e.* to arrange the rocks into a neat fence or the jelly beans into a flag.

In our study of chemistry quite often we come across changes in entropy. We know that while the structure of a solid is usually well ordered and that of a liquid is less ordered, structure of a gas is quite random. From these facts, we can deduce that the change from a solid to a liquid is accompanied by an increase in entropy and the same happens in case of a change from a liquid to a gas. When molecules containing many atoms break apart to form smaller molecules, there is also an increase in entropy. The free energy change ΔG associated with a reaction tells whether or not the reaction will occur under the specified conditions.

3. WHAT PREDICTIONS CAN BE MADE FROM EXAMINATION OF THE FOLLOWING EQUATION?

$$\Delta G = \Delta H - T\Delta S?$$

For a reaction to occur spontaneously (that is, without the net addition of energy) and actually produce the products shown in its equation, the free energy change at the specified conditions must be negative. The three factors that play a role in determining the spontaneity of a reaction (that is, the sign of ΔG) are: (1) whether the reaction as written is exothermic ($\Delta H < 0$), (2) the temperature at which it is proposed to run the reaction and (3) the entropy change ΔS associated with the reaction. These factors are related by the equation: $\Delta G = \Delta H - T\Delta S$.

Careful examination of this equation inables us to make the following predictions:

1. If a reaction is exothermic (ΔH is negative) and the entropy ΔS is positive (more disorder), the free energy change is always negative and the reaction is always spontaneous.
2. If a reaction is endothermic (ΔH positive) and the entropy change ΔS is negative (less disorder), the free energy change is always positive and the reaction is never spontaneous.
3. If the enthalpy changes, ΔH and the entropy change, ΔS both are positive or negative; the spontaneity of the reaction depends on the temperature. These predictions may be summarized in tabular form as under:

Table 12.1: Enthalpy, Entropy, Free Energy and spontaneity of a reaction

Enthalpy	*Entropy*	*Free energy*
Exothemic, $\Delta H < 0$	Increased disorder, $\Delta S > 0$	Spontaneous, $\Delta G < 0$
Exothemic, $\Delta H < 0$	Decreased disorder, $\Delta S < 0$	Spontaneity depends on tempeature
endothemic, $\Delta H > 0$	Increased disorder, $\Delta S > 0$	Spontaneity depends on tempeature
Endothemic, $\Delta H > 0$	Decreased disorder, $\Delta S < 0$	Reaction is never spontaneous, $\Delta G > 0$

The phase changes of water illustrate the dependence of some free energy changes on temperature. When ice melts, the change is endothermic (ΔH is positive) and entropy increases (ΔS is positive) because the water molecules lose ordered arrangement of ice crystals. The $T\Delta S$ factor for melting ice at 298 K is numerically larger than ΔH; the free energy change ΔG is then negative, so melting is spontaneous at that temperature. Similarly, the freezing of water at temperatures below 273K is an example of a spontaneous change that is exothermic and accompanied by a decrease in entropy (an increase in order).

4. HOW ARE ACTIVATION ENERGY, REACTION MECHANISM AND REACTION RATE INTERRELATED?

Activation energy, in chemistry, is the minimum amount of energy needed to cause a chemical reaction. A chemical reaction between two species occurs only when an atom, ion, or molecule of one species collides with an atom, ion, or molecule of the other species. Only a fraction of the total collisions results in a reaction, because usually only a small percentage of the species interacting have the minimum amount of kinetic energy, a molecule must possess for it to react. When the reactants collide, they may form an intermediate product whose chemical energy is higher than the combined chemical energy of the reactants. In order for this transition state in the reaction to be achieved, some energy must enter into the reaction other than the chemical energy of the reactants. This energy is the activation energy.

Once the intermediate product, or activated complex, is formed, the final products are formed from it. The path from reactants through the activated complex to the final products is known as the reaction mechanism. (Reaction mechanisms for complex reactions may involve several steps) Because the heat energy of a substance is not uniformly distributed among its atoms, ions, or molecules, some species may carry enough heat energy to react while others do not. If the activation energy is low, a greater proportion of the collisions between reactants will result in reactions. If the temperature of the system is increased, the average heat energy is increased, a greater proportion of collisions between reactants result in reaction and the reaction proceeds more rapidly. A catalyst increases the reaction rate by providing a reaction mechanism with lower activation energy, so that a greater proportion of collisions result in reaction.

The activation energy and rate of a reaction are related by the equation: $k=A\exp(-Ea/RT)$, where k is the rate constant, A is a temperature-independent constant (often called the frequency factor), exp is the function ex, Ea is the activation energy, R is the universal gas constant and T is the temperature. This relationship was derived by Arrhenius in 1899. Because the relationship of reaction rate to activation energy and temperature is exponential, a small change in temperature or activation energy causes a large change in the rate of the reaction. Activation energies are usually determined experimentally by measuring the reaction rate k at different temperatures T, plotting the logarithm of k against 1/T on a graph and determining the slope of the straight line that best fits the points. A reaction that is spontaneous is always accompanied by the net release of free energy (energy available to do useful work). However, some spontaneous reactions first require added energy to get started. The energy they finally release includes both this added energy and the calculated free energy of the reaction. We all know that paper burns in air (the reaction has a negative free energy), but the reaction does not take place until extra energy is added (the heat from a burning match).

That added energy is called activation energy.

5. WHAT ARE REQUIREMENTS OF A REACTION AT THE MOLECULAR LEVEL?

Let us consider what happens at the molecular level when a reaction takes place. Bonds between the atoms in the reacting molecules break and new bonds form to combine the atoms in a different way. For this reaction to occur, the reacting molecules must collide. Together they must have enough kinetic energy (energy of motion) to overcome the repulsion between the clouds of electrons that surround the molecules. As they collide, the two reacting molecules must be oriented so that those atoms that will be bonded together in the product are next to each other. Without this molecular orientation, the molecules will retreat from the collision without reacting as has been depicted in the following figure. A collision that takes place with enough energy and with the correct molecular orientation is called an effective collision. Needless to say that not all collisions are effective.

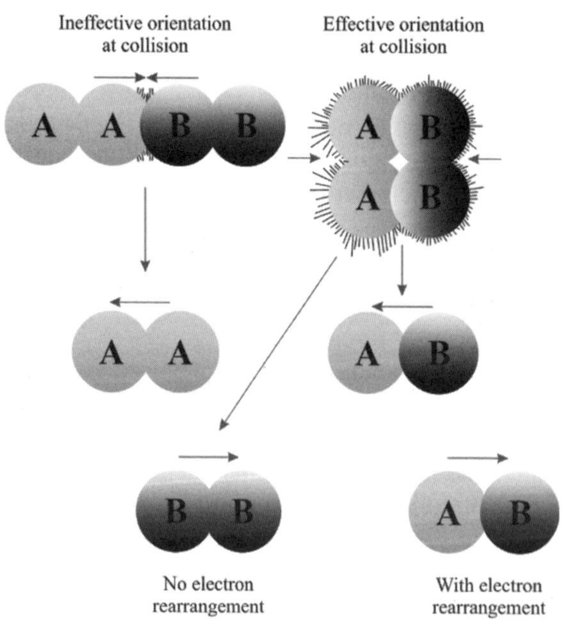

Fig. 12.1: For a reaction between molecules to occur, the molecules must be correctly oriented when they collide.

The Course of a Reaction

Figure 12.2 plots the course of a reaction. The initial average energy of the reactants is indicated at the left side of each graph. If molecules are to collide effectively, they must have more than the average energy. They must have enough energy to overcome the repulsive forces between molecules. This added amount of energy, the activation energy, is the difference between the initial energy and the energy at the peak of each graph. Molecules having sufficient energy can collide and, if they are correctly oriented at collision, their bonds may break and the bonds of the products form. As the new bonds form, energy is released, leaving the product molecules with the average energy shown at the right of the graph.

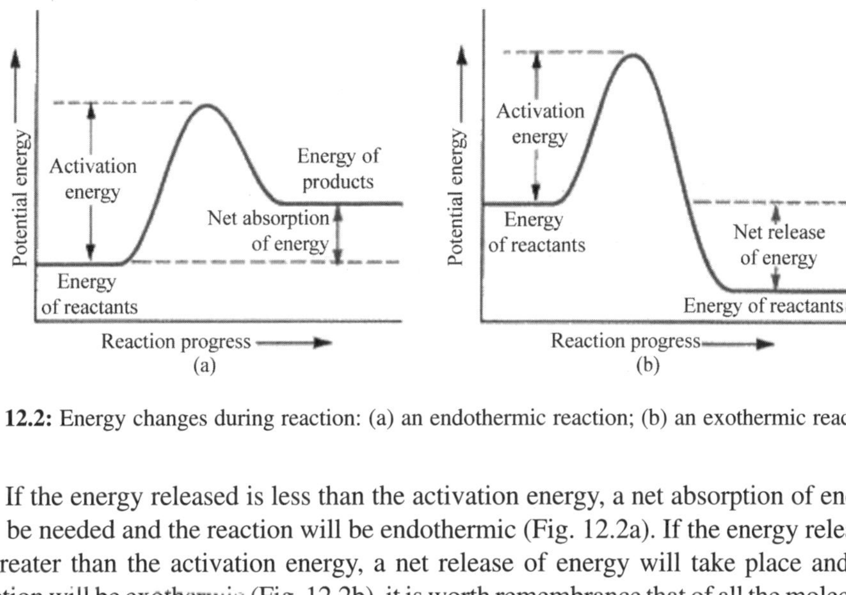

Fig. 12.2: Energy changes during reaction: (a) an endothermic reaction; (b) an exothermic reaction.

If the energy released is less than the activation energy, a net absorption of energy will be needed and the reaction will be endothermic (Fig. 12.2a). If the energy released is greater than the activation energy, a net release of energy will take place and the reaction will be exothermic (Fig. 12.2b). it is worth remembrance that of all the molecules present, only some will collide and of those collisions, only some will be effective and will result in reactions.

This picture of a reaction is analogous to riding a bicycle over a mountain pass. The activation energy of the reaction is comparable to the energy needed to pedal to the top of the pass. The energy released by the rearrangement of bonds is comparable to that gained in coasting down from the top of the pass to the floor of the next valley. If this second valley is higher than the one we started from, the energy gained in coasting down is less than the energy expended in pedaling up. This situation corresponds to an endothermic reaction, which has a net absorption of energy. If the second valley is lower than the one we started from, we gain more energy coasting down than was used pedaling up. This situation corresponds to an exothermic reaction, which results in a net release of energy.

6. WHAT ARE THE WAYS TO CHANGE THE RATE OF A REACTION?

The rate of a reaction measures how fast the concentrations of the reactants decrease or how fast the concentrations of the products increase. The rate of a reaction is different from its spontaneity. In speaking of chemical reactions, spontaneity has no connotation of speed; it refers only to the fact that the reaction will occur with the release of free energy as the equation is written and at the specified temperature. For example, the reaction of hydrogen with oxygen is a spontaneous one.

$$H_2(g) + {}^1/_2O_2(g) \rightarrow H_2O(g);\ \Delta H = -235 \text{ kJ/mol}$$

But a mixture of the two gases can be kept for years at room temperature (298 K) with only imperceptible amounts of water being formed. Only when more energy is added does reaction take place at a measurable rate.

Changing the Rate of a Reaction

A reaction will form products more rapidly if the conditions under which the reaction occurs are changed so that more molecules have enough energy to reach the peak of either of the graphs. There are three ways to increase the size of this set of molecules.

Increase in the concentration of reactant molecules

The more molecules are present in the reaction vessel, the more likely is the probability of a collision. We can increase the number of molecules by increasing the concentration of the reactants. If the reactants are both gases, an increase in pressure decreases the volume and brings the molecules closer together, thus increasing the likelihood of collision.

Increase in the temperature of the reaction

The rate of a reaction will increase if the number of molecules with enough energy to provide the activation energy of the reaction increases. Fig. 12.3 shows the distribution of energies in a collection of molecules at two different temperatures. In this figure, molecules with energy greater than at point A are sufficiently energetic to provide the activation energy necessary for collision. The screened area under each curve represents the number of molecules at that temperature with an energy greater than A. The screened area is much larger under the higher-temperature curve. Therefore, at the higher temperature, more collisions occur and the reaction proceeds faster. At lower temperatures, these results are reversed and the reaction is slower.

7. WHY FOODS STORED IN REFRIGERATORS DO NOT SPOIL?

We store food in a refrigerator because of the effect of temperature on reaction rates. The rates of the reactions that lead to food spoilage are decreased considerably by cooling the food from room temperature to that in a refrigerator. The rates of these

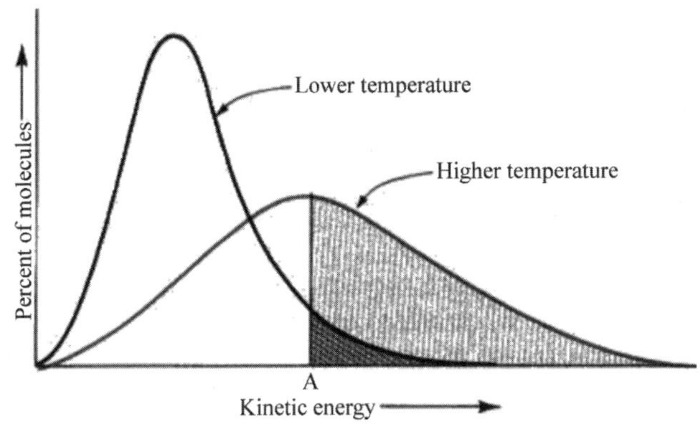

Fig. 12.3: The distribution of energies in a collection of reacting molecules at two different temperatures.

reactions are decreased even further by storing food in a freezer. Recent developments in low-temperature surgery have also resulted from the application of this principle. By cooling the patient, metabolic reactions are slowed and the operation can be performed more deliberately, with fewer traumas to the patient.

8. HOW CAN THE ACTIVATION ENERGY OF A REACTION BE LOWERED?

We know that a certain amount of energy, *viz.* the activation energy, is necessary for a reaction to take place. If the activation energy could be lowered, more molecules would be able to react. How can the activation energy of a reaction be lowered? Just as a pass between the valleys might be lower than that originally used, a pathway for the reaction may have lower activation energy.

A catalyst can provide such an alternative pathway. A catalyst is a substance which, when added to a reaction mixture increases the rate of the overall reaction yet is recovered unchanged after the reaction is complete. Suppose a substance C is added to a reaction mixture. If the formation of the product occurs at a faster rate in the presence of C than in its absence and if C is recovered unchanged, then C is a catalyst for the reaction. Activation energy is still required, but it is less than that of the uncatalyzed reaction. There are many examples of catalysts.

9. WHY VEHICLES SHOULD STOP SPEWING OUT LARGE AMOUNTS OF CARBON MONOXIDE?

Since the mid-1970s, many automobile exhaust systems have been manufactured with catalysts for the reaction:

$$2CO(g) + O_2(g) \rightarrow 2CO_2(g)$$

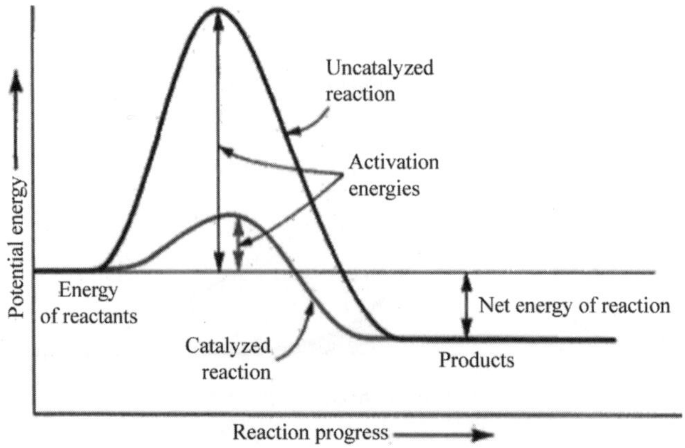

Fig. 12.4: The effect of a catalyst on activation energy.

Since in the absence of a catalyst, the above reaction requires a very high temperature, it does not occur significantly at normal exhaust temperatures. The well-being of the public requires that vehicles stop spewing out large amounts of carbon monoxide. The introduction of a catalyst to the exhaust system of the car makes possible the oxidation of carbon monoxide to carbon dioxide at lower exhaust temperatures, with a considerable improvement in air quality.

10. HOW DO ENZYMES CHANGE THE RATES OF BIOLOGICAL REACTIONS?

The catalysts that trigger biological processes are enzymes that have enormous power to change the rates of chemical reactions. In fact, most of the reactions that occur so readily in the living cell would, in the absence of enzymes, occur too slowly to support life. For example, the enzyme carbonic anhydrase catalyzes the reaction of carbon dioxide and water to form carbonic acid:

$$CO_2 + H_2O \underset{\text{Carbonic anhydrase}}{\rightleftarrows} H_2CO_3$$

Carbonic anhydrase increases the rate of this reaction almost tenfold over that of the uncatalyzed reaction. Red blood cells are especially rich in this enzyme. For this reason, they are able to absorb carbon dioxide as it is produced in the body and transport it back to the lungs, where it is released as one of the waste products of the body. Whether catalysts are inorganic like those in automotive emission-control systems or organic like the enzymes of living system, they are remarkably effective because they provide an alternative pathway for a reaction, one that has a lower activation energy.

11. WHY FLOUR MILLS ARE REQUIRED TO OPERATE UNDER STRICT CONDITIONS?

When a reaction is to take place between reactants in two different physical states, the reaction rate is increased if we increase the surface area of the more-condensed reactant. Such reactions include a gas or a liquid with a solid or a gas with a liquid. We may consider the reaction between oxygen in the air with cellulose, a reaction we call burning. Cellulose is the main component of wood and flour. A match will ignite twigs but will not ignite a large log; the twigs have more surface area. If the cellulose is ground to a fine powder (enormous surface area) as in flour, a spark is sufficient to start a very rapid reaction, that is, an explosion. For this reason, flour mills operate under strict regulations designed to prevent static electricity. Similarly, the reaction between a gas and a liquid will take place more rapidly if the liquid is sprayed in small drops through which the gas passes than if the gas is passed over the surface of a large body of liquid.

Some reactions, classified as photochemical reactions, are very sensitive to light. A mixture of the reactants in such a reaction will be stable in the dark indefinitely. When exposed to light of the correct wavelength, the reaction occurs, often at an explosive rate. The reaction of hydrogen with chlorine is one such reaction.

$$H_2(g) + Cl_2(g) \xrightarrow{\text{Dark}} \text{No reaction}$$

$$H_2(g) + Cl_2(g) \xrightarrow{\text{Light}} 2HCl(g)$$

The decomposition of nitrogen dioxide into nitrogen monoxide and atomic oxygen is another photochemical reaction. Small amounts of nitrogen dioxide are found in the exhaust gases from gasoline engines. Its decomposition on bright sunny days triggers the series of reactions that causes smog.

12. WHY THE DECOMPOSITION OF HYDROGEN PEROXIDE IS CATALYZED BY IODIDE ION?

Because the uncatalyzed reaction is exothermic, the energy of the products will be less than the energy of the reactants. The shape of the graph will be like this:

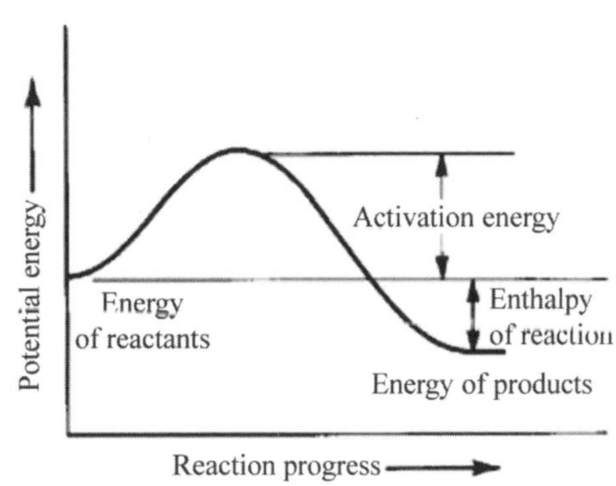

Fig. 12.5(a): Shape of the graph in case of an uncatalyzed reaction

Because the catalyzed reaction is the same reaction as before, the energy of the reactants and the energy of the products will be the same as in the figure above. But, because it is catalyzed, the activation energy will be less as shown here:

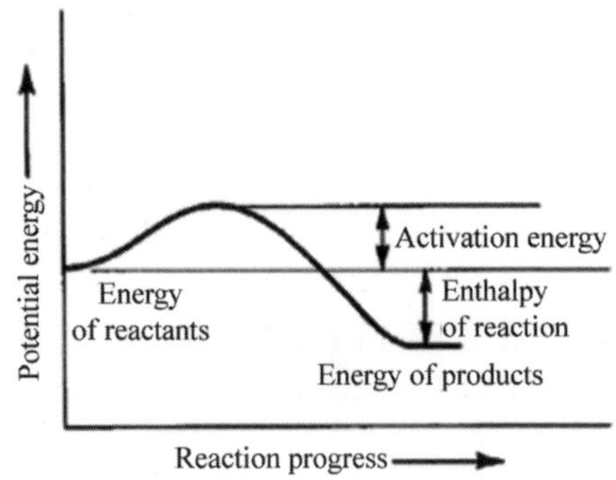

Fig. 12.5(b): Shapes of graphs in case of a catalyzed reaction

13. WHAT HAPPENS WHEN A STATE OF DYNAMIC CHEMICAL EQUILIBRIUM IS REACHED?

Many chemical reactions are reversible; that is, the products of the reaction can combine to re-form the reactants. An example of a reversible reaction is that of hydrogen with iodine to form hydrogen iodide:

$$H_2(g) + I_2(g) \rightleftarrows 2HI(g)$$

We can study this reversible reaction by placing hydrogen and iodine in a reaction vessel and then measuring the concentrations of H_2, I_2 and HI at different times after the reactants are mixed. Fig. 12.6 is a plot of the concentrations of reactants and products of this reaction versus time. The concentration of hydrogen iodide increases very rapidly at first, then more slowly and finally, after the time indicated by the vertical line marked "Equilibrium," remains constant. Similarly, the concentrations of hydrogen and iodine are large at the start of the reaction but decrease, rapidly at first and then more slowly. Finally, they, too, become constant. If this reaction were not reversible, the concentrations of hydrogen and iodine would have continued to decrease and the concentration of hydrogen iodide to increase. This process does not happen. Instead, as soon as any molecules of hydrogen iodide are formed, some decompose into hydrogen and iodine. Two reactions are taking place simultaneously: the formation of hydrogen iodide and its decomposition. When the concentrations of all these components become constant (at the equilibrium point in Fig. 12.6), the rate of the forward reaction ($H_2 + I_2 \rightarrow 2HI$) must be equal to the rate of the reverse reaction ($2HI \rightarrow H_2 + I_2$). A state of dynamic chemical equilibrium has then been reached, one in which two opposing reactions are proceeding at equal rates, with no net changes in concentration.

We have encountered this criterion for equilibrium before. In the equilibrium between a liquid and its vapor, the rate of vaporization is equal to the rate of

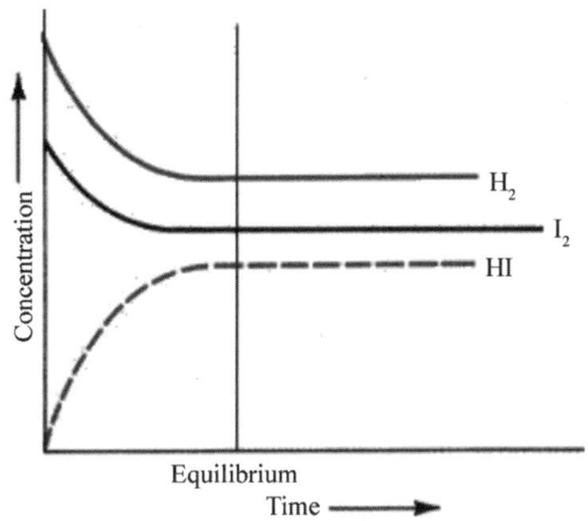

Fig. 12.6: Concentration changes during the reversible reaction $H_2(g) + I_2(g) \rightleftarrows 2HI$ as it proceeds toward equilibrium.

condensation. In the equilibrium of a saturated solution with undissolved solute, the rate of dissolution is equal to the rate of precipitation. In the equilibrium of a weak acid with its ions, the rate of dissociation is equal to the rate of recombination. We may note that none of these reactions is static, rather two opposing changes are occurring at equal rates.

14. WHAT ARE CHARACTERISTICS OF CHEMICAL EQUILIBRIA?

At equilibrium, the rate of the forward reaction is equal to the rate of the reverse reaction and the concentrations of the substances participating in the equilibrium are constant. Although individual reactant molecules may be reacting to form product molecules and individual product molecules may be reacting to re-form the reactants, the concentrations of the reactants and the products remain constant. Free energy or Gibbs free energy is a quantity derived from the relationships between heats and works and used as a measure of the relative stability of a physical or chemical system, *i.e.*, the tendency of the system to react or change. If the change in free energy, ΔG, is negative, the transformation of the system will occur spontaneously, since transitions in which the energy decreases are favored, whereas those in which it increases (ΔG positive) are not. The change in free energy for a given process at a particular temperature depends on three factors, as seen from the equation $\Delta G = \Delta H - T\Delta S$, where ΔH is the change in the enthalpy of the system, T is the temperature in degrees Kelvin and ΔS is the change in entropy. At equilibrium, the free energy change is zero. Neither the forward nor the reverse reaction is spontaneous and neither is favored.

Let consider the ice-water change phenomenon. Above 0°C, ice melts spontaneously to form liquid water; ΔG for this change is negative. Below 0°C, the change from ice to water is not spontaneous; ΔG is positive. At 0°C, the two states are in equilibrium. The rate of melting is equal to the rate of freezing: the amount of ice and water and the amount of liquid water present remain constant and the free energy change is zero as long as no energy is added to or subtracted from the mixture.

15. WHY DO WE OMIT THE CONCENTRATION OF A PURE SOLID OR A PURE LIQUID IN THE EQUILIBRIUM CONSTANT EXPRESSION?

We know the mathematical relationship between the concentrations of the components of an equilibrium, known as the equilibrium constant, K_{eq}. We also know that, for the general equation of a reversible reaction

$$aA + bB \rightleftarrows cC + dD$$

the equilibrium constant expression is

$$K_{eq} = [C]^c[D]^d / [A]^a[B]^b$$

where the brackets mean concentration in mol/L. Therefore, for the reversible reaction

$$H_2 + I_2 \rightleftarrows 2HI$$

the equilibrium constant expression is

$$K_{eq} = [HI]^2 / [H_2]\,[I_2]$$

In each of these expressions, the concentrations of the products of the reaction, each raised to a power equal to the coefficient of that product in the balanced equation for the reaction, are multiplied in the numerator. The concentrations of the reactants in the equation, each raised to a power equal to the coefficient of that reactant in the balanced equation, are multiplied in the denominator.

When we write an equilibrium constant expression, we omit the concentration of a substance that is a pure solid or a pure liquid. The concentration of a pure liquid or solid is so great that it remains essentially constant during a reaction. We tacitly admit this fact by omitting water from the equation for the ionization of a weak acid and from the water ion product expression, K. The equation for ionization of acetic acid is given as:

$$HC_2H_3O_2 \leftarrow\!\!\dashv H^+ + C_2H_3O_2^-$$

$$K_A = [H^+][C_2H_3O_2^-] / [HC_2H_3O]$$

even though water is required for the ionization to occur. Similarly, for the ionization of water:

$$H_2O \rightleftarrows H^+ + OH^- \; K_w = [H^+][OH^-]$$

the concentration of water molecules was not included in the constant K_w. Applying this rule to an equilibrium involving solids:

$$CaCO_3(s) \rightleftarrows CaO(s) + CO_2(g) \; K_{eq} = [CO_2]$$

16. WHAT SHOULD BE THE EQUILIBRIUM CONSTANT EXPRESSION FOR THE REACTION BETWEEN GASEOUS NITROGEN AND HYDROGEN?

1. The numerator of the constant should contain the product NH_3 enclosed in brackets to represent concentration and raised to the second power, because 2 is the coefficient in the equation $[NH_3]^2$
2. The denominator should include the reactants of the equation, N_2 and H_2, enclosed in brackets. The nitrogen term is to the first power; the hydrogen term is raised to the third power: $[N_2][H_2]^3$
3. The complete expression should be

$$Keq = [NH_3]^2/[N_2][H_2]^3$$

17. HOW SHOULD WE WRITE THE EQUILIBRIUM CONSTANT EXPRESSION FOR THE FOLLOWING REACTION?

$$Fe_2O_3(s) + 3H_2(g) \rightleftarrows 3H_2O(g) + 2Fe(s)?$$

Both Fe_2O_3 and Fe are solids, so their concentrations do not appear in the equilibrium constant expression. The numerator of the expression will be $[H_2O]^3$, the denominator will be $[H_2]^3$. The expression is:

$$Keq = [H_2O]^3 / [H_2]^3$$

The value of an equilibrium constant does not depend on how equilibrium was reached. The value of the equilibrium constant does depend on how the equation for the equilibrium is written. The value of equilibrium constant does change with a change in temperature. The equilibrium constant is a very useful concept, for it allows the prediction and calculation of the concentrations of the various species present in a reaction mixture at equilibrium. This calculation is important in determining the pH of a solution, the solubility of a sparingly soluble salt, how far a reaction goes toward completion and other similar data.

18. WHAT IS LE CHATELIER'S PRINCIPLE?

A system in equilibrium is a special situation where everything is in balance. Things rarely stay in balance; changes occur that shift the balance and the equilibrium that was present. We may recall the equilibrium that exists between a liquid and its vapor in a closed container:

Liquid dvapor

At a given temperature, the vapor has a particular pressure; if the temperature is increased, it has a higher pressure. Increasing the temperature causes the equilibrium to shift to the right toward a higher concentration of vapor, but, if the system is maintained at that higher temperature, equilibrium will again be established. We have the idea of the common-ion effect and we know that, when acetate ion is added to a solution of acetic acid, the hydrogen ion concentration decreases. The addition of acetate ion causes the equilibrium:

$$HC_2H_3O_2 \rightleftarrows H^+ + C_2H_3O_2^-$$

to shift to the left. The original equilibrium is upset by the addition of the acetate ion. When a new equilibrium is established, there are more acetate ions, fewer hydrogen ions and more acetic acid molecules. But remember, both before and after the addition of acetate ions, the concentrations of acetic acid, hydrogen ion and acetate ion are related by the equilibrium constant

$$K_a = [H^+][C_2H_3O_2^-]/[HC_2H_3O_2]$$

It is possible to predict how a particular stress or change in conditions will affect an equilibrium. Such predictions are based on a principle first stated by the French chemist Henri Le Châtelier (1850–1936). In 1888, Le Chatelier's Principle was proposed as follows:

"When a stress or change in conditions is applied to a system in equilibrium, the system shifts to absorb the effect of that stress. In considering this principle, it is important to realize that equilibrium is not present while the change is taking place. The sequence is: The system is in equilibrium, the stress is applied and the system changes to absorb the stress and finally equilibrium is again reached.

19. HOW DO CONCENTRATION CHANGES AFFECT EQUILIBRIA?

If the stress applied to a system in equilibrium is a change in the concentration of a component of the equilibrium, the system shifts to counteract that change. If the concentration of a substance is increased, the reaction that consumes that substance is favored and the equilibrium shifts away from that substance. If the concentration of a substance is decreased, the reaction that produces that substance is favored and the equilibrium shifts toward that substance.

It may be elaborated that equilibrium is a combination of two reactions, one forward and one reverse. Changing the concentration of a reactant in a reaction equation, changes the rate of that reaction. When the concentration of a component in equilibrium is increased, the rate of the reaction in which that substance is a reactant is increased and more of the product of that reaction is formed. When the concentration of a component is decreased, the rate of the reaction that uses that substance is decreased and less of its product is formed. In studying the common-ion effect, we may calculate the effect of a concentration change on the equilibrium between acetic acid and its ions:

$$HC_2H_3O_2 \rightleftarrows H^+ + C_2H_3O_2^-$$

In 1.0 M acetic acid, the concentrations of both hydrogen and acetate ions are calculated to be 4.2×10^{-3} M. We then raise the concentration of acetate ions to 1.0 M by the addition of solid sodium acetate. Now question arises as to how this change affects the equilibrium? The reaction that consumes acetate ion is the one toward the left, i.e. toward the formation of acetic acid molecules. Every acetic acid molecule formed uses up a hydrogen ion, which decreases the concentration of hydrogen ions in the newly established equilibrium and increases the concentration of acetic acid molecules. We may calculate the concentrations in the new equilibrium and find that the hydrogen ion concentration has decreased from 4.2×10^{-3} M to 1.8×10^{-5} M. The equilibrium has shifted to absorb the effect of the stress caused by the increase in the concentration of acetate ions. As an illustrative example we may consider the equilibrium:

$PCl_3(g) + Cl_2(g) \rightleftarrows PCl_5(g)$ and write the equilibrium constant expression for this reaction predicting how the equilibrium position will be affected by a) an increase in concentration of PCl_3 and b) a decrease in concentration of Cl_2.

The equilibrium constant expression for this reaction is: $Keq = [PCl_5] / [PCl_3][Cl_2]$.

The reaction that consumes phosphorus trichloride is the forward reaction. Increasing the concentration of PCl_3 will increase the rate of this reaction, decrease the concentration of chlorine and increase the concentration of phosphorus pentachloride. The equilibrium will shift toward right. The reaction that produces chlorine is the one toward the left (the reverse reaction). If the concentration of chlorine is reduced, the rate of the reverse reaction will increase. The equilibrium will shift to produce chlorine and phosphorus trichloride. When equilibrium is reestablished, there will be less phosphorus pentachloride and more phosphorus trichloride present. Thus, decreasing the concentration of chlorine will decrease the rate of the forward reaction and the equilibrium will shift to left.

20. WHAT ARE THE EFFECTS OF PRESSURE CHANGES ON EQUILIBRIA INVOLVING GASES?

A pressure change is a stress to those equilibria that involve gases, that is, those equilibria that have different numbers of gaseous molecules on the left and right sides of the equilibrium equation. Increased pressure favors the reaction that decreases the number of gaseous molecules. In the following equilibrium:

$N_2(g) + 3H_2(g) \rightleftharpoons 2NH_3(g)$, the forward reaction produces two molecules of gas, whereas the reverse reaction produces four molecules of gas. In terms of moles, there are four moles of gas on the left of this equilibrium and two on the right. We know that the volume of a gas is independent of the composition of the gas; that is, at the same temperature and pressure, one mole of any gas behaving ideally occupies the same volume as one mole of any other gas. Thus, the gases on the left occupy a total of four volumes and those on the right occupy two volumes. Increased pressure decreases the volume available to this gaseous equilibrium and favors the forward reaction, because the forward reaction decreases the number of gaseous molecules. Thus an increase in pressure on this equilibrium will favor the formation of more ammonia.

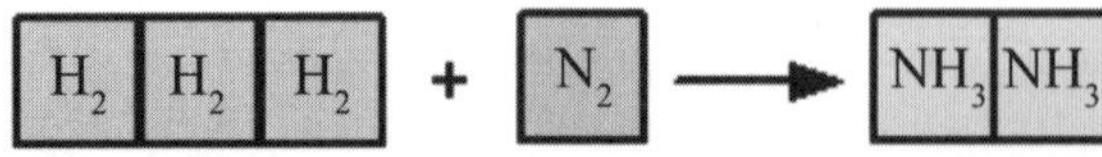

21. HOW DO TEMPERATURE CHANGES AFFECT EQUILIBRIA?

A change in temperature is in fact a stress on a system in equilibrium. It changes the rate of both forward and reverse reactions and at the same time changes the value of the equilibrium constant. In every equilibrium, two reactions, one forward and one reverse, proceed simultaneously. One of these is endothermic ($\Delta H > 0$) and one is exothermic ($\Delta H < 0$). When the equilibrium is shown as an equation, the enthalpy term ΔH refers to the forward reaction. For example, in the hydrogen iodide equilibrium, the forward reaction is exothermic, as shown by the equation:

$$H_2(g) + I_2(g) \rightleftarrows 2HI(g) \ \Delta H = -51.0 \text{ kJ}$$

When the temperature of an equilibrium mixture is increased, the rates of both reactions increase, but the rate of the endothermic reaction (the reaction that absorbs the added energy) is increased more. For the hydrogen iodide equilibrium, an increase in temperature favors the endothermic reverse reaction. When the system returns to equilibrium, the hydrogen iodide concentration will be smaller and the concentrations of hydrogen and iodine will be larger. The equilibrium constant will be changed:

At 445°C, $K_{eq}=[HI]^2/[H_2][I_2] = 64$ and at 490°C, $K_{eq} = [HI]^2/[H_2][I_2] = 45.9$

As an illustrative example we may consider as to how an increase in temperature will affect the following equilibrium?

$$PCl_3(g) + Cl_2(g) \rightleftarrows PCl_5(g) \ \Delta H = -92.5 \text{ kJ}$$

Since in this equilibrium, the forward reaction (to form phosphorus pentachloride) is exothermic and the reverse reaction (to consume phosphorus pentachloride) is endothermic, an increase in temperature favors the endothermic reaction and the equilibrium will shift to the left to absorb the added energy and produce more phosphorus trichloride.

22. HOW DO CATALYSTS AFFECT EQUILIBRIA?

A catalyst changes the rate of a reaction by providing an alternative pathway with a lower activation energy. The lower-energy pathway is available to both the forward and the reverse reactions of the equilibrium. The addition of a catalyst to a system in equilibrium does not favor one reaction over the other. Instead, it increases equally the rates of both the forward and the reverse reactions. The rate at which equilibrium is reached is increased, but the relative concentrations of reactants and products at equilibrium and hence the equilibrium constant, are unchanged.

23. WHAT IS USEFULNESS OF SOLUBILITY CONSTANT?

When a sparingly soluble salt is added to water, a small amount dissolves until the solution is saturated at that temperature. Equilibrium is then established between the undissociated ions of the solid and the ions present in solution. For example, when a saturated solution of calcium sulfate is prepared, the undissolved solute is in equilibrium with dissolved ions: $CaSO_4(s) \rightleftarrows Ca^{2+} + SO_4^{2-}$

The equilibrium constant for this equation is

$$K_{sp} = [Ca^{2+}][SO_4^{2-}] = 2.45 \times 10^{-5}$$

We notice that solid calcium sulfate is not included in the equilibrium constant expression for obvious reasons. To emphasize this omission, as well as the fact that this constant is for an ion product rather than a true equilibrium constant expression, the constant is known as the solubility product constant, K_{sp}.

If the solubility of a substance is known, its solubility product constant can be calculated. For example, we can calculate the solubility product constant for silver chloride knowing that the solubility of silver chloride is 6.56×10^{-4}g/L. For silver chloride, Ksp is $[Ag^+][Cl^-]$, where the brackets mean concentration in mol/L. For every mole of silver chloride dissolved, one mole of silver ion and one mole of chloride ion are found in solution. Therefore, the first step in solving the problem is to convert the solubility from g/L to mol/L. The formula weight of silver chloride is 107.8 + 35.4 = 143.3. The solubility of AgCl is 6.56×10^{-4}g /1 Lx 1 mol /143.3 g = 4.58 $\times 10^{-6}$ mol/L. Therefore, $[Ag^+] = [Cl^-] = 4.58 \times 10^{-6}$ mol/L. Substituting this value into the solubility product expression gives: $K_{sp} = [Ag^+][Cl] = (4.58 \times 10^{-6})(4.58 \times 10^{-6}) = 20.9 \times 10^{-12} = 2.09 \times 10^{-11}$

Similarly if the solubility product constant of a sparingly soluble salt is known, its solubility can be calculated. Since the value of the solubility product constant of a salt depends on the solubility of the salt, its usefulness is shown by the fact that any sparingly soluble salt will precipitate from a solution in which the product of the concentration of its ions, each raised to the power shown in the solubility product constant expression, is greater than the solubility product constant. For example, the solubility product constant of silver bromide is 5.2×10^{-13}. Whenever $[Ag+][Br^-]$ is greater than 5.2×10^{-13} in a solution, silver bromide will precipitate. The two concentrations need not be equal. Silver bromide will precipitate out when

$$[Ag^+] = 0.1 \text{ M}, [Br^-] = 10^{-11} \text{ M because } [Ag^+][Br^-] = 10^{-12}$$

$$[Ag+] = [Br^+] = 7.4 \times 10^{-7} \text{ M because } [Ag^+][Br^-] = 5.4 \times 10^{13}$$

$$[Ag^+] = 10^{-10} \text{ M}, [Br^-] = 6 \times 10^{-2} \text{ M because } [Ag^+][Br^-] = 6 \times 10^{-12}$$

In each case, $[Ag^+][Br^-]$ is greater than 5.2×10^{-13}. In a saturated solution of silver bromide, $[Ag^+][Br^-] = 5.2 \times 10^{-13}$ and $AgBr(s) \rightleftarrows Ag^+Br^-$

If more bromide ions are added to the solution, the equilibrium will shift to the left, forming more precipitate and decreasing the amount of silver ion in solution. Conversely, if bromide ions are removed from solution, the precipitate will redissolve until the product of $[Ag^+]$ and $[Br^-]$ again equals 5.2×10^{-13}.

24. HOW EQUILIBRIA OF WEAK ACIDS RESPOND TO CONCENTRATION CHANGES IN BUFFER SOLUTIONS?

We know about the equilibria between the molecules and ions of weak acids and how such equilibria respond to the stress of changing concentrations. Another interesting and important example of how such equilibria respond to concentration changes comes from the study of buffers. A buffer is a system that resists change. A chemical buffer system is one that resists change in hydrogen ion concentration (pH). Buffers are important because many chemical reactions, particularly those in biological systems, proceed best at a particular pH. If the reaction takes place in a solution that remains at that pH throughout the reaction, the most satisfactory results will be obtained.

A chemical buffer system that is designed to resist changes in hydrogen ion concentration will contain a weak acid and its conjugate base, both in such concentrations as will give the solution the desired pH. For example, if a buffer of pH 4.74 is needed, a solution that is 0.1 M in both acetic acid and acetate ion would be suitable. This buffer of pH 4.74 contains the following equilibrium:

Notice that the concentrations of both the acid and its conjugate base are much larger than that of the hydrogen ion. From our study of equilibria, we know that, if an acid (H^+) is added to this solution, the equilibrium will shift to the left because the reverse reaction consumes the added hydrogen ions. If, on the other hand, hydroxide ion or some other strong base is added to the solution, that base will react with the hydrogen ions present and more acid will ionize to replenish the supply and keep the system in equilibrium. In the buffer we describe, if these added amounts are small, the concentrations of the acetic acid molecules and of acetate ion are sufficiently large that the pH of the solution will remain essentially constant. Since the addition of a small amount of H^+ to a very dilute solution of a strong acid causes a large change in pH, in choosing the buffer for a particular reaction, the following criteria should be considered.

$HC_2H_3O_2$	$\rightleftarrows$	H^+	+	$C_2H_3O_2^-$
0.1 M		1.8×10^{-5} M		0.1 M

1. ***pH:*** Buffering is most effective if the pH required for the reaction is close to the pK_a of the weak acid of the buffer. An effective buffer has a pH that equals the pK_a of the weak acid ±1 unit. For acetic acid, the pK_a is 4.74; therefore, a solution of acetic acid and sodium acetate will function as a buffer within a pH range of approximately 3.74–5.74.

2. ***Concentration*:** The concentration of a buffer refers to the total concentration of both the weak acid and the anion of the weak acid. A 0.1 M acetic acid-acetate ion buffer may be made up of 0.025 mol acetic acid and 0.075 mol sodium acetate in a liter of solution, or any other combination in which $[HC_2H_3O_2] + [C_2H_3O_2^-] = 0.1$ M.
3. ***Capacity*:** The capacity of a buffer, or its effectiveness, is the amount of hydrogen or hydroxide ion that the buffer can absorb without a significant change in pH. Capacity depends on both concentration and pH. The most effective buffer is one with equal concentrations of a weak acid and its salt, that is, one in which the pH of the solution is equal to the pK_a of the weak acid. Although a solution of acetic acid and sodium acetate will buffer within the pH range 3.74–5.74, it is most effective as a buffer at, or very near, pH 4.74. Some common buffers and the pH at which each is most effective may be listed in a tabular form as under.

Some common buffers and the pH at which each is most effective

Name of acid	*Formula*	K_a	*Anion*	*pH of buffer when (Anion)/[A] = 1*
Phosphoric	H_3PO_4	7.5×10^{-3}	H_2PO_4	2.12
Formic	H_2CO_2	1.8×10^{-4}	HCO_2	3.74
Acetic	$HC_2H_3O_2$	1.8×10^{-5}	$C_2H_3O_2$	4.74
Carbonic	$CO_2 + H_2O$	4.3×10^{-7}	HCO_3	6.37
Dilhydrogen phosphate ion	$H_2PO_4^-$	6.2×10^{-8}	HPO_4^{2-}	7.21
Bicarbonate ion	HCO_3^-	5.6×10^{-11}	CO_3^{2-}	10.25
Monohydrogen phosphate ion	HPO_4^{2}	2.2×10^{-13}	PO_4^{3-}	12.66

25. HOW CAN WE SUMMARIZE ALL ASPECTS OF A CHEMICAL REACTION?

In order to react, two species must collide and their valence electrons concerned should rearrange themselves in a way that old bonds are broken and new bonds formed. For a reaction to occur spontaneously, the enthalpy of the reaction, the temperature at which it is to occur and the entropy change must be such that the free energy of the reaction as calculated by the equation $\Delta G = \Delta H - T\Delta S$, is less than zero. Often activation energy must be added to initiate the reaction. The course of a chemical reaction can be plotted against the energy of the reactants and products. This plot shows the activation energy necessary for an effective collision between reactants, as well as the exothermic or endothermic nature of the reaction. The number of collisions between reactants can be increased by increasing their concentrations. Increasing the temperature of a reaction will increase the number of molecules that are sufficiently energetic to collide effectively. The addition of a catalyst provides a reaction pathway that requires lower activation energy. Catalysts in biological systems (enzymes) are particularly important as they allow complex reactions to occur at body temperature.

Many chemical reactions are reversible. Allowed to proceed spontaneously, they reach an equilibrium state in which the rate of the forward reaction equals the rate of the reverse reaction. At equilibrium, a relationship exists between the concentrations of the components of the reaction known as the equilibrium constant. Ionic equilibria may involve sparingly soluble salts. The equilibrium constant for the equilibrium between a sparingly soluble salt and its ions in solution is known as the solubility constant, K_{sp}. The concentration of the undissolved salt does not appear in this constant. Ionic equilibria also involve weak electrolytes. A buffer solution resists change in properties concerned. An acid-base buffer resists changes in pH. Buffers, prepared from weak acids or bases and their salts, are particularly effective within one pH unit of the pK_a of the acid.

Chapter 13

ATTRACTIVE FORCES BETWEEN CHEMICAL SPECIES

1. WHAT ARE POSTULATES OF KINETIC MOLECULAR THEORY?

Following are the five postulates of the kinetic molecular theory of gases:

1. The volume occupied by a gas is largely empty space; by comparison, the real volume of the gas molecules is negligible.
2. There are no appreciable attractive forces between the molecules of a gas.
3. The gaseous molecules are in constant, random, rapid, straight-line motion.
4. The molecules collide frequently with one another and with the walls of the container, but no energy is lost by these collisions.
5. The average kinetic energy of the molecules is proportional to the temperature of the sample.

However, these postulates do not hold completely for liquids and solids. We also know that when the average kinetic energy of a gas sample decreases, the sample eventually reaches a temperature at which this kinetic energy is no longer great enough to overcome the attractive forces between the molecules and they start to interact with one another. These forces become more effective as the temperature lowers and the sample becomes first a liquid and then, at a lower temperature, a solid. In these states, the sample has different properties.

For condensed states, the kinetic molecular theory must be modified. The properties of molecules in these states can be described as follows:

1. Instead of negligible volume, the molecules of a liquid or a solid occupy most of the volume of the sample. This fact is apparent from the lack of compressibility of matter in these two states.
2. There are appreciable attractive forces between the molecules and these forces keep the liquid or the solid molecules together.
3. Although not moving as rapidly as in a gas, the molecules are still in constant, rapid motion.
4. In a liquid this movement can be seen by gently placing a drop of ink (or dye) on the surface of a large amount of water. Without any stirring, the colored ink

will be rapidly dispersed throughout the entire container as the colored molecules are buffeted about by the constantly moving water molecules.

5. While, in a liquid the molecules collide without loss of kinetic energy, in a solid the particles are held so rigidly in place that they do not collide.
6. Just as in gases, each molecule has some kinetic energy in condensed states too and the average of these energies is proportional to the temperature of the sample. There is a distribution of kinetic energies in the sample similar to that in a gas sample.

Fig. 13.1: When a drop of a dye is placed on the surface of some water (as shown in this figure), the molecules of dye are buffeted by the moving molecules of water and dispersed throughout.

2. WHAT IS IMPORTANCE OF ATTRACTIVE FORCES BETWEEN PARTICLES?

The magnitude of the attraction of one particle for another is important in determining whether the substance containing those particles is a solid, a liquid, or a gas under normal conditions. These attractive forces also represent a part of the potential or stored energy of a sample. We know that when the kinetic energy of a sample changes, its temperature changes. However, the temperature does not change when the potential energy of a sample changes. Instead the energy that is added or subtracted breaks or forms bonds. For example, when 6.02 kJ of energy is added to 1 mol (18 g) of ice at 0°C, there is no change in temperature, only a change in state from solid to liquid takes place. The added energy counteracts the forces that held the water molecules in the rigid ice structure. It does not break the bonds within molecules; rather it weakens the bonds only between molecules.

$$H_2O(s) + \text{energy} \rightarrow H_2O(l)$$

The bonds between the atoms, ions, or molecules in a substance range from very strong to comparatively weak. The melting and boiling points of substances of similar molecular weight are a measure of the relative strength of their intermolecular, interatomic, or interionic bonds. A higher-melting substance contains stronger bonds between its particles and a lower-melting substance has comparatively weaker interparticle bonds.

3. HOW CAN WE ILLUSTRATE THE EXISTENCE OF INTERMOLECULAR FORCES SUCH AS HYDROGEN BONDING IN LIQUIDS?

Low-melting solids and compounds that are liquids at room temperature are usually covalent and ionic compounds. They almost always melt well above room temperature. (The melting point of sodium chloride, a typical ionic solid, is 801°C.) In samples of covalent compounds that are either low-melting solids or liquids, three types of intermolecular forces: dispersion forces, dipole-dipole interactions and hydrogen bonding are possible.

Hydrogen bonding is a special kind of dipole-dipole interaction. It represents the attraction between a small partially negative atom (usually nitrogen, oxygen, or fluorine) and partially positive hydrogen (usually bonded to another very electronegative atom like nitrogen, oxygen, or fluorine) in another molecule. We can illustrate the existence of hydrogen bonding by comparing the boiling points of the hydrides of several elements. (A hydride of an element is the compound which the element forms with hydrogen; for example, ammonia is the hydride of nitrogen). We know that the boiling point of a substance is a measure of its intermolecular forces. The following figure (Fig. 13.2) is a plot of the boiling points of the hydrides of the elements in Groups 14/IV, 15/V, 16/VI and 17/VII of the periodic table.

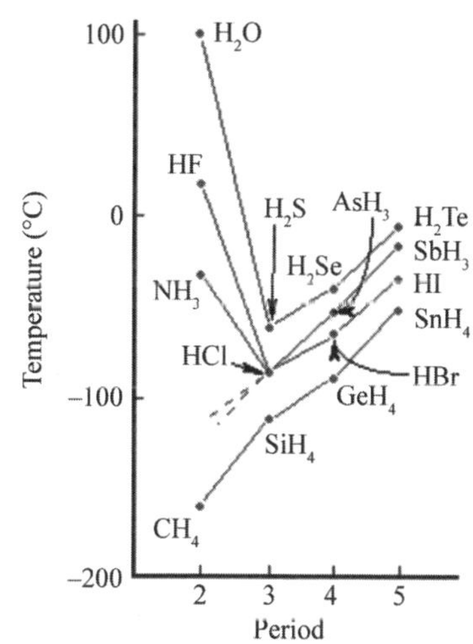

Electronegativities

IV	V	VI	VII
C **2.5**	N 3.0	O 3.5	F 4.0
Si 1.8	P 2.1	S 2.5	Cl 3.0
Ge 2.0	As 2.2	Se 2.6	Br 2.8
Sn 1.8	Sb 2.0	Te 2.1	I 2.5
Pb 2.3	Bi 2.0	Po -	

Fig. 13.2: Molecular weights versus boiling points of the hydrides of Groups 14/IV, 15/V, 16/VI and 17/VII. The solid line shows the actual boiling points and the dashed line shows the predicted boiling point of the hydride of the lightest member of each group if there was no hydrogen bonding.

Elements of Group 14 show the regular increase in boiling point that would be expected for a regular increase in molecular weight. For Groups 15, 16 and 17, the hydrides of the three heaviest members of each group show a regular increase in boiling point, but the boiling point of the hydride of the lightest member of each group is much higher than would be predicted. Their expected boiling points are shown by the continuation of the dashed line plotting boiling points against molecular weight. For Group 16, the graph shows that the boiling point of water is much higher than would be predicted on the basis of molecular weight. Ammonia and hydrogen fluoride also boil at higher temperatures than expected ones. We postulate that the abnormally high boiling points of ammonia, water and hydrogen fluoride are due to strong interactions between the molecules, so strong that these compounds behave more as aggregates of molecules than as single molecules. These aggregates are held together by hydrogen bonds. Within this group, the hydrogen bond strength is greatest in H-F, less in H-O and even less in H-N bonds. We notice that this order corresponds to the order of decreasing electronegativity of F, O and N. Let us consider the nature of hydrogen bonding in water at the molecular level. We know that the O-H bond is polar covalent wherein the oxygen atom bears a partial negative charge.

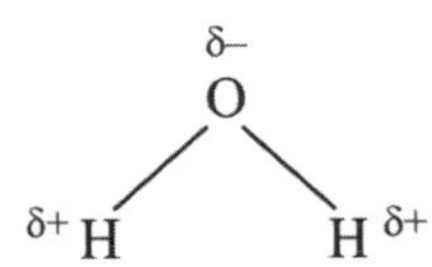

When two water molecules come close to each other, a partially positive hydrogen atom of one water molecule interacts with the partially negative oxygen atom of the other to form a hydrogen bond. The hydrogen atom is apparently bonded to oxygen atoms in both molecules. We say that the two molecules of water are held together by a hydrogen bond. The following figure (Fig. 13.3) shows several water molecules bonded together. The hydrogen bonds between neighboring molecules are shown as dashed lines. The overall effect is a network of bonds with no separate molecules. In closing this discussion of hydrogen bonding, we must emphasize that, even though hydrogen bonds are stronger than dipole-dipole interactions or dispersion forces, they are still much weaker than intermolecular or interionic bonds. Hydrogen bonds play an important role in chemistry, particularly in the chemistry of biochemical molecules such as proteins and nucleic acids.

4. WHAT INTERPARTICLE FORCES EXIT IN SOLIDS?

High-melting solids are either metals or ionic compounds, or network covalent compounds. This latter group includes, the hardest materials known such as diamonds, quartz, silica etc. The interparticle forces in high-melting solids also fall into three categories:

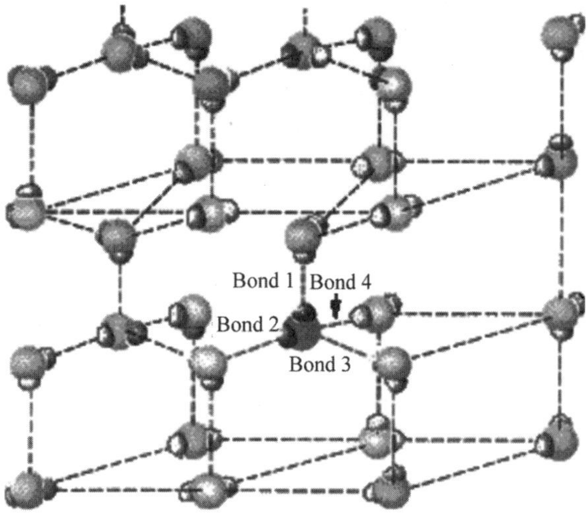

Fig. 13.3: Hydrogen bonding in ice. Each water molecule is bonded to four others, two hydrogen atoms are attracted to the oxygen atoms in two other water molecules and the oxygen atom attracts hydrogens in two more water molecules.

Network Covalent Solids

Most of covalent bonds in small molecules are low-melting, but the interatomic bonds in these small molecules are very strong. Network covalent solids (Fig. 13.4a) can be thought of as huge covalent molecules of enormous molecular weight. In these molecules every atom is covalently bonded to several other atoms so as to form a network of covalent bonds (hence the name). Such compounds have very high melting points. Diamonds (mp 3550°C) and quartz (SiO_2; mp 1610°C) are typical network covalent solids. When these compounds melt, a regular covalent bond of shared electrons is broken, the same kind of bond that is broken when water is decomposed to hydrogen and oxygen.

Ionic Bonds

Ionic compounds are not atoms or molecules but ions. In the solid state, the ions are held in a regular, rigid structure (Fig. 13.4b) by the electrostatic forces of attraction between oppositely charged particles. Ionic solids usually have very high melting points, indicating the great strength of the electrostatic forces between ions.

Metallic Bonding

The bonding in metals differs from that in ionic or covalent solids. We know that metals typically have one, two, or three valence electrons. A picture of a metallic solid shows these valence electrons as a fluid within which float the nuclei surrounded by

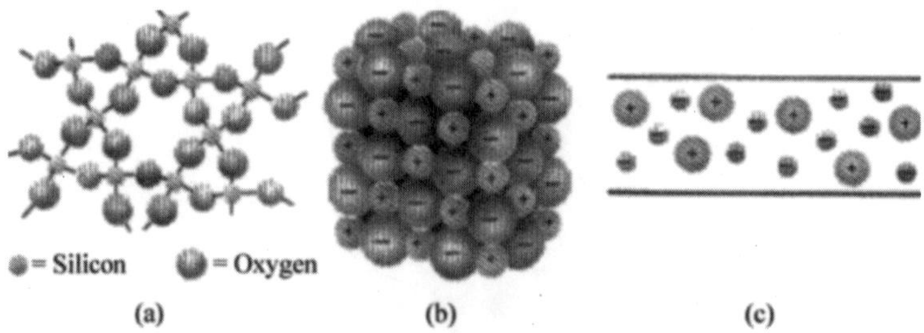

Fig. 13.4: Bonding in solids: (a) a network covalent solid (notice how closely the structure resembles the structure of hydrogen bonding in ice); (b) an ionic solid; (c) a metal.

their inner electrons (Fig. 13.4c). The conductivity of metals is due to the movement of this "sea" of electrons. Hammering metal into a thin sheet spreads out the fluid; similarly, drawing a metal into thin wire is a rearrangement of the sea of electrons containing the nuclei into a thin stream. Metal fatigue, a problem in the aircraft industry and in nuclear power plants, is associated with metals losing their "fluid" nature and assuming a rigid structure. The melting points of metals vary over a wide range. Mercury is the lowest-melting (–39°C) and tungsten is one of the highest-melting metals (3410°C).

5. WHAT IS CORRELATION BETWEEN EVAPORATION AND CONDENSATION?

The following figure (Fig. 13.5) refers to the distribution of kinetic energies in a liquid.

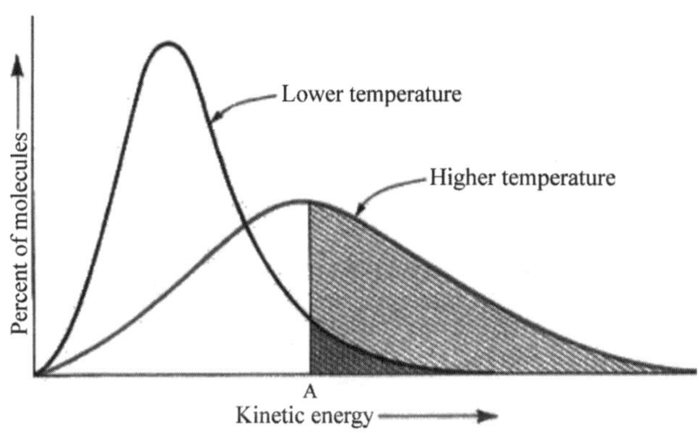

Fig. 13.5: Distribution of kinetic energies among the molecules of a liquid at two different temperatures.

Let us assume that a molecule with kinetic energy greater than 'A' in Fig. 13.5 has enough energy to overcome the attraction of its neighboring molecules. If the molecule is on the surface of the liquid, it can escape from the body of the liquid and become a gaseous molecule. The number of molecules that have at least this much energy is represented by the area under the curve to the right of A. When these molecules escape the liquid, we say that they have evaporated or vaporized; the process is called evaporation. We can show this process by the equation:

$$\text{Liquid} + \text{energy} \rightarrow \text{gas}$$

If these gaseous molecules are confined in the space above the liquid, that is, if the liquid sample is in a closed container, some of them, in their random gaseous motion, strike the surface of the liquid and are recaptured by it. This recapturing is called condensation. When the rate of condensation equals the rate of evaporation, the molecules are in a state of dynamic equilibrium. When a system is in a state of dynamic equilibrium, two opposing processes are going on at the same rate. In this equilibrium, the opposing processes are evaporation and condensation and Rate of evaporation = Rate of condensation.

For this equilibrium to be established, the system (the closed flask containing both liquid and vapor) must be at constant pressure. The molecules that have escaped are in the atmosphere over the liquid and, like all gases, exert a pressure. The partial pressure they exert is known as the vapor pressure of the liquid. One should note that this equilibrium can take place only in a closed container or closed system. The vapor must accumulate enough to allow a normal distribution of energies before the rate of condensation will equal the rate of vaporization. In an open container, the vapor can escape and equilibrium will not be reached. We recognize this fact when we store liquids in closed containers. Even ethyl ether, with a boiling point (b.p 34.5°C) slightly above room temperature, can be stored at room temperature in a tightly closed container. Fig. 13.6 shows the two processes of evaporation from an open container and equilibrium in a closed container.

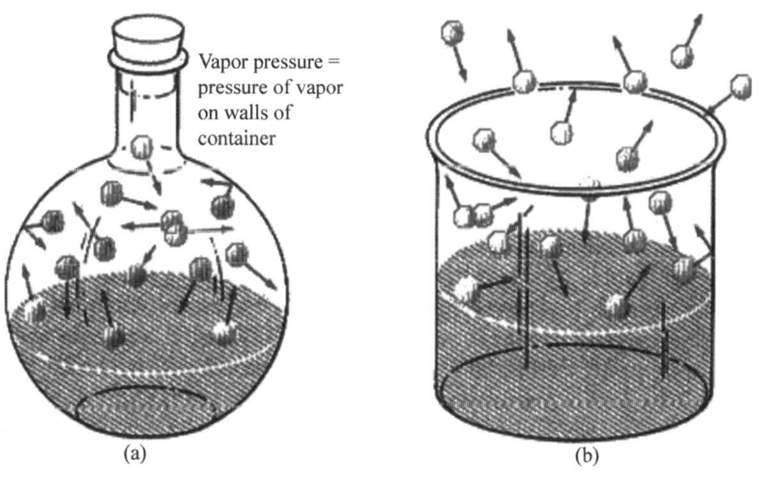

Fig. 13.6: (a) Equilibrium between vapor and liquid in a closed container. (b) Non-equilibrium (evaporation) in an open container; equilibrium cannot be established because the vapor does not collect.

Vapor pressure increases with temperature. At higher temperatures, more molecules have sufficient energy to escape the body of the liquid. Fig. 13.7 illustrates how the vapor pressure of a liquid changes with temperature. In this figure, the vapor pressure of water (b.p 100°C) and of carbon tetrachloride (b.p 78°C) are plotted against temperature (°C). Notice how rapidly the vapor pressure increases. Notice too that, for each, the vapor pressure equals 760 torr at the boiling point. This fact then defines the normal boiling point of a liquid: the temperature at which the vapor pressure of a liquid equals 1 atm.

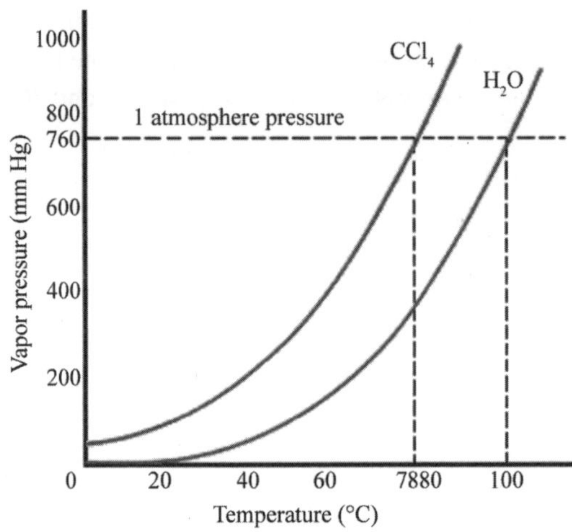

Fig. 13.7: A plot of vapor pressure versus temperature for carbon tetrachloride (CCl_4) and for water (H_2O). Notice that the vapor pressure of these two liquids equals 760 torr (1 atm) at their normal boiling points (78°C for carbon tetrachloride and 100°C for water).

6. WHY DO LIQUIDS BOIL AT LOWER TEMPERATURES AT HIGH ALTITUDES AND WHY ARE FOODS COOKED IN SHORTER TIME IN PRESSURE COOKERS?

Although the normal boiling point of a substance is defined as the temperature at which its vapor pressure equals 1 atm, a liquid will boil whenever its vapor pressure equals the gaseous pressure in the space over the liquid. At high altitudes, such as on a mountain, liquids boil at lower temperatures than normal because the atmospheric pressure is less than 1 atm. Foods take longer times to cook because the water boils at a lower temperature. Conversely, in a pressure cooker, the vapor is confined and builds up a pressure greater than that of the atmosphere. The boiling point of water in a pressure cooker is above 100°C because of this greater pressure, which means that the water can be hotter than 100°C and still remain a liquid. Foods cook in a shorter time because they are cooking in water that is at a temperature above 100°C.

7. HOW IS SPECIFIC HEAT OF A SAMPLE CALCULATED?

At the molecular level, the temperature of a liquid is proportional to the average kinetic energy of the molecules within the liquid. Any change in temperature corresponds to a change in the average kinetic energy of the molecules. The specific heat of a liquid is the amount of energy necessary to change the temperature of a one-gram sample by one degree Celsius. The amount of energy necessary to cause this change differs from liquid to liquid which means that each liquid has a unique specific heat. Given data for the mass of the sample, its temperature changes (ΔT) and the energy added, the specific heat of the sample can be calculated as:

$$\text{Specific heat} = \text{energy change}/(\text{mass of sample})\ (\Delta T)$$

As an illustrative example let us calculate the specific heat of ethyl alcohol if 402 J are required to change the temperature of a 9.63 g sample from 21°C to 38°C.

Here we want to calculate the specific heat (in J/g°C) of ethyl alcohol when the given data is: 9.63 g alcohol, 402 J, ΔT = 17°C (temperature change from 21°C to 38°C) Equation:

$$\text{Specific Heat} = \frac{402\ \text{J}}{(9.63\ \text{g})\ (17^{\circ}\text{C})}$$

Answer: 2.46 J/g°C

As yet another example let us calculate the final temperature of a sample supposing that 655 J of energy are added to 16.5 g water at 25.0°C.

We want the final temperature which will be 25.0°C + ΔT, therefore we must calculate ΔT. The given data is: 16.5 g water, 655 J

Conversion factors: Specific heat of water = 4.184 J/g°C

Specific heat = (energy)/(mass × ΔT)

Rearranging this equation gives:

ΔT = (energy)/(mass × specific heat)

Equation

$$\Delta T = \frac{655\ \text{J}}{(16.5\ \text{g})\ (4{,}184\ \text{J/g}^{\circ}\text{C})} = 9.49^{\circ}\text{C}$$

8. HOW THE MOLAR HEAT OF VAPORIZATION OF A LIQUID CAN BE CALCULATED?

When energy is supplied to a liquid, its temperature rises slowly and finally to the boiling point at a rate that is dependent on its specific heat. Energy added to a liquid at its boiling point does not change its temperature. Instead, this added energy counteracts the intermolecular forces in the liquid and the molecules break apart one by one to vaporize and become gaseous. The amount of energy required to vaporize one mole of a liquid at its boiling point is its molar heat of vaporization (ΔH_{vap}). The following Table (Table 13.1) shows the molecular weight, boiling point, specific heat and molar heat of vaporization for several liquids. The liquids are listed in order of increasing polarity. Notice that this order is independent of molecular weight.

Table 13.1: Physical properties of some liquids

Liquid	*Molecular weight*	*B.P. (°C)*	*ΔH_{vap} (kJ/mol)*	*Specific heat (J/g°C)*
Carbon disulfide (CS_2)	76.1	46°	28.4	1.000
Chloroform ($CHCl_3$)	119.4	61.7°	31.4	0.966
Ethyl alcohol (C_2H_5OH)	46.1	78.5°	40.4	2.45
Water (H_2O)	18	100°	40.7	4.18

The molar heat of vaporization of a liquid can be calculated from experimental data. As an illustrative example let us calculate the molar heat of vaporization of benzene when 7.47 kJ are required in order to vaporize 13.6 g benzene at its boiling point,

We want the molar heat of vaporization in joules per mole (ΔHvap =? J/mol)

It is given that: 7.47 kJ are required to vaporize 13.6 g benzene

Conversion factor: 1 mol benzene = 78.1 g benzene

Equation

$$? \text{ J/mol} = \frac{7.47 \text{ kJ}}{13.6 \text{ g}} \times \frac{78.1 \text{ g benzene}}{1 \text{ mol benzene}}$$

Answer: 42.9 kJ/mol

9. WHAT IS CORRELATION BETWEEN SURFACE TENSION AND VISCOSITY?

The surface tension of a liquid is the property that causes the liquid to have the minimum possible surface area. Within a liquid, attractive forces operate all around a molecule; on the surface, attraction is only into the body of the liquid as has been depicted in the Fig. 13.8. This attraction into the liquid causes the molecules on the surface to minimize surface area, becoming more tightly bound together than those within the liquid.

Surface tension measures this surface binding. Between different liquids of similar molecular weight, the difference in surface tension can be observed by comparing the different-sized drops formed by the liquids being compared. Large drops mean large intermolecular forces and large surface tension. Surface tension decreases as the temperature and kinetic energy of the molecules increase. The resistance to flow, or the

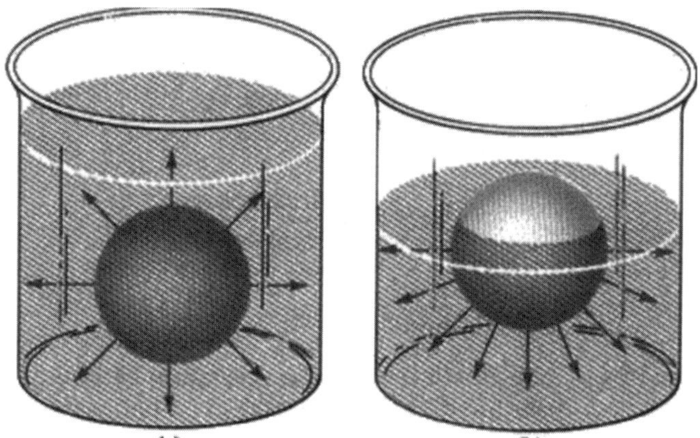

Fig. 13.8: Attractive forces (a) within a liquid and (b) at its surface.

viscosity of a liquid, is another measure of the strength of its intermolecular forces. All liquids flow; some (like gasoline) flow easily, others (like diesel oil) flow very slowly. Resistance to flow means high viscosity and strong intermolecular forces just as happen in high surface tension. Viscosity decreases as temperature increases which we may recall from we may the difference in rate of flow between cold and warm diesel oil.

10. WHY DO SOLIDS VAPORIZE AND SUBLIME?

The physical properties of solids indicate that the particles (ions, molecules, or atoms) in a solid occupy fixed positions from which they cannot easily move. This orderly arrangement is called the crystal structure or crystal lattice of the solid. Strong attractive forces between the particles keep them in this arrangement. Even so, as suggested in the Fig. 13.9), each particle in a solid has some kinetic energy and is in constant motion within its space in the solid (unless the solid is at a temperature of absolute zero, at which all motion ceases). The ions, atoms, or molecules of which the solid is composed, vibrate, rotate and even move around within their assigned space in the crystal structure.

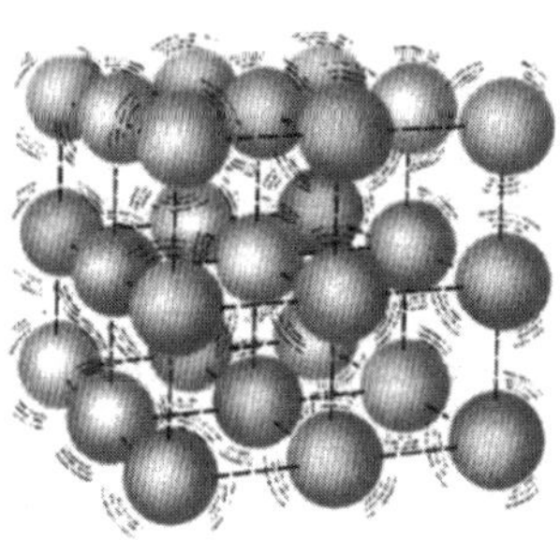

Fig. 13.9: The crystal structure of a solid, showing the movement of the component ions, atoms, or molecules within their assigned space.

Similar to gas and liquid samples, particles in a solid have a distribution of kinetic energies that depends on the temperature of the sample. At every temperature, some particles have enough energy to overcome the forces that hold them in place and escape from the solid as a vapor. Hence each solid, like each liquid vaporizes and has a vapor pressure that increases as the temperature increases. The process by which atoms or molecules go directly from the solid state to the gaseous state is called sublimation. Solids that sublime have unusually high vapor pressures. Dry ice (solid carbon dioxide) is a familiar example of a solid that sublimes. Solid water (ice) sublimes. Those who live in a cold, dry climate, may have noticed that ice disappears (actually it sublimes to a colorless gas) at temperatures below 0°C. Moth balls sublime and disappear slowly after protecting woolens during a hot summer.

11. WHAT HAPPENS DURING FREEZING AND MELTING OF SOLIDS?

As a solid is heated, the change in its temperature depends on its specific heat. The solution of specific heat problems is the same whether for a liquid or a solid. A solid melts (changes to a liquid) when the average kinetic energy of its particles is high enough to overcome the attractive forces between them. By definition, the melting point of a solid is the temperature at which the liquid and solid states are in equilibrium at a pressure of 1 atm. We show this equilibrium in equation form by writing "solid" as a reactant and "liquid" as a product and connecting them with a double (equilibrium) arrow:

$$\text{solid} \underset{\text{freezing}}{\overset{\text{melting}}{\rightleftharpoons}} \text{liquid}$$

During melting, both solid and liquid are present until all the solid is converted to liquid. Conversely, during freezing, solid and liquid are present until all the liquid is converted to solid. The processes are indicated by writing "melting" over the forward equilibrium arrow and "freezing" under the reverse equilibrium arrow. Because freezing is the reverse of melting, the freezing point of a substance is the same as its melting point. The molar heat of fusion, ΔH_{fus}, of a substance is the amount of heat that must be supplied to convert one mole of that substance from a solid to a liquid at its melting point. Calculations involving heats of fusion are similar to those involving heats of vaporization.

As an illustrative example let us find out how many joules must be supplied to 34 g solid carbon tetrachloride at is melting point to change the sample to a liquid when the molar heat of fusion of carbon tetrachloride (CCl_4) is 3.26 kJ per mol.

***Solution*:** We want to calculate the energy needed in Joules when it has been given that the molar heat of fusion of carbon tetrachloride (CCl_4) is 34 g CCl_4

The conversion factor and equation may be written as:

$$\Delta H_{fus} = \frac{3.26 \text{ kJ}}{\text{mol}}$$

$$1 \text{ mol } CCl_4 = 153.8 \text{ g } CCl_4$$

Equation

$$? \text{ J} = 34 \text{ g } CCl_4 \times \frac{1 \text{ mol } CCl_4}{153.8 \text{ g } CCl_4} \times \frac{32 \text{ k J}}{1 \text{ mol } CCl_4}$$

Answer: 0.72 kJ

12. HOW DO TRANSITIONS FROM THE SOLID THROUGH THE LIQUID TO THE GASEOUS STATE TAKE PLACE?

We have discussed the properties of matter in its three states: solid, liquid and gas. In each case, we talk about the average kinetic energy of the molecules in that state and about changes in average kinetic energy as the temperature is changed. We have also talked about changes from one state to another. The energy necessary to transform a substance from a solid below its melting point to a gas above its boiling point can be determined by calculating the energy required for each change involved and summing these energies. In the following figure the energy required for such a series of changes is plotted. In this change, pressure has been held constant at 1 atm, so the melting and boiling points of the substance are the normal ones.

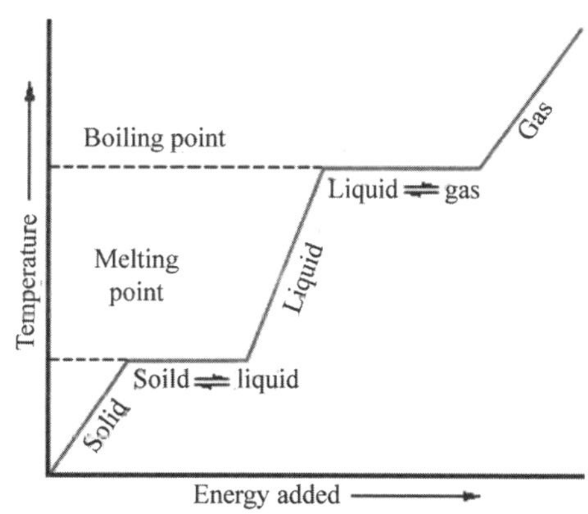

Fig. 13.10: The change in temperature of a sample as heat energy is added and the sample changes from a solid to a liquid to a gas

We may note carefully the shape of the resulting graph. Although the length and slopes of the segments differ between compounds, the general shape will remain the same. The temperatures for the horizontal segments are the melting point and the boiling point of the substance and the lengths of these segments depend on the molar heat of fusion and molar heat of vaporization. The slopes of the other segments of the graph depend on the specific heats of the substance: the first segment on the specific heat of the solid, the second on the specific heat of the liquid and the third on the specific heat of the gas.

It may be noted that the sloping segments represent a change in temperature and therefore a change in kinetic energy, whereas the horizontal segments represent no change in temperature and therefore no change in kinetic energy. These segments do represent a change in the potential energy of the sample as bonds are broken. The amount of energy required for changing a sample from one state and temperature to another state and temperature can be calculated if the various physical constants are known for that substance.

As an illustrative example let us calculate the amount of energy (in joules) needed to change 55.0 g ice at –5°C to steam at 110°C when the given data is:

Specific heat of ice = 2.05 J/g°C; Specific heat of liquid water = 4.184 J/g°C

Specific heat of steam = 2.01 J/g°C; Molar heat of fusion of water = 6.02 kJ/mol
Molar heat of vaporization of water = 40.7 kJ/mol.

The answer to this problem is obtained by calculating the energy change for each segment of the graph and then finding the total energy change. In this solution, the letter of each step corresponds with a lettered segment of the figure.

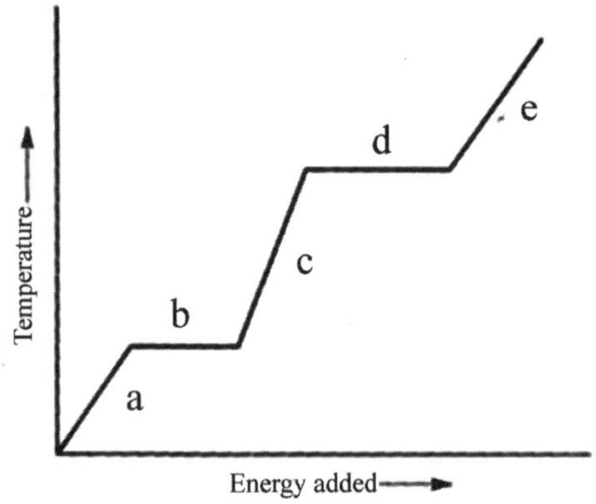

a. To raise the temperature of ice from –5°C to 0°C, a 5-degree change in the temperature of the solid, the equation is

$$? \text{ J} = 55.0 \text{ g} \times \frac{2.05 \text{ J}}{\text{g°C}} \times 5°\text{C} = 0.564 \text{ kJ}$$

b. To melt the ice at 0°C:

$$? \text{ J} = 55.0 \text{ g} \times \frac{1 \text{ mol}}{18.0 \text{ g}} \times \frac{6.02 \text{ kJ}}{1 \text{ mol}} = 18.4 \text{ kJ}$$

c. To raise the temperature of liquid water form 0°C to 100°C:

$$? \text{ J} = 55.0 \text{ g} \times \frac{4.184 \text{ J}}{\text{g°C}} \times 100\text{°C} = 23.0 \text{ kJ}$$

d. To vaporize liquid water at 100°C:

$$? \text{ J} = 55.0 \text{ g} \times \frac{1 \text{ mol}}{18.0 \text{ g}} \times \frac{40.7 \text{ kJ}}{1 \text{ mol}} = 124.4 \text{ kJ}$$

e. To heat water vapor (steam) from 100°C to 110°C:

$$? \text{ J} = 55.0 \text{ g} \times \frac{2.01 \text{ J}}{\text{g°C}} \times 10\text{°C} = 1.1 \text{ kJ}$$

The sum of these energy changes is the total amount of energy required by the change:

$$?\text{J} = 0.564 \text{ kJ} + 18.4 \text{ kJ} + 23.0 \text{ kJ} + 124.4 \text{ kJ} + 1.1 \text{ kJ} = 167.5 \text{ kJ}$$

Correcting to three significant figures, the required amount of heat is 168 kJ.

13. WHY WATER IS THE BEST SOLVENT FOR IONIC COMPOUNDS?

We know that water, H_2O, is a polar covalent compound of hydrogen and oxygen with the H-O-H bond angle of 104.4°. Due to the charge separation the molecule is highly a polar one, having the positive end toward the hydrogens and the negative end toward the oxygens. More accurately it is a quadrupole with two occupied nonbonding orbitals on the oxygen in a tetrahedral arrangement with the hydrogens. Because of the polar nature of the molecule it is a particularly good solvent for ionic compounds, the ions being stabilized in solution by oriented water dipoles. It has a very high dielectric constant, (80 at room temperature). Because of the high dielectric constant the force of attraction and repulsion of electric charges are very weak. Two opposite charges will attract each other with 1/80 (one 80th) of the force that would exist in a vacuum. This along with the polarity and hydration sphere formed around ions makes it a very good solvent for ionic compounds (electrolytes). It boils at 100°C and freezes at 0°C. The solid form, ice, is unusual in that it is less dense than the liquid. Thus lakes freeze over on the top.

14. WHAT THE TERM 'pH' STAND FOR?

We know that water ionizes to a small extent and the concentrations of hydrogen ions H^+ and hydroxide ions OH^- at room temperature are 1×10^{-7} moles/liter. Addition of acid increases the hydrogen ion concentration and decreases the hydroxide ion concentration. The hydrogen ions don't exist by themselves but combine with water molecules to form the hydronium ion H_3O^+. This exerts a structural influence on more water molecules in the vicinity. Addition of bases will increase the hydroxide ion concentration and decrease the hydrogen ion concentration. The product of the two will always equal 1×10^{-14}. Water is called neutral when the concentration of each hydrogen ion and hydroxide ion is 1×10^{-7}M. It is neither acidic nor alkaline (basic) since the concentration of (1×10^{-7}M). The term "*pH*" means the power of the hydrogen ion concentration. pH 7 means that the hydrogen ion concentration is 1×10^{-7}M. At pH 5 the hydrogen ion concentration is 1×10^{-5}M and the hydroxide ion concentration is 1×10^{-9}M. The solution with pH 5 is acidic.

15. WHY DO METHANE AND HYDROGEN SULFIDE HAVE LIMITED USES AS SOLVENTS?

Before opining on the question let us consider briefly the properties of water summarized below in the tabular form (Table 13.3). The same table also lists the properties of methane, a compound of similar molecular weight (therefore expected to be similar to water in many physical properties) and of hydrogen sulfide (the hydride of sulfur), an element in the same column of the periodic table as oxygen. Our study of the periodic table showed that the compounds: water and hydrogen sulfide can be expected to be quite similar. We notice that both methane and hydrogen sulfide are gaseous under normal conditions. Water, on the other hand, is a liquid under normal conditions and a solid or gas at temperatures not far from normal.

Table 13.2: A comparison of Properties of Water with Methane and Hydrogen sulfide

Property	*Methane, CH_4*	*Water, H_2O*	*Hydrogen sulfide, H_2S*
Formula weight	16.0	18.0	34.1
Melting point	–182°C	0°C	–85°C
Boiling point	–16°C	100°C	–61°C
Density: solid (g/cm^3)	–	0.999	–
: liquid (g/mL)	–	1.000	–
: gas (g/L)	0.71	0.80	1.5
Specific heat: solid (J/g°C)	–	2.05	–
: liquid (J/g°C)	–	4.18	–
: gas (J/g°C)	2.1	2.01	–
Heat of fusion (kJ/mol)	0.934	6.02	2.37
Heat of vaporization (kJ/mol)	8.90	40.7	18.7
Toxicity	High	None	Very high
Odor	Bad	None	Bad
Usefulness as solvent	none*	Excellent	none*

*Both methane and hydrogen sulfide would be useful solvents below their boiling points but, as both melt well below room temperature and are highly toxic; they have limited use as solvents.

The physical state of water at normal temperature can by and large be attributed to the polarity of its molecules and the strength of its hydrogen bonds. Methane is neither polar nor hydrogen bonded. Hydrogen sulfide is polar, but much less so than water and is not hydrogen bonded.

16. WHAT IS UNIQUE WITH WATER?

The answer to this question is too obvious to go in details. We know that water is an excellent solvent. It dissolves ionic compounds, polar compounds and low-molecular-weight nonpolar compounds. The implications of this property, particularly in biological processes, are enormous. The specific heat and heat of vaporization of water keep most of our planet temperate. We may consider how much less the daily temperature fluctuates near large bodies of water than in deserts. Our bodies evaporate water (sweat) in order to maintain their constant temperature. The small difference in density between ice and water is of far-reaching importance. Ice is slightly less dense and floats on water. Lakes freeze from the top down, not only allowing a heaven for fish in freezing water but allowing the ice to melt in summer.

17. HOW CAN WE SUMMARIZE THE INTERMOLECULAR FORCES AND THEIR CONSEQUENCES?

Substances are liquids or solids if the attractive forces between their particles are strong enough to overcome their motion (their kinetic energy). These forces, in order of increasing strength, are: dispersion forces (London forces), dipole-dipole interaction and hydrogen bonds. In addition to these forces, the particles in solids may be bonded to one another by ionic bonds, network covalent bonds, or metallic bonds. A liquid has a vapor pressure that increases with temperature. When the vapor pressure of a liquid equals atmospheric pressure, the liquid boils. When the pressure of the atmosphere is 1 atm, the temperature at which a liquid boils is its normal boiling point. Vapor pressure in the solid state is less than that in the liquid state. When the vapor pressures are equal the solid melts. Within a solid, the particles are joined by rigid bonds that define the crystal lattice of the solid. Each substance has a specific heat for each physical state in which it occurs. Specific heat measures the rate of change of kinetic energy within the sample. Each substance also has a heat of fusion and a heat of vaporization that measure the strength of the attractive forces between the particles. The energy necessary to cause a change in temperature and/or a change in state can be calculated.

Chapter 14

BASIC CONCEPTS OF BONDING IN ORGANIC CHEMISTRY

1. WHAT IS UNIQUENESS OF CARBON AND ITS COMPOUNDS?

Answer to the question 'why carbon is such a unique element' may be found by looking at its chemistry in a little detail. Carbon appears in the second row and fourteenth group of the periodic table and has four bonding electrons in its valence shell. Similar to other non-metals, carbon needs eight electrons to satisfy its valence shell. Carbon therefore forms four bonds with other atoms (each bond consisting of one of carbon's electrons and one of the bonding atom's electrons). Every valence electron participates in bonding, thus a carbon atom's bonds will be distributed evenly over the atom's surface. These bonds form a tetrahedron (a pyramid with a spike at the top).

Organic chemistry, at the most basic level, is the study of reactions and properties of compounds of carbon. Organic compounds are vitally important to life. The proteins that serve as the structural and functional basis of living organisms, the drug molecules that are used to treat diseases and the petroleum compounds that serve as an energy source are all organic compounds. To understand life as we know it, we must first understand a little bit of organic chemistry. Organic molecules contain both carbon and hydrogen. Though many organic chemicals also contain other elements, it is the carbon-hydrogen bond that defines them as organic. Organic chemistry defines life. Just as there are millions of different types of living organisms on this planet, there are millions of different organic molecules, each with different chemical and physical properties. There are organic chemicals that make up our hair, our skin, our fingernails and so on. The diversity of organic chemicals is due to the versatility of the carbon atom.

Organic chemicals get their diversity from the many different ways carbon can bond to other atoms. In fact chemical bond is the most basic concept in all of chemistry. The chemical bond is essentially the sharing of electrons between two atoms, a sharing which holds or bonds the atoms together. In order to understand the chemical bond, atoms and electrons must first be understood. The simplest organic chemicals, called hydrocarbons, contain only carbon and hydrogen atoms; the simplest hydrocarbon (called methane) contains a single carbon atom bonded to four hydrogen atoms. But carbon can bond to other carbon atoms in addition to hydrogen, as is the case in the molecule ethane in which there is a carbon-carbon bond also. In fact, the uniqueness of carbon comes from the fact that it can bond to itself in many different ways. Carbon atoms can form long chains (Hexane, a 6-carbon chain); branched chains (Isohexane); and ring (Cyclohexane, a ringed hydrocarbon). There appears to be almost no limit to the number of different structures that carbon can form. To add to the complexity of organic chemistry, neighboring carbon atoms can form double and triple bonds in addition to single carbon-carbon bonds. Since each carbon atom forms four bonds, as the number of bonds between any two carbon atoms increases, the number of hydrogen atoms in the molecule decreases.

2. WHAT IS PROBABILITY OF FINDING AN ELECTRON IN ATOMIC ORBITALS?

Atoms have three components: protons, neutrons and electrons. Protons have a positive charge of +1; neutrons have zero charge and electrons have a negative charge of -1. The protons and neutrons occupy the center of the atom, called the nucleus. The electrons revolve in "orbitals" surrounding the nucleus. In the olden times, it was believed that the electrons orbited the nucleus in the same way that planets orbit the sun. Though it is not exactly true, it explains the origin of the term orbital to describe the location of the electron.

In reality, it is impossible to tell the precise trajectory of an electron in the way that the path of a planet can be followed. The best that can be achieved is to describe the probability of locating the electron in a region of space. The distribution of this probability over space is called an orbital. Let us treat the simplest case, in which the nucleus is surrounded by just one electron (for example, the hydrogen atom). In this case, the probability of finding an electron in its lowest energy, or most stable, state is distributed in a spherically symmetric way around the nucleus. The probability of finding the electron is highest at the nucleus and decreases as the distance from the nucleus increases (Fig. 14.1). The density of dots in a region represents the probability of locating the electron around/at the nucleus. Such a location may seem counterintuitive, since the electron is known to generally be located some distance from the nucleus, not directly at it. Question arises as to how can this apparent contradiction be explained?

Although the probability of finding the electron is greatest at the nucleus, the volume of space occupied by the nucleus is extremely small compared to the rest of the atom. The probability of finding the electron at a point decreases farther from the nucleus, but the volume in which it can be found becomes much, much larger. So the electron is more likely to be found some distance from the nucleus. It is the product of the volume in which the electron can be found and the probability of finding it that gives the probability of finding the electron as a function of distance from the nucleus. The radial distribution function, which is a product of the probability and the volume element, gives the probability of finding the electron in a spherical shell some distance- r, from

Fig. 14.1: Is orbital: The lowest energy, spherically symmetric orbital is called the 1s orbital.

the nucleus. It is clear that the electron is most likely to be found in a shell some distance from the nucleus rather than directly at the nucleus. In fact, since the volume element is almost zero directly at the nucleus, the electron is very unlikely to be found there.

3. WHAT IS DIFFERENCE BETWEEN AN 's' ORBITAL AND A 'p' ORBITAL AND HOW THESE ORBITALS LOOK LIKE?

The 1s orbital is the lowest energy orbital that an electron can occupy, but several higher energy orbitals are significant in chemistry. The next lowest energy orbital that an electron can occupy is the 2s orbital, which looks much like the 1s orbital except that the electron is more likely to be found farther from the nucleus. The third lowest energy orbital is called the 2p orbital. The major and highly important difference between a p orbital and an s orbital is that the p orbital is not spherically symmetric (Fig. 14.2).

Fig. 14.2: 2p Orbital image of 2p orbital

Instead, the p orbital is oriented along a specific axis in space. There are three p orbitals, oriented along the x, y and z axes (Fig. 14.3).

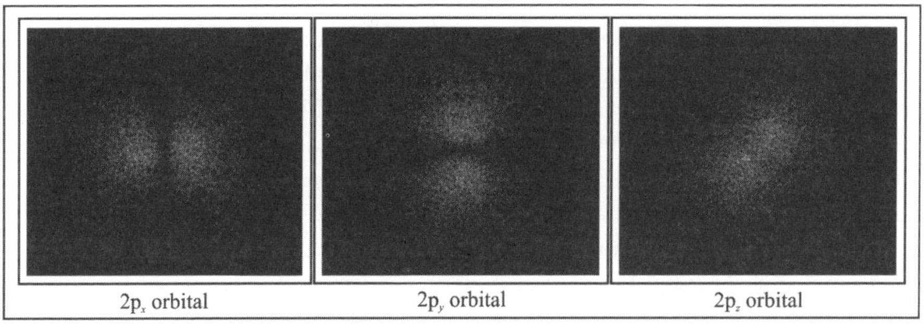

Fig. 14.3: Images of the three 2p orbitals

The images of s and p orbitals depict the actual distribution over space of the probability of finding an electron in atoms with one electron. However, after a second electron is added, the problem cannot be solved exactly due to the electron-electron repulsion. Instead, approximations have to be made. Fortunately, use of the exact orbitals of the one-electron atom as a basis for the orbitals in many-electron atoms is an excellent approximation. For example, in carbon, the energy of the electron is correct to within 1% using this approximation. Thus, orbitals in many-electron atoms look essentially like the one-electron atom orbitals.

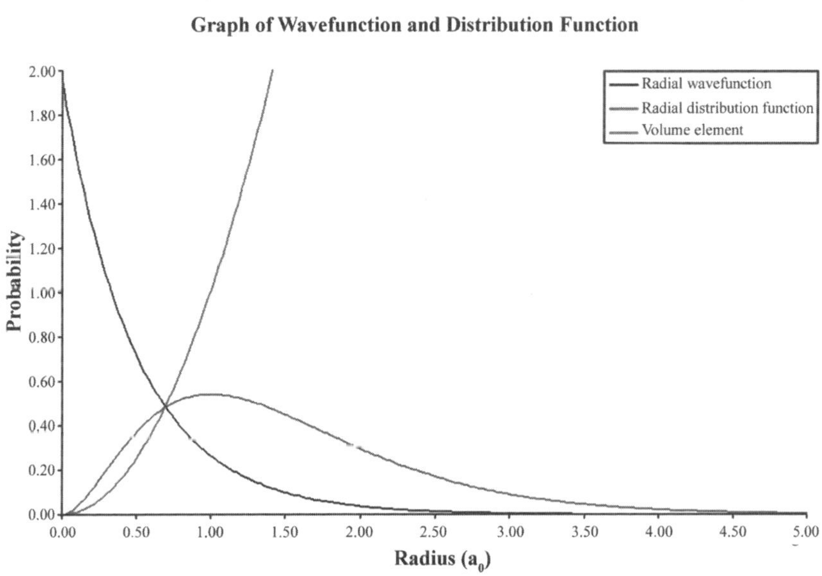

Fig. 14.4: Graph of probability density, radial distribution function and volume as a function of radius for the 1s electron.

4. WHAT HAPPENS WHEN TWO ATOMIC NUCLEI COMBINE?

We know that a chemical bond is essentially the sharing of electrons between two atoms. Since electrons are negatively charged and exert an attractive force on nuclei, they serve to hold the atoms together if they are located between two nuclei. One may ask how this notion of a chemical bond relates to orbitals describing electron density. When two atoms approach each other, their atomic orbitals overlap. The overlapped atomic orbitals can add together to form molecular orbitals. This phenomenon is called the linear combination of atomic orbitals (LCAO). The area of greatest overlap between the original atomic orbitals represents the chemical bond that is formed between them. Since the sharing of electrons is the basis of the chemical bond, the molecular orbitals formed represent chemical bonds. Let us take the simplest example of hydrogen. We can imagine two 1s orbitals gradually coming closer and closer together until there is a good deal of overlap between them (vide Fig. 14.5).

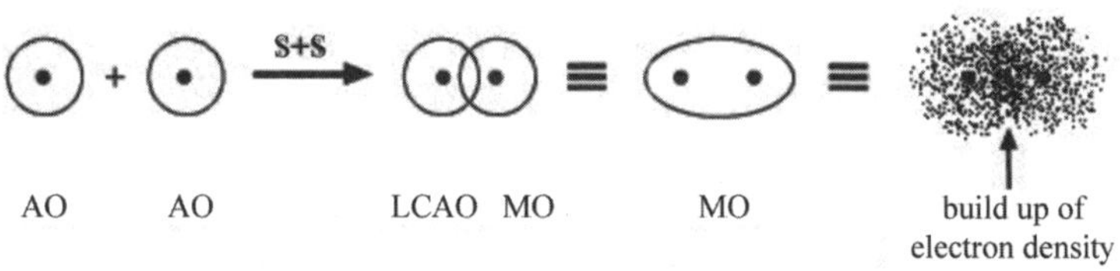

Fig. 14.5: Two 1s atomic orbitals of hydrogen atoms overlap to form a hydrogen molecular orbital

At this point, the area in space of greatest electron density will be between the two nuclei, which themselves were at the center of the original atomic orbitals. This electron density, now part of a new molecular orbital, represents the chemical bond. When the area of greatest overlap occurs directly between the two nuclei on internuclear axis or axis containing the nuclei of both atoms, the bond is called a sigma (б) bond.

There are two final points to be considered when atomic orbitals combine. First, more than one atomic orbital from a single atom can be used to form new molecular orbitals. For example, a 2s orbital and a 2p orbital from one atom might add together and overlap with one or more orbitals from a second atom to form new molecular orbitals. Second, parts of orbitals can possess a sign (+ or -). The s orbital has the same sign throughout, while in the p orbitals, one lobe is + and the other lobe is -. Signs do not matter with respect to electron density, but they must be taken into account when orbitals are added or subtracted. If two orbitals of the same sign are added, electron density will increase, while if two orbitals of opposite signs are added, the shared electron density will cancel out.

5. WHAT ORBITAL COMBINATIONS MAKE sp^3 BONDS?

In order to illustrate this point we can treat bonding in carbon. Carbon has six electrons out of which only two electrons can occupy an s orbital at a time. The first two electrons

Fig. 14.6: Tetrahedral geometry

in carbon occupy the 1s orbital. The next two occupy the higher-energy, but similarly shaped 2s orbital. The final two electrons occupy the 2p orbitals. In carbon, the electrons in the 1s orbital are too low in energy to form bonds. Thus, electrons used to form bonds must come from the 2s and 2p orbitals. Carbon, in general makes four bonds and when it does so, these bonds are arranged in a way that they are as far away from each other as possible. This arrangement is referred to as tetrahedral (Fig. 14.6).

It is clear that the individual 2s and 2p orbitals cannot form bonds in this arrangement due to their geometry. The 2s orbital is completely symmetric, while the 2p orbitals are aligned along specific axes. None of these orbitals is well-equipped to form bonds in the tetrahedral geometry alone. However, we have stated that a chemical bond doesn't have to be formed from individual atomic orbitals, but can be formed from a combination of several atomic orbitals from the same atom. For each bond that is made in the tetrahedral geometry, a part of the 2s and a part of each of the 2p orbitals will contribute, resulting in a tetrahedral arrangement. To achieve this geometry, both the 2s and all three of the 2p orbitals ($2p_x$, $2p_y$ and $2p_z$) must contribute. The new bonds that are formed are called sp^3 bonds, since one s and 3 p orbitals are used to form them.

Note that when sp^3 bonds are formed, they all point in separate directions as far away from each other as possible, leading to 109.5° angles between the bonds. This is the tetrahedral geometry as shown in Fig. 14.6. When carbon makes four sp^3 bonds to hydrogen, the resulting molecule is tetrahedral in shape and looks like Fig. 14.6. Carbon can also make a bond to another carbon atom. Two carbon atoms, both of which have sp^3 bonds oriented tetrahedrally, can bond to each other if each uses one of its sp^3 bonds (Fig. 14.7).

Fig. 14.7: sp^3 bond between two carbon atoms

sp^2 bonds

Carbon sometimes makes just three bonds. In this case, not all of the 2p orbitals combine with the 2s orbital to form bonds. Instead, a combination of the 2s orbital and two of the 2p orbitals make three sp^2 bonds, while the other p orbital does not participate in this combination and can make a fourth bond on its own.

Like the sp^3 bonds, the sp^2 bonds are oriented such that they are as far away from each other as possible. This geometry is called the trigonal planar geometry. It is called trigonal planar because each of the bonds points to one of the vertices of a triangle, but all three bonds are located in the same plane. Here we may ask what about the orphan or lone 2p orbital, the one which did not add to make sp^2 bonds? This orbital still exists and it is perpendicular to the plane in which the sp^2 bonds form. It too is able to form bonds and it does so independently of the sp^2 bonds.

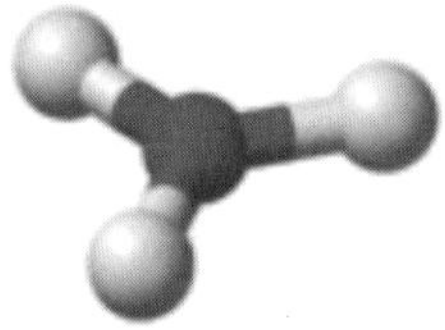

Fig. 14.8: Trigonal planar geometry

When two carbon atoms with sp^2 orbitals form a bond to each other using their sp^2 orbitals, a sigma (ó) bond is formed between them. Moreover, the extra p orbitals which exist above and below each carbon atom also overlap with each other. This overlap between p orbitals leads to the formation of a second bond in addition to the ó bond formed between the sp^2 orbitals. This second bond that does not occur directly between the nuclei on the internuclear axis, but rather above and below the internuclear axis, is called a pi bond (ð bond). When a sigma (ó) bond and a pi (ð) bond form together between two atoms, a double bond is said to have formed.

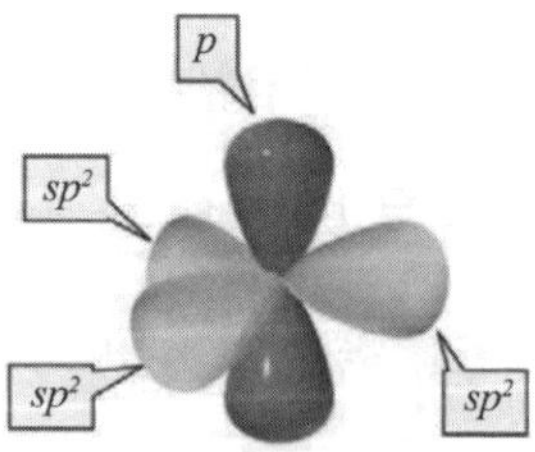

Fig. 14.9: Carbon atom with sp^2 orbitals and a 2p orbital

Figure 14.10: A ó bond is formed from two sp^2 orbitals and a ð bond is formed from two 2p orbitals in ethane molecule and together they give a double bond.

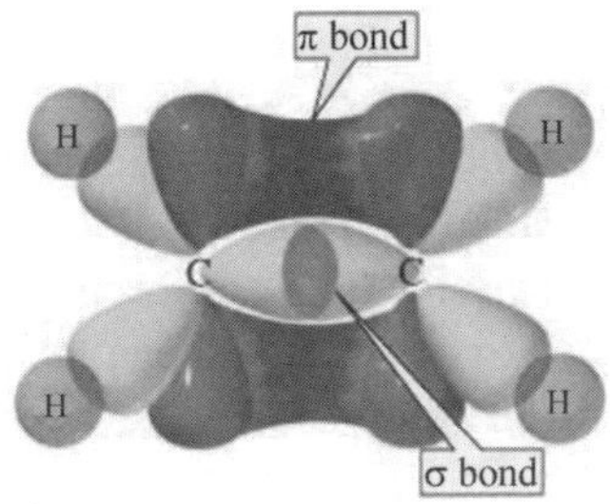

Fig. 14.10

6. HOW ARE SIMPLE HYDROCARBONS NAMED AND THEIR FORMULAS EXPRESSED?

Hydrocarbons, *i.e.* compounds containing just carbon and hydrogen atoms are the most basic compounds encountered in organic chemistry. The simple hydrocarbons come in three varieties depending on the type of carbon-carbon bonds that occur in the molecule. Alkanes are the first category of simple hydrocarbons and contain only carbon-carbon single bonds. The alkanes are named by combining a prefix that describes the number of carbon atoms in the molecule with the root ending "ane". The names and prefixes for the first ten alkanes are given in the following table (Table 14.1).

Table 14.1: Name and Prefixes of first ten Alkanes.

No. of carbon atoms	*Prefix*	*Alkane name*	*Chemical formula*	*Structural formula*
1	Meth	Methane	CH_4	CH_4
2	Eth	Ethane	C_2H_6	CH_3CH_3
3	Prop	Propane	C_3H_8	$CH_3CH_2CH_3$
4	But	Butane	C_4H_{10}	$CH_3CH_2CH_2CH_3$
5	Pent	Pentane	C_5H_{12}	$CH_3CH_2CH_2CH_2CH_3$
6	Hex	Hexane	C_6H_{14}	–
7	Hept	Heptane	C_7H_{16}	–
8	Oct	Octane	C_8H_{18}	–
9	Non	Nonane	C_9H_{20}	–
10	Dec	Decane	$C_{10}H_{22}$	–

The chemical formula for any alkane is given by the expression C_nH_{2n+2}. The structural formula, shown for the first five alkanes in the table, shows each carbon atom and the elements that are attached to it. This structural formula is important when we begin to discuss more complex hydrocarbons. The simple alkanes share many properties in common. All enter into combustion reactions with oxygen to produce carbon dioxide and water vapor. In other words, alkanes are flammable which makes them good fuels. For example, methane is the main component of natural gas and butane is common lighter fluid.

On the most basic level, hydrocarbons can broadly be divided into three groups: those containing just single bonds, those containing one or more double bonds and those containing one or more triple bonds. Before discussing how to name these compounds, it is instructive to examine how they are represented by chemists. The second category of simple hydrocarbons, the alkenes, consists of molecules that contain at least one double-bonded carbon pair. Alkenes follow the same naming convention used for alkanes. A prefix (to describe the number of carbon atoms) is combined with the ending "ene" to denote an alkene. Ethene, for example is the two-carbon molecule that contains one double bond. The chemical formula for the simple alkenes follows the expression C_nH_{2n}. Because one of the carbon pairs is doubly bonded, simple alkenes have two hydrogen atoms less than alkanes.

Alkynes are the third category of simple hydrocarbons and are molecules that contain at least one triple-bonded carbon pair. Like the alkanes and alkenes, alkynes are named by combining a prefix with the ending "yne" to denote the triple bond. The chemical formula for the simple alkynes follows the expression C_nH_{2n-2}. Since carbon atoms can bond in so many different ways; a single molecule can have different bonding configurations. Let us consider the following two molecules:

$$C_6H_{14}\ (CH_3CH_2CH_2CH_2CH_2CH_3) \text{ and } C_6H_{14}\ (CH_3CH_2CHCH_3CH_2CH_3)$$

Both molecules have identical chemical formulas but different structural formulas and thus some of the chemical properties are different. These two molecules are called isomers. Isomers are molecules that have the same chemical formula but different structural formulas.

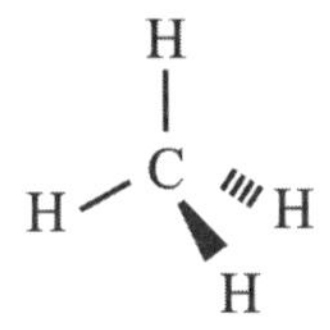

Fig. 14.11: Sketch of two dimensional representation of methane

7. HOW IS TETRAHEDRAL GEOMETRY OF HYDROCARBONS REPRESENTED IN TWO DIMENSIONS?

We may recall that when carbon forms four bonds, it adopts the tetrahedral geometry in which only two bonds can simultaneously occupy one plane simultaneously. The other two bonds point in back and in front of the plane. In order to represent the tetrahedral geometry in two dimensions, solid wedges are used to represent bonds pointing out of the plane of the drawing toward the viewer and dashed wedges are used to represent bonds pointing out of the plane of the drawing away from the viewer. Let us consider the following representation of the molecule methane:

In this drawing, the two hydrogens connected by solid lines to the carbon atoms in the center of the molecule, exist in a plane (specifically, the plane of the computer monitor of the piece of paper, etc.). The hydrogen atom connected to the carbon atom by a solid wedge points out of this plane toward the viewer and the hydrogen atom connected to the carbon atom by the dashed wedge points behind this plane and away from the viewer.

8. WHAT IS MEANT BY A SKELETAL STRUCTURE?

In drawing structures of hydrocarbons, it can be time-consuming to write out each atom and bond individually. Hence hydrocarbons can be represented in a short hand notation called a skeletal structure. In a skeletal structure, only the bonds between carbon atoms are represented. Neither individual carbon and hydrogen atoms nor the bonds between carbon and hydrogen atoms are drawn. In the case that the molecule contains just single bonds (sp^3 bonds), these bonds are drawn in a "zigzag" fashion. This is because in the tetrahedral geometry all bonds point as far away from each other as possible and the structure is not linear. We may consider the following representations of the propane molecule:

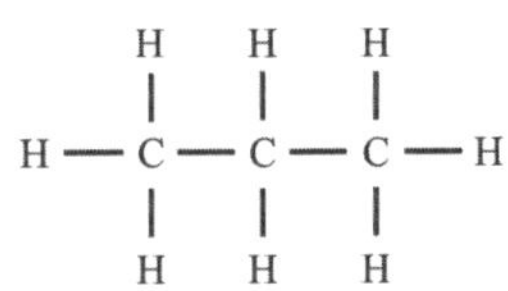

(Full structure of propane)

(Skeletal structure of propane)

Only the bonds between carbons atoms have been drawn and these have been drawn in a "zigzag" manner. Note that there is no evidence of hydrogens in a skeletal structure. Since, in the absence of double or triple bonds, carbon makes total four bonds the presence of hydrogens is implicit. Whenever an insufficient number of bonds to a carbon atom are specified in the structure, it is assumed that the rest of the bonds are made to hydrogens. For example, if the carbon atom makes only one explicit bond, there are three hydrogens implicitly attached to it. If it makes two explicit bonds, there are two hydrogens implicitly attached, etc. Note also that two lines are sufficient to represent three carbon atoms. It is the bonds only that are being drawn out and it is understood that there are carbon atoms (with three hydrogens atoms attached) at the terminal ends of the structure.

9. WHAT ARE RULES FOR NAMING SUBSTITUTED ALKANES?

When hydrocarbons contain only single bonds, they are called alkanes. Alkanes are named using a prefix for the number of carbon atoms they contain, followed by the suffix -ane. The following table lists the rules for naming alkanes.

Alkane nomenclature is straightforward; difficulties come only when one of the hydrogen or carbon atoms on the molecule is replaced by another atom or group. When this takes place, the group which replaces the hydrogen or carbon is called a substituent. Since the only kinds of molecules we've discussed so far are alkanes, let's consider the

Table 14.2: Name of Hydrocarbons.

Number of carbon atoms	*Prefix*	*Compound name*	*Image*
1	Meth-	Methane	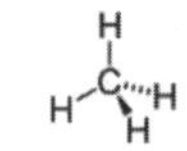
2	Eth-	Ethane	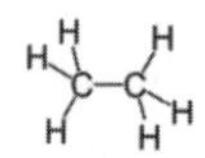
3	Prop-	Propane	
4	But-	Butane	
5	Pent-	Pentane	
6	Hex-	Hexane	
7	Hept-	Heptane	
8	Oct-	Octane	
9	Non-	Nonane	

situation in which one of the hydrogen atoms on an alkane has been replaced by another alkane. Let us consider the molecule, 3-methypentane; depicted as:

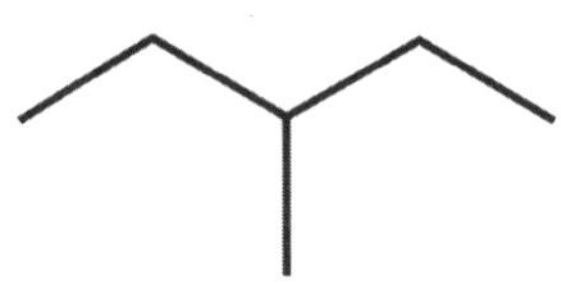

Skeletal structure of 3-methylpentane

Consider the long chain of five carbon atoms at the top of the image. If this were all that composed the molecule, it would simply be called pentane. However, one of the hydrogens on the carbon third from the end has been replaced with an alkane, specifically methane. How are we to name this molecule?

1. First, we identify the longest chain of carbon atoms and name this alkane. It will serve as the root name for the molecule. In the example above, the root name is pentane.
2. Next, we number the carbon atoms, starting at the end that gives the substituent the lowest number. In the example above, we can count from either end and arrive at 3 for the substituent.
3. Next, we name the substituent as if it were an independent alkane. However, we replace suffix -ane with -yl. This name will serve as the prefix. In the example above, methane is the substituent, so we call it methyl.

4. The compound is named in accordance with "number-prefix-root name". In the example above, the name is 3-methylpentane.

10. WHAT HAPPENS IF AN ALKANE HAS MORE THAN ONE SUBSTITUENT?

In this case, the rules concerning nomenclature of alkanes are followed and the carbons on the longest chain are numbered to give the lowest number possible to one of the substituents. The substituents are then all named in the prefix (*e.g.* 2-ethyl,3-methyl). If more than one substituent is attached to the same carbon atom, the number of that carbon atom is repeated to indicate the number of substituents and the prefixes di- (2) or tri- (3) are used. If there are more than one substituents on different carbon atoms, the prefixes are ordered alphabetically (*e.g.* ethene before methane). The prefixes di- and tri- are ignored when considering alphabetical order. For illustration we way consider the following compound:

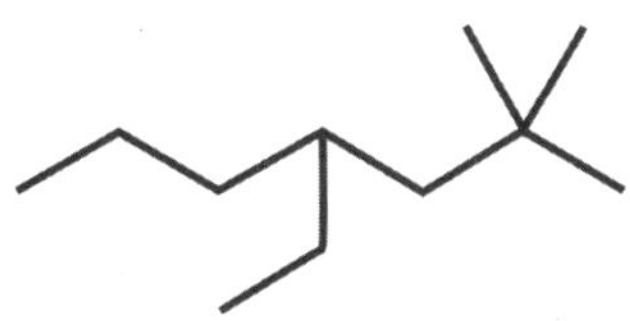

Skeletal structure of 4-ethyl-2,2-dimethylheptane

The longest carbon chain has seven carbon atoms, so the root name is heptane. Numbering from the right gives the lowest number to the first substituent. There are two methyl substituents at the second carbon atom, so we use the prefix 2,2-dimethyl. There is another substituent on the fourth carbon atom, so we use the prefix ethyl. Ethyl comes before methyl alphabetically, so we name the compound 4-ethyl-2,2-dimethylheptane.

11. WHAT IS SPECIAL WITH NAMING OF ALKENES AND ALKYNES?

Alkenes are hydrocarbons containing one or more double bonds. They are named using the same general naming rules that are used for alkanes, except that the suffix is now -ene. There are a few other small differences as outlined below:

- The main chain of carbon atoms must contain both carbons in the double bond and the main chain is numbered so that the double bond gets the smallest number.
- Before the root name, the number of the carbon atom at which the double bond starts (the smaller number) is written.
- If more than one double bond is present, the prefixes di-, tri-, tetra-, etc. are used before the prefix -ene and (strangely) the letter "a" is added after the prefix for the number of carbon atoms.

Alkynes are hydrocarbons containing one or more triple bonds. They are named using the same general procedures that are used for alkenes, replacing the suffix –ene with -yne. If a molecule contains both a double and a triple bond, the carbon chain is numbered so that the first multiple bond gets a lower number. If both bonds can be assigned the same number, the double bond takes precedence. The molecule is then named "n-ene-n-yne", with the double bond root name preceding the triple bond root name (*e.g.* 2-hepten-4-yne).

12. HOW ARE FUNCTIONAL GROUPS NAMED?

Since carbon-carbon and carbon-hydrogen bonds are among the most stable bonds in chemistry, alkanes are extremely unreactive and they seldom undergo reactions. Thus, alkanes alone are not typically used in reactions in organic chemistry. Instead, they serve as a backbone or template on which unreactive carbon or hydrogen atoms can be replaced by substituents consisting of more reactive atoms or groups of atoms.

In addition to carbon and hydrogen, hydrocarbons can also contain other elements. In fact, many common groups of atoms can occur within organic molecules, these groups of atoms are called functional groups. Generally a substituent consisting of an atom or group of atoms other than carbons and hydrogens is called a functional group. One good example is the hydroxyl functional group. The hydroxyl group consists of a single oxygen atom bound to a single hydrogen atom (-OH). The group of hydrocarbons that contain a hydroxyl functional group is called alcohols. The alcohols are named in a fashion similar to the simple hydrocarbons; a prefix is attached to a root ending (in this case "anol") that designates the alcohol.

The existence of a functional group completely changes the chemical properties of the molecule. Ethane, the two-carbon alkane, is a gas at room temperature; ethanol, the two-carbon alcohol, is a liquid. Ethanol, common drinking alcohol, is the active ingredient in "alcoholic" beverages such as beer and wine. Functional groups are significant because they are the part of the molecule that undergoes reactions. One functional group may change into another one, or a functional group might react with a separate molecule to build up a larger structure. Functional groups are the essential "reacting units" in organic chemistry.

We generally encounter two functional groups: the double bonds in alkenes and the triple bonds in alkynes. They are able to undergo reactions that the single bonds in alkanes cannot. Thus, double and triple bonds are the first and simplest examples of functional groups. There are several other significant functional groups, summarized in the table below. Note that in organic chemistry, the letter "R" represents any alkane and the letter "X" represents any halogen.

Alkyl halides (R-X)

One of the simplest functional groups is the alkyl halide. In an alkyl halide, one of the hydrogen atoms in an alkane has been replaced by a halogen. Alkyl halides are easy to

The Important Functional Group and their Structured Representatives

Functional group	*Structure*
Alkyl halide	R-X
Alcohol	R-OH
Ether	R-O-R
Amine	NR_3

name. The name of the alkane is preceded by the number of the carbon on which the halogen is substituted and the name of the halogen, modified so that -ine is replaced by -o (*e.g.* 2-bromopropane). If a molecule also contains a multiple bond, numbers are assigned to give the lowest number to the first functional group. In the event of a tie, the lowest number goes to the multiple bonds.

Alcohols (R-OH)

The alcohol is a very common functional group and a very easy one to name. The molecule is named as if it were an alkane (or alkene or alkyne), except that the suffix -ane is replaced by -ol and the number of the carbon atom on which the -OH group is located is placed before the name of the compound (*e.g.* 2-butanol). The alcohol functional group takes precedence over alkyl substituents, multiple bonds and halides and always gets the lowest number.

Ethers (R-O-R)

An ether is a molecule consisting of two alkyl groups connected to an oxygen atom. Ethers are named by considering one alkyl group (the shorter one) plus the oxygen atom to be a substituent and the other alkyl group (the longer one) to be an alkane. The alkyl group plus oxygen atom is called an "alkoxy" substituent and is named by replacing -ane suffix from the alkane with -oxy (*e.g.* methane becomes methoxy). The alkoxy substituent gets priority over alkyl and halide substituents, but not over alcohols, which will get the lower number.

Amines (NR_3)

An amine is a derivative of the molecule ammonia, NH_3, in which one or more of the hydrogens has been replaced by an alkyl substituent (R group). Amines are named by treating the amino group as a substituent and giving it the name "amino" (*e.g.* 2-aminobutane). If multiple hydrogens have been replaced by alkyl substituents, then these alkyl substituents are stated before the word "amino" (*e.g.* 2-dimethylaminobutane).

13. HOW ARE ISOMERS CLASSIFIED?

Several methods are used to distinguish between organic molecules. They can be classified as alkanes, alkenes, or alkynes depending on the presence of single, double and triple bonds and their functional groups can be identified. A systematic method for naming compounds has also been presented, allowing different compounds to be differentiated. Sometimes, two different molecules can possess the same chemical formula. Two distinct molecules with the same chemical formula are called isomers.

Constitutional Isomers

Constitutational isomers or structural isomers are molecules with the same chemical formula but different structures of atoms and bonds. For example, both 3-methylpentane and hexane have the same chemical formula, C_6H_{14}, yet they clearly have different structures:

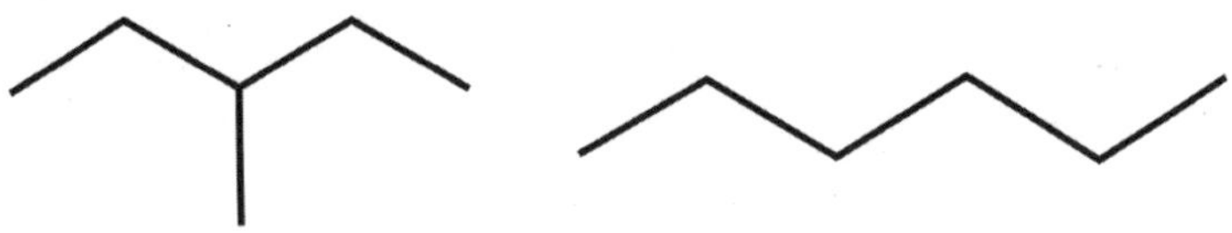

3-methylpentane ***hexane***

Another example involves functional groups. Methoxy methane, an ether and ethanol, an alcohol, both have the chemical formula C_2H_6O:

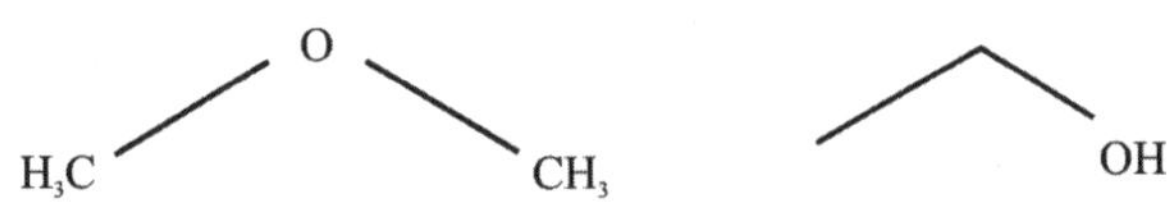

Methoxy methane ***Ethanol***

Enantiomers

Sometimes, two molecules can have the same chemical formula and even the same structure of atoms bonded to each other but can still be distinctly different molecules (isomers). We may ask as to how can this be so? Not only the arrangements in which atoms are bonded to each other, but also the actual geometry of those bonds in space, determine the properties of a molecule. Stereoisomers are distinct molecules with the same sequence of bonded atoms but a different orientation of these atoms in space. Consider the following representations of the molecule bromochlorofluromethane:

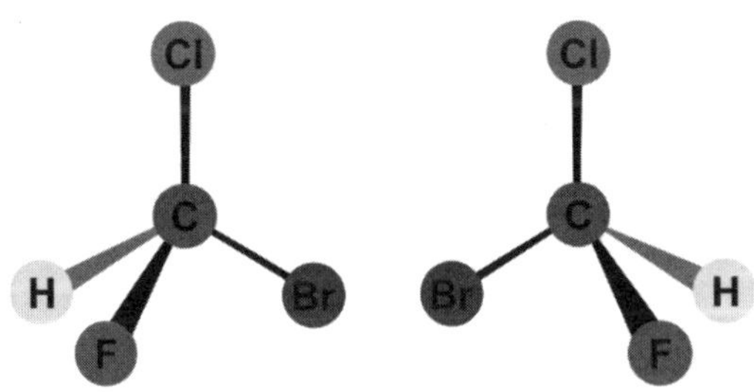

S enantiomer ***R enantiomer***

Fig. 14.12: Enantiomeric representations of bromochlorofluromethane

The two representations clearly depict the same structure of atoms bonded to each other. Yet they are clearly different molecules. No rotation will cause the molecule on the right to adopt the same arrangement of bonds in space as the molecule on the left. If bromine is rotated to adopt the same location as the molecule on the left, then the hydrogen and fluorine atoms will have switched places. Clearly the two molecules are different. In fact they are mirror images of each other. What is special about bromochlorofluromethane is that its mirror image can never be manipulated to produce the original molecule; that is, the mirror image can never be superimposed on the original.

Two molecules which are non-superimposable mirror images of each other are called enantiomers. One special property of enantiomers is that despite being distinct molecules, they have the same physical properties. The only exception is that enantiomers will rotate plane polarized light in different directions. This property is used to make a distinction between them. The amino acids that compose proteins in living organisms are enantiomers, but their only one form exists in most living organisms. Thus, often only one enantiomer of a drug molecule has an effect in the human body.

The simplest ones of enantiomers involve four different groups bonded to a central carbon atom. When the four groups are distinct, the molecule is an enantiomer and the carbon atom is called a chiral center or stereogenic center. Two enantiomers can be distinguished by assigning a letter to the chiral center, either R or S. First, the three substituents are numbered according to atomic mass. The atom with the highest atomic mass is assigned highest priority (1), the second heaviest is assigned next highest priority (2) and so on. If two atoms are the same, the atoms bonded to them are examined and so on until the first point of difference, at which point the assignment of priority is made. If a rotation through the numbers 1, 2 and 3 is clockwise, the enantiomer is labeled R; if it is counterclockwise, it is labeled S. Let's assign R or S configurations to the above enantiomers. In these molecules =, bromine is the heaviest, so it gets the highest priority (1), Chlorine comes next (2) and Fluorine comes third (3). In the molecule on the left, if we follow along passing from 1 to 2 to 3, we find that we have made a rotation in the counterclockwise direction; thus the left molecule is the S enantiomer. If we now follow from 1 to 2 to 3 on the right, we find that we have moved in a clockwise direction; the molecule on the right is the R enantiomer. The system of assigning R or S labels to enantiomers, called the Cahn–Ingold–Prelog system is a simple way of distinguishing enantiomers.

Diastereomers

Diastereomers are stereoisomers that are not enantiomers; that is, they are distinct molecules with the same structural arrangement of atoms that are non-superimposable, non-mirror images of each other. Let us consider the following representations of tartaric acid:

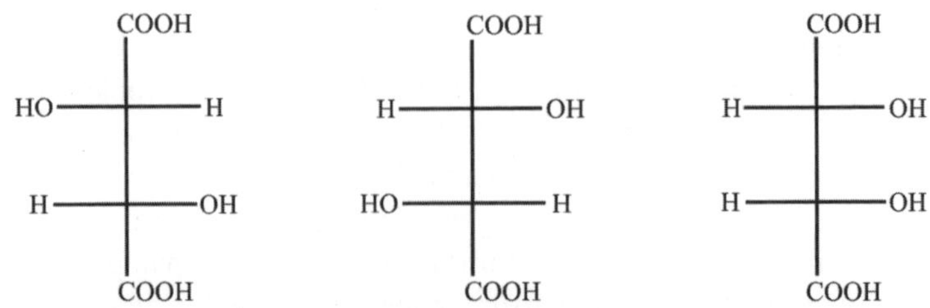

(2S,3S)-tartaric acid ***(2R,3R)-tartaric acid*** ***(2R,3S)-tartaric acid***

Fig. 14.13: Enantiomeric representations of tartaric acid

In tartaric acid, both of the central carbon atoms are stereogenic centers. Thus, they both must be assigned R or S configurations according to the Cahn–Ingold–Prelog system. The first two representations of this molecules shown above, (2S,3S)-tartaric acid and (2R,3R)-tartaric acid, are clearly enantiomers of each other since they are mirror images. But what about the third representation of the molecule, (2R,3S)-tartaric acid? It is clearly distinct from the first two, yet the same atoms are bonded to each other. It is also very clearly not a mirror image of either of the first two. Thus, as a non-superimposable, non-mirror image with the same arrangement of atoms, (2R,3S)-tartaric acid is a diasteromer of the first two molecules.

Another type of diasteromer can occur in the presence of a double bond. Consider the following representations of the molecule 2-butene:

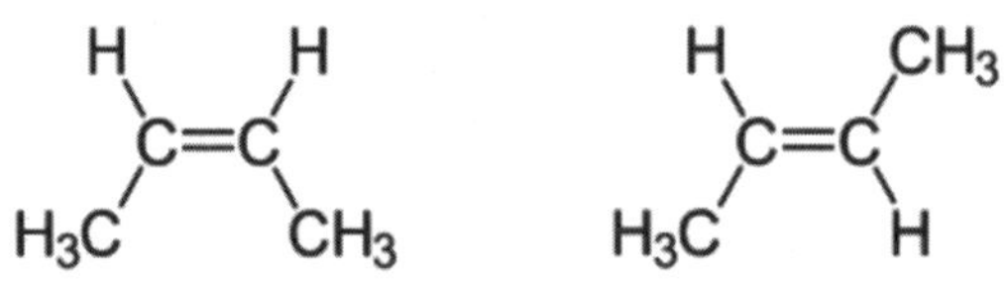

cis-2-butene ***trans-2-butene***

Fig. 14.14: Diastereomeric representations of 2-butene

The molecules are clearly distinct; they clearly have the same arrangement of atoms structurally; and they clearly are not mirror images. Thus, these molecules are diastereomers. Diastereomers such as these can be distinguished using the Cahn–Ingold–Prelog system. If there are only two substituents, as above, the molecule is simply inspected to determine whether the higher priority substituents are on the same side of the double bond. If they are, the molecule is designated cis. If not, it is designated trans. If more than two different substituents are present on the double bond, then the highest priority substituent on each carbon atom composing the double bond is identified. If these highest priority substituents are on the same side of the double bond, the molecule is designated Z; if not, it is designated E.

14. WHAT THE TERMS: CONFORMATIONAL ANALYSIS AND NEW MAN PROJECTION STAND FOR?

We have seen that a different geometric arrangement of bonds in space can produce distinct molecules, stereoisomers, despite the same bonding structure. Even within the same molecule, however, different spatial arrangements are available to a molecule due to rotations about bonds. Different spatial arrangements due to rotations about bonds are called conformations. A conformational analysis is a study of the energetics of different spatial arrangements of atoms relative to rotations about bonds. Conformational analyses are assisted greatly by a representation of molecules in a manner different to skeletal structures. A representation of a molecule in which the atoms and bonds are viewed along the axis about which rotation occurs is called a Newman projection (as has been depicted below).

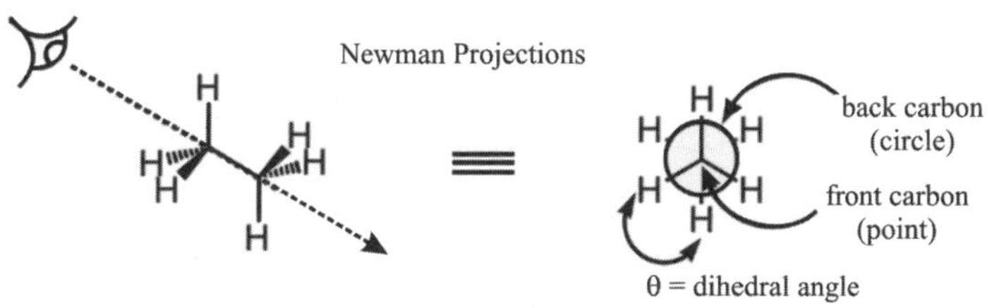

Fig. 14.15: Newman Projections

15. WHAT ARE NEWMAN PROJECTIONS OF ETHANE?

In a Newman projection, the molecule is viewed along an axis containing two atoms bonded to each other and the bond between them, about which the molecule can rotate. In a Newman projection, the "substituents" of each atom composing the bond, be they hydrogens or functional groups, can then be viewed both in front of and behind the carbon-carbon bond. Specifically, one can observe the angle between a substituent on the front atom and a substituent on the back atom in the Newman projection, which is called the dihedral angle or torsion angle.

In ethane specifically, we can imagine two possible "extreme" conformations. In one case, the dihedral angle is 0° and the hydrogens on the first carbon line up with or eclipse the hydrogens on the second carbon. When the dihedral angle is 0° and the hydrogens line up perfectly, ethane has adopted the eclipsed conformation. The other extreme occurs when the hydrogens on the first carbon are as far away as possible from those on the second carbon; this occurs at a dihedral angle of 60° and is called the staggered conformation.

The staggered conformation of ethane is more stable, a lower energy conformation than the eclipsed conformation because the eclipsed conformation involves unfavorable interactions between hydrogen atoms. Specifically, the negatively charged electrons in the bonds repel each other most when the bonds line up. Thus, ethane spends most of its time in the more stable staggered conformation.

staggered
θ = 0°

eclipsed
θ = 0°

Fig. 14.16: A sketch of conformations of ethane: Staggered and Eclipsed

16. WHAT CONFORMATIONS ARE POSSIBLE IN BUTANE?

In butane, two of the substituents, one on each carbon atom being viewed, is a methyl group. Since methyl groups are much larger than hydrogen atoms, when an eclipsed conformation occurs in butane, the interactions are especially unfavorable. There are four possible "extreme" conformations of butane: 1) Fully eclipsed, when the methyl groups eclipse each other; 2) Gauche, when the methyl groups are staggered but next to each other; 3) Eclipsed, when the methyl groups eclipse hydrogen atoms; and 4) Anti, when the methyl groups are staggered and as far away from each other as possible.

The fully eclipsed conformation is clearly the highest in energy and least favorable as the largest methyl groups are interacting directly with each other. As the molecule rotates, it adopts the relatively stable gauche conformation. As it continues to rotate, it encounters less favorable eclipsed conformation in which a methyl group eclipses a hydrogen atom. As rotation continues, the molecule comes to the anti conformation, which is the most stable since the substituents are staggered and the methyl groups are as far away from each other as possible.

fully eclipsed
(eclipsed)
°Me/Me = 0°

gauche
(staggered)
°Me/Me = 60°

eclipsed
(eclipsed)
°Me/Me = 120°

anti
(staggered)
°Me/Me = 180°

Fig. 14.17: A sketch of conformations of butane: Fully eclipsed, Gauche, Eclipsed and Anti

17. WHAT IS AN ENERGY DIAGRAM AND WHY ALKANES DO NOT INTERCONVERT?

An energy diagram is a graph which represents the energy of a molecule as a function of some changing parameter. An energy diagram can be made as a functionof dihedral angle for butane.

Clearly, the anti and gauche conformations are significantly more stable than the eclipsed and fully eclipsed conformations. Butane spends most of its time in the anti and gauche conformations. There is one final, very important point. At room temperature,

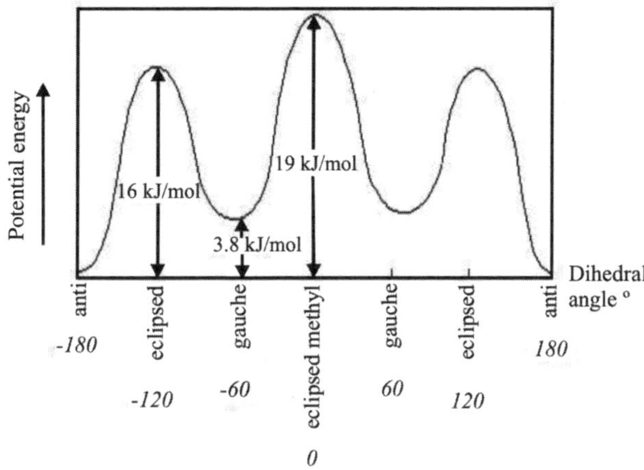

Conformer potentials of butane about central C-C bond

Fig. 14.18: Energy diagram for conformations of butane as a function of dihedral angle

approximately 84 kJ/mol of thermal energy is available to molecules. Thus, if the barrier to a rotation is less than 84 kJ/mol, the molecule will rotate. In ethane and butane, the barriers to rotation are significantly less than 84 kJ/mol. Therefore, even though the eclipsed conformations are unfavorable, the molecules are able to adopt them. In reality, since these conformations are not as stable, the molecules will quickly pass through them at room temperature and return a staggered conformation. Molecules are constantly converting between different staggered conformations all the time, quickly passing through eclipsed conformations in between. Thus, in alkanes, no single "true" conformation exists all the time; the molecule instead constantly converts between conformations, spending more time in those that are more stable. This constant conversion lies in stark contract to alkenes, which adopt the cis- or trans- (E or Z-) conformations and retain them at room temperature; they do not interconvert because the barrier to rotation is too high.

The alkanes we have studied so far have been of two types: linear and branched. There is a third type of alkane in which the molecule does not have ends but instead forms a ring. These molecules are called cycloalkanes.

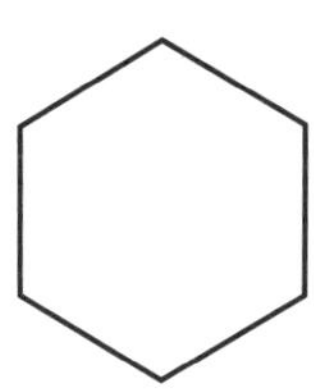

Fig. 14.19: Skeletal structure of a cyclohexane

18. WHAT THE TERMS 'ANGLE STRAIN, TORSIONAL STRAIN, RING STRAIN AND RING FLIP' STAND FOR?

Stable cycloalkanes cannot be formed with carbon chains of just any length. Recall that in alkanes, carbon adopts the sp^3 tetrahedral geometry in which the angles between bonds are 109.5°. For some cycloalkanes to form, the angle between bonds must deviate from this ideal angle, an effect known as angle strain. Additionally, some hydrogen atoms may come into closer proximity with each other than is desirable (become eclipsed), an effect called torsional strain. These destabilizing effects, angle strain and torsional strain are known together as ring strain. The smaller cycloalkanes, cyclopropane and cyclobutane, have particularly high ring strains because their bond angles deviate substantially from 109.5° and their hydrogens eclipse each other. Cyclopentane is a more stable molecule with a small amount of ring strain, while cyclohexane is able to adopt the perfect geometry of a cycloalkane in which all angles are the ideal 109.5° and no hydrogens are eclipsed; it has no ring strain at all. Cycloalkanes larger than cyclohexane have ring strain and are not commonly encountered in organic chemistry.

Most of the time, cyclohexane adopts the fully staggered, ideal angle chair conformation. In the chair conformation, if any carbon-carbon bond were examined, it would be found to exist with its substituents in the staggered conformation and all bonds would be found to possess an angle of 109.5°.

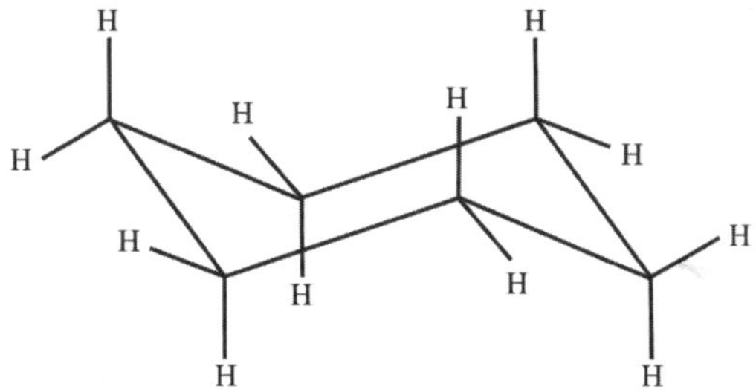

Fig. 14.20: A sketch of cyclohexane in the chair conformation

In the chair conformation, hydrogen atoms are labeled according to their location. Those hydrogens which exist above or below the plane of the molecule are called axial. Those hydrogens which exist in the plane of the molecule are called equatorial. Although the chair conformation is the most stable conformation that cyclohexane can adopt, there is enough thermal energy for it to also pass through less favorable conformations before returning to a different chair conformation. When it does so, the axial and equatorial substituents change places. The passage of cyclohexane from one chair conformation to another, during which the axial substituents switch places with the equatorial substituents, is called a ring flip.

19. WHAT CHAIR CONFORMATIONS CAN BE ADOPTED BY METHYL-CYCLOHEXANE?

Methylcyclohexane is cyclohexane in which one hydrogen atom is replaced with a methyl group substituent. Methylcyclohexane can adopt two basic chair conformations: one in which the methyl group is axial and one in which it is equatorial. Methylcyclohexane strongly prefers the equatorial conformation. In the axial conformation, the methyl group comes in close proximity to the axial hydrogens, an energetically unfavorable effect known as a 1, 3-diaxial interaction. Thus, the equatorial conformation is preferred for the methyl group. In most cases, if the cyclohexane ring contains a substituent, the substituent will prefer the equatorial conformation.

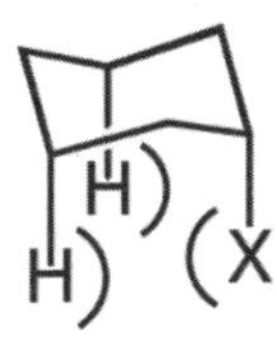

Fig. 14.21: 1,3-diaxial interaction

Sometimes, multiple chemical structures in which electrons are shared between just two atoms can be used to represent a single molecule. In these cases no single structure is correct and the molecule exhibits special stability. This phenomenon imparts special chemical properties, especially for certain cyclic systems.

20. WHAT IS MEANT BY RESONANCE STABILIZATION?

In order to define resonance stabilization let us consider the following representations of an acetate anion:

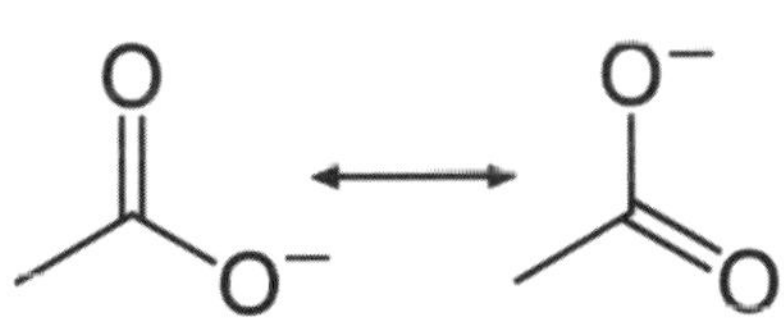

Fig. 14.22: Resonance structures of the acetate anion

The bonding in acetate may be described with a double bond made to the first oxygen atom and a negative charge placed on the second, or it may be represented with a negative charge on the first and a double bond to the second. Which structure is correct? Experimental evidence indicates that the bond lengths for the both carbon-oxygen bonds in acetate are actually in between the lengths of single and double bonds. When two chemical structures involving electrons shared between just two atoms may be drawn that represent a molecule, the molecule's true structure is intermediate between them. The contribution from multiple hypothetical structures to the true structure of a molecule is called resonance. The two structures are called resonance structures and the intermediate is called a resonance hybrid. Resonance often results in the presence of electrons which cannot be assigned to a single atom or bond; the electrons are said to be delocalized. A resonance hybrid is always more stable than the resonance structures that contribute to it, a phenomenon known as resonance stabilization.

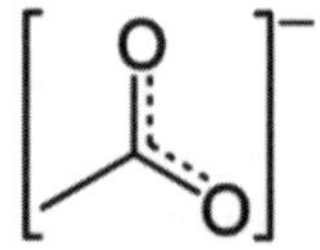

Fig. 14.23: Resonance hybrid of the acetate anion; the dashed line represents a delocalized electron

21. WHAT IS UNDERLYING BASIS OF AROMATICITY AND HÜCKEL'S RULE?

A conjugated system is a chemical system, such as a series of alternating single and double bonds, in which there is a p orbital on each atom. Owing to resonance, in a conjugated system of alternating bonds, the double and single bonds are able to switch places, producing an overall more stable structure. Conjugated systems can also exist in cyclic molecules. A conjugated cyclic system which is stabilized by resonance is called an aromatic system. The classic example of an aromatic system involves a six-membered ring. There are two possible chemical structures for a conjugated six-membered ring.

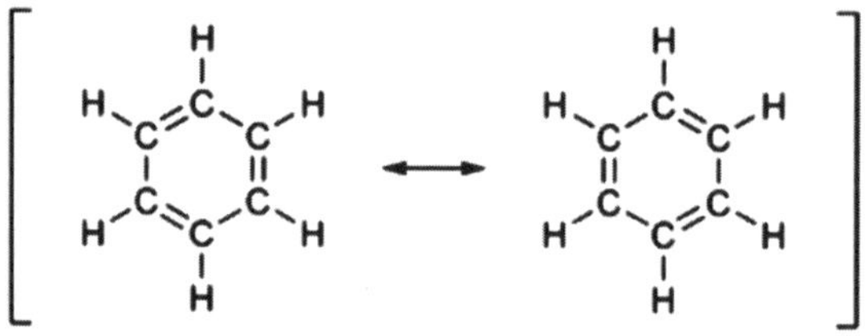

Fig. 14.24: Resonance structures of benzene

A conjugated six-membered ring is called benzene. In an aromatic system like benzene, each atom has a p orbital, the electrons of which are delocalized about the system. Benzene and other aromatic compounds exhibit chemistry very different from ordinary, non-aromatic hydrocarbons. To make distinction between them, non-aromatic compounds are labeled aliphatic. Benzene and other aromatic compounds can have substituents. When benzene itself is a substituent, it is called a phenyl group. Benzene is typically drawn as under in such a way that the hybrid between the resonance structures is emphasized.

We have stated that a conjugated cyclic system stabilized by resonance is described as aromatic. But not every conjugated cyclic system is aromatic since not all are

stabilized by resonance, mainly due to differences in filling molecular orbitals with electrons. A system of rules was developed by chemist Erich Hückel to determine whether a system is aromatic:

- The system must be conjugated (a p orbital must be present on each atom).
- The system must be cyclic.
- The system must be planar.
- The system must contain a total of 4n $_{+2}\pi$ electrons (n is any integer).
- The last rule is known as Hückel's Rule. The π electrons can come from double bonds (2 electrons per double bond) or negative charges (2 per negative charge). Benzene, for example, has 6 π electrons.

22. HOW THE PARENT NAME IS DETERMINED IN CASE OF CYCLIC ALKANES?

Hydrocarbons are the simplest organic compounds that contain only two elements, hydrogen and carbon. A saturated hydrocarbon or alkane is a hydrocarbon in which all of the carbon-carbon bonds are single bonds. Each carbon atom forms four bonds and every hydrogen atom forms a single bond to a carbon. The bonding around each carbon atom is tetrahedral, so all bond angles are 109.5°. As a result, the carbon atoms in higher alkanes are arranged in zigzag rather than linear patterns. The general formula for an alkane is C_nH_{2n+2} where n is the number of carbon atoms in the molecule. There are two ways of writing a condensed structural formula. For example, butane may be written as $CH_3CH_2CH_2CH_3$ or $CH_3(CH_2)_2CH_3$.

The parent name of the molecule is determined by the number of carbons in the longest chain. In the case where two chains have the same number of carbons, the parent is the chain with the most substituents. The carbons in the chain are numbered starting from the end nearest the first substituent. In the case where there are substituents having the same number of carbons from both ends, numbering starts from the end nearest the next substituent. When more than one of a given substituent is present, a prefix is applied to indicate the number of substituents. Besides using di- for two, tri- for three, tetra- for four, etc. one should use the number assigned to the carbon to indicate the position of each substituent.

Branched substituents are numbered starting from the carbon of the substituent attached to the parent chain. From this carbon, count the number of carbons in the longest chain of the substituent. The substituent is named as an alkyl group based on the number of carbons in this chain. Numbering of the substituent chain starts from the carbon attached to the parent chain. The entire name of the branched substituent is placed in parentheses, preceded by a number indicating which parent-chain carbon it joins. Substituents are listed in alphabetical order. To alphabetize, ignore numerical (di-, tri-, tetra-) prefixes (*e.g.*, ethyl would come before dimethyl), but don't ignore positional prefixes such as iso and tert (*e.g.*, triethyl comes before tertbutyl).

In case of cyclic alkanes the parent name is determined by the number of carbons in the largest ring (*e.g.*, cycloalkane such as cyclohexane). In the case where the ring is attached to a chain containing additional carbons, the ring is considered to be a substituent on the chain. A substituted ring that is a substituent on something else is named using the rules for branched alkanes. When two rings are attached to each other, the larger ring is the parent and the smaller is a cycloalkyl substituent. The carbons of the ring are numbered such that the substituents are given the lowest possible numbers.

Chapter 15

ORGANIC COMPOUNDS

1. WHY ORGANIC CHEMISTRY IS OF VITAL IMPORTANCE AND WHAT MADE THE VITAL FORCE THEORY OF ORGANIC COMPOUNDS OBSOLETE?

We now know that organic chemistry is the branch of chemistry that deals only with the compounds of carbon which in spite of being only the fourteenth most common element on earth, forms by far the greatest number of different compounds. We also know that all carbon compounds except a few inorganic carbon compounds are organic. Organic chemistry is of vital importance to the petrochemical, pharmaceutical and textile industries, where our prime concern is the synthesis of new organic molecules and polymers. In general, a particular type of organic compound, such as an alcohol, aldehyde, ether, or ketone, is identified by the presence of a characteristic functional group of atoms.

The functional group is not only the part of the molecule but also the most responsible component for its particular chemical nature. Organic compounds containing nitrogen are of great importance in biochemistry. They generally contain the amine group (NH_2). Molecules containing both the NH_2 and COOH groups are called amino acids and are the building blocks of proteins. Inorganic carbon compounds include the oxides of carbon, the bicarbonates and carbonates of metal ions, the metal cyanides and a few others.

There are over six million organic compounds characterized, including the foods we eat, (made of carbohydrates, lipids, proteins and vitamins), furs and feathers, hides and skins and the organisms they came from not to mention plastics, synthetic and natural fibers, dyes and drugs, insecticides and herbicides, ingredients in perfumes and flavoring agents and petroleum products. The name organic chemistry came from the word organism. Prior to 1828, all organic compounds had been obtained from organisms or their remains. The belief then was that the synthesis of organic compounds from inorganic compounds in the laboratory was impossible. All efforts had failed and scientists became convinced that some "vital force" that living organisms had, was necessary to make an organic compound. The synthesis of urea from inorganic substances in 1828 made this vital force theory obsolete.

2. WHAT ARE DIFFERENT FORMS OF CARBON?

Carbon, an interesting and versatile element with atomic number 6 and symbol C, has been known since prehistoric times. It gets its name from carbo, the Latin word for charcoal, which is almost pure carbon. In various forms, carbon is found not only on Earth, but in the atmospheres of other planets, in the Sun and stars, in comets and in some meteorites. On Earth, carbon can be considered to be the most important of all the chemical elements for obvious reasons. Carbon is the fourth most abundant element in the known universe but not nearly as common on the earth, despite the fact that living organisms contain significant amounts of the element. Common carbon compounds in the environment include the gaseous carbon dioxide (CO_2) and methane (CH_4).

Carbon exists in several allotropic forms. Diamond is one form with a very strong crystal lattice, known as a precious gem from the most ancient records. Graphite is another allotrope in which the carbon atoms are arranged in planes which are loosely attracted to one another (hence its use as a lubricant). The recently discovered fullerenes are yet another form of carbon. Most elemental carbon is taken from the ground in the form of coal but certainly diamonds should not be ignored. Carbon has a very high melting and boiling point and rapidly combines with oxygen at elevated temperatures. In small amounts it is an excellent hardener for iron, yielding the various steel alloys upon which so much of modern construction depends. An important (but rare) radioactive isotope of carbon, C-14, is used to date ancient objects of organic origin. It has a half-life of 5730 years but there is only 1 atom of C-14 for every 10^{12} atoms of C-12 (the usual isotope of carbon). The great number of carbon compounds is possible because of the ability of carbon to form strong covalent bonds to each other while also holding the atoms of other nonmetals strongly. Chains of carbon atoms can be thousands of atoms long, as in polyethylene.

3. WHAT ARE ISOMERS AND WHY THE NUMBER OF POSSIBLE ISOMERS FOR ANY GIVEN FORMULA BECOMES VERY LARGE?

An isomer, in chemistry, is one of two or more compounds having the same molecular formula but different structures (arrangements of atoms in the molecule) and isomerism is the occurrence of such compounds. Isomers have the same number of atoms of each element in them and the same atomic weight but differ in other properties. For example, there are two compounds with the molecular formula C_2H_6O, one is ethanol (also called ethyl alcohol), CH_3CH_2OH, a colorless liquid alcohol; the other is dimethyl ether, CH_3OCH_3, a colorless gaseous ether. Among their different properties, ethanol has a boiling point of 78.5°C and a freezing point of –117°C; dimethyl ether has a boiling point of –25°C and a freezing point of –138°C. Ethanol and dimethyl ether are isomers because they differ in the way the atoms are joined together in their molecules.

Isomers are classified as structural isomers, which have the same number of atoms of each element and molecular weight but different bonding patterns, or as stereoisomers, which have the same number of atoms of each element, molecular weight and bonding pattern but in which the atoms have different spatial relationships. Tautomers are structural isomers that readily convert from one isomeric form to another and therefore exist in equilibrium. Stereoisomers are important in metabolism; in many cases only one of several isomeric forms of a compound can take part in biochemical reactions. For example, there are 16 stereoisomers of a simple sugar whose molecular formula is $C_6H_{12}O_4$. Of these, only d-glucose is readily utilized in human metabolism. Isomerism is another reason why there are so many organic compounds. Depending on how they are arranged, isomers may have similar or different properties. Because the number of carbon atoms per molecule increases as the compound gets more complex, the number of possible isomers for any given formula becomes very, very large.

Formula	*Number of isomers*
C_8H_{18}	18
$C_{10}H_{22}$	75
$C_{20}H_{42}$	366
$C_{40}H_{82}$	6.25×10^{13} (approx.)

4. WHY FUNCTIONAL GROUPS OF ORGANIC FAMILIES ARE IMPORTANT?

Since organic compounds are huge in number, they require organization. They are sorted into organic families defined by functional groups. Functional group, in organic chemistry, is group of atoms within a molecule that is responsible for certain properties of the molecule and reactions in which it takes part. Organic compounds are frequently classified according to the functional group or groups they contain. For example, methanol, ethanol and isopropanol are all classified as alcohols since each of them contains a functional hydroxyl group. Functional groups are small structural units within molecules at which most of the compound's chemical reactions occur. For example, two of the most important families are the alcohols and the carboxylic acids. Their functional groups, the alcohol group and the carboxyl group, respectively, distinguish them from the rest of the other types of organic compounds. The type of properties associated with a compound is different depending on the family it is in. For example, the alkanes have just C-C and C-H single bonds. Since C and H are so alike in electronegativity, they are least able to attract ions or polar molecules and least able to interact with them.

When a polar group of atoms, like the OH group or NH group are attached to a carbon atom, the molecule has a polar site. It may undergo chemical reactions when it attracts polar and ionic reactants, but it will usually be near this functional group. This is why compounds with the same functional group have similar chemical properties, or kinds of reactions. For example, in the amine group, both methylamine and ethylamine give the same kinds of reactions. Only the NH_2 group of the amines change in the previous two reactions and it changes the same way in both.

5. WHAT THE SYMBOL 'R' IN STRUCTURAL FORMULAS OF ORGANIC COMPOUNDS STAND FOR?

To save space and time, condensed structures may be used to simplify the drawing and writing of structural formulas of organic compounds. While C-H bonds are obviously understood, CH_3 means that three hydrogen atoms are bonded to a carbon atom. The symbol R stands for the word radical and in structural formulas of organic compounds or in organic chemistry in general, represents alkane-like groups. It is sometimes used to simplify equations or to summarize them. The R takes care of any alkane group that we may substitute in. It takes care of all the equations, including methylamine, ethylamine, propylamine, butylamine, etc. It gives a simpler view of how one group, not just the individual compound, changes into another and how each group affects its properties.

6. WHY THERE ARE SO MANY ISOMERS FOR ORGANIC COMPOUNDS AND WHAT CONVENTIONS ARE FOLLOWED FOR DRAWING CONDENSED VERSION OF THE STRUCTURAL FORMULAS?

In organic chemistry, there are a few basic structural shapes that we generally encounter. They are chains and rings. There are also two types of chains: a straight chain and a branched chain. In a straight chain, one carbon atom holds no more than two other carbon atoms. As its name implies, the straight chain is a straight link of carbon, atoms sometimes oxygen or nitrogen atoms in structural formula. Because of twisting and contouring, the chain may have several conformations. Branched chains have at least one carbon holding more than two other carbon atoms. It will, as its name implies, have branches of other chains coming off another chain. Branching is one of the reasons why there are so many isomers for each compound. Rings (or cyclic compounds) are composed of rings of carbon and sometimes oxygen or nitrogen. In the fully condensed version of the structural formula of cyclohexane, or any cyclic compound for that matter, the conventions for drawing them are as under:

- A carbon atom occurs at every corner unless Oxygen or Nitrogen (or some other atom) is explicitly written at the corner.
- A line connecting two corners is a covalent bond between adjacent ring atoms.
- Remaining bonds are understood to hold H atoms (C normally has 4 total bonds). Double bonds are always explicitly shown.

The atom that is not carbon is called the heteroatom. The heteroatom's bonds must be shown or clearly understood. For example, nitrogen normally forms three bonds, so three bonds must be indicated for it. Oxygen normally forms two, so its two bonds must be shown. Rings may have more than one heteroatom.

7. WHY ALKANES ARE INSOLUBLE IN WATER?

It may be mentioned that piperidine and tetrahydropyran are both soluble in water, but cyclohexane and other alkanes are insoluble. The reason is that the alkanes have only C-C and C-H bonds, which are relatively nonpolar and do not tend to form hydrogen bonds. Hydrogen bonds usually form when the H is attached to O or N atoms, two of the three most electronegative atoms. The - charge (relatively negative charge) of the O and the N atoms attract the + charge of the H, since there is a sizable difference in their electronegativities. Since piperidine and tetrahydropyran have heteroatoms of N or O, the water molecules (and other polar fluids) tend to cluster around each molecule. When they cluster around them, the partially + charge of the H in water tends to form hydrogen bonds with the partially - charge in the N or O. Therefore, the molecules of piperidine and tetrahydropyran become dissolved in the water. Because alkanes are relatively nonpolar, they do not dissolve in water, a polar molecule, because hydrogen bonds do not form and water does not cluster around each alkane molecule.

8. CHOLESTEROL MOLECULES ARE EXPECTED TO DISSOLVE IN WATER BUT PRACTICALLY THEY ARE INSOLUBLE, EXPLAIN WHY.

Hydrocarbons are organic compounds that consist of only C and H atoms. They include the alkanes, alkenes, alkynes and aromatic hydrocarbons. Because of their relatively nonpolarity, all hydrocarbons are insoluble in water. When hydrocarbons burn in sufficient oxygen, carbon dioxide and water are the sole products. The main structural difference among hydrocarbon families is the presence of double or triple bonds between carbon atoms. The alkanes are saturated organic compounds, or those with only single bonds. Unsaturated organic compounds are those which have double or triple bonds. To predict the properties from molecular structure is one of the goals of organic chemistry. Functional groups, heteroatoms and molecular size influence physical properties in ways that permit some broad generalizations. A molecule of cholesterol, which has an OH group, is expected to dissolve in water, because it is polar due to its OH group. But it is practically insoluble, because of the effect of the large size of the alkane-like part of the molecule; and the polarity of the OH group is overwhelmed by the huge hydrocarbon group. It is more soluble in nonpolar solvents.

9. WHAT ARE CHARACTERISTICS OF SATURATED OPEN-CHAIN AND CYCLIC HYDROCARBONS?

Hydrocarbons may be considered as the parent substances from which all other organic compounds are derived. The hydrocarbons are conveniently classified into two major groups: open-chain and cyclic ones. In open-chain compounds, containing more than one carbon atom, the carbon atoms are attached to each other to form an open chain; the chain may carry one or more side branches. In cyclic compounds the carbon atoms form one or more closed rings. The two major groups are subdivided according to their chemical behavior into saturated and unsaturated compounds.

The saturated open-chain hydrocarbons form a homologous series called the paraffin (Latin parum affinis, "little affinity") series or the alkane series. The composition of each of the members of the series corresponds to the formula C_nH_{2n+2}, where n is the number of carbon atoms in the molecule. Among the members of the series are methane CH_4; ethane, C_2H_6; propane, C_3H_8; and butane, C_4H_{10}. All the members of the series are unreactive; that is, they do not react readily at ordinary temperatures with such reagents as acids, alkalies, or oxidizers. The first four members of the series are gases at ordinary temperature and pressure; intermediate members are liquids; and the heavier members are semisolids or solids. Petroleum contains a great variety of saturated hydrocarbons and such petroleum products as gasoline, kerosene, heavy fuel oil, lubricating oils and petroleum jelly. The unsaturated open-chain hydrocarbons include the alkene or olefin series, the diene series and the alkyne series. The alkene series is made up of chain hydrocarbons in which a double bond exists between two carbon atoms. The general formula for the series is C_nH_{2n}, where n stands for the number of carbon atoms. As in the paraffin series, the lower members are gases, intermediate compounds are liquids and the higher members of the series are solids. The alkene series compounds are more active chemically than the saturated compounds. They easily react with substances such as halogens, adding atoms at the double bonds. They are not found to any extent in natural products, but are produced in the destructive distillation of complex natural substances, such as coal and are formed in large amounts in petroleum refining, particularly in the "cracking" process. The first member of the series is ethylene, C_2H_4. The dienes contain two double bonds between pairs of carbon atoms in the molecule. They are related to the complex hydrocarbons in natural rubber and are important in the manufacture of synthetic rubber and plastics; important members of this series are butadiene, C_4H_6 and isoprene, C_5H_8.

10. HOW ARE ORGANIC COMPOUNDS NAMED ACCORDING TO THE IUPAC SYSTEM?

A saturated hydrocarbon or alkane is a hydrocarbon in which all of the carbon-carbon bonds are single bonds. Each carbon atom forms four bonds and every hydrogen atom forms a single bond to a carbon. The bonding around each carbon atom is tetrahedral, so all bond angles are 109.5°. As a result, the carbon atoms in higher alkanes are arranged in zigzag rather than linear patterns. There are two ways of writing a condensed structural formula. For example, butane may be written as $CH_3CH_2CH_2CH_3$ or $CH_3(CH_2)_2CH_3$. Open-chain alkanes (without rings) all have the general formula C_nH_{2n+2}, where n equals the number of carbon atoms.

The IUPAC (International Union of Pure and Applied Chemistry) is composed of chemists representing the national chemical societies of several countries. One committee of the IUPAC, the Commission on Nomenclature of Organic Chemistry, has set a system for naming organic compounds. The last syllable in the name of a compound designates the family to which it belongs. The alkanes all end in -ane. The name ending for all alkanes (and cycloalkanes) is -ane. The parent chain is the longest continuous chain of carbons atoms in the structure. A prefix that specifies the number of carbon atoms in the parent chain is attached to the name ending, -ane. The carbon atoms of the parent chain are numbered starting from whichever end of the chain gives the location of the first branch the lower of two possible numbers. When two or more groups are attached to the parent, each is named and located with a number. The alkyl substituent names are assembled in alphabetical order. Hyphens are always used when separating numbers from words. When two or more substituents are identical, prefixes such as di- (2), tri- (3) and tetra- (4) are used and the location number of every group is specified. Always separate a number from another number in a name by a comma. When identical groups are on the same carbon, repeat the number of the carbon in the name as 2,2-dimethylpentane. Sometimes, we may need to go around corners and zigzag to find the longest (parent) chain.

The parent name of the molecule is determined by the number of carbons in the longest chain. In the case where two chains have the same number of carbons, the parent is the chain with the most substituents. The carbons in the chain are numbered starting from the end nearest the first substituent. In the case where there are substituents having the same number of carbons atoms from both ends, numbering starts from the end nearest the next substituent. When more than one of a given substituent is present, a prefix is applied to indicate the number of substituents. Besides using di- for two, tri- for three, tetra- for four, etc., one should use the number assigned to the carbon to indicate the position of each substituent.

Branched substituents are numbered starting from the carbon of the substituent attached to the parent chain. From this carbon, count the number of carbons in the longest chain of the substituent. The substituent is named as an alkyl group based on the number of carbons in this chain. Numbering of the substituent chain starts from the carbon attached to the parent chain. The entire name of the branched substituent is placed in parentheses, preceded by a number indicating which parent-chain carbon it joins. Substituents are listed in alphabetical order. To alphabetize, we should ignore numericals (di-, tri-, tetra-) prefixes (*e.g.*, ethyl would come before dimethyl), but we must not ignore the positional prefixes such as iso and tert (*e.g.*, triethyl comes before tertbutyl).

The parent name is determined by the number of carbons in the largest ring (*e.g.*, cycloalkane such as cyclohexane). In the case where the ring is attached to a chain containing additional carbons, the ring is considered to be a substituent on the chain. A substituted ring that is a substituent on something else is named using the rules for branched alkanes. When two rings are attached to each other, the larger ring is the parent and the smaller is a cycloalkyl substituent. The carbons of the ring are numbered such that the substituents are given the lowest possible numbers.

Alkyl group, in chemistry, is the group of carbon and hydrogen atoms derived from an alkane molecule by removing one hydrogen atom. The name of the alkyl group is derived from the name of its alkane by replacing the -ane suffix with -yl, *e.g.*, methyl, CH_3, from methane, CH_4 and ethyl, C_2H_5, from ethane, C_2H_6. In some cases different alkyl groups can be formed from the same alkane by removing different hydrogen atoms; the alkyl groups are then distinguished by adding a prefix, *e.g.*, 1-propyl or n-propyl, $CH_2CH_2CH_3$ and 2-propyl or isopropyl, $CH(CH_3)_2$, both formed from propane, $CH_3CH_2CH_3$. When a functional group is joined with an alkyl group, replacing the hydrogen that was removed, a compound is formed whose characteristics depend largely on the functional group. An alkyl is basically an alkane minus one of its hydrogen atoms.

11. WHAT ARE CHARACTERISTICS OF ALKANES?

Alkanes are usually stable at room temperature toward not only reactants such as concentrated aqueous acids or bases but even with the most reactive metals. Fluorine, however, attacks virtually all organic compounds, including the alkanes, to give mixtures of products. Hot nitric acid, chlorine and bromine can also react with alkanes. The chlorination of methane can give the following compounds depending on the relative quantities of reactants used:

CH_3Cl : Methyl chloride
CH_2Cl_2 : Methylene chloride
$CHCl_3$: Chloroform
CCl_4 : Carbon tetrachloride

When heated at high temperatures in the absence of air, alkanes can "crack," meaning that they break up into smaller molecules. The cracking of methane gives finely powered carbon and hydrogen gas.

$$CH_4 \xrightarrow{\text{High temperatures}} C + 2H_2$$

The controlled cracking of ethane gives ethene ("ethylene"), which is an important raw material in the organic chemicals industry, used to make polyethylene plastics, ethyl alcohol and ethylene glycol (an antifreeze).

$$\underset{\text{Ethane}}{CH_3CH_3} \xrightarrow{\text{High temperatures}} \underset{\text{ethene}}{CH_2{=}CH_2} + H_2$$

An alkene is any of a group of aliphatic hydrocarbons whose molecules contain one or more carbon-carbon double bonds. Alkenes with only one double bond have the general formula C_nH_{2n}. In the International Union of Pure and Applied Chemistry (IUPAC) system of chemical nomenclature, the name of an alkene is derived from the name of the corresponding alkane by replacing the -ane alkane suffix with -ene and, if necessary, adding a prefix to indicate the location of the double bond in the molecule. The IUPAC name of the simplest alkene, $H_2C{=}CH_2$, is ethene, which is derived from ethane. Propene is related to propane. Two alkenes: 1-butene and 2-butene are related to butane; these two compounds, which differ in the location of the double bond in their molecules, are structural isomers. In addition to these IUPAC names, many of the alkenes have common names, *e.g.*, ethene and propene are called as ethylene and propylene respectively. The alkenes as a group are sometimes called the ethylene series. Since the carbon-carbon double bond is sometimes called an olefinic linkage, the alkenes are sometimes called the olefins. Many of the reactions in which alkenes take part involve the cleavage of half the carbon-carbon double bond and subsequent formation of two single bonds, one to each of the adjacent carbon atoms. Such reactions include hydrogenation, with the formation of an alkane and hydration, with the formation of an alcohol.

The rules for naming alkenes are basically the same as those of alkanes but with two differences. The parent chain must include the double bond even if it makes it shorter than the others. And the parent alkene chain must be numbered from whichever end gives the first carbon of the double bond the lower of two possible numbers. Also, the location number should be given as to where the double bond is (except ethene or propene, where the location will always be 1). Alkenes which have two double bonds are dienes, those with three are trienes and so forth. Each double bond has to be located by a number.

An alkyne is any of a group of aliphatic hydrocarbons whose molecules contain one or more carbon-carbon triple bonds. Alkynes with one triple bond have the general formula C_nH_{2n-2}. In the International Union of Pure and Applied Chemistry (IUPAC) system of chemical nomenclature, the name of an alkyne is derived from the name of the corresponding alkane by replacing the -ane alkane suffix with -yne and, if necessary, adding a prefix to indicate the location of the triple bond in the molecule. The IUPAC name of the simplest alkyne, HC -triple bond -CH, is thus ethyne, which is derived from cthanc. Ethync is morc commonly known as acetylene; it is an extremely important starting material in commercial chemical synthesis. The next simplest alkyne is propyne, CH_3-Ctriplebond-CH. There are two butynes: 1-butyne and 2-butyne that are structural isomers which differ in the location of the triple bond in their molecule. The alkynes are sometimes referred to as the acetylene series, the higher members of the series being named as derivatives of acetylene, *e.g.*, propyne as methyl acetylene, 1-butyne as ethyl acetylene and 2-butyne as dimethylacetylene. The usefulness of the alkynes in chemical synthesis is due both to the reactions of the triple bond itself and to the relative acidity of a hydrogen atom bonded to a triply bonded carbon.

12. WHY DO ALKENES ACT AS LEWIS BASES?

As there is no free rotation of a double bond, many alkenes exhibit geometric isomerism. For example, cis-2-butene and trans-2-butene are geometric isomers. Cis means "on the same side," while trans means "on opposite sides." They refer to the orientation of the structure presented. Generally, the differences in physical properties are measurable, but their chemical properties are very similar. Alkenes are Lewis bases (electron pair donors) because the pi bond of the carbon-carbon double bond is projected outward where electron-seeking reactants are able to get it. They will react very readily with Lewis acids (electron pair acceptors) and strong Brønsted acids (proton donors). Ethene readily reacts with hydrogen chloride to make 1-chloroethane. Another reaction involves ethene and water, giving ethanol (ethyl alcohol), while sulfuric acid acts like a catalyst.

13. WHY AND HOW ARE VEGETABLE OILS AND COAL HYDROGENATED?

When an alkene is hydrogenated, it becomes an alkane. It requires a catalyst, powdered platinum, for example and often high heat and pressure. The term 'Hydrogenation' is used for the reaction that involves the combination of hydrogen with unsaturated organic compounds. Unsaturated organic compounds have at least one pair of carbon atoms connected by a double or triple bond. When an unsaturated compound is treated with hydrogen at a suitable temperature and in the presence of a catalyst, such as finely-divided nickel, platinum, or palladium, the multiple bonds between the carbon atoms are broken and a hydrogen atom attaches itself to each carbon atom. For example, when ethylene (C_2H_4) is hydrogenated, the product is ethane (C_2H_6). Hydrogenation is also used with more complicated molecules, yielding a great variety of synthetic products that are important in the laboratory and for industry.

The hydrogenation reaction is applied on a large industrial scale in a number of processes, the most important of which is the hydrogenation of vegetable oils to produce many edible fats, such as margarine. Oils are esters of fatty acids containing one or more double bond linkages between carbon atoms; solid fats are saturated compounds. To convert the oils, which often have an unpleasant taste and odor, to fats with odor and taste sufficiently innocuous so that they can be used for cooking, the oils are hydrogenated. The reaction is performed at a temperature of about 200°C (about 392°F) in the presence of finely-divided nickel and under a pressure of 3 to 4 atmospheres of hydrogen. Large volumes of cheap and plentiful vegetable oils, such as soybean and cottonseed oil, are hydrogenated sufficiently to yield creamy solids that resemble lard at room temperatures. Such products are widely used in foods. Low-grade oils, such as fish oils, are hydrogenated and used in the manufacture of soap and wax.

The hydrogenation process is also applied in the production of synthetic gasoline. The Bergius process, named after the German chemist Friedrich Bergius, is used on a large scale in many parts of the world where petroleum resources are low. It utilizes coal and coal tar as a starting material. The coal, mixed with heavy oil, is ground to a fine paste and heated with hydrogen, under pressure, in the presence of a catalyst composed of metallic sulfides. The resulting oil is further hydrogenated and a third hydrogenation yields gasoline. One ton of coal yields about 304 liters (about 80 gallons) of gasoline. The Fischer-Tropsch process, named after its developers, the German chemists Franz Fischer and Hans Tropsch, was used extensively in Germany in the 1930s to produce synthetic petroleum and diesel fuel. It uses a mixture of carbon monoxide and hydrogen gases with a catalyst containing nickel, cobalt, or modified iron. The process is currently used to produce the raw materials for manufacturing synthetic fats and soaps.

14. WHAT ARE CHARACTERISTICS OF ALCOHOLS?

Whereas the term 'Alcohol' stands for any of a class of organic compounds with the general formula R-OH, where R represents an alkyl group made up of carbon and hydrogen in various proportions and -OH represents one or more hydroxyl groups, in common usage the term usually refers to ethanol. The class of alcohols also includes methanol; the amyl, butyl and propyl alcohols; the glycols; and glycerol. An alcohol is generally classified by the number of hydroxyl groups in its molecule. An alcohol that has one hydroxyl group is called monohydric which include methanol, ethanol and isopropanol. Glycols have two hydroxyl groups in their molecules and so are dihydric. Glycerol, with three hydroxyl groups, is trihydric. The monohydric alcohols are further classified as primary, secondary, or tertiary according to the number of carbon atoms bonded to the carbon atom to which the hydroxyl group is bonded.

Alcohols are generally less volatile, have higher melting points and are more soluble in water than the corresponding hydrocarbons (in which the -OH group is replaced with hydrogen). For example, at room temperature methanol is a liquid, while methane is a gas. The four most common alcohols are:

			OH | $CH_3CH_2CH_2OH$
	CH_3OH	CH_3CH_2OH	CH_3CHCH_3
Methanol	Ethanol	1-propanol	2-propanol
(methyl alcohol)	(ethyl alcohol)	(propyl alcohol)	(isopropyl alcohol)

Many of the properties and reactions characteristic of alcohols are due to the electron charge distribution in the C-O-H portion of the molecule. Chemical reactions involving the hydroxyl group in an alcohol molecule include: those in which the hydroxyl group is replaced as a whole, *e.g.*, reaction of ethanol with hydrogen iodide to form ethyl iodide and water; those in which only the hydrogen in the hydroxyl group is replaced, *e.g.*, the reaction of ethanol with sodium, an active metal, to form sodium ethoxide and hydrogen; and those in which the carbon-oxygen bond becomes a double bond to form an aldehyde or ketone depending on whether it is a primary or secondary alcohol. Ethanol, when fermented from sugar, is the alcohol in beverages. It can also be made from ethene by the addition of water for nonbeverage use, like an additive to gasoline to make "gasohol." 2-Propanol (better known as isopropyl alcohol) is in (with some water) rubbing alcohol. It is also used in gasoline to prevent freezing of the gas line in automobiles by keeping excess moisture dissolved in the gasoline.

15. WHY IS THE CONCENTRATION OF ETHANOL OBTAINED BY FERMENTATION LIMITED TO ABOUT 20 PROOF AND POISONOUS OR NAUSEATING SUBSTANCES ARE ADDED TO IT?

Ethanol or ethyl alcohol, CH_3CH_2OH, is a colorless liquid with characteristic odor and taste; commonly called grain alcohol or simply alcohol. Ethanol is the alcohol of beer, wines and liquors. It can be prepared by the fermentation of sugar (*e.g.*, from molasses), which requires an enzyme catalyst that is present in yeast; or it can be prepared by the fermentation of starch (*e.g.*, from corn, rice, rye, or potatoes), which requires, in addition to the yeast enzyme, an enzyme present in an extract of malt. The concentration of ethanol obtained by fermentation is limited to about 10% (20 proof) since at higher concentrations ethanol inhibits the catalytic effect of the yeast enzyme. (The proof concentration of an alcoholic beverage is numerically double the percentage concentration.) For non-beverage uses ethanol is more commonly prepared by passing ethylene gas at high pressure into concentrated sulfuric or phosphoric acid to form the corresponding ester; the acid-ester mixture is diluted with water and heated, forming ethanol by hydrolysis and the alcohol is then removed from the mixture by distillation, usually with steam.

Ethanol is used extensively as a solvent in the manufacture of varnishes and perfumes; as a preservative for biological specimens; in the preparation of essences and flavorings; in many medicines and drugs; as a disinfectant and in tinctures (*e.g.*, tincture of iodine); and as a fuel and gasoline additive. Many U.S. automobiles manufactured since 1998 have been equipped to enable them to run on either gasoline or E85, a mixture of 85% ethanol and 15% gasoline. E85, however, is not yet widely available. Denatured, or industrial, alcohol is ethanol to which poisonous or nauseating substances have been added to prevent its use as a beverage; a beverage tax is not charged on such alcohol, so its cost is quite low. Medically, ethanol is a soporific, *i.e.*, sleep-producing; although it is less toxic than other alcohols, death usually occurs if the concentration of ethanol in the bloodstream exceeds about 5%. Behavioral changes, impairment of vision, or unconsciousness occur at lower concentrations.

16. WHAT MAJOR REACTIONS ARE EXHIBITED BY ALCOHOLS?

Being much reactive, the alcohols exhibit a variety of reactions, some of which may be stated as under:

Oxidation Reactions

To study the oxidation of alcohols, they must be subclassified into the primary, secondary and tertiary alcohols. Only primary and secondary alcohols are oxidized by oxidizing agents. When an alcohol is oxidized, an H attached to the alcohol carbon is removed as H^- and the H atom of the OH group leaves as H^+. The two H become part of a water molecule with an O provided by the oxidizing agent. Primary alcohols are oxidized to aldehydes. The net ionic equation for this when a dichromate ion is used as an oxidizing agent is:

$$3RCH_2OH + Cr_2O_7^{2-} + 8H^+ ==> 3RCH{=}O + 2Cr^{3+} + 7H_2O$$

For example, this is the oxidation of 1-propanol into propanal:

$$3CH_3CH_2CH_2OH + Cr_2O_7^{2-} + 8H^+ ==> 3CH_3CH_2CH{=}O + 2Cr^{3+} + 7H_2O$$

Since the boiling point of propanal is much lower than 1-propanol, it is boiled out of the solution as it forms. If it is not permitted to leave, it will further be oxidized into propanoic acid (CH_3CH_2COOH) since its tendency to be oxidized is greater than alcohol's.

Secondary alcohols are oxidized to ketones. For example, the oxidation of 2-propanol gives propanone (more commonly called acetone):

$$\begin{array}{l} \;CH_3 \qquad\qquad\qquad\quad\; CH_3 \\ \;\;| \qquad\qquad\qquad\qquad\;\; | \\ 3C\text{-}OH + Cr_2O_7^{2-} + 8H^+ ==> 3C{=}O + 2Cr^{3+} + 7H_2O \\ \;\;| \qquad\qquad\qquad\qquad\;\; | \\ \;CH_3 \qquad\qquad\qquad\quad\; CH_3 \end{array}$$

Since ketones resist oxidation, they do not have to be removed as they form. Tertiary alcohols are not oxidized except by chain-breaking reactions, because they have no removable H atom on the alcohol carbon.

Dehydration Reactions

In a strong acid and heat, alcohols can undergo dehydration, losing a water molecule and leaving behind a carbon-carbon double bond. For example:

Ethanol	==>	ethene + water
Cyclohexanol	==>	cyclohexene + water

Not only do chemists know that it works, they also want to find out and know how it works. The examples above were elimination reactions. The thing that makes them possible is the proton-accepting ability of the oxygen atom of the OH group and therefore, will react with concentrated strong acids.

Substitution Reactions

In acidic conditions, the OH group of an alcohol can be replaced by a halogen atom. For example:

$CH_3CH_2OH + HI$	==>	$CH_3CH_2I + H_2O$
Ethanol + hydrogen iodide	==>	iodoethane + water
		(Ethyl iodide)

Once the OH group has been protonated (added H^+), the bond between the carbon and oxygen is weakened, allowing a halide ion to displace the H_2O.

17. WHAT ARE CHARACTERISTICS OF ETHERS?

Ethers are any of a number of organic compounds whose molecules contain two hydrocarbon groups joined by single bonds to an oxygen atom. The most common of these compounds is ethyl ether, $CH_3CH_2OCH_2CH_3$, often called simply ether, a colorless, volatile liquid with a distinctive odor; its IUPAC name is ethoxyethane. Ethyl ether boils at 34.5°C and is extremely flammable. It is insoluble in water but mixes with many organic solvents and is widely used as a solvent itself, *e.g.*, for fats and oils. Its most familiar historical application is as an anesthetic. Ethers are basically two alkyl groups joined to one oxygen atom. For example:

CH_3OCH_3: Dimethyl ether

$CH_3CH_2OCH_2CH_3$: Diethyl ether

$CH_3OCH_2CH_3$: Methyl ethyl ether

The simple ethers have very low boiling points because hydrogen bonds do not exist between neighboring molecules. Because of this, the boiling points of ethers are much lower than comparable alcohols. For example, 1-butanol ($CH_3CH_2CH_2CH_2OH$) boils at 117°C, much higher than its isomer, diethyl ether, which boils at 34.5°C.

Ether such as ethyl ether in which both hydrocarbon groups are identical is said to be simple, or symmetrical, ether. An ether in which the two groups differ (*e.g.*, methyl ethyl ether, $CH_3OCH_2CH_3$) is said to be a mixed, or unsymmetrical, ether. Ethers are often prepared commercially by heating an alcohol with sulfuric acid; the reaction is one of dehydration. In the laboratory ethers are often prepared by reaction of an alkyl halide with a sodium alkoxide (a method called the Williamson synthesis). Ethers are usually chemically unreactive but can be cleaved (broken apart) at high temperatures by concentrated hydrogen halides; initially an alkyl halide and an alcohol are formed. Epoxides are a special class of cyclic ethers. Dimethyl ether was the "ether" once used as an anesthetic. It was first used as an anesthetic to reduce pain by W.T.G. Morton (1819–1868), a Boston dentist. Morton had learned about ether from a chemist named Charles T. Jackson (1805–1880). Eventually Morton convinced Dr. J.C. Warren (1778–1856) to let him use ether as an anesthetic on one of his patients.

18. WHY AMINES ACT AS WEAK BASES AND BOIL AT LOWER TEMPERATURES THAN ALCOHOLS?

The amino group, in chemistry, is a functional group that consists of a nitrogen atom attached by single bonds to hydrogen atoms, alkyl groups, aryl groups, or a combination of these three. An organic compound that contains an amino group is called an amine. Amines are derivatives of the inorganic compound ammonia, NH_3. When one, two, or all the three of the hydrogens in ammonia are replaced by an alkyl or aryl group, the resulting compound is known as a primary, secondary, or tertiary amine, respectively. Like ammonia, the amines are weak bases because the unshared electron pair of the nitrogen atom can form a coordinate bond with a proton. Amines will react with a mineral acid to form an amine salt, *e.g.*, with hydrochloric acid to form an amine hydrochloride. A water-insoluble amine can be made to dissolve by adding acid to form its water-soluble amine salt. Amines react similarly with alkyl halides to form alkyl ammonium salts. Amines can be synthesized by reacting ammonia with an alkyl halide and neutralizing the resulting alkyl ammonium salt with an alkali, *e.g.*, sodium hydroxide. This procedure yields a mixture of primary, secondary and tertiary amines that is easily separated into its three components by fractional distillation. Amines can also be prepared by the reaction of ammonia with an alcohol or by the reduction of any of a variety of compounds containing nitrogen in a higher oxidation state. Amines take part in many kinds of chemical reactions; in particular, they can react with an acid chloride, acid anhydride, or ester to form an amide. All reactions of amines involve bonding of an electron-deficient atom to the amino nitrogen through its unshared electron pair. One of the most important amines is aniline, an aromatic amine. The simplest amines are:

H	H	H	CH_3
|	|	|	|
H-N-H	CH_3-N-H	CH_3-N-CH_3	CH_3-N-CH_3
Ammonia	Methylamine	Dimethylamine	Trimethylamine

The NH bond is not as polar as the OH bond in alcohols, so amines boil at lower temperatures than alcohols of comparable molecular masses. Amines which have low molecular masses are soluble in water because of hydrogen bonding. This is also why trimethylamine has a lower boiling point than dimethylamine (even though it is bigger), because there is no N–H bond to cause significant hydrogen bonding.

19. WHEN DO AMINES ACT AS WEAK BRØNSTED ACIDS?

Like ammonia, amines are Brønsted bases (proton acceptors) because of presence of nitrogen. In water, this equilibrium will exist as:

```
        H          H
        |          |+
$CH_3CH_2$-N-$CH_3$ + $H_2O$ <=====> $CH_3CH_2$-N-$CH_3$ + $OH^-$
                   |
                   H
```

Ethylmethylamine + water <====> ethylmethylammonium ion + hydroxide ion

Because of reactions similar to above, aqueous solutions of amines are basic and therefore have pH values above 7. When mixed with a strong proton donor like hydrochloric acid, amines change almost entirely into its protonated form by accepting a proton.

$$CH_3CH_2\text{-}N(H)\text{-}CH_3 + HCl \Longrightarrow CH_3CH_2\text{-}N^{+}(H)_2\text{-}CH_3 + Cl^-$$

When water-insoluble amines give this reaction, they can become more soluble because they have been changed into ions that can be hydrated by water. Once amines have become protonated, they act as weak Brønsted acids and can neutralize a strong base by reversing the protonation.

20. WHAT REACTIONS CARBONYL COMPOUNDS UNDERGO?

Carbonyl group, in chemistry, is a functional group that consists of an oxygen atom joined by a double bond to a carbon atom which in turn is joined to the remainder of the molecule by two single bonds or one double bond. If the carbonyl group is joined only to alkyl or aryl groups, the compound is a ketone; if it is joined to at least one hydrogen atom, the compound is an aldehyde. The chemical reactivity of aldehydes and ketones is primarily due to the difference in electronegativity between carbon and oxygen. Because oxygen has the greater affinity for electrons, it acquires a partial negative charge, becoming electron-rich; the carbon atom of the carbonyl group thus becomes electron-deficient, acquiring a partial positive charge. One major type of reaction of aldehydes and ketones involves the addition of an electron-rich chemical species to the electron-deficient carbon atom of the carbonyl group. Another type of reaction is due to the tendency of the electron-deficient carbon atom of the carbonyl group to partially attract electrons from carbon atoms adjacent to it in the molecule, thus increasing the acidity of hydrogen atoms that are bonded to the adjacent carbon.

21. WHAT ARE CHARACTERISTICS OF ALDEHYDES AND KETONES AND HOW ARE THEY NAMED?

Aldehyde is any of a class of organic compounds that contain the carbonyl group and in which the carbonyl group is bonded to at least one hydrogen; the general formula for an aldehyde is RCHO, where R stands for hydrogen or an alkyl or aryl group. Aldehydes are formed by partial oxidation of primary alcohols and form carboxylic acids when they are further oxidized. The common name for an aldehyde is often derived from the name of the acid it forms; the IUPAC name is usually derived from the name of the alcohol from which it is formed. Low molecular weight aldehydes, *e.g.*, formaldehyde and acetaldehyde, have sharp, unpleasant odors and higher molecular weight aldehydes, *e.g.*, benzaldehyde and furfural, have pleasant, often flowery, odors and are found in the essential oils of certain plants. Aldehydes are used for the manufacture of synthetic resins, *e.g.*, Bakelite and for making dyestuffs, flavorings, perfumes and other chemicals. Some are used as preservatives and disinfectants. Aldehydes are the most easily oxidized of all organic compounds, though ketones resist oxidation strongly. They can also be slowly oxidized by oxygen in the air. In this example, potassium dichromate, together with hydrogen sulfate, easily oxidizes ethanal to ethanoic acid.

The parent chain is the longest chain that includes the aldehyde group. Its name is made by replacing the ending -e of parent alkane name with -al. The numbering of the chain always starts with the carbon of the aldehyde group being 1. For example: (parent chain will be bold for visual clarity).

			CH_3
			$\vert$
H	CH_3	CH_3CH_2	CH_3CH
$\vert$	$\vert$	$\vert$	$\vert$
C=O	C=O	C=O	C=O
$\vert$	$\vert$	$\vert$	$\vert$
H	H	H	H
Methanal	Ethanal	Propanal	2-methylpropanal

The term 'Ketone' is used for any of a class of organic compounds that contain the carbonyl group, C=O and in which the carbonyl group is bonded only to carbon atoms. The general formula for a ketone is RCOR', where R and R' are alkyl or aryl groups. The simplest ketone, where R and R' are methyl groups, is acetone; this is one of the most important ketones used in industry. Low-molecular-weight ketones are used chiefly as solvents. Ketones may be prepared by several methods, including the oxidation of secondary alcohols and the destructive distillation of certain salts of organic acids. Ketones are related to the aldehydes but are less active chemically. In naming ketones, the parent chain includes the carbonyl group and is numbered so that the carbonyl location is the lowest number. The number of the location must be part of the name whenever there would be an uncertainty.

		CH_3
		$\vert$
CH_3	CH_3CH_2	CH_3CH
$\vert$	$\vert$	$\vert$
C=O	C=O	C=O
$\vert$	$\vert$	$\vert$
CH_3	$CH_2CH_2CH_3$	$CH_2CH_2CH_3$
Propane (aceton)	3-hexanone	2-methyl-3-hexanone

22. WHAT TYPES OF REACTIONS ARE EXHIBITED BY CARBOXYLIC ACIDS?

In general carboxylic acids are compounds whose molecules contain a carboxyl group that is joined to a hydrogen atom, an alkyl group, or an aryl group by a single bond to its carbon atom. Dicarboxylic acids, compounds that contain two carboxyl groups, are important in a number of industrial processes. There are four main types of reactions of carboxylic acids that are chiefly due to either the weak acidity of the hydroxyl hydrogen or to the difference in electronegativity between carbon and oxygen. One type involves cleavage of the hydroxyl oxygen-hydrogen bond, *e.g.*, reaction with an alcohol to form an ester or reaction with an alkali to form a water-soluble salt. A second type involves addition of an electron-rich species to the electron-deficient carbon atom of the carboxyl group. A third type is characterized by the joining of a carbon atom directly to the carboxyl group. A fourth type involves the loss of carbon dioxide (decarboxylation). The second and third types are similar to reactions of the carbonyl group; the carboxyl group may be thought of as a carbonyl group joined to a hydroxyl group. The parent chain must include the carboxyl carbon, which is given position number 1. The name of the alkane attached is changed by replacing the -e with -oic acid. For example:

HCOOH	CH_3COOH	$CH_3CH_2CH_2COOH$
Methanoic acid	Ethanoic acid (acetic acid)	Butanoic acid

Because carboxylic acids have both a lone oxygen and an OH group they are strongly hydrogen-bonded to each other, therefore having high boiling points. Since the carboxyl group is weakly acidic, all carboxylic acids neutralize OH^-.

$$RCOOH + NaOH \Longrightarrow RCOONa + H_2O$$

Carboxylic acids are used to synthesize two important derivatives, esters and amides. In esters, the OH of the carboxyl group is replaced by OR. To name esters, first have the name of the akyl group attached to the O, followed by the name of the parent carboxylic acid after its name is altered by changing -ic acid to -ate. For example: $CH_3COO(CH_2)_7CH_3$ (Octyl ethanoate).

Esters are responsible for mimicing many pleasant fragrances in nature.

Ester	*Similar aroma*
$HCOOCH_2CH_3$	Rum
$HCOOCH_2CH(CH_3)_2$	Raspberries
$CH_3COO(CH_2)_4CH_3$	Bananas
$CH_3COO(CH_2)_2CH(CH_3)_2$	Pears
$CH_3COO(CH_2)_7CH_3$	Oranges
$CH_3(CH_2)_2COOCH_2CH_3$	Pineapples
$CH_3(CH_2)_2COO(CH_2)_4CH_3$	Apricots

In a mixture with carboxylic alcohol, an equilibrium reaction forms ester and water in the presence of heat and a strong acid catalyst.

RCOOH + HOR' <==> RCOOR' + H_2O

For example:

$CH_3CH_2CH_2COOH + HOCH_2CH_3$ <==> $CH_3CH_2CH_2COOCH_2CH_3 + H_2O$

Butanoic acid + ethanol <==> ethyl butanoate + water

23. WHAT ARE CHARACTERISTICS OF ESTERS?

Ester is an organic functional group that forms many sweet-smelling compounds. The chemical structure of an ester is represented by the general formula, R-CO-OR', where a central carbon atom that has a double bond to an oxygen atom (the carbonyl group), C=O, a single bond to another carbon atom represented by R and a single bond to an oxygen atom connected to a carbon atom represented by R'. For example, when ethanol and acetic acid react, ethyl acetate (an ester) and water are formed; the reaction is called esterification. Ethyl acetate is used as a solvent. Methyl acetate, formed by the reaction between methanol and acetic acid, is a sweet-smelling liquid used in making perfumes, extracts and lacquers. Esters react with water (hydrolysis) under basic conditions to form an alcohol and an acid. When heated with a hydroxide certain esters decompose to yield soap and glycerin; the process is called saponification. Common fats and oils are mixtures of various esters, such as stearin, palmitin and linolein, formed from the alcohol glycerol and fatty acids. Naturally occurring esters of organic acids in fruits and flowers give them their distinctive odors. Esters perform important functions in the animal body; *e.g.*, the ester acetylcholine is a chemical transmitter of nerve stimuli. An ester is hydrolyzed to its parent acid and alcohol when it is heated together with a stoichiometric excess of water (with acid catalyst). Saponification is the reaction of an ester with a strong base, like sodium hydroxide. For example:

$CH_3COOCH_2CH_3$ + NaOH ==> CH_3 COO^- + Na^+ + $HOCH_2CH_3$

Ethyl ethanoate + sodium hydroxide ==> ethanoate ion + sodium ion + ethanol

_______________________/

|

Sodium ethanoate salt

The mixtures of salt of long-chain carboxylic acids are the ingredients of ordinary soap.

24. IN WHAT RESPECTS AMIDES DIFFER FROM AMINES?

In amides, the OH of the carboxyl group is replaced by nitrogen holding any combination of H atoms or hydrocarbon groups. The names of simple amides are devised by writing the name of the parent carboxylic acid and replacing the ending, -oic acid, with -amide

For example: CH_3CONH_2 (Ethanamide (acetamide). The preparation of simple amides is similar to those of esters, but they are in an excess of ammonia. For example:

$$RCOOH + NH_3 \Longrightarrow RCONH_2 + H_2O$$
Carboxylic acid + ammonia ==> simple amide + water

When amides are in water, they are changed back to the carboxylic acid and ammonia. Unlike amines, the amides are nonbasic even though they have the NH_2 group in simple amides. This is because of the presence of O atom in the carbonyl group, which is very electronegative. This tightens the electrons on N so that it is unable to accept a proton.

25. WHAT ARE AROMATIC COMPOUNDS?

Aromatic Compounds are a large group of organic chemical compounds that usually contain closed rings of carbon atoms. Some aromatic rings, however, may also contain an oxygen or a nitrogen atom. The term was originally restricted to the coal tar product benzene and its derivatives, but it now includes about half of all organic compounds, the remaining half being classified as aliphatic compounds. Important aromatic compounds include all hormones and vitamins except vitamin C; virtually all flavorings, perfumes and organic dyes, synthetic and natural; those alkaloids that are not alicyclic (aliphatic bases such as putrescine are sometimes incorrectly classified as alkaloids); some substances of military importance, such as TNT and tear gas; many drugs; and a wide variety of other substances. Certain derivatives of aromatic compounds have special properties and special names. For example, the addition of the hydroxyl group to an aromatic hydrocarbon yields a phenol. The aromatic hydrocarbons include benzene and the alkyl derivatives of benzene, a six sided ring drawn with alternating single and double bonds or with a circle in the middle.

26. WHAT ARE CHARACTERISTICS OF BENZENE?

Benzene, C_6H_6, the most important aromatic organic compound, is a clear, colorless, highly flammable liquid with a pronounced characteristic and a pleasant aromatic odor. It is a flammable and toxic hydrocarbon liquid which boils at 80.1°C, solidifies at 5.5°C and has a density of 0.8787 g/mL. The simplest picture of the benzene molecule, proposed by the German chemist Friedrich Kekulé (1865), is a hexagon of six carbon atoms joined by alternating single and double bonds and each bearing one hydrogen atom, symbolized by benzene;. However, modern studies have shown that the six carbon-carbon bonds are all of equal strength and distance; thus the double-bond electrons do not belong to any particular bonds but rather are delocalized about the ring, with the result that the strength of each bond is between that of a single bond and that of a double bond.

Benzene is the parent substance of the aromatic compounds, a large and important group of organic compounds. It is the first of a series of hydrocarbons known as the benzene series. The second member of the series is toluene, $C_6H_5CH_3$, from which trinitrotoluene is derived and the third member is xylene, $C_6H_4(CH_3)_2$, a solvent. In xylene and other benzene derivatives in which two of the hydrogens have been replaced, there are three possible arrangements of the substitution groups; in the ortho (o) configuration the groups are on adjacent carbon atoms, in the meta (m) configuration the groups are separated by one carbon atom (alternate) and in the para (p) configuration the groups are on opposite sides of the ring. Benzene can be found in the air by several different ways of transmission. It can be produced for use with plastic or produced through the burning of fossil fuels. Benzene can also be found in the soil as well as some areas of polluted groundwater. Exposure to benzene has been linked to leukemia.

27. WHY A RESONANCE STRUCTURE HAS BEEN PROPOSED FOR THE BENZENE RING?

Any compound with the benzene ring is an aromatic compound. The term "aromatic" was coined because of the fact that the earlier compounds found with the rings had pleasant fragrances, but it turns out that the ring has nothing to do with smell. An aromatic compound like aspirin is entirely odorless. Since the benzene ring looks very much like a triene (alkene with three double bonds), it should be expected that its properties would be like one, but it is not. It does not undergo the addition reactions that alkenes do and it is unusually stable to oxidizing agents. The reason is that the benzene ring is a resonance structure. The carbon atoms, in fact, are not connected by alternating single and double bonds, but they all are connected by hybrid partial double bonds. Overlapping p orbitals cause delocalization of pi bonds, giving stability to the ring system. If benzene undergoes an addition reaction, that would break the stabilizing electron network. Instead, benzene prefers to react by substitution. Benzene will react by substitution with chlorine, bromine and nitric acid, provided that a Lewis acid catalyst is present.

28. WHAT ARE ADVANTAGES AND DISADVANTAGES OF IUPAC SYSTEM OF NOMENCLATURE?

International Union of Pure and Applied Chemistry (IUPAC), is an international organization that was established in 1919 to advance the chemical sciences and contribute to the application of chemistry to the service of humanity. IUPAC is an association of national chemistry associations that is responsible for the standardization of weights, measures, symbols and terminology. It also publishes definitive atomic and chemical data and establishes standard methods for use in various laboratories.

IUPAC nomenclature is a system of naming chemical compounds and of describing the science of chemistry in general. It is developed and kept up to date under the auspices of the International Union of Pure and Applied Chemistry (IUPAC). The rules for naming organic and inorganic compounds are contained in two publications, known as the Blue Book and the Red Book respectively. A third publication, known as the

Green Book, describes the recommendations for the use of symbols for physical quantities (in association with the IUPAP), while a fourth, the Gold Book, contains the definitions of a large number of technical terms used in chemistry. Similar compendia exist for biochemistry (the White Book, in association with the IUBMB), analytical chemistry (the Orange Book), macromolecular chemistry (the Purple Book) and clinical chemistry (the Silver Book). These "colour books" are supplemented by shorter recommendations for specific circumstances which are published from time to time in the journal 'Pure and Applied Chemistry'. This discussion treats the system of nomenclature in general, notably its aims and historical development.

The primary function of chemical nomenclature is to ensure that the person who hears or reads a chemical name is under no ambiguity as to which chemical compound it refers to: each name should refer to a single substance. It is considered less important to ensure that each substance should have a single name, although the number of acceptable names is limited. It is also preferable that the name convey some information about the structure or chemistry of a compound. CAS numbers form an extreme example of names which do not perform this function: each refers to a single compound but none contain information about the structure. The form of nomenclature which should be used depends on the public to which it is addressed: as such there is no single correct form, but rather different forms which are more or less appropriate in different circumstances.

A common name will often suffice to identify a chemical compound in a particular set of circumstances. To be more generally applicable, the name should indicate at least the chemical formula. To be more specific still, the three-dimensional arrangement of the atoms may need to be specified. In a few specific circumstances (such as the construction of large indices), it becomes necessary to ensure that each compound has a unique name: this requires the addition of extra rules to the standard IUPAC system (the CAS system is the most commonly used in this context), at the expense of having names which are longer and less familiar to most readers. Another system gaining popularity is the International Chemical Identifier, while InChI symbols are not human-readable; they contain complete information about substance structure. That makes them more general than CAS numbers.

The IUPAC system is often criticized for the above failures when they become relevant (for example in differing reactivity of sulfur allotropes which IUPAC doesn't distinguish). While IUPAC has a human-readable advantage over CAS numbering, it would be difficult to claim that the IUPAC names for some larger, relevant molecules (such as rapamycin) are human-readable and so most researchers simply use the informal names.

29. UNDER WHAT CIRCUMSTANCES AND WHEN DID IUPAC COME INTO BEING?

Though the nomenclature of alchemy is rich in description, it does not effectively meet the aims it was meant for. Opinions differ whether this was deliberate on the part of the early practitioners of alchemy or whether it was a consequence of the particular (and often esoteric) theoretical framework in which they worked. While both explanations are probably valid to some extent, it is remarkable that the first "modern" system of chemical nomenclature appeared at the same time as the distinction (by Lavoisier) between elements and compounds, in the late eighteenth century. The French chemist Louis-Bernard Guyton de Morveau published his recommendations in 1782, hoping that his "constant method of denomination" would "help the intelligence and relieve the memory". The system was refined in collaboration with Berthollet, de Fourcroy and Lavoisier and promoted by the latter in a textbook which would survive long after his death at the guillotine in 1794. The project was also espoused by Jöns Jakob Berzelius, who adapted the ideas for the German-speaking world.

The recommendations of Guyton covered only what would be today known as inorganic compounds. With the massive expansion of organic chemistry in the mid-nineteenth century and the greater understanding of the structure of organic compounds, the need for a less adhoc system of nomenclature was felt just as the theoretical tools became available to make this possible. An international conference was convened in Geneva in 1892 by the national chemical societies, from which the first widely accepted proposals for standardization arose. A commission was set up in 1913 by the Council of the International Association of Chemical Societies, but its work was interrupted by World War I. After the war, the task passed to the newly formed International Union of Pure and Applied Chemistry, which first appointed commissions for organic, inorganic and biochemical nomenclature in 1921 and continues to do so to this day.

Chapter 16

MACROMOLECULES (POLYMERS)

1. WHAT ARE ORGANIC POLYMERS AND WHAT IS DIFFERENCE BETWEEN ADDITION AND CONDENSATION POLYMERS?

Almost all of the molecules, chemists generally study or come across, are relatively small with low molecular masses. But in nature and the world of synthetics, there are many substances with molecular masses in thousands and hundred thousands, containing hundreds or thousands of atoms. These molecules are called macromolecules. Macromolecular substances are almost everywhere in nature and in synthetics. Synthetic macromolecules are examples of how chemists take ordinary substances in nature, like coal, oil, air and water and make new materials with useful applications. Some macromolecules have more structural order than others. In fact a polymer is a macromolecule in which all of the molecules have a small characteristic structural feature that repeats itself again and again. Polymers do not consist of molecules identical in size but identical in kind.

The reaction that makes a polymer out of a monomer is called polymerization. Sometimes during polymerization, a long chain may develop branches which themselves are long. Sometimes two monomers are mixed together and polymerized, providing a unit 'a' and a unit 'b'. For example, saran is a copolymer of $CH_2{=}CCl_2$ and $CH_2{=}CHCl$ and its molecules have the structure:

$$\text{etc-}CH_2\text{-}CCl_2\text{-}CH_2\text{-}CHCl\text{-etc}$$

Chemicals called promoters initiate polymerizations and pieces of their molecules become joined to the ends of the polymer systems (where "etc." has been written).

Polymerizations that make addition polymers generally join the monomer molecules end to end. For example, in the addition polymerization of ethylene:

$$\text{———} + CH_2{=}CH_2 + CH_2{=}CH_2 + CH_2{=}CH_2 + CH_2{=}CH_2 + CH_2{=}CH_2 + \text{———},$$

through addition polymerization gives:

$$\text{———}\ \text{-}CH_2\text{-}CH_2\text{-}CH_2\text{-}CH_2\text{-}CH_2\text{-}CH_2\text{-}CH_2\text{-}CH_2\text{-}CH_2\text{-}CH_2\text{-}\ \text{———}$$

Condensation polymers are copolymers, which have monomers having functional groups that can react with each other, usually spitting out water. The reaction between a carboxylic acid and an alcohol creates an ester. If the carboxylic acid and the alcohol were the monomers of the polymer, during polymerization, they would create polyester and spit out water in the process. The polymerization of a carboxylic acid and an amine similarly creates polyamides.

2. WHAT FACTORS AND PROPERTIES OF POLYMERS DECIDE THEIR USEFULNESS?

Polymers are very useful because chemically they are very inert. They are not attached by air or its pollutants. They are stable toward water. They must keep these properties under high temperatures, as Teflon does on a nonstick frying pan. Polyesters and polyamides have functional groups that water can attack, but this does not happen due to the enormous size of the molecule. They are completely insoluble in water and therefore make a good storing container for food but bad for waste disposal since it does not decay.

Physical properties of polymers are the ones that are most sought-after for their practical use. For example, Teflon (a polyolefin) has slipperiness toward almost anything. Nylon (a polyamide) isn't eaten by moths and has superior tensile strength and the ability to be made into fibers and fabrics. Dacron (a polyester) does not mildew and when made into fibers, it is superior to cotton, with greater strength and lower mass and it doesn't stretch as much. The best fiber-forming polymers have molecules with shapes that let the molecules align side by side and be twisted into cables. Because the molecules are large, substantial London forces of attraction occur between them. Hydrogen bonds are also present in some polymers, like the polyamides.

3. WHAT GIVES PLASTICS THEIR MAIN CHARACTERISTICS?

The term Plastic is used for any organic material with the ability to flow into a desired shape when heat and pressure are applied to it and to retain the shape when they are withdrawn. A plastic is made up principally of a binder together with plasticizers, fillers, pigments and other additives. The binder gives a plastic its main characteristics and usually its name. Thus, polyvinyl chloride is both the name of a binder and the name of a plastic into which it is made. Binders may be natural materials, *e.g.*, cellulose derivatives, casein, or milk protein, but are more commonly synthetic resins. In either case, the binder materials consist of very long chainlike molecules called polymers. Cellulose derivatives are made from cellulose, a naturally occurring polymer; casein is also a naturally occurring polymer. Synthetic resins are polymerized, or built up, from small simple molecules called monomers.

Plasticizers are added to a binder to increase flexibility and toughness. Whereas fillers are added to improve particular properties, *e.g.*, hardness or resistance to shock, pigments are used to impart various colors. Virtually any desired color or shape and many combinations of the properties of hardness, durability, elasticity and resistance to heat, cold and acid can be obtained in a plastic.

4. WHAT ARE DIFFERENCES BETWEEN THERMOPLASTICS AND THERMOSETS?

There are two basic types of plastic: thermosets, which cannot be re-softened after being subjected to heat and pressure; and thermoplastics, which can be repeatedly softened and remolded by heat and pressure. When heat and pressure are applied to a thermoplastic binder, the chainlike polymers slide past each other, giving the material "plasticity". However, when heat and pressure are initially applied to a thermosetting binder, the molecular chains become cross-linked, thus preventing any slippage if heat and pressure are reapplied. Thus thermoplastics are materials that soften or melt when heated and harden when cooled. Thermoplastic polymers consist of long polymer molecules that are not linked to each other. *i.e.*, have no cross-links. They are often supplied as granules and heated to permit fabrication by methods such as molding or extrusion. Thermoplastics include polyethylene, polypropylene, polystyrene, polyester, polyvinyl chloride, acrylics, nylons, spandex-type polyurethanes and cellulosics. Thermosets are materials that can not be softened on heating. In thermosetting polymers, the polymer chains are joined (or cross-linked) by intermolecular bonding. Thermosets are usually supplied as partially polymerized or as monomer-polymer mixtures. Cross-linking is achieved during fabrication using chemicals, heat, or radiation; this process is called curing or vulcanization. Important thermosets include phenolics, ureas, melamines, epoxies, polyesters, silicones, rubbers and polyurethanes.

5. WHAT ARE POLYACRYLICS?

Polyacrylics or acrylic plastics are a group of thermoplastics that are transparent and highly decorative. They are derivatives of acrylic acid, $H_2C{=}CH\text{-}COOH$. The best-known acrylic plastic, sold under the trade names Plexiglas and Lucite, is polymethyl methacrylate. It takes a high polish, is clear and colorless and is transparent to visible and ultraviolet light. Since it is a thermoplastic, it can be shaped while hot to form a number of objects, such as windshields for airplanes and transparent ornamental objects. Other esters of acrylic acid and methylacrylic acid similarly polymerize and copolymerize to transparent thermoplastics, differing somewhat in hardness and in softening temperatures.

6. WHAT ARE MAIN CHARACTERISTICS OF POLYETHYLENE AND POLYSTYRENE?

Polyethylene is a widely used plastic. It is a polymer of ethylene, $CH_2{=}CH_2$, having the formula $(\text{-}CH_2\text{-}CH_2\text{-})_n$ and is produced at high pressures and temperatures in the presence of any one of several catalysts, depending on the desired properties of the finished product. Polyethylene is resistant to water, acids, alkalis and most solvents. Its many applications include films or sheets for packaging, shower curtains, unbreakable bottles, pipes, pails, drinking glasses and insulation for wire and cable.

Polystyrene, a widely used plastic and a polymer of styrene, is a colorless and transparent thermoplastic that softens slightly above 100°C (212°F) and becomes a viscous liquid at around 185°C (365°F). It is resistant to acids, alkalis, oils and alcohols. It is produced either as a solid or as a foamed plastic marketed under the trade name Styrofoam. Its many uses include electrical and thermal insulation, translucent window panels, storage-battery cases and toilet articles.

7. WHAT ARE USES OF POLYURETHANES?

Polyurethanes, a group of plastics that may be either thermosetting or thermoplastic, can be made into both flexible and rigid foams. The flexible foam is often used in furniture and automobile cushions, in mattresses and for carpet backings. The rigid foam is used for the thermal insulation of refrigerators, trucks and buildings. In the furniture industry the rigid foam is molded into mirror frames, chair shells and other parts that were formerly made from wood. Some polyurethanes are highly elastic materials that are resistant to chemical attack and to abrasion. They are used in such things as solid rubber tires and shoe heels. Lycra, a fiber used in stretch clothing, is polyurethane. Polyurethanes are also used as decorative and protective coatings, exhibiting great hardness and toughness.

8. WHAT ARE POLYESTERS AND POLYAMIDES?

Polyesters and polyamides are condensation polymers. Polyesters are created using two monomers: dicarboxylic acids (two carboxyl groups on each end) and diols (two alcohol groups on each end). As they create esters during polymerization, water molecules are spit out. A similar reaction happens with polyamides using diamines and dicarboxylic acids.

9. HOW YARN IS MADE FROM RAYON?

Rayon is an artificial textile material, composed of cellulose obtained from cotton linters or from the pulp of trees such as spruce. Since its introduction in about 1900, rayon has been used in many textile fields. At first rayon was called artificial silk because, in its filament form, it somewhat resembles silk; however, this comparison is misleading because the chemical composition of rayon is entirely different from that of silk. Rayon can be made by either the viscose process or the cuprammonium process; both produce fiber classified by the U.S. Federal Trade Commission as rayon. In the viscose process purified cellulose is treated with sodium hydroxide, then with carbon disulfide, to form a viscous yellow liquid called viscose. In the cuprammonium process purified cellulose is treated with cuprammonium liquor, then with sodium hydroxide, to form viscose. The manufacture of rayon filaments and all manufactured fibers, is done by means of an extrusion process called spinning. In this procedure the fiber-forming liquid is forced through tiny holes in a nozzle or spinneret into a liquid bath containing chemicals that produce filaments of pure cellulose, which can be spun into yarn. The filaments are drawn together to form both fibers and yarn in a single, continuous process.

10. WHAT ARE CHARACTERISTICS AND IMPORTANT USES OF RUBBER?

The term rubber stands for any solid substance that upon vulcanization becomes elastic; the term includes both natural rubber (caoutchouc) and synthetic rubber. The term elastomer is sometimes used to designate synthetic rubber only and is sometimes extended to include caoutchouc as well. All rubberlike materials are polymers, which are high molecular weight compounds consisting of long chains of one or more types of molecules, such as monomers. Vulcanization (or curing) produces chemical links between the loosely coiled polymeric chains; elasticity occurs because the chains can be stretched and the crosslinks cause them to spring back when the stress is released. Natural rubber is a polyterpene, *i.e.*, it consists of isoprene molecules linked into loosely twisted chains. The monomer units along the backbone of the carbon chains are in a cis- arrangement and it is this spatial configuration that gives rubber its highly elastic character. In gutta-percha, which is another natural polyterpene, the isoprene molecules are bonded in a trans- configuration leading to a crystalline solid at room temperature. Unvulcanized rubber is soluble in a number of hydrocarbons, including benzene, toluene, gasoline and lubricating oils.

Rubber is water repellent and resistant to alkalis and weak acids. Rubber's elasticity, toughness, impermeability, adhesiveness and electrical resistance make it useful as an adhesive, a coating composition, a fiber, a molding compound and an electrical insulator. In general, synthetic rubber has many advantages over natural rubber: better aging and weathering, more resistance to oil, solvents, oxygen, ozone and certain chemicals and resilience over a wider temperature range. The advantages of natural rubber are less buildup of heat from flexing and greater resistance to tearing when hot.

It may be asked by vehicle-owners and users as to why be rubber so good for our car tires? The obvious and quick answer may be that it is long wearing, heat resistant, elastic, water-resistant and air tight due to which a tire is not worn away quickly by constant contact with the road and its environment. Neither will it burn away from the constant friction. And, as we move along the shock absorbent quality of the rubber in our tires keeps us cushioned from the bumps in the road. Yes, our life would hardly be the same if we didn't have rubber as our trusty servant. Natural rubber contains monomers of a diene called isoprene (2-methyl-1,3-butadiene). It is part of a large family of polymers called elastomers because of its ability to recover its shape after being deformed. Natural rubber contains several alkene groups and makes it easily attacked by ozone. This problem has largely been solved by adding antioxidant compounds to rubber products when they are made.

11. HOW IS NATURAL RUBBER OBTAINED?

Natural rubber is obtained from the milky secretion (latex) of various plants, but the only important commercial source of natural rubber (sometimes called Pará rubber) is the tree Hevea brasiliensis. The only other plant under cultivation as a commercial rubber source is guayule (Parthenium argentatum), a shrub native to the arid regions of Mexico and the SW United States. To soften the rubber so that compounding ingredients can be added, the long polymer chains must be partially broken by mastication, mechanical shearing forces applied by passing the rubber between rollers or rotating blades. Thus, for most purposes, the rubber is ground, dissolved in a suitable solvent and compounded with other ingredients, *e.g.*, fillers and pigments such as carbon black for strength and whiting for stiffening; antioxidants; plasticizers, usually in the form of oils, waxes, or tars; accelerators; and vulcanizing agents. The compounded rubber is sheeted, extruded in special shapes, applied as coating or molded, then vulcanized. Most Pará rubber is exported as crude rubber and prepared for market by rolling slabs of latex coagulated with acid into thin sheets of crepe rubber or into heavier, firmly pressed sheets that are usually ribbed and smoked.

During the two World Wars, demand for rubber for automotive use sky-rocketed. Unable to meet this demand through natural sources, scientists began the search for a synthetic alternative. Today, rubber is made from such things as petroleum and coal. Some of these man-made rubbers are actually better for specific usages than the natural variety. Neoprene, for example, is highly resistant to oil, gasoline, sunlight and ozone. Every year rubber wear leaves about 80,000 tons of rubber on United States roads. But, surprisingly, very little trace of this rubber fall-out can be found. It appears that a large number of micro-organisms that live on the roadside find the product quite tasty and do a good job of cleaning off the surface of the highway.

12. HOW ARE CARBOHYDRATES CLASSIFIED?

There are three main classes of carbohydrates: monosaccharides, disaccharides and polysaccharides.

Monosaccharides

Carbohydrates that are unable to react with water and usually are the units making up larger carbohydrates are called monosaccharides. The most common is glucose, a pentahydroxyaldehyde and it is the most widely occurring structural unit in the entire living world. It is the building unit for many important polysaccharides such as cellulose and starch. Fructose is another monosaccharide that makes sugar (sucrose) when combined with glucose. The open-chain forms account for less than 0.1% of all the molecules. Monosaccharides usually come in cyclic forms.

Cyclic Forms of Monosaccharides

When dissolved in water, molecules of carbohydrates exist in equilibrium involving more than one structure. Glucose exists as two cyclic forms and one open chain form (rare) in water. The equilibrium shifts to supply more of any of its members when a specific reaction occurs to only one. Galactose, another monosaccharide, differs by the projection of the OH group at carbon 4.

Disaccharides

Disaccharides are carbohydrates composed of two monosaccharide molecules. They are created through dehydration synthesis (removal of water) and broken apart through hydrolysis (addition of water). For example, these are molecules of sucrose and lactose:

$$\text{Glucose} + \text{fructose} \rightarrow \text{sucrose} + H_2O$$
$$\text{Galactose} + \text{glucose} \rightarrow \text{lactose} + H_2O$$

Polysaccharides

Polymers of carbohydrates are called polysaccharides and make up some of the most important naturally occurring compounds. They have thousands of monosaccharide units linked to each other by oxygen bridges. They include starch, glycogen and cellulose, all three of which yield only glucose when completely hydrolyzed. Starch occurs naturally in plants, which use it to store glucose units for energy. It is often found in seeds and tubers (*e.g.*, potatoes). It consists of two kinds of polymers of glucose. The simpler kind is called amylose and it makes up about 20% of starch. It is basically a chain of glucose units linked by oxygen bridges. The average amylose molecules have over 1000 units. During digestion, the oxygen bridges are hydrolyzed and the glucose units are broken up.

$$\text{Amylose} + nH_2O \rightarrow n\ \text{glucose}$$

Here n is very large, usually over 1000. The other part of starch, the part that makes up most of it, is made up of amylopectin. It is even larger than amylose. Basically, it consists of several amylose molecules linked and cross-linked with oxygen bonds.

13. WHY SHOULD WE PREFER COMPLEX CARBOHYDRATES AND ROUGHAGE IN OUR DIETS?

Carbohydrates are naturally occurring compounds composed of carbon, hydrogen and oxygen. The carbohydrate group includes sugars, starches, cellulose and a number of other chemically related substances. For the most part, these carbohydrates are produced by green plants through the process known as photosynthesis. Carbohydrates are important as foods; they supply energy and are used in the production of fats. They are also used in various forms in industry and commerce. There are three main classes of carbohydrates. Monosaccharides are the simple sugars, *e.g.*, fructose and glucose; they have the general formula $(CH_2O)_n$, in which n is an integer larger than 2. Disaccharides include lactose, maltose and sucrose. Upon hydrolysis, a disaccharide molecule yields two monosaccharide molecules. Most disaccharides have the general formula $C_n(H_2O)_{n-1}$, with n larger than 5. Polysaccharides include such substances as cellulose, dextrin, glycogen and starch; they are polymeric compounds made up of the simple sugars and can be hydrolyzed to yield simple sugars. The disaccharides are sometimes grouped with the simpler polysaccharides (usually those made up of three or four simple sugar units) to form a class of carbohydrates called the oligosaccharides.

Though carbohydrates (starches and sugars) provide a readily available energy source, surplus carbohydrates are also converted by the body to glycogen and fat, the storage forms of calories for energy and to some of the amino acids used in protein synthesis. Most health professionals recommend that carbohydrates comprise 50% to 60% of the dietary calories, of which most (c.80% of all carbohydrates eaten) should be complex carbohydrates, such as cereals and vegetables. Complex carbohydrates are preferred because the fast-acting simple carbohydrates, such as honey and sugar, are difficult for the body (especially the pancreas) to handle in large doses. Simple carbohydrates also lack the vitamins, minerals, proteins and fiber that generally accompany foods rich in complex carbohydrates. Cereals, fruits, vegetables, legumes and pasta are good sources of complex carbohydrates.

Glycogen is used to store glucose units for energy in animals. Its structures are much like amylopectin. Any unused glucose molecules in our bloodstream are put into the liver and muscles to be stored as glycogen until needed. Cellulose is also a polymer of glucose. It is much like amylose, but it has the oxygen bridges differently oriented

It makes up cell walls in plants. Humans lack the enzyme to hydrolyze it and plant materials made with it, like leafy vegetables, are not digested and used as "roughage" or fiber. Some animals have bacteria living in their digestive tracts that can hydrolyze cellulose for them. Lipids are largely hydrocarbon like and therefore do not dissolve in water but in nonpolar solvents like diethyl ether and benzene. The lipid family is very large and diverse. It includes cholesterol, hormones and the edible fats and oils we eat.

14. HOW ARE FATS CLASSIFIED?

Fats and Oils are oily, greasy, or waxy substances that, in their pure state, are normally tasteless, colorless and odorless. Fats and oils are lighter than water and are insoluble in it; they are slightly soluble in alcohol and are readily dissolved in ether and other organic solvents. Fats are soft and greasy at ordinary temperatures, whereas fixed oils, as distinct from essential oils and petroleum, are liquid. Some waxes, which are hard solids at ordinary temperatures, are chemically similar to fat.

The fats normally occur in animal and vegetable tissue as a mixture of several pure fats plus free fatty acids. The most common of such fats are palmitin, the ester of palmitic acid; stearin, the ester of stearic acid; and olein, the ester of oleic acid. These pure chemical compounds are contained in different proportions in various natural fats and oils and they determine the physical characteristics of each of these substances.

Fats are designated as saturated or unsaturated, depending on whether the chemical bonds between the carbon atoms of the fat molecules contain all the hydrogen atoms they are capable of holding (saturated) or have capacity for additional hydrogen atoms (unsaturated). While saturated fats are generally solid at room temperature, unsaturated and polyunsaturated fats are liquids. Unsaturated fats may be converted to saturated fats by adding hydrogen atoms in a process known as hydrogenation. This process creates saturated fats, called trans-fatty acids.

Vegetable fats are usually obtained by pressure extraction from seeds and fruit. Animal fats are often extracted from animal tissue by boiling fat-containing tissue in water and then allowing it to cool. The heat melts the fat out of the tissue; it then rises to the surface of the water because of its low specific gravity and can be skimmed off. The largest consumption of fats and oils is in the form of food. Certain natural fats, such as butter fat and lard, are used as food with little or no preparation. Certain unsaturated oils, such as cottonseed and peanut oils, are often partly hydrogenated to increase their melting points and make them suitable for use as shortening and cooking fats. Tallow, the animal fats and oils from sheep and cattle, is used for candle making and in margarine. Natural oils that contain esters of unsaturated acids are known as drying oils and possess the property of forming a permanent dry film when exposed to air. Linseed oil and other oils of this type are extensively used in the production of paints. Fats also serve as a raw material in the manufacture of soap.

15. WHAT ARE CHARACTERISTICS OF EDIBLE OILS?

Fats and oils are group of organic substances that form an important part of the diet and also are useful in many industries. The fats are usually solid and the oils generally liquid at ordinary room temperatures. Some tropical products, liquids in their sites of origin, become solids in cooler climates; in commerce these often retain the name originally given to them, *e.g.*, palm oil and coconut oil. Chemically fats and oils are either simple or mixed glyceryl esters of organic acids belonging to the fatty-acid series. Fats and oils are derived from both plant and animal sources.

Among the vegetable oils of greatest commercial importance are cottonseed, linseed, olive, palm, corn, peanut, soybean and castor oils. The method of obtaining the oils is similar for all: the fruits or seeds after being cleaned are crushed and pressed cold to obtain the highest grade of oil and then pressed warm, yielding a grade suitable for industrial use. Sometimes solvents are used to remove the remaining oil from the crushed mass. Edible oils are those used in foods and for these the highest grade is utilized; these must be pale in color, free from disagreeable odor and taste and wholesome. The lower grades are suitable for making soap and for other industrial purposes. The chemical property that makes fats solid and oils liquid is the amount of saturation in the ester.

Animal fats are esters of saturated fatty acids; vegetable oils are esters of unsaturated fatty acids.

16. WHY DO MOST FATS BECOME RANCID UPON STANDING?

Conversion of liquid vegetable oils into solid fats is an important chemical industry. This process, sometimes called hardening, involves hydrogenation of the unsaturated fatty-acid portion of the oil molecule by heating the oil with hydrogen in the presence of a metal catalyst and by controlling the extent of hydrogenation, various products can be obtained. For example, controlled hydrogenation of cottonseed oil produces a solid vegetable cooking fat. Most fats become rancid upon standing; since a major factor leading to rancidity is air oxidation of double bonds (to form foul-smelling aldehydes); saturated fats are much more resistant to rancidity than unsaturated fats.

17. WHAT IS CORRELATION BETWEEN LEVELS OF FAT AND TRIGLYCERIDES IN THE BLOOD STREAM TO HEART ATTACKS IN HUMAN?

A fat is a solid triester of glycerol. It is formed when a molecule of glycerol (an alcohol with three hydroxyl groups), reacts with three molecules of fatty acids, a long-chain aliphatic carboxylic acid. The more correct name for a fat is a triglyceride. The three fatty acid fragments in a fat may be all the same (a simple triglyceride) or they may be different from each other. Fats in the diet provide a concentrated source of energy; 1 gram of fat supplies about 9 calories as opposed to only 4 calories per gram of carbohydrates and protein. Fats in the body, in addition to acting as a source of stored energy, supply physical protection and insulation for tissues and form important portions of cell membrane structure. Fats also aid in the absorption of the fat-soluble vitamins (vitamins A, D, E and K) from the intestine. Milk, butter, meat and oils are important sources of fat.

In a simple triglyceride such as palmitin or stearin, all three fatty-acid groups are identical. In a mixed triglyceride, two or even three different fatty-acid groups are present; most fats and oils contain mixed triglycerides. The edible fats and oils we eat are triacylglycerols (commonly called triglycerides). Fats exist in foods and are usually stored in the body as Triglycerides. Recent research relating levels of triglycerides in the blood stream to heart attacks in human presents a sometime confusing picture but mounting level evidence suggests that, along with other indicators, triglyceride levels can be used to predict heart attack risk, especially in women and diabetics.

18. WHY SHOULD WE AVOID INTAKE OF SATURATED FATS?

Animal fats used in foods include butter, lard, chicken fat and suet. Cod-liver oil and some other fish oils are used therapeutically as sources of vitamins A and D. Nutritionally fats and oils are valued as a source of energy. Because they contain less oxygen than other nutrients, they oxidize more readily and release more energy. Fats are digested in the human body mainly by the enzyme lipase (in the pancreatic juice) aided by the bile. There are several theories to explain the method of absorption of fats; favored by many is the view that they are absorbed by the epithelial cells of the lining of the small intestine in the form of the fatty acids and glycerol into which they are split by digestion and that a recombination to re-form the fat occurs within the cells. Most of the fat then enters the lymphatic system through the villi in the lining of the small intestine, although some is probably absorbed directly by the blood vessels of the villi. Medical research indicates the possibility that saturated fats in the diet contribute to the incidence of arteriosclerosis; such fats may raise the blood's level of cholesterol, which is deposited in the arteries.

19. WHY DO HEART ASSOCIATIONS RECOMMEND RESTRICTING THE TOTAL FATS IN A DIET?

Research focusing on heart disease and other circulatory disorders indicates that excessive consumption of foods high in fats, including lipids and sterols, increases the risk of developing these diseases. Various studies have shown that as much as 50 per cent of the caloric intake in a modern diet consists of fats. Other studies demonstrate that the likelihood of heart disease decreases when the intake of saturated fats is reduced. Recent research indicates that excessive consumption of trans fats, such as those produced in the hydrogenation of vegetable oil to make margarine, may also increase the risk of heart disease.

When saturated solid fats are added to the diet, the amount of cholesterol in the blood increases, but when liquid, unsaturated fats or oils (particularly the polyunsaturated type) replace solid fat, the amount of cholesterol decreases. High cholesterol levels in the blood appear to promote the deposition of hard, fatty materials in the arteries, causing the arteries eventually to clog. When the coronary artery around the heart becomes clogged in this manner, blood supply to the heart is interrupted and a heart attack occurs. The Heart Associations recommend restricting the total fats in a diet, reducing the consumption of cholesterol-rich foods, such as milk and butter, replacing saturated fats with unsaturated fats and limiting the intake of trans fats. Scientists are continuing to seek a thorough understanding of how the body handles fatty materials and why cholesterol levels rise to the point that fat is deposited on artery walls.

20. WHAT ARE CHARACTERISTICS OF COMMON FATTY ACIDS?

Fatty Acids, is the common name for a group of organic acids that includes the saturated (hydrogenated), straight-chain acids, with a single carboxyl (COOH) group, that are produced by the hydrolysis of fats, due to which this name has been dubbed. The group also includes all other saturated straight-chain acids and acids with a branched chain or cyclic structure. Formic acid, HCOOH and acetic acid, CH_3COOH, are the simplest fatty acids. Both have sour taste, irritate the skin and have a sharp smell. Of more complicated structure are butyric, caproic, caprylic and capric acids, all of which have unpleasant odors. Stearic, palmitic, oleic and naphthenic acids are greasy materials with little odor. A growing source of fatty acids is tall oil, a by-product of the pinewood used to make paper pulp. Fatty acids are useful in preparing biodegradable detergents, thickeners for paints and lubricants. Stearic acid is used to combine rubber with other substances, such as pigments, or materials that control the flexibility of rubber products; it is also used in the polymerization of styrene and butadiene in making artificial rubber. New uses for fatty acids include ore flotation and the manufacture of disinfectants, varnish driers and heat stabilizers for vinyl resins. Fatty acids are also used in plastic products, such as coatings for wood and metal and in automobile parts ranging from air-cleaner housings to upholstery.

Table 16.1: Common Fatty Acids

Fatty acid	*No of carbon atoms*	*Structure*
Myristic acid	14	$CH_3(CH_2)_{12}COOH$
Palmitic acid	16	$CH_3(CH_2)_{14}COOH$
Stearic acid	18	$CH_3(CH_2)_{16}COOH$
Oleic acid	18	$CH_3(CH_2)_7CH{=}CH(CH_2)_7COOH$
Linoleic acid	18	$CH_3(CH_2)_4CH{=}CHCH_2CH{=}CH(CH_2)_7COOH$
Linolenic acid	18	$CH_3CH_2CH{=}CHCH_2CH{=}CHCH_2CH{=}CH(CH_2)_7COOH$

21. WHY ARE VEGETABLE OILS HYDROGENATED?

Vegetable oils (liquids) are generally less expensive to produce than animal fats (solids). Using hydrogenation, chemists can break some of the carbon-carbon double bonds and replace them with hydrogen to make them chemically identical to the triacylglycerols in animal fats. Breaking one of the three double bonds in the molecule can make the product melt above room temperature. The hydrolysis of triacylglycerol in the presence of sufficient sodium hydroxide is called saponification. The fatty acids are released as sodium salts and the mixture of these salts of long-chain fatty acids is soap.

22. WHAT ARE NONESSENTIAL AND ESSENTIAL AMINO ACIDS AND HOW DO THEY MAKE POLYPEPTIDES?

The term amino acid stands for any one of a class of simple organic compounds containing carbon, hydrogen, oxygen, nitrogen and in certain cases sulfur. These compounds are the building blocks of proteins. They are characterized by the presence of a carboxyl group (COOH) and an amino group (NH_2) attached to the same carbon at the end of the compound. The 20 amino acids commonly found in animals are: alanine, arginine, asparagine, aspartic acid, cysteine, glutamic acid, glutamine, glycine, histidine, isoleucine, leucine, lysine, methionine, phenylalanine, proline, serine, threonine, tryptophan, tyrosine and valine. More than 100 less common amino acids also occur in biological systems, particularly in plants. Every amino acid except glycine can occur as either of two optically active stereoisomers, d or l; the more common isomer in nature is the l-form. When the carboxyl carbon atom of one amino acid covalently binds to the amino nitrogen atom of another amino acid with the release of a water molecule, a peptide bond is formed. Amino acids are released in the intestinal tract by the digestion of food proteins and are then carried in the bloodstream to the body cells, where they are used for growth, maintenance and repair. Cellular catabolism breaks amino acids down into smaller fragments. Many of the amino acids necessary in metabolism can be synthesized in the human or animal body when needed; these are called nonessential. Others cannot be synthesized in sufficient quantities; these are termed essential and must be provided in the diet.

When amino acids are joined together, they make polypeptides. The carboxyl group of one amino acid becomes joined to the amino group of another by a peptide bond (amide bond). Glycylalanine is a dipeptide because it has two amino acids linked by a peptide bond. A different dipeptide could be made using the same two amino acids, but putting them in a different orientation. This would make alanylglycine (Ala-Gly). The three letter abbreviations can be used to represent the structural formulas of the amino acids. Also, because the molecules produced are still dipolar ions, more amino acids can be put together. Proteins have hundreds, thousands, or sometimes even millions of these amino acids.

23. WHAT ARE PROTEINS AND HOW ARE THEY CLASSIFIED?

Proteins are a group of highly complex organic compounds found in all living cells and comprising the most abundant class of all biological molecules. Protein comprises approximately 50% of cellular dry weight. Hundreds of protein molecules have been isolated in pure, homogeneous form; many have been crystallized. All contain carbon, hydrogen and oxygen and nearly all contain sulfur as well. Some proteins also incorporate phosphorous, iron, zinc and copper. Proteins are large molecules with high molecular weights (from about 10,000 for small ones [of 50–100 amino acids] to more than 1,000,000 for certain forms); they are composed of varying amounts of the same 20 amino acids, which in the intact protein are united through covalent chemical linkages called peptide bonds. The amino acids linked together, form linear unbranched polymeric structures called polypeptide chains; such chains may contain hundreds of amino-acid residues; these are arranged in specific order for a given species of protein.

A protein molecule that consists of but a single polypeptide chain is said to be monomeric; proteins made up of more than one polypeptide chain, as many of the large ones are, are called oligomeric. Based upon chemical composition, proteins are divided into two major classes: simple proteins, which are composed of only amino acids and conjugated proteins, which are composed of amino acids and additional organic and inorganic groupings, certain of which are called prosthetic groups. Conjugated proteins include glycoproteins, which contain carbohydrates; lipoproteins, which contain lipids; and nucleoproteins, which contain nucleic acids.

Classified by biological function, proteins include the enzymes, which are responsible for catalyzing the thousands of chemical reactions of the living cell; keratin, elastin and collagen, which are important types of structural, or support, proteins; hemoglobin and other gas transport proteins; ovalbumin, casein and other nutrient molecules; antibodies, which are molecules of the immune system; protein hormones, which regulate metabolism; and proteins that perform mechanical work, such as actin and myosin, the contractile muscle proteins. Every protein molecule has a characteristic three-dimensional shape, or conformation. Fibrous proteins, such as collagen and keratin, consist of polypeptide chains arranged in roughly parallel fashion along a single linear axis, thus forming tough, usually water-insoluble, fibers or sheets. Globular proteins, *e.g.*, many of the known enzymes, show a tightly folded structural geometry approximating the shape of an ellipsoid or sphere.

Because the physiological activity of most proteins is closely linked to their three-dimensional architecture, specific terms are used to refer to different aspects of protein structure. The term primary structure denotes the precise linear sequence of amino acids that constitutes the polypeptide chain of the protein molecule. Automated techniques for amino-acid sequencing have made possible the determination of the primary structure of hundreds of proteins. The physical interaction of sequential amino-acid subunits results in a so-called secondary structure, which often can either be a twisting of the polypeptide chain approximating a linear helix (α-configuration), or a zigzag pattern (β-configuration). Most globular proteins also undergo extensive folding of the chain into a complex three-dimensional geometry designated as tertiary structure. Many globular protein molecules are easily crystallized and have been examined by X-ray diffraction, a technique that allows the visualization of the precise three-dimensional positioning of atoms in relation to each other in a crystal.

24. HOW ARE BIOCHEMICALS AND BIOCHEMICAL REACTIONS CORRELATED?

Biochemistry is the branch of biosciences that studies the molecular basis of life. The study of biochemistry includes the knowledge of the structure and function of molecules found in the biological world and an understanding of the precise biochemical pathways by which organic molecules are either put together or broken down. Biochemistry is the science concerned chiefly with the chemistry of biological processes; it attempts to utilize the tools and concepts of chemistry, particularly organic and physical chemistry, for elucidation of the living system. The science has been variously referred to as physiological chemistry and as biological chemistry. In fact biochemistry is the study of the chemicals of living systems and their interactions. Alone, these chemicals are not living, but together, their interactions work together to create and sustain a living organism. The reactions between them obey all of the known laws of chemistry. To sustain a living system, materials, energy and information is needed. The substances that supply these three necessities are the basic materials which, together with water and a few kinds of ions, make up cells and tissues.

Proteins are the main source of the materials. Lipids and carbohydrates are the major sources of the chemical energy needed to maintain function. And information is carried in a genetic code by molecules of nucleic acid. This section will not go in depth because many aspects extend into the field of biology. Biochemists study such things as the structures and physical properties of biological molecules, including the proteins, the carbohydrates, the lipids and the nucleic acids; the mechanisms of enzyme action; the chemical regulation of metabolism; the molecular basis of genetic expression; the chemistry of vitamins; chemiluminescence; biological oxidation; and energy utilization in the cell. The study of the chemistry of the immune response offers the possibility of treatment and cure for such diseases as AIDS and lupus.

A vast array of biomolecules is present in the cell. The structure of each biomolecule determines in what chemical reactions it is able to participate and hence what role it plays in the cell's life processes. Among the most important classes of biomolecules are nucleic acids, proteins, carbohydrates and lipids. Nucleic acids are responsible for storing and transferring genetic information. There are enormous molecules made up of long strands of subunits, called bases that are arranged in a precise sequence. These are "read" by other components of the cell and used as a guide in making proteins. Proteins are large molecules built up of small subunits called amino acids. Using only 20 different amino acids, a cell constructs thousands of different proteins, each of which has a highly specialized role in the cell.

25. HOW WAS PROTEIN NATURE OF ENZYMES ESTABLISHED AND WHY ENZYMES ARE TEMPERATURE SPECIFIC?

The proteins of greatest interest to biochemists are the enzymes, which are the "worker" molecules of the cell. These enzymes serve as promoters, or catalysts, of chemical reactions. A very important function that proteins can serve is as catalysts. In fact enzymes are biological catalysts, agents which increase the rate of chemical reactions without being used up in the reaction. They are proteins which possess special binding sites for a certain biochemicals. Weak binding interactions with the biochemical reactants allow enzymes to accelerate specific reaction rates millions of times. Enzyme kinetics is the study of enzyme reactions and mechanisms. The term enzyme comes from zymosis, the Greek word for fermentation, a process accomplished by yeast cells and long known to the brewing industry, which occupied the attention of many 19th-century chemists.

Louis Pasteur recognized in 1860 that enzymes were essential to fermentation but assumed that their catalytic action was inextricably linked with the structure and life of the yeast cell. Not until 1897 was it shown by German chemist Edward Büchner that cell-free extracts of yeast could ferment sugars to alcohol and carbon dioxide; Büchner denoted his preparation of zymase. This important achievement was the first indication that enzymes could function independently of the cell. The first enzyme molecule to be isolated in pure crystalline form was urease, prepared from the jack bean in 1926 by American biochemist J.B. Sumner, who suggested, contrary to prevailing opinion, that the molecule was a protein. In the period from 1930 to 1936, pepsin, chymotrypsin and trypsin were successfully crystallized; it was confirmed that the crystals were protein and the protein nature of enzymes was thereby firmly established.

Enzymes speed up reactions inside an organism. The molecule which an enzyme catalyzes is called a substrate. Enzymes can act only on the substrate that they were designed for. This is again because of protein shape. If the substrate molecule's shape matches the enzyme's active site, it undergoes the reaction specified. This is called the lock-and-key theory of enzyme action.

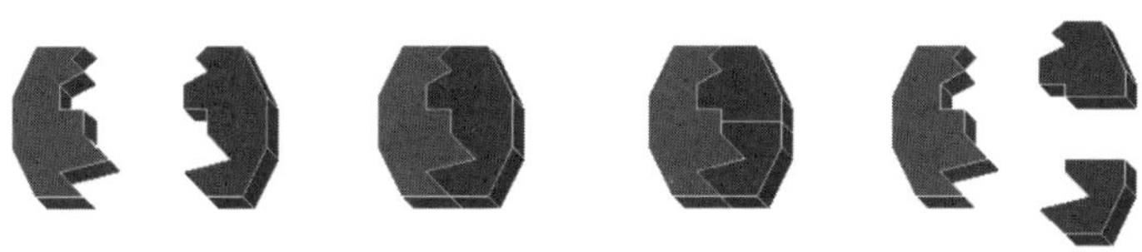

Enzymes can either break or put together substrates. And they can enhance the rate of reactions to over half-a-million molecules per second. Because of the lock-and-key theory, enzymes must retain their shape to keep their function. If the temperature is too high, the enzyme will change its shape, becoming almost like a random mass of coils, in a process called denaturation and thus losing its function. Some dangerous poisons work by deactivating enzymes by changing their shapes.

26. WHAT SPECIES AND FACTORS ACCOUNT FOR AN ENZYME'S ACTIVITY?

Like all catalysts, enzymes accelerate the rates of reactions while experiencing no permanent chemical modification as a result of their participation. Enzymes can accelerate, often by several orders of magnitude, reactions that under the mild conditions of cellular concentrations, temperature, pH and pressure would proceed imperceptibly (or not at all) in the absence of the enzyme. The efficiency of an enzyme's activity is often measured by the turnover rate, which measures the number of molecules of compound upon which the enzyme works per molecule of enzyme per second. Carbonic anhydrase, which removes carbon dioxide from the blood by binding it to water, has a turnover rate of 106. That means that one molecule of the enzyme can cause a million molecules of carbon dioxide to react in one second.

Most enzymatic reactions occur within a relatively narrow temperature range (usually from about 30°C to 40°C), a feature that reflects their complexity as biological molecules. Each enzyme has an optimal range of pH for activity; for example, pepsin in the stomach has maximal reactivity under the extremely acid conditions of pH 1–3. Effective catalysis also depends crucially upon maintenance of the molecule's elaborate three-dimensional structure. Loss of structural integrity, which may result from such factors as changes in pH or high temperatures, almost always leads to a loss of enzymatic activity. An enzyme that has been so altered is said to be denatured. Consonant with their role as biological catalysts, enzymes show considerable selectivity for the molecules upon which they act (called substrates). Most enzymes will react with only a small group of closely related chemical compounds; many demonstrate absolute specificity, having only one substrate molecule which is appropriate for reaction.

Numerous enzymes require for efficient catalytic function the presence of additional atoms of small nonprotein molecules. These include coenzyme molecules, many of which only transiently associate with the enzyme. Nonprotein components tightly bound to the protein are called prosthetic groups. The region on the enzyme molecule in close proximity to where the catalytic event takes place is known as the active site. Prosthetic groups necessary for catalysis are usually located there and it is the place where the substrate (and coenzymes, if any) bind just before reaction takes place. The side-chain groups of amino acid residues making up the enzyme molecule at or near the active site participate in the catalytic event. For example, in the enzyme trypsin, its complex tertiary structure brings together a histidine residue from one section of the molecule with glycine and serine residues from another. The side chains of the residues in this particular geometry produce the active site that accounts for the enzyme's reactivity.

27. WHAT ARE NUCLEIC ACIDS AND ON WHAT ASPECTS AN ORGANISM'S HEREDITARY INFORMATION RESIDES?

All the proteins of an organism are made under the chemical direction of nucleic acids. These are of two types: RNA (ribonucleic acid) and DNA (deoxyribonucleic acid). DNA forms the chemical basis through which genctic information is carried. All physical and biological characteristics are inherited through it. DNA molecules direct their own duplications and guide the assembly of amino acid units into a unique polypeptide sequence. Both DNA and RNA have similar structures. They consist of a chain of deoxyribose or ribose sugars linked by phosphate groups with side chain bases. One structural difference between DNA and RNA is that the sugar in DNA, deoxyribose, is lacking an oxygen atom where ribose has an OH group. The side chain bases are all heterocyclic amines. Their molecular shapes give the nucleic acids their function. They are represented by single letters—A for adenine, T for thymine, U for uracil, G for guanine and C for cytosine. A, T, G and C occur in DNA. A,U, G and C occur in RNA.They make up the "letters" of the genetic alphabet.

Genes, the individual units of heredity, are unique in the length of the backbone and the sequence in which the bases occur. All of an organism's hereditary information resides on these aspects of DNA. In DNA, there are two backbones (or strands) of nucleic acids and they are intertwined like a spiral staircase called the DNA double helix. A always pairs with T and C always pairs with G in DNA. A will pair with U, but U occurs in RNA only, which is important in the work of RNA. If an A is found on one strand, a T will be found on the other. The same goes for C and G.

Because of the limitation of base pairings, replication is very accurate. When replicating, the two strands divide apart and the complementary nucleotides (monomers of DNA, consists of phosphate, sugar and base) will join with the two strands, creating two separate molecules. Nucleotides are made by cells and are always present in the cellular "soup," ready to be used for duplication or other processes. DNA replication is used in cell reproduction, where the two identical copies are carried to the two resulting cells. Each polypeptide in a cell is made under the direction of its own gene. To make the polypeptide, a few basic steps are followed:

Transcription Translation

DNA =============> mRNA ===========> tRNA ==> polypeptide

In transcription, a type of RNA called messenger RNA (mRNA) is used to read the code of a gene in DNA. The gene in the DNA strand opens up, allowing RNA nucleotides to bond in a certain sequence complementary to the bases on the DNA (though U is used instead of T in RNA). Once completed, the mRNA goes out of the cell nucleus to a ribosome (manufacturing center for polypeptides) and the DNA reforms its original shape. At the ribosome, transfer RNA (tRNA) reads the mRNA in a process called translation. The tRNA has an open faced part (anticodon) for reading the mRNA consisting of three letters (codon) a time. Each tRNA is attached to a specific amino acid and as it reads the mRNA, the amino acids are put together in a certain sequence. For example, the anticodon CCA on tRNA matches up with the codon GGU on mRNA. Since CCA is always with glycine, the glycine will become part of the growing chain. If the next codon is GCU, it matches up with anticodon CGA, which is alanine. So, the alanine will be attached to the glycine. Through this process, long chains of polypeptides are formed.

28. WHY CHEMISTS ARE HAUNTING FOR INORGANIC POLYMERIC MATERIALS?

Polymers are usually large, organic molecules that make up substances such as proteins, rubber and plastics. Most plastics consist of organic polymers made up of extremely long carbon chains. Inorganic chemists are making rapid advances in the development of new inorganic polymers. Current research has produced inorganic polymers known as polyphosphazenes, which consist of long chains of alternating nitrogen and phosphorus atoms. Polyphosphazenes may eventually be used in the medical field to provide materials for artificial blood vessels, limbs and joints. Chemists have found that changing the side groups of atoms attached to these nitrogen-phosphorus chains forms plastics that possess unique properties, such as the ability for a plastic pill capsule to time-release ingested drugs into the circulatory system. Another inorganic polymer, polysulfurnitride, consists of alternating sulfur and nitrogen atoms. This polymer conducts electricity and becomes a superconductor at the temperature of –273°C (–460°F). However, because polysulfurnitride is unstable, it is not currently used in practical applications.

Chapter 17

METALLURGY

1. WHAT IS MEANT BY EXTRACTIVE METALLURGY?

Metallurgy is the science and technology of metals and their alloys. Modern metallurgical research is concerned with the production of radioactive metals, for obtaining metals economically from low-grade ores, for obtaining and refining rare metals hitherto not used and for the formulation of alloys. Powder metallurgy deals with the manufacture of ferrous and nonferrous parts by compacting elemental metal or alloy powders in a die. The resultant materials are then heated in a controlled-atmosphere furnace to bond the particles so that the part will retain the shape at normal temperatures and pressures. Welding and soldering are techniques for joining metals metallurgically. Extractive metallurgy is the study and practice of separating metals from their ores and refining them to produce a pure metal. We may discuss the extraction of metals in general terms, but methods for the treatment of ores are quite diverse.

2. WHAT THE TERMS: "ADIT, ALLOY AND CARBOLOY" STAND FOR?

Adit, in mining, is underground passage excavated nearly horizontally, with one end open to the earth's surface, usually used to service a mine. The adit end is the furthermost end from the surface, *i.e.*, the location where miners work. The adit collar is the area where an adit opens to the surface and must be reinforced against any surface weakness.

An alloy is a substance with metallic properties and consisting of a metal fused with one or more metals or nonmetals. Alloys may be a homogeneous solid solution, a heterogeneous mixture of tiny crystals, a true chemical compound, or a mixture of these. Alloys are used more extensively than pure metals because they can be engineered to have specific properties. For example, they may be poorer conductors of heat and electricity, harder, or more resistant to corrosion. Alloys of iron and carbon include cast iron and steels; brass and bronze are important alloys of copper; amalgams are alloys that contain mercury; and chromium is an important additive in stainless steel. Because pure gold and silver are soft, they are often alloyed with one another or with other metals. New alloys are being engineered for use in new technology, including materials for the space program. Metallic glasses and crystalline alloys have also been developed and metal alloys are sometimes bonded with ceramics, graphites and organic materials as composites. Carboloy (portmanteau word from carbon and alloy], is an alloy containing cobalt, tungsten and carbon. This alloy is extremely hard, harder than steel; it is used to cut steel, porcelain, quartz and other materials. Its hardness is little affected by heat and it retains a sharp cutting edge even at red heat.

3. WHY IS AMALGAMATION PROCESS CRITICIZED?

Amalgam is an alloy containing mercury. The alloy may be liquid or solid, depending on the proportion of mercury, although all naturally occurring amalgams, *i.e.*, those of gold and silver, are solid. Amalgams are widely used. Silver, gold and copper amalgams are used in dentistry and tin amalgam is used in making mirrors. Amalgamation process, in particular, is a method to extract gold and silver from ores. The ore is crushed and treated with mercury, in which the metal dissolves. When the amalgam is heated, the mercury evaporates, leaving pure gold or silver. The process has been criticized as an environmental and occupational threat due to the extreme toxicity of mercury.

4. WHAT ARE ANNEALING AND ASSAYING PROCESSES?

Annealing is the process in which glass, metals and other materials are treated to render them less brittle and more workable. Annealing consists of heating the material and then cooling it very slowly and uniformly; the time and temperatures required in the process are set according to the properties desired. Annealing increases ductility and lessens the possibility of a failure by relieving internal strains. The process, also called hot working, was known to the ancients. Assaying, in metallurgy, is the process of determining the specific metallic content of an ore, alloy, or other substance, especially one containing precious metals. It is concerned, in some cases, with subjecting of the substance to complete chemical analysis and, in others, with simply determining the quantity present of one or more of the metal constituents. An accurate assay depends first upon procuring a representative sample of the ore in question. Since distribution of the ore's components is not uniform, a common method employed in obtaining this sample is to procure several samples, crush and mix them together and from the final mass take the sample to be assayed. Assays are said to be gravimetric when the weight of the metal is determined and volumetric when the analysis involves the volume of the metal in solution as compared to that in a standardized solution. A wet assay (one which involves the use of liquid reagents) is generally used in a determination of weight. In a dry assay the ore is fused and the metal is finally obtained in a pure state.

5. WHAT SPECIES ARE DUBBED AS 'BABBITT METALS AND BRITANNIA METAL'?

Babbitt metal is an antifriction metal first produced by Isaac Babbitt in 1839. In present-day usage, the term is applied to a whole class of silvery-white metals, or "white metals." These alloys usually consist of relatively hard crystals embedded in a softer matrix, a structure important for machine bearings. They are composed primarily of tin, copper and antimony, with traces of other metals added in some cases and lead substituted for tin in others. Britannia metal is a silvery-white alloy of tin with antimony, copper and sometimes bismuth and zinc. It is very similar in appearance to pewter, but is harder. It is used widely for the manufacture of tableware.

6. WHAT ARE ADVANTAGES OF BASIC OXYGEN PROCESS?

Basic oxygen process is the method of producing steel from a charge consisting mostly of pig iron. The charge is placed in a furnace similar to the one used in the Bessemer process of steelmaking except that pure oxygen, instead of air, is blown into the charge to oxidize the impurities present. One desirable feature of this process is that it takes less than an hour and is thus much faster than the open-hearth process, another important method of steelmaking. A second advantage is that a major byproduct is carbon monoxide, which can be used as a fuel or in producing various chemicals, such as acetic acid. The basic oxygen process also produces less air pollution than methods using air.

7. WHAT IS BAUXITE?

Bauxite, an important ore of aluminum, is the mixture of hydrated aluminum oxides, usually containing oxides of iron and silicon in varying quantities. A noncrystalline substance formerly thought to be a mineral; bauxite is claylike and earthy and ranges in color from white to deep brown or red according to the nature and quantity of its components. Among the specific aluminum-bearing minerals in bauxite are gibbsite, $Al(OH)_3$ and diaspore, $HAlO_2$. Bauxite is a soft mineral with hardness varying from 1 to 3 and specific gravity from 2 to 2.55. Bauxite usually occurs in aggregates in lumps about the size of a pea. Generally bauxite occurs characteristically in pisolitic form, *i.e.*, composed of small, round concretions. Although aluminum is the commonest of all metals on the surface of the earth, it most frequently occurs strongly bound in silicate minerals, which cannot be used commercially as ores because of the difficulty of extracting the aluminum. Bauxite, on the other hand, forms by the rapid weathering of granite rocks in warm, humid climates and can be easily purified and converted directly into either alum or metallic aluminum. Its composition varies, alumina constituting from about 50% to about 70%. First discovered in Les Baux, France, bauxite is widely distributed, with important deposits occurring in Africa, South America, Russia, the West Indies and the United States (Alabama, Arkansas and Georgia). It is mined in large quantities in Australia, Guinea and Jamaica. Bauxite is the main source of aluminum and of its compounds, including alumina, alums and alundum. It is used in the preparation of abrasives and as a refractory for spark plugs and furnace linings.

8. HOW PURE ALUMINUM IS OBTAINED BY BAYER PROCESS?

Bayer process is a procedure for obtaining alumina from the aluminum ore bauxite. The alumina can then be used for various industrial purposes or smelted to provide aluminum. The first step in the process is the mixing of ground bauxite into a solution of sodium hydroxide. By applying steam and pressure in tanks containing the mixture, the bauxite slowly dissolves. The alumina released reacts with the sodium hydroxide to form sodium aluminate. After the contents of the tank have passed through other vessels where the pressure and temperature are reduced and impurities removed, the solution of sodium aluminate is placed in a special tank where the alumina is precipitated out. The precipitate is removed from the tank, washed and heated in a kiln to drive off any water present. The residue is a commercially pure alumina.

9. HOW IS STEEL OBTAINED BY BESSEMER PROCESS?

Bessemer process [for Sir Henry Bessemer], is a industrial process for the manufacture of steel from molten pig iron. The principle involved is that of oxidation of the impurities in the iron by the oxygen of air that is blown through the molten iron; the heat of oxidation raises the temperature of the mass and keeps it molten during operation. The process is carried on in a large container called the Bessemer converter, which is made of steel and has a lining of silica and clay or of dolomite. The capacity is from 8 to 30 tons of molten iron; the usual charge is 15 or 18 tons. The converter is egg-shaped. At its narrow upper end it has an opening through which the iron to be treated is introduced and the finished product is poured out. The wide end, or bottom, has a number of perforations through which the air is forced upward into the converter during operation. The container is set on pivots (trunnions) so that it can be tilted at an angle to receive the charge, turned upright during the "blow," and inclined for pouring the molten steel after the operation is complete. As the air passes upward through the molten pig iron, impurities such as silicon, manganese and carbon unite with the oxygen in the air to form oxides; the carbon monoxide burns off with a blue flame and the other impurities form slag. Dolomite is used as the converter lining when the phosphorus content is high; the process is then called basic Bessemer. The silica and clay lining is used in the acid Bessemer, in which phosphorus is not removed. In order to provide the elements necessary to give the steel the desired properties, another substance (often spiegeleisen, an iron-carbon-manganese alloy) is usually added to the molten metal after the oxidation is completed. The converter is then emptied into ladles from which the steel is poured into molds; and the slag is left behind. The whole process is completed in 15 to 20 min. The Bessemer process was superseded by the open-hearth process.

10. HOW IS CAST IRON PRODUCED BY BLAST FURNACE?

Blast furnace is a structure used chiefly in smelting. The principle involved in this means of extracting metals is that of the reduction of the ores by the action of carbon monoxide, *i.e.*, the removal of oxygen from the metal oxide in order to obtain the metal. Blast furnaces differ in construction. The one used in the production of iron consists of a chimney like structure (usually 80–100 ft/24–30 m high) made of iron or steel and lined with firebrick. It is narrow at the top, increasing in diameter downward, but narrowing again suddenly almost at the bottom, to form the hearth or crucible.

There the fine molten products are caught. The furnace is fed from the top with a charge of definite quantities of ore, coke and a flux, mostly limestone. Preheated compressed air is introduced at the bottom through pipes (tuyères) entering just above the hearth. The air passes upward through the charge. The coke is oxidized to carbon dioxide, which changes to carbon monoxide at the high temperature. The carbon monoxide then reduces the ores and, taking on oxygen, reverts to carbon dioxide. This gas, together with unused carbon monoxide, nitrogen and other constituents of the air originally introduced, is led off through a pipe from the top of the furnace and, being still at a high temperature, is employed to heat the stoves into which fresh air for the process is brought. As the operation proceeds, the mass in the furnace becomes molten and descends into the crucible. The iron sinks to the bottom; impurities, called the slag, being lighter, float on top. The slag is drained through a pipe in the upper portion of the crucible. The iron is tapped from below and run into sand molds to harden. The product is known as pig iron or cast iron. Efforts to increase production rates have led to the addition of pure oxygen and steam and the sizing of ore to obtain better gas-solid contact. Flux and ore are sometimes combined into pellets. Pig iron prepared in the blast furnace is converted into steel by the Bessemer process. Copper ore treated in a blast furnace yields a copper matte, from which only a part of the impurities are removed. It is usually further refined by electrolytic methods.

11. WHAT IS DIFFERENCE BETWEEN 'BRASS AND BRONZE'?

Brass

Brass is an alloy having copper (55%–90%) and zinc (45%–10%) as its essential components. The properties of brass vary with the proportion of copper and zinc and with the addition of small amounts of other elements, such as aluminum, lead, tin, or nickel. In general, brass can be forged or hammered into various shapes, rolled into thin sheets, drawn into wires and machined and cast. Its ductility reaches a maximum with about 30% zinc and its tensile strength with 45% although this property varies greatly with the mechanical and heat treatment of the alloy. Cartridge brass (70% copper, 30% zinc) is used for cartridge cases, plumbing and lighting fixtures, rivets, screws and springs. Aluminum brass (not exceeding 3% aluminum) has greater resistance to corrosion than ordinary brass. Brass containing tin (not exceeding 2%) is less liable to corrosion in seawater; it is sometimes called naval brass and is used in naval construction. Dutch metal (80%–85% copper, 15%–20% zinc) is used as a substitute for gold leaf. When iron is added to brass it produces hard, tough alloys. One of these is delta metal (55% copper, 41% zinc, 1%–3% iron and fractional percentages of tin and manganese), which can be forged, rolled, or cast and is used for bearings, valves and ship propellers.

Bronze

Bronze, in metallurgy, is an alloy of copper, tin, zinc, phosphorus and sometimes small amounts of other elements. Bronzes are harder than brasses. Most are produced by melting the copper and adding the desired amounts of tin, zinc and other substances. The properties of the alloy depend on the proportions of its components. Aluminum bronze has high strength and resists corrosion; it is used for bearings, valve seats and machine parts. Leaded bronze, containing from 10% to 29% lead, is cast into heavy–duty bushings and bearings. Silicon bronze is used for telegraph wires and chemical containers. Phosphor bronze is used for springs. Bronze is used for coins, medals, steam fittings and gunmetal and was formerly employed for cannon. Because of its particularly sonorous quality, bell metal, containing from 20% to 24% tin, is used for casting bells. Bronze has long been used in art, *e.g.*, for castings, engravings and forgings.

12. WHAT IS DIFFERENCE BETWEEN CALCINATION AND DRYING?

Calcination, in metallurgy, is the process of heating solid material to drive off volatile chemically combined components, *e.g.*, carbon dioxide. It is sometimes a step in the extraction of metals from ores. Calcination is distinguished from drying, in which mechanically held water is driven off by heating and from roasting, in which a material is heated in the presence of air to oxidize impurities. Originally calcination meant the method of obtaining lime (calcium oxide) from limestone by heating it to drive off carbon dioxide.

13. HOW IS CASTING DONE?

Casting or founding is shaping of metal by melting and pouring into a mold. Most castings, especially large ones, are made in sand molds. Sand, mixed with a binder to hold it together, is pressed around a wooden pattern that leaves a cavity in the sand. Molten metal is poured into the cavity and allowed to solidify. Permanent metal molds are used to make many small, simple parts; shell molding gives greater accuracy for a large volume of semi-precision parts. A two-step process, investment casting, produces small, complex shapes. Wax or plastic replicas of the parts are molded in accurate metal molds. These replicas are covered with sand in a box to make the final mold. When the whole mold is heated, the replica melts, leaving behind a cavity into which metal is poured. Large numbers of small, precise parts of metals that have a low melting point, such as zinc, are made by die-casting; in an automatic process, molten metal is forced under pressure into metal molds. Cast iron and cast steel are more brittle than forged iron and forged steel.

14. HOW IS COAL MINING DONE AND HOW ARE MINERS AND EQUIPMENTS PROTECTED IN UNDERGROUND COAL MINES?

Coal mining is the physical extraction of coal resources to yield coal. It is also the business of exploring for, developing, mining and transporting coal in any form. Strip mining is the process in which the overburden (earth and rock material overlying the coal) is removed to expose a coal seam or coal bed. Excavators either dispose off the overburden or store the waste material for replacement after the coal has been extracted. Once exposed, the coal is usually removed in a separate operation. Surface soil is often stripped separately and spread back onto the reclaimed surface. The environment can also be protected by seeding or planting grass or trees on the fertilized restored surface of a strip mine. The term strip mining is most often used in reference to coal mining, although the process may also be used to extract certain metallic ores as well. Sometimes the terms open-pit, open-cast, or surface mining are used in the same sense, although they usually refer to metalliferous mining or the mining of other minerals. Underground coal mining is the extraction of coal from below the surface of the earth. The coal is worked through tunnels, passages and openings that are connected to the surface for the purpose of the removal of the coal. Mechanical equipment breaks the coal to a size suitable for haulage. Alternatively, the coal is drilled and the resultant holes are loaded with explosives and blasted in order to break the coal to the desired size.

In order to protect the miners and equipment in an underground coal mine, much attention is paid to maintain and support a safe roof or overhead ceiling for the extraction openings. Long-wall mining is a method of underground mining believed to have been developed in Shropshire, England, near the end of the 17th century. A long face, or working section, of coal, some 600 ft (180 m) in length, is operated at one time. The miners and machinery at the working face are usually protected by hydraulic jacks or mechanical props which are advanced as the coal is extracted. The excavated, or gob, area is either allowed to cave in, or is filled in by waste material called stowing. The Anderton shearer is a widely used coal cutter and loader for long-wall mining. It shears coal from the face, as it moves in one direction and loads coal onto an armored conveyor as it travels back in the opposite direction. It is ordinarily used for coal seams greater than 3.5 ft (9.1 cm) in thickness.

15. WHAT IS BRAZING AND WHAT ARE ITS ADVANTAGES?

Brazing is a method of joining metal parts using nonferrous filler metals such as copper, silver and aluminum alloys that have high melting points. Brazing differs from soldering in the respect that it uses a higher temperature; and unlike welding, the parts are not melted in it. Brazing is best for dissimilar or thinner metal parts and for parts difficult to weld or solder. Brazing joins two pieces of base metal when a melted metallic filler flows across the joint and cools to form a solid bond. Similar to soldering, brazing creates an extremely strong joint, usually stronger than the base metal pieces themselves, without melting or deforming the components. Two different metals, or base metals such as silver and bronze, are perfect for brazing. We use this method to make a bond that is invisible, resilient in a wide range of temperatures and can withstand jolting and twisting motion. The process of brazing is the same as soldering, although metals and temperatures differ. We can braze pipes, rods, flat metals, or any other shape as long as the pieces fit neatly against each other without large gaps. Brazing handles more unusual configurations with linear joints, whereas most welding makes spot welds on simpler shapes.

First, We must clean the entire area to be joined or else the melted braze mixture will clump instead of flowing, making an inconsistent joint. Wash the surface and then apply melted flux. Flux removes oxides, prevents more oxidation during brazing and smoothes the surface so that braze "flows" evenly across the joint. Next, We gather our torch and braze alloy. The torch uses fuels like acetylene and hydrogen to create an extremely high temperature, often between 800°F and 2000°F (430–1100°C). The temperature must be low enough that the base metals don't melt, yet high enough to melt the braze. Torches have sensitively controls to reach the proper temperature depending on the associated melting points. Finally, We complete the joint by applying the braze. Braze, like solder, comes in a stick, disc, or wire, depending on our preference or the shape of the joint. After the base metals near the joint have been heated with the torch, bring the wire to the hot pieces so the braze melts, flowing around the joint. By "flow," brazers mean it penetrates the joint, working into every cavern. If the brazing was performed correctly, when the bond cools and solidifies, it is nearly unbreakable.

Brazing offers many advantages over spot welding or soldering. For instance, a brazed joint is smooth and complete; creating an airtight and watertight bond for piping that can be easily plated so the seam disappears. It also conducts electricity like the base alloys. Only brazing can join dissimilar metals, such as bronze, steel, aluminum, wrought iron and copper, with different melting points.

16. HOW ARE ORES CONCENTRATED?

This simply means getting rid of as much of the unwanted rocky material as possible before the ore is converted into the metal. In some cases this is done chemically. For example, pure aluminium oxide is obtained from bauxite by a process involving a reaction with sodium hydroxide solution. When an ore has a low percentage of the desired metal, a method of physical concentration must be used before the extraction process begins. In one such method, the ore is crushed and placed in a machine where, by shaking, the heavier particles containing the metal are separated from the lighter rock particles by gravity. One such method is the flotation process, used commonly for copper sulfide ores. In certain cases (as when gold, silver, or occasionally copper occur "free," *i.e.*, uncombined chemically in sand or rock), mechanical or ore dressing methods alone are sufficient to obtain relatively pure metal. Waste material is washed away or separated by screening and gravity; the concentrated ore is then treated by various chemical processes. Some copper ores can be converted into copper (II) sulphate solution by leaving the crushed ore in contact with dilute sulphuric acid for a long time. Copper can then be extracted from the copper (II) sulphate solution. But, in many cases, it is possible to separate the metal compound from unwanted rocky material by physical means. A common example of this involves froth flotation.

17. HOW ARE DAMPS DETECTED?

Damp, in mining, is any mixture of gases in an underground mine, especially oxygen-deficient or noxious gases. The term damp probably is derived from the German dampf, meaning fog or vapor. Several distinct types of damp are recognized. Firedamp is methane and other flammable gases, often mixed with air; it results from the decomposition of coal or other carbonaceous materials. Explosive mixtures of firedamp with air usually contain from 1% to 14% methane. The mixture of gases that remains after a firedamp explosion is called afterdamp; it consists chiefly of carbon dioxide and nitrogen. Chokedamp is any mixture of oxygen-deficient mine gases that causes suffocation. (In England, carbon dioxide is called chokedamp). Several methods are used for detection of damps. The Davey safety lamp is one of the earliest detection devices. The color and height of the lamp flame indicate the amount of firedamp present; if the flame is extinguished, chokedamp is present. Canaries were formerly kept in mines; the birds are overcome by relatively small quantities of noxious gases and their death warned the miners of the presence of damps. Special colorimetric detectors are now used. The methanometer is a special portable instrument used to detect firedamp.

18. HOW IS A DIE MADE?

Die is any of various devices used for drawing wire and for blanking, bending, cutting, machine forging and embossing. Dies used for striking, or stamping, coins and medals are cut in intaglio, one for the front, another for the back, of the coin. Such dies were used as early as c.800 B.C. in Greece. Die-making, or die-sinking, formerly entirely a hand process in which the graver (a cutting tool), riffler (a file) and chisel were employed, has been accelerated in modern times by the use of die-making machines supplemented by hand finishing. A punch, or male die, is commonly made as the counterpart in relief of the original die, or matrix; both are preserved as models and duplicates are made from them for working dies. Sheet metal or other material is blanked (cut) out, shaped, or embossed between the dies by power-operated levers or drop hammers, or by die-casting. The die used for drawing wire or extruding rods is made of hard metal with a hole or a series of progressively smaller holes through which the metal is forced. For making screws or threading pipe a hollow hard metal die with internal threading is used.

19. WHAT IS DIE-CASTING?

Die-casting is a process by which molten metal is forced by a plunger or compressed air into a metallic die and the pressure maintained until the metal has solidified. Die castings are accurate, are sharply outlined, have a good surface finish and can be made in complicated designs. Zinc, aluminum and magnesium alloys are the principal metals used. The high cost of the die usually limits the process to large-scale, high-speed production. Typical products are carburetor bodies and zippers. Type-casting machines are specialized die-casting machines.

20. WHAT IS DOWSER?

Divining rod or dowser is a stick used in searching for underground water or minerals. This form of divination is still in common use in many parts of the world. The instrument is typically a forked twig. The operator holds the forked ends of the twig close to his body, with the stem pointing forward. When he walks over a spot under which water or the desired mineral lies, the stem of the divining rod is supposedly pulled down. Impartial research, however, has indicated that successes in this method result mostly from chance and possibly also from a heightened sensitivity to visual cues of which the diviner is unaware.

21. WHAT IS DRAINAGE?

Drainage, in mining, is removal of water seeping into shafts and other underground mine workings from the surrounding ground. Unless seeping water is removed continually, it may endanger haulage and mining equipment, weaken supporting structures and, in some instances, flood the mine completely. Water in a mine is drained into sumps, or reservoirs, usually excavated below the lowest working level of the mine and are then removed by pumps. Methods to minimize seepage include the sealing of visible fissures through which water enters and the injection of concrete into the ground surrounding the shafts.

22. WHAT IS DIFFERENCE BETWEEN 'DUCTILITY AND MALLEABILITY'?

Ductility is ability of a metal to plastically deform without breaking or fracturing, with the cohesion between the molecules remaining sufficient to hold them together. Ductility is important in wire drawing and sheet stamping. The metal must neither break nor be scraped off during these processes. Platinum, steel, copper and tungsten have high ductility. Ductility is a focus of rheology, the study of how materials deform and flow in response to force.

Malleability is the property of a metal describing the ease with which it can be hammered, forged, pressed, or rolled into thin sheets. Metals vary in this respect; pure gold is the most malleable. Silver, copper, aluminum, lead, tin, zinc and iron are also very malleable. Some heating usually increases malleability. Zinc, for example, at ordinary temperatures is very brittle, but is malleable in the temperature range from about 120°C to 150°C. Impurities adversely affect the malleability of metals.

23. WHAT IS DIFFERENCE BETWEEN 'DURALUMIN AND MAGNALIUM'?

Duralumin is an alloy of aluminum (over 90%) with copper (about 4%), magnesium (0.5%–1%) and manganese (less than 1%). Before a final heat treatment the alloy is ductile and malleable; after heat treatment, a reaction between the aluminum and magnesium, produces increased hardness and tensile strength. Because of its lightness and other desirable physical properties, duralumin is widely used in the aircraft industry.

Magnalium is an alloy of aluminum and about 5% magnesium. Although weak and soft in the elemental state, magnesium alloys with aluminum, manganese, zinc, tin, zirconium and cerium to produce alloys useful in engineering materials. The strength and hardness of aluminum increases when it is alloyed with such substances as magnesium, manganese, nickel, chromium, zinc, iron, copper and silicon. Magnalium, lighter and more workable than aluminum, is used in making metal mirrors and scientific instruments.

24. HOW ARE METALS EXTRACTED FROM THEIR ORES?

We may look at the various factors which influence the choice of a method for extracting metals from their ores by reduction by carbon, reduction by a reactive metal (like sodium or magnesium) and by electrolysis.

Reducing the Metal Compound to the Metal

At its simplest, where we are starting from metal oxides, the ore is being reduced because oxygen is being removed. However, if we are starting with a sulphide ore, for example, that's not of any help. It is much more helpful to use the definition of reduction in terms of addition of electrons. To a reasonable approximation, we can think of these ores as containing positive metal ions. To convert them to the metal, we need to add electrons. In some compounds the metal may not literally be present as a positive ion. Instead, it may be part of a covalent bond, but will always be the least electronegative element present and so will carry some degree of positive charge. That means that its oxidation state will always be positive. Reducing that oxidation state to zero (in the raw element) will always involve adding electrons.

Choosing a Method of Reduction

There are various economic factors we need to think about in choosing a method of reduction for a particular ore. What follows is a quick summary. We need to consider: the cost of the reducing agent; energy costs and the desired purity of the metal. There may be various environmental considerations as well, some of which will have economic considerations.

Carbon Reduction

Carbon (as coke or charcoal) is cheap. It not only acts as a reducing agent, but it also acts as the fuel to provide heat for the process. However, in some cases (for example with aluminium) the temperature needed for carbon reduction is too high to be economic, so a different method has to be used. Carbon may also be left in the metal as an impurity. Sometimes this can be removed afterwards (for example, in the extraction of iron); sometimes it can't (for example in producing titanium) and a different method would have to be used in cases like this.

Reduction Using a More Reactive Metal

Titanium is produced by reducing titanium(IV) chloride using a more reactive metal such as sodium or magnesium and this is the only way of producing high purity metal.

The more reactive metal sodium releases electrons easily as it forms its ions. These electrons are used to reduce the titanium(IV) chloride. This is a good example of a reduction in which the metal isn't originally present as an ion. Titanium(IV) chloride is a covalent liquid. The reduction is from titanium in the +4 oxidation state to the metal in the zero oxidation state. The downside of this is expense. We have first to extract (or to buy) the sodium or magnesium. The more reactive the metal is the more difficult and expensive the extraction becomes. That means that we have to use a very expensive reducing agent to extract the titanium. In case of titanium extraction, there are other problems in its extraction which also add to the cost.

Reduction by Electrolysis

This is a common extraction process for the more reactive metals, for example, for aluminium and metals above it in the electrochemical series. We may also come across it in other cases such as one method of extracting copper and in the purification of copper. During electrolysis, electrons are being added directly to the metal ions at the cathode (the negative electrode). The downside (particularly in the aluminium case) is the cost of the electricity. An advantage is that it can produce very pure metals.

25. WHY MATERIALS USED IN CONSTRUCTION ARE TESTED FOR THEIR FATIGUE STRENGTH?

Fatigue, in engineering, is microscopic cracking of materials, especially metals, after repeated applications of stress. Fissures may be formed within pieces of metal during their manufacture when, while cooling from the molten state, they shrink and tensile stresses arise. Once a crack has started it spreads under repeated stress until the metal ruptures. Examples of fatigue are found in steel rails, beams and girders. Metallic fatigue resulted in the catastrophes encountered by many of the Liberty ships built during World Wars I and II and the crashes of a number of the earliest jet aircraft constructed. Materials used in construction are tested for fatigue strength, or endurance limit, by being subjected mechanically to cyclic applications of stress. Steel parts are sometimes treated by shot blasting to increase their fatigue resistance.

26. WHAT HAPPENS IN FLOTATION PROCESS?

In mineral treatment and mining, Flotation process is the process for concentrating the metal-bearing mineral in an ore. Crude ore is ground to a fine powder and mixed with water, frothing reagents and collecting reagents. When air is blown through the mixture, mineral particles cling to the bubbles, which rise to form a froth on the surface. The waste material (gangue) settles to the bottom. The froth is skimmed off and the water and chemicals are distilled or otherwise removed, leaving a clean concentrate. The process, also called the froth-flotation process, is used for a number of minerals, especially silver.

27. HOW ARE METALS FORGED?

Forging is shaping metal by heating it and then hammering or rolling it. Forging is the method by which metal was first worked when it came into use about 4000 B.C. in Egypt and Asia. Modern forging is done with a power-driven hammer; Dies are usually used. These are steel blocks hollowed out or carved in relief in the shape of the desired part. One die is stationary, the other is attached to the underside of a hammer or press ram. A piece of metal is then hammered or squeezed until it takes on the shape of the die cavity. Sometimes there are several sets of dies that form the metal in stages into the final shape. Smaller parts may be forged cold; larger parts, hot. Maximum pressure exerted is about 100 tons per square inch. Variations of the forging process are called sizing, heading and coining. Some industrial machines for forging are the drop hammer, steam hammer and hydraulic press. Forging toughens iron and steel, while casting makes them brittle.

28. WHAT HAPPENS IN 'FRASCH AND FROTH FLOTATION PROCESSES'?

Frasch Process

Frasch process [for Herman Frasch, the German-American chemist who devised it], is a process for the extraction of sulfur from subsurface deposits. Three pipes, one inside another, are sunk to the bottom of the sulfur bed. Water heated under pressure to a temperature well above the melting point of the sulfur is conducted down the outer pipe and air under pressure down through the innermost pipe. The heated water melts the sulfur and the compressed air forces it through the middle pipe to the surface.

Froth Flotation

In this process ore is first crushed and then treated with something which will bind to the particles of the metal compound that we want and make those particles hydrophobic. "Hydrophobic" literally means "water fearing". In concentrating copper ores, for example, pine oil is often used. The pine oil binds to the copper compounds, but not to the unwanted rocky material. The treated ore is then put in a large bath of water containing a foaming agent (a soap or detergent of some kind) and air is blown through the mixture to make a lot of bubbles. Because they are water-repellent, the coated particles of the metal compound tend to be picked up by the air bubbles, float to the top of the bath and are allowed to flow out over the sides. The rest of the rocky material stays in the bath.

29. WHAT ARE ADVANTAGES OF GALVANIZATION?

Galvanizing is the process of coating a metal, usually iron or steel, with a protective covering of zinc. Galvanized iron is prepared either by dipping iron, from which rust has been removed by the action of sulfuric acid, into molten zinc so that a thin layer of the zinc remains on the surface of the iron upon removal or by a method of electroplating. Iron is also coated with zinc by a method in which the iron is first covered with the zinc dust and then baked; an alloy is formed at the surface, the resulting product being known as sherardized iron. Sheets of pure iron, copper iron and various steels, as well as wire and netting, are often galvanized, since the zinc coating resists oxidation and the action of moisture very successfully. When the coating is broken or pierced some protection is still afforded, since the zinc reacts with the corroding agent first.

30. WHAT IS DIFFERENCE BETWEEN 'GERMAN SILVER AND GUNMETAL'?

Discovered in early 19th century by a German industrial chemist, E. A. Geitner, German silver is the name for various alloys of copper, zinc and nickel, sometimes also containing lead and tin. They were originally named for their silver-white color, but use of the term silver is now prohibited for alloys not containing that metal. German silver varies in composition, the percentage of the three elements ranging approximately as: copper, from 50% to 61.6%; zinc, from 19% to 17.2% and nickel, from 30% to 21.1%. The proportions are always specified in commercial alloys. German silver is extensively used because of its hardness, toughness and resistance to corrosion for articles such as tableware (commonly silver plated), marine fittings and plumbing fixtures. Because of its high electrical resistance it is used also in heating coils.

Gunmetal is a bronze that is an alloy of copper, tin and a small amount of zinc. Although originally used extensively for making guns (from which it received its name), it has been superseded by steel and it is now chiefly employed in casting machine parts. The so-called 88–10–2 (copper-tin-zinc) alloy is the "government bronze," composed of 88% copper, 10% tin and 2% zinc. The percentages of the three elements are varied slightly in gunmetals produced for different purposes. The metal commonly called gunmetal today is very often steel treated to simulate the bronze alloy. In other cases, copper and tin are used alone; in still others, copper, tin and lead are used.

31. WHY AND HOW ARE METALS HARDENED?

Hardening, in metallurgy, is treatment of metals to increase their resistance to penetration. A metal is harder when it has small grains, which result when the metal is cooled rapidly. Sometimes small areas on the surface of a casting are given a fine-grain structure by chill hardening; metal pieces (chills) are inserted in the wall of a sand mold. The area next to the chill cools faster and becomes harder than the surface next to the sand. Metals worked cold, as by being rolled into thinner pieces, become hardened, partly by reducing grain size and partly by distorting the shape of the grains so that they increasingly resist further distortion. Alloying may harden a metal by changing its chemical composition. In hardening by precipitation, one constituent of a supersaturated solid solution separates from the solution. Usually the process is carried out at above room temperature. At room temperature the process takes longer; it is then known as age-hardening. Aluminum-copper alloys are hardened by precipitation. Iron-carbon alloys, steel and cast iron, for example, respond well to heat treatments. By varying the percentage of carbon and the rate of cooling from a high temperature, many gradations of hardness, softness, toughness and other properties are achieved. To impart hardness the metal is rapidly cooled from a high temperature by quenching in water, oil, or molten salt. Later heat treatment by tempering or annealing modifies the metal slightly to give other desirable qualities. Steels with a low percentage of carbon can be given a hard surface by increasing the amount of carbon at the surface so that they will respond to heat treatment, a process known as carburizing, or casehardening. One way to do this is to pack steel in charcoal and then heat it. Another way is to heat the metal in a furnace with a hydrocarbon gas atmosphere; still another is to heat the metal in molten-salt bath containing potassium and sodium cyanides. If the salt bath cited is of a lower temperature, the steel surface will also pick up nitrogen, which helps harden it; the process is then called cyaniding. At even lower temperatures the steel picks up only nitrogen and is nitrided.

32. WHAT IS MEANT BY LEACHING?

Leaching is a method of extraction in which a solvent is passed through a mixture to remove some desired substance from it. A simple example of leaching is the passage of boiling water through ground coffee to dissolve and carry out the chemicals necessary for producing the beverage. Another example is the removal of sugar from sugar beets using water as the solvent. Leaching is also used to remove metals from their ores. In one procedure certain crushed ores of copper are placed into a series of tanks. When a solvent, such as sulfuric acid, is pumped into the first tank, it dissolves the copper from the ore. Eventually overflowing the first tank, the solution passes into the second, where more copper is dissolved. When this tank overflows, the process is repeated in the third tank and so on. The copper is ultimately removed from the solution by chemical or other treatment.

33. WHAT METHODS ARE USED FOR SEPARATION OF METALS?

Processes for separating the metal from the impurities 'it is found with' or the other elements with which it is combined, depend upon the chemical nature of the ore to be treated and upon the properties of the metal to be extracted. Gold and silver are often removed from the impurities associated with them by treatment with mercury, in which they are soluble. Another method for the separation of gold and silver is the so-called cyanide process. The Parkes process, which is based on silver being soluble in molten zinc while lead is not, is used to free silver from lead ores. Since almost all the metals are found combined with other elements in nature, chemical reactions are required to set them free. These chemical processes are classified as pyrometallurgy, electrometallurgy and hydrometallurgy.

Pyrometallurgy, or the use of heat for the treatment of an ore, includes smelting and roasting. If the ore is an oxide, it is heated with a reducing agent, such as carbon in the form of coke or coal; the oxygen of the ore combines with the carbon and is removed in carbon dioxide, a gas. The waste material in the ore is called gangue; it is removed by means of a substance called a flux which, when heated, combines with it to form a molten mass called slag. Being lighter than the metal, the slag floats on it and can be skimmed or drawn off. The flux used depends upon the chemical nature of the ore; limestone is usually employed with a siliceous gangue. A sulfide ore is commonly roasted *i.e.*, heated in air. The metal of the ore combines with oxygen of the air to form an oxide and the sulfur of the ore also combines with oxygen to form sulfur dioxide, which, being a gas, passes off. The metallic oxide is then treated with a reducing agent. When a carbonate ore is heated, the oxide of the metal is formed and carbon dioxide is given off; the oxide is then reduced.

Electrometallurgy includes the preparation of certain active metals, such as aluminum, calcium, barium, magnesium, potassium and sodium, by electrolysis: a fused compound of the metal, commonly the chloride, is subjected to an electric current, the metal collecting at the cathode. Hydrometallurgy, sometimes called leaching, involves the selective dissolution of metals from their ores. For example, certain copper oxide and carbonate ores are treated with dilute sulfuric acid, forming water-soluble copper sulfate. The metal is recovered by electrolysis of the solution. If the metal obtained from the ore still contains impurities, special refining processes are required.

34. WHAT IS MEANT BY MINING?

Mining is extraction of solid mineral resources from the earth. These resources include ores, which contain commercially valuable amounts of metals, such as iron and aluminum; precious stones, such as diamonds; building stones, such as granite; and solid fuels, such as coal and oil shale. The search for and discovery of mineral deposits is called prospecting, or exploration. When a mineral deposit is found, it is studied to determine if it can be mined profitably. If so, the deposit can be worked or extracted by a variety of mining methods.

35. WHO MADE MOTHER LODE FAMOUS?

Mother Lode is belt of gold-bearing quartz veins, central California, along the western foothills of the Sierra Nevada. The term is sometimes limited to a strip c.70 mi (110 km) long and from 1 to 6.5 mi (1.6–10.5 km) wide, running NW from Mariposa. Popularly it is used to mean the gold-bearing area East of the Sacramento and San Joaquin rivers and West of the Sierra Nevada. The discovery of alluvial gold on the South Fork of the American River led to the 1848 gold rush. Mark Twain and Bret Harte helped make the Mother Lode famous.

36. WHAT MINERALS ARE DUBBED AS 'ORE AND GANGUE MINERALS'?

An ore is any naturally-occurring source of a metal that we can economically extract the metal from. Aluminium, for example, is the most common metal in the Earth's crust, occurring in all sorts of minerals. However, it isn't economically worthwhile to extract it from most of these minerals. Instead, the usual ore of aluminium is bauxite - which contains from 50–70% of aluminium oxide. We always treat bauxite as if it was aluminium oxide for chemistry purposes, although it is actually more complicated than that in reality. Copper is much rarer, but fortunately can be found in high-grade ores (ones containing a high percentage of copper) in particular places. Because copper is a valuable metal, it is also worth extracting it from low-grade ores as well. Ores are commonly oxides, for example: Bauxite Al_2O_3, Hematite Fe_2O_3 and Rutile TiO_2 Or sulphides, for example: Pyrite FeS_2 and Chalcopyrite $CuFeS_2$ and a whole lot of other things as well which we won't actually come across as a part of the foundation level purposes.

Nearly all rock deposits contain some metallic minerals, but in many cases the concentration of metal is too low to justify mining the ore. Ores are usually found concentrated in deposits with a definite gradation of metal concentration from the ore to the surrounding rock. The ore often occurs in veins, which are ore-filled fissures in the rock. The veins vary in thickness from only fractions of an inch to several hundred feet. Minerals with no commercial value, called gangue minerals, are usually found mixed with the ore in the vein. Some veins are buried deep within the ground, but others lie close to the surface. Veins of ore exposed to weathering are often eroded and redeposited in placers and alluvial deposits, *e.g.*, ore-bearing stream and lake gravels or beach sands. Some ores are simple chemical compounds, while others are chemically complex minerals. Important ores of aluminum, iron, manganese and tin are oxides; important ores of antimony, copper, lead, mercury, nickel, silver and zinc are sulfides. Some metals, called native metals, *e.g.*, copper, gold, platinum and silver, occur uncombined in nature. The recovery of metals from their ores is one area of the field of metallurgy.

37. WHAT ARE ADVANTAGES OF PLATING AND HOW IS IT DONE?

Plating is application of a plate, or coat, of metal to a surface for decoration, reflection of light, protection against corrosion, or increased wearing quality. The practice is of ancient origin: gilding was developed early; the Romans soldered silver plates to articles of baser metals; and in the 5th century a process was described where iron weapons were coated with copper by dipping them in a copper solution. In the 18th century the plating of copper or brass with silver by fusion was originated in England. Plating by electrolysis, or electroplating, is the most common method because it permits the control of the thickness of the plating. Cadmium, zinc, silver, gold, tin, copper, nickel and chromium are the most frequently used plating metals. Any of the common metals and some nonmetals, *e.g.*, plastics, with suitably prepared surfaces can be used as a base.

In electroforming, a mold made of specially prepared wax is electroplated. The wax is then melted and the plating itself constitutes the finished product. The part to be plated may be dipped in molten metal. For example, molten zinc is used in galvanizing; lead-tin alloy, in terneplating; and tin, in tinning. Without electricity, some metals can be precipitated from chemical solutions onto the surface of plastics or metals. A powdered, liquid, or gaseous plating metal can be made to diffuse into a heated base-metal surface. Sherardizing is the diffusion coating of zinc on steel; chromizing, chromium on steel; and calorizing, aluminum on steel. The application by welding or brazing of a sheet or plate to the base metal is called cladding. Decorative metal coatings are applied to plastic by vapor deposition. The metal, usually aluminum and the plastic parts are put in a vacuum chamber.

When the aluminum is vaporized by a surge of electricity, the particles settle on any exposed surface. Worn parts can be sprayed with molten metal to build them up to their original dimensions. Iron or steel is best protected from corrosion by a coating of cadmium or zinc, which oxidizes, creating a protective coat for the underlying material. Parts such as automobile bumpers usually have an undercoating of copper, a protective layer of nickel and a thin, shiny surface layer of chromium for decoration. In the manufacture of phonograph records an original recording is made on a metal disk coated with lacquer, which is then plated. From the plated master a die is made to be used in a hydraulic press. The silvering of mirrors is a type of plating in which silver is precipitated chemically on glass.

38. WHY AND HOW IS PROSPECTING DONE?

Prospecting is search for mineral deposits suitable for mining. Modern prospecting has replaced earlier methods based on chance or superstition (*e.g.*, use of the divining rod) with others based on a scientific knowledge of modern geology and mineralogy. Surface indications of deposits are confirmed by extensive sampling, *e.g.*, by examination and analysis of material taken from holes drilled at regular intervals. Modern geophysical methods of prospecting use instruments that measure variations in the earth's magnetic or gravitational field, or in the direction, nature and velocity of waves set up in the ground by underground explosions. Electrical methods employ instruments that indicate relative electrical conductivity between points in the earth's surface or electromotive forces generated by large ore bodies. Geochemical prospecting involves the chemical or spectrographic analysis of soil, plant and water samples. Scintillometers or Geiger counters are used to locate radioactive materials. A portable radioactive source is the basis of the berylometer, useful in locating beryl. Some minerals fluoresce in the presence of ultraviolet light; they are sought in the dark with portable ultraviolet lamps. Aerial and satellite photography and airborne instruments have proved useful for preliminary prospecting in unexplored territories.

39. WHAT 'QUARRYING METHOD' WAS USED IN ANCIENT TIMES AND HOW QUARRYING OF 'ORNAMENTAL STONE' IS DONE TODAY?

Quarrying is open or surface excavation of rock used for various purposes, including construction, ornamentation, road building and as an industrial raw material. Rock that has been quarried is commonly called stone. Quarrying methods depend chiefly on the desired size and shape of the stone and its physical characteristics. For industrial use (*e.g.*, limestone for preparing cement), as the aggregate in concrete, or for road beds, the rock is shattered. Explosives are detonated in a series of holes drilled in the rock in a pattern designed to yield the greatest amount of fracturing. The rock fragments may be further reduced in crushing machines and sorted according to size by screening. For building stone, rocks that do not shatter are separated by blasting; for softer rocks or when explosives cannot be used (*e.g.*, because they would disturb adjacent workings), a process known as broaching, or channeling, is used. In this process a line of holes is drilled perpendicular to the joints or cleavage planes of a formation; wedges are inserted into the holes and hammered until the stone splits off. This method was probably used in ancient times, notably by the Incas and the Egyptians.

Much quarrying of ornamental stone today is done by using pneumatically operated channelers. After the vertical cuts have been made, gadding machines (working on the same principle) are used to make horizontal cuts. Wedges are then used to split off the long blocks, which are subdivided and removed. Wire saws are also used; these consist of several pulleys over which passes an endless steel wire. Holes are drilled in the rock, each hole being made large enough to accommodate a pulley and the shaft to which it is attached. The wire, extending from one pulley to another, presses down against the rock between them. As the cut is deepened by the constantly moving wire the pulleys are continuously lowered into the holes.

40. WHAT IS MEANT BY REFINING?

Refining is any of various processes for separating impurities from crude or semi finished materials. It includes the finer processes of metallurgy, the fractional distillation of petroleum into its commercial products and the purifying of cane, beet and maple sugar and many other substances. The nature of the refining process depends on such factors as the type of material involved, the value of the end product and the degree of purity necessary. The purification of a metal is based upon physical or chemical differences between the metal and its accompanying impurities, including density, melting point, magnetic properties and reaction to certain chemicals. It may be accomplished by a continuation of a process used in separating the metal from its ore. Electrolysis is much used in the refining of various metals (*e.g.*, copper); the open hearth process and the basic oxygen process are used in the refining of iron; the amalgamation process and the cyanide process are chemical means of metal refining.

41. WHAT IS A SAFETY LAMP?

Safety lamp is an oil lamp designed for safe use in mines and other places where flammable gases such as firedamp may be present. Its invention (c.1816) is usually attributed to Sir Humphry Davy. The Davy lamp is based upon the principle that to be ignited a substance must first be heated to its kindling temperature and that if such heating is prevented combustion will not occur. The flame in the lamp is surrounded by a metal-gauze screen that distributes the heat over a large area so that the maximum temperature of the screen is below the ignition temperature of the flammable gas mixture (*e.g.*, firedamp). Improvements devised by K.G. Bischof and others include special locks to prevent the accidental opening of the lamp and devices to permit the lamp to be held upside down without danger. If firedamp or related gas mixtures are present in a mine, the Davy lamp flame burns higher and with a blue halo; the height of the flame and color of the halo indicate the amount of combustible gas in the air. If the mine air is deficient in oxygen, the lamp flame is extinguished. Coal miners often place the safety lamp close to the ground to detect gases, *e.g.*, carbon dioxide, which are denser than air and thus collect in poorly ventilated depressions in the mine. Other gas-detecting devices are now more commonly used, *e.g.*, the methanometer; as a light source the Davy lamp has been replaced by electric lighting.

42. HOW IS SHAFT SINKING ACCOMPLISHED?

Shaft sinking is excavation from the surface of an opening in the earth. Shafts, which are generally vertical, are usually distinguished from tunnels, which are horizontal. Little difficulty is experienced in shaft sinking through solid rock, which contains little water. When loose, water-bearing strata have to be contended with, careful shoring and sealing of the shaft lining become necessary and pumping facilities are needed. Shafts are usually circular or rectangular and are generally lined with wood, masonry, concrete, steel, or cast iron. Shafts sunk in loose water-bearing soils, where there is great external pressure on the shaft sides, are nearly always circular; rectangular shafts with wood lining are often used in mining work, as the shafts are frequently of a temporary nature. Shaft sinking through rock is generally accomplished by blasting. When the loose surface material has been removed, holes are drilled and the charges are placed and are fired by electricity. The broken rock is removed and the process is repeated. In an ordinary rectangular shaft the lining consists of timbers 8 or 12 in. (20 or 30 cm) square placed horizontally around the shaft. Shafts of a more permanent nature are generally circular in form and lined with cast iron or with concrete masonry 1 to 2 ft (30–61 cm) thick, built in sections as the work advances. When excessive quantities of water are met with, cast-iron tubbing is sometimes used. This consists of heavy cast-iron rings made in segments, with flanges for connecting and bolted together in place. Cement grout is forced into the space between the outside of the tubbing and the surrounding earth to form a seal. Shaft sinking by the freezing process in very watery soil is accomplished by sinking pipes in the area to be excavated and circulating brine at low temperature in them until the earth is frozen and hard so that it can be excavated in the same manner as rock. In the grouting method, liquid cement is forced into the water-bearing earth under very high pressure. On mixing with the water, the cement solidifies the adjacent area and it is removed by drilling and blasting as with rock.

43. WHAT IS MEANT BY 'SHAPE MEMORY'?

Shape memory is a property possessed by certain alloys that allows them to return, when heated, to their original shape after having been deformed. This effect results because, as the alloy is deformed, it undergoes a martensitic (or athermal) transformation, a solid-state transition that rapidly changes the crystalline structure of the alloy without thermal activation, that is readily reversed once an appropriate amount of heat is applied. Among the alloys are copper-aluminum-nickel, copper-zinc-aluminum, nickel-titanium and iron-manganese-silicon. Applications of shape memory include heat-activated fasteners, switches, eyeglass frames, orthopedic devices, teeth braces and blood clot filters.

44. WHAT IS DIFFERENCE BETWEEN 'SINTERING AND SMELTING'?

Sintering is a process of forming objects from a metal powder by heating the powder at a temperature below its melting point. In the production of small metal objects it is often not practical to cast them. Through chemical or mechanical procedures a fine powder of the metal can be produced. When the powder is compacted into the desired shape and heated, *i.e.*, sintered, for up to three hours, the particles composing the powder join together to form a single solid object. Smelting, in metallurgy, is any process of melting or fusion, especially to extract a metal from its ore. Smelting processes vary in detail depending on the nature of the ore and the metal involved, but they are typified in the use of the blast furnace.

45. WHAT PROCESSES ARE USED IN MODERN STEELMAKING AND HOW IS STEEL CLASSIFIED?

Steel is an alloy of iron, carbon and small proportions of other elements. Iron contains impurities in the form of silicon, phosphorus, sulfur and manganese; steelmaking involves the removal of these impurities, known as slag and the addition of desirable alloying elements. Steel was first made by cementation, a process of heating bars of iron with charcoal in a closed furnace so that the surface of the iron acquired a high carbon content. The crucible method, originally developed to remove the slag from cementation steel, melts iron and other substances together in a fire-clay and graphite crucible. The famous blades of Damascus and Tolėdo, Spain, were made by the cementation and crucible techniques.

The Bessemer process, the open-hearth process and the basic oxygen process are more widely used in modern steelmaking. The open-hearth uses a type of furnace called a regenerative furnace; instead of a firebox at one end and a flue at the other, it has devices at each end for the intake and outflow of both fuel and air. The air is preheated by a system of current reversals that causes very high temperatures. This process, developed c.1866 by Sir William Siemens, uses iron ore and pig iron. In the basic oxygen process, or Linz-Donawitz process, developed in the 1950s, the design of the furnace is changed and oxygen added to the air intake permits more rapid refining of the charge (material in the furnace). The electric-arc furnace is another modern development; it provides a means of making large quantities of high-grade steel, with the advantages of positive temperature control, freedom from contamination of the product by the fuel and simultaneous deoxidation and desulfurization actions.

Steel is shaped for commercial use in rolling mills, where successive passages of the red-hot ingot between variously shaped rollers give it the desired form. Pittsburgh, one of the world's great steel centers, built its first rolling mill in 1811; Bessemer steel rails were rolled in Chicago as early as 1865. Steel is often classified by its carbon content: a high-carbon steel is serviceable for dies and cutting tools because of its great hardness and brittleness; low or medium-carbon steel is used for sheeting and structural forms because of its amenability to welding and tooling. Alloy steels, now most widely used, contain one or more other elements to give them specific qualities. Aluminum steel is smooth and has a high tensile strength. Chromium steel finds wide use in automobile and airplane parts on account of its hardness, strength and elasticity, as does the chromium-vanadium variety. Nickel steel is the most widely used of the alloys; it is nonmagnetic and has the tensile properties of high-carbon steel without the brittleness. Nickel-chromium steel possesses a shock resistant quality that makes it suitable for armor plate. Wolfram (tungsten), molybdenum and high-manganese steel are other alloys. Stainless steel, which was developed in England, has a high tensile strength and resists abrasion and corrosion because of its high chromium content.

46. WHAT IS TACONITE?

Taconite is a low-grade iron ore, a flint like rock usually containing less than 30% iron. Resistant to drilling and to the extraction of its contained metal, the rock was long considered worthless. Experiments begun in 1912 by the American scientist Edward W. Davis and continued by him for nearly 40 years produced the pelletizing method for upgrading the ore. The development of the jet piercer (a high-temperature flame thrower) provided penetration speeds of up to 40 ft (12 m) an hour for blasting holes in the rock. In 1956 exploitation of the vast reserves of taconite in the Mesabi Range of Minnesota was begun by some of the largest steel companies in the United States.

47. WHY AND HOW ARE METALS TEMPERED?

Tempering is the process involving slow and moderate heating to increase the hardness and toughness of metals that have undergone previous heat treatment. Metals are usually hardened by being heated to high temperatures and quenched rapidly. This treatment causes brittleness, which is reduced by tempering. Steel is notably responsive to tempering and makers of tools, weapons, armor and other articles of steel have long had great skill in the process. Tempering is not necessary for such products as razors and files, in which hardness is sought but brittleness is not a serious disadvantage. Other products, *e.g.*, swords and saws, require tempering for toughness. In the handicraft process of tempering, the condition of the steel during heating is judged by its color, caused by an oxide film. A desired hardness can be achieved by plunging the steel into a bath when it has cooled to the right shade of yellow or brown or blue. To secure a bath of the right temperature, various liquids are used, *e.g.*, pure water, saltwater, oil and molten metal. The process of softening steel that is harder than desired is called annealing. In modern mass production the processes of tempering are guided by scientific tests in place of the artisan's skill. Comparable to tempering is the process of hastening the cooling of a surface of a casting to increase the hardness of the part so "chilled."

48. WHY IS 'TYPE METAL' SO NAMED?

Type metal, so named because of its one time extensive use for making printing type, is an alloy of lead with antimony, tin and sometimes copper. Expanding upon solidification, the alloy takes a fine and clear impression of the mold in which it hardens. It has a low melting point. Once used extensively for type, it is also employed in making the metal parts of various musical instruments and for ornaments of intricate design and pattern. The percentages of the metals in the alloy vary, according to the use to which it is to be put.

49. WHY AND HOW ARE METALS WELDED?

Welding is the process for joining separate pieces of metal in a continuous metallic bond. While cold-pressure welding is accomplished by the application of high pressure at room temperature, forge welding (forging) is done by means of hammering and with the addition of heat. In most processes in common use, the metal at the points to be joined is melted; additional molten metal is added as a filler and the bond is allowed to cool. In the Thomson process, resistance to an electric current, passed through the sections to be joined, causes them to melt. Other notable methods of welding include the thermite process, oxyacetylene, electric arc, oxyhydrogen and the atomic hydrogen flame. In this last-named method, molecules of hydrogen gas passing through an electric arc are broken up into atoms of hydrogen by absorbing energy; when outside the arc, the atoms reunite into molecules, yielding in the process enough heat to weld the material. Another process, the argon-arc method, is widely used with metals such as stainless steel, aluminum, magnesium and titanium, which require an inert atmosphere for successful welding. The use of argon prevents slag from forming in the weld and greatly increases the speed of the welding.

Chapter 18

SYSTEMS OF MEASUREMENT

1. WHAT IS MEANT BY THE TERMS: STANDARD AND UNIT DESIGNATION'?

A unit of measurement is some specific quantity that has been chosen as the standard against which other measurements of the same kind are made. For example, the meter is the unit of measurement for length in the metric system. When an object is said to be 4 meters long, that means that the object is four times as long as the unit standard (1 meter). The term standard refers to the physical object on which the unit of measurement is based. For example, for many years the standard used in measuring length in the metric system was the distance between two scratches on a platinum-iridium bar kept at the Bureau of Standards in Sèvres, France. A standard serves as a norm against which other measuring devices of the same kind are made. The meter stick in a school classroom or home is thought to be exactly one meter long because it was made from a permanent model kept at the manufacturing plant that was originally copied from the standard meter in France. All measurements consist of two parts: a scalar (numerical) quantity and the unit designation. In the measurement 8.5 meters, the scalar quantity is 8.5 and the unit designation is meters.

2. WHAT PROBLEMS ARE POSED BY THE BRITISH SYSTEM OF UNITS?

The need for units and standards developed at a point in human history when people needed to know how much of something they were buying, selling, or exchanging. A farmer might want to sell a bushel of wheat, for example, for ten dollars, but he or she could do so only if the unit "bushel" were known to potential buyers. Furthermore, the unit "bushel" had to have the same meaning for everyone who used the term. The measuring system that most Americans know best is the British system, with units including the foot, yard, second, pound and gallon. The British system grew up informally and in a disorganized way over many centuries. The first units of measurement probably came into use shortly after the year 1215. These units were tied to easily obtained or produced standards. The yard, for example, was defined as the distance from King Henry II's nose to the thumb of his outstretched hand. (Henry II of England reigned from 1154 to 1189).

The British system of measurement consists of a complex, irrational (meaning, in this case, not sensibly organized) collection of units whose only advantage is its familiarity. As an example of the problems it poses is that the British system has three different units known as the quart. These are the British quart, the U.S. dry quart and the U.S. liquid quart. The exact size of each of these "quarts" differs. In addition, a number of different units are in use for specific purposes. Among the units of volume in use in the British system (in addition to those mentioned above) are the bag, barrel (of which there are three types: British and U.S. dry, U.S. liquid and U.S. petroleum), bushel, butt, cord, drachm, firkin, gill, hogshead, kilderkin, last, noggin, peck, perch, pint and quarter.

3. WHAT IS THE MOST RECENT STANDARD ADOPTED FOR THE METER?

In an effort to bring some rationality to systems of measurement, the French National Assembly established a committee in 1790 to propose a new system of measurement with new units and new standards. That system has come to be known as the metric system and is now the only system of measurement used by all scientists and in every country of the world except the United States and the Myanmar Republic. The units of measurement chosen for the metric system were the gram (abbreviated g) for mass, the liter (L) for volume, the meter (m) for length and the second (s) for time. A specific standard was chosen for each of these basic units. The meter was originally defined as one ten-millionth the distance from the North Pole to the equator along the prime meridian. As a definition, this standard is perfectly acceptable, but it has one major disadvantage: a person who wants to make a meter stick would have difficulty using that standard to construct a meter stick of his or her own.

As a result, new and more suitable standards were selected over time. One improvement was to construct the platinum-iridium bar standard. Manufacturers of measuring devices could ask for copies of the fundamental standard kept in France and then make their own copies from those. As we can imagine, the more copies of copies had to be made, the less accurate the final measuring device would be. The most recent standard adopted for the meter solves this problem. In 1983, the International Conference on Weights and Measures defined the meter as the distance that light travels in $^{1}/$ 299,792,458 second. The standard is useful because it depends on the most accurate physical measurement known, the second and because anyone in the world is able, given the proper equipment, to determine the true length of a meter.

4. WHAT FUNDAMENTAL UNITS MAKE UP THE S.I. SYSTEM?

In 1960, the metric system was modified somewhat with the adoption of new units of measurement. The modification was given the name of Le Système International d'Unités, or the International System of Units. This system is more commonly known as the SI system. Nine fundamental units make up the SI system. These are the meter (abbreviated m) for length, the kilogram (kg) for mass, the second (s) for time, the ampere (A) for electric current, the kelvin (K) for temperature, the candela (cd) for light intensity, the mole (mol) for quantity of a substance, the radian (rad) for plane angles and the steradian (sr) for solid angles.

5. WHAT ARE THE MOST FAMILIAR DERIVED UNITS?

Many physical phenomena are measured in units that are derived from SI units. For example, frequency is measured in a unit known as the hertz (Hz). The hertz is the number of vibrations made by a wave in a second. It can be expressed in terms of the basic SI unit as s^{-1}. Pressure is another derived unit. Pressure is defined as the force per unit area. In the SI system, the unit of pressure is the pascal (Pa) and can be expressed as kilograms per meter per second squared, or $kg/m/s^2$. Even units that appear to have little or no relationship to the nine fundamental units can, nonetheless, be expressed in these terms. The absorbed dose, for example, indicates that amount of radiation received by a person or object. In the metric system, the unit for this measurement is the gray. One gray can be defined in terms of the fundamental units as meters squared per second squared, or m^2/s^2.

Many other commonly used units can also be expressed in terms of the nine fundamental units. Some of the most familiar are the units for area (square meter: m^2), volume (cubic meter: m^3), velocity (meters per second: m/s), concentration (moles per cubic meter: mol/m^3), density (kilogram per cubic meter: kg/m^3), luminance (candela per square meter: cd/m^2) and magnetic field strength (amperes per meter: A/m). A set of prefixes is available that makes it possible to use the fundamental SI units to express larger or smaller amounts of the same quantity. Among the most commonly used prefixes are milli- (m) for one-thousandth; centi- (c) for one-hundredth; micro- (ì) for one-millionth; kilo- (k) for one thousand times; and mega- (M) for one million times. Thus, any volume can be expressed by using some combination of the fundamental unit (liter) and the appropriate prefix. One million liters, using this system, would be a megaliter (ML) and one millionth of a liter would be a microliter (ì L).

6. WHAT ARE NATURAL UNITS?

One characteristic of all of the derived units is that they have been selected arbitrarily (by individual preference or convenience rather than by law). The committee that established the metric system could, for example, have defined the meter as one one-hundredth the distance between Paris and Sèvres. It was completely free to choose any standard it wanted to. Some measurements, however, suggest "natural" units. In the field of electricity, for example, the charge carried by a single electron would appear to be a natural unit of measurement. That quantity is known as the elementary charge (e) and has the value of $1.6021892 \times 10^{-19}$ coulomb. Other natural units of measurement include the speed of light (c: 2.99792458×10^{8} m/s), the Planck constant, the mass of an electron (m_e: 0.9109534 × 10^{-30} kg) and the mass of a proton (m_p: 1.6726485 × 10^{-27} kg). As one can see, each of these natural units can be expressed in terms of SI units, but they often are used as basic units in specialized fields of science.

7. HOW ARE UNITS CONVERTED BETWEEN SYSTEMS?

For many years, an effort has been made to have the metric system, including SI units, adopted worldwide. As early as 1866, the U.S. Congress legalized the use of the metric system. More than a hundred years later, in 1976, Congress adopted the Metric Conversion Act, declaring it the policy of the nation to increase the use of the metric system in the United States. In fact, little progress has been made in that direction. Indeed, elements of the British system of measurement continue in use for specialized purposes throughout the world. All flight navigation, for example, is expressed in terms of feet, not meters. As a consequence, it is still necessary for an educated person to be able to make conversion from one system of measurement to the other. In 1959, English-speaking countries around the world met to adopt standard conversion factors between British and metric systems. To convert from the pound to the kilogram, for example, it is necessary to multiply the given quantity (in pounds) by the factor 0.45359237. A conversion in the reverse direction, from kilograms to pounds, involves multiplying the given quantity (in kilograms) by the factor 2.2046226. Other relevant conversion factors are 1 inch equals 2.54 centimeters and 1 yard equals 0.9144 meter.

8. HOW ANCIENT RULES OF THUMB GAVE WAY TO MORE CAREFULLY DEFINED UNITS?

The metric system of measurement for expressing the amounts of various quantities was developed in France during the Napoleonic reign of France in the 1790's. Nearly all early units of size were based on the always-handy human body. The metric system of units defines seven basic units: one each for length, mass, time, electric current, temperature, amount of substance and luminous intensity. The metric system of measurement is an internationally agreed-upon set of units for expressing the amounts of various quantities such as length, mass, time and temperature. The metric system has several advantages over the English system which is still in place in the U.S. However the scientific community has adopted the metric system almost from its inception. In fact, the metric system missed being nationalized in US by one vote in the Continental Congress in the late 1700's or early 1800's. As of 1994, every nation in the world has adopted the metric system, with only four exceptions: the United States, Brunei, Burma and Yemen (which use the English units of measurement). Because of its convenience and consistency, scientists have used the metric system of units for more than 200 years. Originally, the metric system was based on only three fundamental units: the meter for length, the kilogram for mass and the second for time. Today, there are more than 50 officially recognized units for various scientific quantities.

Nearly all early units of size were based on the always-handy human body. In the Middle Ages, the inch is reputed to have been the length of a medieval king's first thumb joint. The yard was once defined as the distance between English king Henry I's nose and the tip of his outstretched middle finger. The origin of the foot as a unit of measurement is obvious. Eventually, ancient "rules of thumb" gave way to more carefully defined units. The metric system was adopted in France in 1799.

9. WHAT BASIC QUANTITIES ARE IN EVERYDAY USE OF NONSCIENTISTS?

The metric system of units defines seven basic units: one each for length, mass, time, electric current, temperature, amount of substance and luminous intensity. (While, amount of substance refers to the number of elementary particles in a sample of matter, luminous intensity has to do with the brightness of a light source.) But only four of these seven basic quantities are in everyday use by nonscientists: length, mass, time and temperature. Their defined units are the meter for length, the kilogram for mass, the second for time and the degree Celsius for temperature. (The other three basic units are the ampere for electric current, the mole for amount of substance and the candela for luminous intensity).

10. HOW ARE METRIC UNITS OF LENGTH, MASS AND TIME DEFINED NOW?

The meter was originally defined in terms of Earth's size; it was supposed to be one ten-millionth of the distance from the equator to the North Pole. Since Earth is subject to geological movements, this distance does not remain the same. The modern meter, therefore, is defined in terms of how far light will travel in a given amount of time when traveling at the speed of light. The speed of light in a vacuum, 186,282 miles (299,727 kilometers) per hour, is considered to be a fundamental constant of nature that will never change. The standard meter is equivalent to 39.3701 inches. The kilogram is the metric unit of mass, not weight. Mass is the fundamental measure of the amount of matter in an object. Unfortunately, no absolutely unchangeable standard of mass has yet been found on which to standardize the kilogram. The kilogram is therefore defined as the mass of a certain bar of platinum-iridium alloy that has been kept since 1889 at the International Bureau of Weights and Measures in Sèvres, France. The kilogram is equivalent to 2.2046 pounds.

The metric unit of time is the same second that has always been used, except that it is now defined in a very accurate way. It no longer depends on the wobbly rotation of our planet (1/86,400th of a day), because Earth is slowing down. Days keep getting a little longer as Earth grows older. So the second is now defined in terms of the vibrations of a certain kind of atom known as cesium-133. One second is defined as the amount of time it takes for a cesium-133 atom to vibrate in a particular way 9,192,631,770 times. Because the vibrations of atoms depend only on the nature of the atoms themselves, cesium atoms will presumably continue to behave exactly like cesium atoms forever. The exact number of cesium vibrations was chosen to come out as close as possible to what was previously the most accurate value of the second. The metric unit of temperature is the degree Celsius, which replaces the English system's degree Fahrenheit. It is possible to convert between Celsius and Fahrenheit simply by multiplying or dividing by 1.8, however, because the scales start at different places. That is, their zero-degree marks have been set at different temperatures.

11. WHAT IS UNIQUE WITH SMALLER AND LARGER METRIC UNITS?

In the metric system, there is only one basic unit for each type of quantity. Smaller and larger units of those quantities are all based on powers of ten (unlike the English system that invents different-sized units with completely different names based on different conversion factors). To create those various units, the metric system simply attaches a prefix to the name of the unit. Latin prefixes are added for smaller units and Greek prefixes are added for larger units. The basic prefixes are: kilo- (1000), hecto- (100), deka- (10), deci- (0.1), centi- (0.01) and milli- (0.001). Therefore, a kilometer is 1,000 meters. Similarly, a millimeter is one-thousandth of a meter.

12. WHAT NONSTANDARD METRIC UNIT ARE PREVALENT?

Minutes are permitted to remain in the metric system even though they don't conform strictly to the rules. The minute, hour and day, for example, are so customary that they're still defined in the metric system as 60 seconds, 60 minutes and 24 hours, not as multiples of ten. For volume, the most common metric unit is not the cubic meter, which is generally too big to be useful in commerce, but the liter, which is one-thousandth of a cubic meter. For even smaller volumes, the milliliter, one-thousandth of a liter, is commonly used. And for large masses, the metric ton is often used instead of the kilogram. A metric ton (often spelled tonne) is 1,000 kilograms. Because a kilogram is about 2.2 pounds, a metric ton is about 2,200 pounds: 10 percent heavier than an American ton of 2,000 pounds. Another often-used, nonstandard metric unit is the hectare for land area. A hectare is 10,000 square meters and is equivalent to 0.4047 acre.

13. WHY IS IT EASIER TO UNDERSTAND MEASUREMENTS MADE IN S. I. SYSTEM?

Measurements are quantitative observations which are very important in science in general and in chemistry in particular. While measurements in science have been posing a big problem to its users from very beginning, it is true even today because there are two different systems in use, namely English System and Metric System. Americans use a system called the English System that uses units like inches, feet, gallons and miles and there is no easily remembered relationship between the units, so a world-wide system of measurement, called the Metric System, or SI system was established. The Système International d'Unités, or SI, is a set of units for scientific measurements which is based on the metric system. This system is based on multiples of ten. For example, if we have 1 meter, we have 100 centimeters, 1000 millimeters and .001 kilometers. This is obviously easier to understand. Many students who are more comfortable with one system do not like the other system. This is largely because they have to do too many conversions from one system to other. The real value of the SI system is in converting units within the system. Calculating energy and pressure from SI base units is much easier than calculating these using base units from the English system.

14. WHAT S.I. SYSTEM-BASED 'BASE UNITS' ARE COMMONLY USED IN CHEMISTRY?

The base units are the fundamental units representing the most straightforward measurements. Details of the SI system based units may be given in a tabular form (Table-18.1) as under:

Table 18.1: S.I.-based Units

Measurement	*Unit*	*Remarks*
Mass	Kilogram, kg	A one liter bottle of water has one kilogram of mass excluding the mass of the bottle.
Length	Meter, m	If we extend our one arm to one side at shoulder height, the length of our hand, arm and across our chest is about one meter.
Amount of Species	Mole, mol	This unit is very important in chemistry. The amount of matter in a mole varies a lot depending on what it is.
Time	Second, s	If we don't have a watch, we can estimate time with slow, deep breaths or by chanting "some prayer."
Temperature	Kelvin, K	Zero on the Kelvin scale is the temperature at which molecular motion stops. Water freezes at 273K and boils at 373K. Room temperature is 298K.

The S.I. base units can be scaled up (or down) using multiples (or factors) of ten. The scale is indicated by prefixes that are familiar to those who have worked with the metric system.

15. WHAT METRIC SYSTEM PREFIXES ARE COMMONLY USED IN CHEMISTRY LABS?

The Metric System Prefixes commonly used in Chemistry Labs are given in the tabular form (Table-18.2) as under:

Table 18.2: Metric System Prefixes

Prefix	*Scale*	*Example*
Kilo-, K	1000	1000 grams = 1 kilogram
Centi-, c	10^{-2}	100 centimeters = 1 meter
Milli-, m	10^{-3}	1000 milliliters = 1 liter
Micro-, μ	10^{-6}	1 micrometer = 0.001 millimeter = 1×10^{-6} meter
Nano-, n	10^{-9}	1 nanometer = 1×10^{-6} millimeter = 1×10^{-9} meter

16. WHAT DERIVED UNITS ARE USED IN CHEMISTRY LABS?

Derived units are more complex units that require more than one dimension or type of measurement. Length is an example of a base unit. Volume is an example of a derived unit, because it is a three dimensional representation of length. The Derived Units used in Chemistry Labs are as under (Table-18.3).

Table 18.3: Derived units are used in Chemistry Labs

Measurement	*Unit*	*Remarks*
Mass	Gram, g	A gram is 1/1000 of a kilogram. A typical solid chemical sample, the size of a pea, has a mass of about a gram. One milliliter of water has a mass of one gram.
Length	Centimeter, cm	A centimeter is 1/100 of a meter. Our thumbnail is about one centimeter wide.
Length	Angstrom, ?	An Angstrom is 1×10^{-10} meter. This unit is used to measure the radii of atoms.
Volume	Milliliter, mL	A milliliter is 1/1000 of a liter. It is the volume that would occupy a cube with sides one cm long. A milliliter is 22 drops from an eyedropper or buret.
Heat	Calorie, cal	One calorie is the amount of thermal energy needed to raise the temperature of one gram of water by one degree Kelvin. A nutritional calorie is one thousand calories.
Energy	Joule, J	The Joule is the SI unit of energy and can be related to base SI units ($1\ J = 1\ kgm^2/s^2$). One calorie is equal to 4.184 J.
Force	Newton, N	One Newton is the force needed to accelerate one kilogram of mass by one meter per second squared ($1\ N = 1\ Kg\ m/s^2$). This unit is not often used in chemistry.
Pressure	Pascal, Pa	One Pascal is the pressure exerted by one Newton of force acting on an area of one square meter ($1\ Pa = 1\ N/m^2$). This unit is not used as often as the other pressure units.
Pressure	Atmosphere, atm	The typical air pressure at sea level is 1 atmosphere. 1 atm = 101.325 kPa
Pressure	Torr	One Torr is the air pressure needed to support a column of mercury one millimeter high (1 Torr = 1 mmHg). This unit is based on devices used to measure air pressure. One atmosphere is 760 Torr.

17. WHAT MAKES THE S.I. SYSTEM EASIER THAN THE ENGLISH SYSTEM?

In the S.I. system, there are base units for measurement. Examples of these units are meter for length, liter for volume and gram for mass. What makes this system even easier than the English system is that the prefixes are standard for all the base units. So if we have a kilogram of water, we have 1000 grams of water and if we have traveled a kilometer of road, we have traveled 1000 meters of road. Details of prefixes, their symbols, their meanings and exponentials are given below (Table-18.4).

Table 18.4: Prefix Notation

Prefix notation	*Symbol*	*Meaning*	*Exponential*
Mega	M	1,000,000	10^6
Kilo	k	1,000	10^3
Hecta	h	100	10^2
Deka	da	10	10^1
–	–	1	10^0
Deci	d	0.1	10^{-1}
Centi	c	0.01	10^{-2}
Milli	m	0.001	10^{-3}
Micro	u	0.000001	10^{-6}

18. WHAT IS RELATIONSHIP BETWEEN ENGLISH AND 'S.I.' SYSTEMS MEASUREMENTS?

Measurements in the scientific world and increasingly, in the nonscientific world too are now made in S.I. (Système International) units. The system was established in order to allow comparison of measurements made in one country with those made in another. SI units and their relative values were adopted by an international association of scientists meeting in Paris in 1960. It is worth noting that metric units are part of this system. The system still in common for nonscientific use in the United States is called the English system, even though England, like most other developed countries, now uses metric units. Anyone using units from both the English and SI systems needs to be aware of a few simple relationships between the two systems. These relationships are given in the following Table (Table-18.5) also.

Table 18.5: Units of the SI system, units derived from the SI system and their relationship to several English units.

Property being measured	*Basic S.I. unit*	*Derived units*	*Relationship to English unit*
Length	meter (m)	Kilometer (km) 1 km = 1000 m centimeter (cm) 1 cm = 0.01 m	1 m = 39.37 in. 1.61 km = 1 mi 2.54 cm = 1 in.

Table 18.5: *(Contd...)*

Table 18.5: (*Contd...*)

Property being measured	***Basic S.I. unit***	***Derived units***	***Relationship to English unit***
Mass	kilogram (kg)	gram (g) 1 g = 0.001 kg	1 kg = 2.204 lb 453.6 g = 1 lb
Volume	cubic meter (m^3)	liter (L) 1 L = 0.001 m^3 cubic centimeter (cm^3, cc) 1 cm^3 = 0.001 L milliliter (mL) 1 mL = 1 cm^3	1 L = 1.057 qt 946 mL = 1.0 qt
Temperature	Kelvin (K)	Celsius(C) K = °C + 273.15	Fahrenheit (F) °F–32° C = 1.8 = $^5/_9$ (°F–32)
Energy	joule (J)	calorie(cal) 1 cal = 4.184 J kilocalorie (kcal) 1 kcal = 1000 cal	

Two features of the SI system make it easy to use. First, it is a base-10 system; that is, the various units of a particular dimension vary by multiples of ten. Once a base unit is defined, units larger and smaller than the base unit are indicated by prefixes added to the name of the base unit. Some of these SI prefixes, along with the abbreviation for each and the numerical factor relating it to the base unit are as under (Table-18.6a).

Table 18.6(a): Prefixes used in the SI system

Prefix	***Symbol***	***Base unit multiplied by***
Mega-	M	1,000,000, or 10^6*
Kilo-	k	1,000, or 10^3
Deci-	d	0.1, or 10^{-1}
Centi-	c	0.01, or 10^{-2}
Milli-	m	0.001, or 10^{-3}
Micro-	μ	0.000001, or 10^{-6}
Nano-	n	0.000000001, or 10^{-9}
Pico-	p	0.000000000001, or 10^{-12}

* A positive exponent means "raised to a power." Thus 10^2 is 10 × 10 or "10 squared." A negative exponent means "divided by that power of 10," thus 10^{-2} is $^1/_{100}$ or 0.01.

The second feature that increases the usefulness of the SI system is the direct relationship between the base units of different dimensions. For example, the unit of volume (cubic meter) is the cube of the unit of length (meter). The base unit of length in the SI system is the meter (m). The meter, approximately 10% longer than a yard, is equivalent to 39.37 inches, or 1.094 yards. The metric units of length most commonly used in chemistry are listed below.

Table 18.6(b): Units of length in the SI system

Unit of length	*Relationship to base unit*
Kilometer (km)	1 km = 1000 m
Meter (m)	1 meter
Decimeter (dm)	10 dm = 1 m
Centimeter (cm)	100 cm = 1 m
Millimeter (mm)	1000 mm = 1 m
Micrometer (μm)	10^6 μm = 1 m
Nanometer (nm)	10^9 nm = 1 m

The base unit of volume in the SI system is the cubic meter (m^3). Other commonly used units of volume are the liter (L), the cubic centimeter (cm^3 or cc) and the milliliter (mL). One liter has a volume equal to 0.001 m^3. The nearest unit of comparable volume in the English system is the quart (1.000 L = 1.057 qt). The SI units of volume are summarized in Table 18.7 and illustrated in Figure 18.1. Note particularly that the volume of 1 cm^3 is the same as the volume of 1 mL.

Table 18.7: Volume Relationship

Unit of volume	*Relationship to liter*
Liter (L)	
Milliliter (mL)	1000 milliliters = 1 liter
Cubic centimeter (cm^3, cc)	1000 cubic centimeters = 1 liter
Micro liter (μL)	10^6 micro liters = 1 liter

19. WHAT IS MASS –WEIGHT RELATIONSHIP?

In discussing SI units we have used the term mass rather than the more familiar term weight. Mass is a measure of the amount of matter in a particular sample. The mass of a sample does not depend on its location; it is the same whether measured on Earth, on the moon, or anywhere in space. Weight is a measure of the pull of gravity on a sample and depends on where the sample is weighed. The standard of mass in the SI system is the kilogram (kg). A safe in Sèvres, France, holds a metal cylinder with a mass of exactly 1 kg. The mass of that cylinder is the same as the mass of 1000 mL (1L) of water at 4°C, thereby relating mass to volume. Notice that the base unit of mass in the SI system is the gram (g), even though the standard of mass in this system is the kilogram.

Astronauts traveling in space and landing on the moon have experienced the difference between mass and weight. In Earth's gravitational field at sea level, a particular astronaut may weigh 198 pounds. On the surface of the moon, this astronaut still has the same mass, but his weight (33 pounds) is only one-sixth of what it is on Earth, because the gravitational field of the moon is much weaker (one sixth) than that of the Earth. In outer space the astronaut is weightless, but his mass remains unchanged.

20. WHEN DO WE MEASURE WEIGHT AND WHEN MASS?

Weight and mass are measured on different instruments. Mass is measured on a balance. An object of unknown mass is put at one end of a straight beam and objects of known mass are added to the other end until their mass exactly balances that of the object whose mass is being measured. Because both ends of the beam, at the moment of balancing, are the same distance from the center of the Earth, this measurement is independent of gravity. Weight, on the other hand, is measured on a scale, which determines weight by measuring the distortion of a spring. Such a measurement depends on the pull of gravity. We will weigh less at the top of a mountain than we do in a valley, because the pull of gravity decreases as we move further from the Earth's center. Nevertheless, our mass is the same in both places. Despite the clear difference in meaning between the terms mass and weight, measuring the mass of an object is often called "weighing" and the terms mass and weight are frequently and incorrectly used interchangeably. One must remember that the correct way to describe the amount of matter in a sample is to state its mass.

21. HOW SHOULD WE RECORD OUR MEASUREMENTS?

It is important to understand that in science, when we take measurements the last number is always estimated. For example, if the ruler we use to measure a piece of paper is calibrated to centimeters (meaning the smallest marked units were centimeters) and the length came out to 20 centimeters, this answer would not be acceptable because the answer has to have an uncertain value which would be the millimeter's value. (Measurements must always be one decimal place more exact than the smallest marked units on the measuring device). In the example above, 20.0, 20.3, or 19.8 would all be acceptable values.

22. WHAT ARE SYSTEMATIC AND RANDOM ERRORS?

In any measurement, chaos, instrumental limitations and human failings prevent perfection. This lack of perfection in measurements is called error. Some errors are random errors, meaning that they cause some of the measurements to be a little high and some to be a little low. These are generally due to chaos and, while they can be minimized, they can never be eliminated. Systematic errors are ones that cause readings to be always high or always low. These are generally due to mistakes, equipment malfunction and poor laboratory technique. For example, reading the volume of a graduated cylinder from the top of the meniscus rather than the bottom will cause all of the volume measurements to be slightly high.

23. WHY CHEMISTRY LABORATORY INSTRUCTORS SHOULD GIVE STUDENTS A WORD OF WARNING REGARDING 'ACCURACY AND PRECISION'?

Accuracy and precision are two indicators of the validity of measurements. Accuracy is the agreement of a measurement with an actual value (how close it is to what it is supposed to be). This is the best indicator of how 'good' a measurement is. Unfortunately, if something is truly worth measuring, we don't know its actual value. Accuracy is important in the chemistry lab during experiments in which the lab instructor knows what the value actually is and the students try to get a result as close as possible to this 'true' value. Scientists routinely check their instruments for accuracy by using them to measure a standard for which the values are well known. If the instrument's readings match the accepted values for the standard, it is probably working correctly.

Poor accuracy, Poor precision

Poor accuracy, good precision

Good accuracy, good precision

Fig. 18.1: Accuracy and precision: Accurate measurements are close to the true value while precise measurements are close to one another.

An accurate measurement is one that is close to the actual value of the property being measured. The accuracy of a measurement depends on the calibration of the tool used to make the measurement. For example, if we are measuring the distance between two cities by driving between them, an accurate measurement requires that the odometer in our car reads 1.00 km for each kilometer driven. If it reads only 0.95 km for each kilometer driven, the accuracy of our measurement will be reduced. Precision, on the other hand, is an indication of how close several measurements are to one another. Systematic errors tend to have an additive effect.

Chemistry is an exact science; its development has been based on careful measurements of properties of matter and careful observations of changes in these properties. Measurements in chemistry must be both accurate and precise. If a student has poor laboratory techniques, his or her measurements will tend to fluctuate more than those of another student with good techniques. Laboratory instructors will tend to have students perform measurements several times then calculate an average with a standard deviation. If the instructor knows what the result should be, the student's average indicates the accuracy and the standard deviation indicates the precision.

A precise measurement is one that can be reproduced. For example, the driving distance between two important cities of India, Delhi and Luchnow, is about 493 km. An accurate odometer, one that reads 1.00 km for every kilometer traveled measures the distance between these two cities as 493 km. However, an odometer that reads 1.00 km for every 0.95 km traveled will measure the distance as 519 km. Each time it is used on the trip, the inaccurate odometer records the same value of 519 km. The odometer reading is precise because it can be reproduced time after time. However, it is not accurate because the odometer itself is not properly calibrated. We may note that accuracy requires precision, but precision does not guarantee accuracy. Most often the measurements we make are not final answers in themselves. Rather, they are used in further calculations involving addition, subtraction, multiplication, or division. These calculations cannot improve the accuracy of the measurements but must record them.

24. WHY INSTRUMENTS USED IN MEDICAL LABORATORIES ARE CALIBRATED EVERY DAY?

The importance of obtaining measurements that are both accurate and precise is rarely greater than in a medical laboratory. Patients and doctors alike want to be certain that instruments give readings that are both precise and accurate. For this reason, instruments used in medical laboratories are calibrated each day (and often at the beginning of each shift) against samples of known concentrations. Periodically, accrediting agencies send the laboratories' samples to analyze. The results obtained on these samples must be accurate within the range allowed. Precision alone is not enough; the determinations must also be accurate.

25. WHY DO WE USE EXPONENTIAL NOTATION?

Measurements in chemistry often involve very large or very small numbers. An example of a very small number is the mass of a hydrogen atom while at the opposite extreme is the mass of the Earth.

Mass of a hydrogen atom = 0.000 000 000 000 000 000 000 001 67 g
Mass of the Earth = 5,976,000,000,000,000,000,000,000,000 g

Such numbers are hard to deal with, difficult to copy without making mistakes and almost impossible to name. To simplify the writing and tabulating such numbers, we use exponential notation (also called scientific notation). When exponential notation is used, the measurement is expressed as a number between 1 and 10 multiplied by a power of 10. When exponential notation is used to express a measurement greater than 1, the original number is expressed as a number between 1 and 10 multiplied by 10^n, where n is the number of places the decimal point was moved to the left. If the number to be expressed in exponential notation is less than 1, the original number is expressed as a number between 1 and 10 multiplied by 10^{-n}, where n equals the number of places the decimal point was moved to the right.

26. HOW MUCH SIGNIFICANT AND IMPORTANT ARE 'SIGNIFICANT FIGURES' AND HOW IS UNCERTAINTY IN A MEASURE-MENT RECORDED?

One of the most tedious parts of Chemistry is dealing with Significant Figures that are more commonly called as Sig Fig's. Often we have heard someone saying 'a chain is only as strong as its weakest link' this is what happens with Sig Fig's. An answer can only be as exact as its least exact number, so if two and two are multiplied together the answer is obviously four, but what if we measured one value at 4.1 (remember that .1 is estimated) and multiplied it by another value that we measured at 5.526. The answer can't be 22.6566, because how can we have our answer more exact that the least exact value that we have. Because 4.1 is the least exact value, the final value can only be 23 because of the rules that the scientific community came up with to deal with this problem called the Rules of Sig Fig's.

Because all measurements have errors, scientists have developed methods of representing numbers such that they can indicate a degree of trustworthiness. Measurements have uncertainty (the lowest decimal place that can be measured with estimation). If we use a ruler to measure length, we read the centimeters directly from the scale. We count the millimeters marks directly. We estimate the length to the tenth of a millimeter based on where it falls between marks, halfway is 0.5 mm, for instance. The uncertainty in that ruler is to the tenth of a millimeter. If we try to use that instrument to measure something very tiny, we can't expect to get very good results. If we perform mathematical operations with the numbers that results from that measurement, we need to round our numbers so that they don't imply less uncertainty than our instruments warrant.

Each time we make a measurement of length, volume, mass, or any other physical quantity, the measurement has some degree of uncertainty. Suppose we have a quantity of liquid whose volume we wish to measure. We are given three different containers in which we might make the measurement, a 50-mL beaker, a 50-mL graduated cylinder and a 50-mL buret. The following Figures (Figures: 1(a), 1(b) and 1(c) shows these containers, each holding an identical volume of liquid.

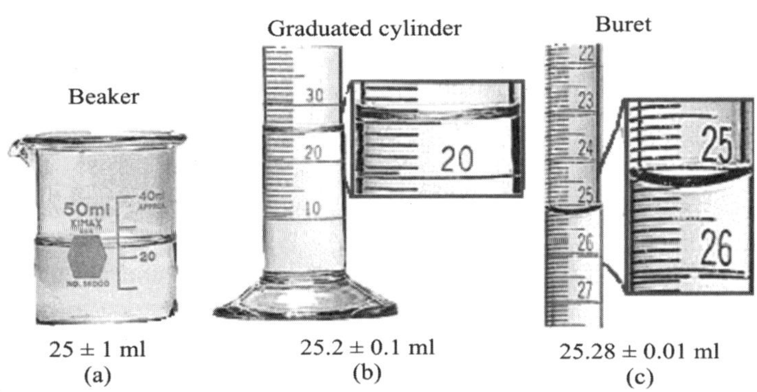

Fig 18.2: Experimental uncertainties in measuring volume

Let us look first at the 50-mL beaker. It has divisions or calibrations every 10 mL. We can see that the level of the liquid in the beaker is between the 20-mL and 30-mL marks. If we look more closely, we can see that the level of liquid is approximately midway between the two marks. We estimate that the volume is 25 mL; however, there is some uncertainty. The volume could be as little as 24 mL or it could be as much as 26 mL. If we record this volume as

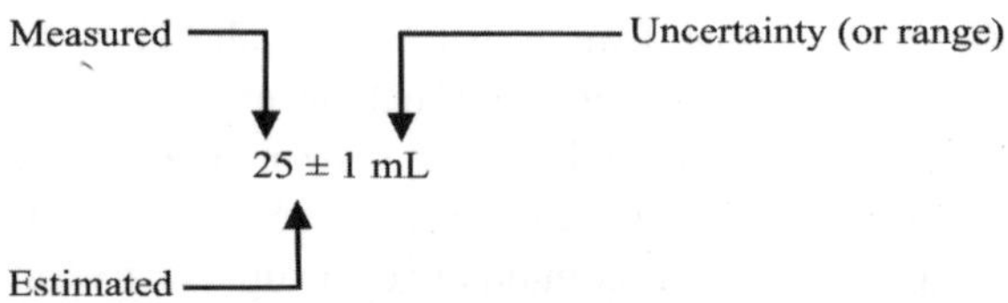

we can show the number we are certain of (20 mL), the number we think is the best estimate (5 mL) and the range within which we are certain the number falls (1 mL), called the uncertainty or range of the reading. In the above figure (b), the same volume of liquid has been placed in a 50-mL graduated cylinder. Divisions on the cylinder are marked every 1 mL. We can read that the volume is between 25 mL and 26-mL and estimate that it is about 0.2 mL above the 25-mL mark. However, it could be as little as 25.1 mL or as much as 25.3 mL. Therefore, we should record the volume of the liquid as

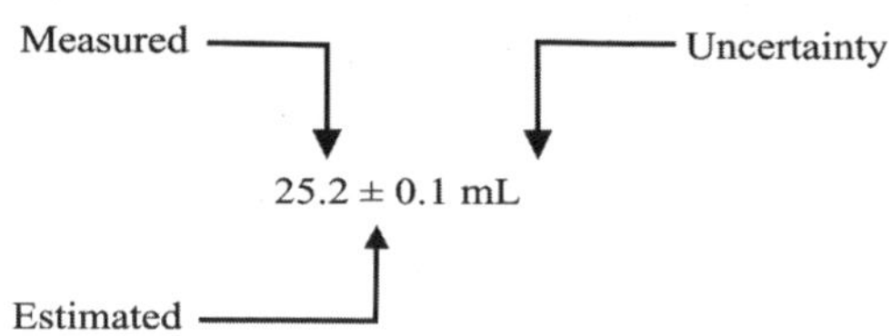

Finally, we measure the liquid in the 50-mL buret [Figure (c)]. Calibration marks on the buret are 0.1 mL apart. We can read that the volume is between 25.2 mL and 25.3 mL and estimate that it is 0.08 mL above the 25.2-mL mark. Therefore, we should report the volume of the liquid as

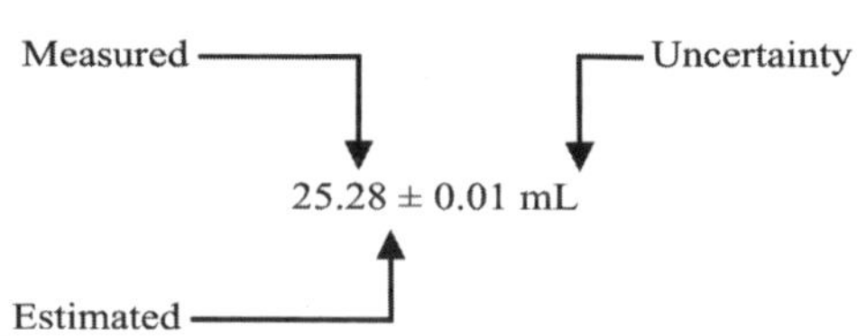

To summarize, the uncertainty of any measurement is assumed to be ±1 in the last recorded digit. This uncertainty is rarely shown but is understood to be present. For example, if we write a measurement as 372, we understand that the uncertainty is ±1; if we write 0.017, we understand that the uncertainty is ±0.001. Uncertainty in measurements is indicated by the number of significant figures used. Significant figures (or significant digits) are all those figures measured plus one that is estimated. Using our volume measurements taken from the above figure, we count the significant figures as follows:

25 mL	Contains two significant figures
25.2 mL	Contains three significant figures
25.28 mL	Contains four significant figures

27. WHEN IS ZERO TAKEN AS A SIGNIFICANT FIGURE AND WHAT IS CORRELATION BETWEEN SIGNIFICANT FIGURE AND EXPONENTIAL NOTATIONS?

A zero that serves only to locate the decimal point is not significant; zeros that are not needed to locate the decimal point are significant, for they report a measurement. If the measurements ml 25 ml, 25.2 ml and 25.28 ml were given in terms of liters, they would be 0.025 L, 0.0252 L and 0.02528 L. The number of significant figures in each measurement is the same as before; the zeros have been added only to show the location of the decimal point.

Suppose we had reported the volume of liquid in a buret as 30.50 mL, or 0.03050 L. Are any of these zeros significant? The zero to the left of the 3 is not significant, for its purpose is to locate the decimal point. The zero between the 3 and the 5 is significant because it shows that the measured volume in that place is 0. The zero after the 5 is also significant as it does not locate the decimal point, rather, it reports a measurement. The use of exponential notation clarifies the significant figures. Any zero that disappears when a number is expressed exponentially is not significant. For example, the mass of a hydrogen atom has been given as: 000 000 000 000 000 000 001 67 g. In exponential notation this number becomes 1.67×10^{-24}g. Because the zeros in the number have disappeared, we know that they merely showed the location of the decimal point and the magnitude of the number, they were not significant. Similarly, the mass of the Earth expressed exponentially is 5.976×10^{27}g. The following Table (Table 18.8) gives further examples of significant figures and exponential notations.

Table 18.8: Significant figures and exponential notations

Number	*Exponential expression*	*Number of significant figures*
560,000	5.6×10^5	Two
560,000	5.60000×10^5	Six (The decimal point in the original number shows that all the zeros are significant).
30,290	3.029×10^4	Four (The first zero is between two digits and is significant.
0.0160	1.60×10^{-2}	Three (The first two zeros show the location of the decimal; they are not significant. The last one does not show the location of the decimal point; it reports a measurement and therefore is significant).

A problem arises when a zero shows both a measurement and the location of the decimal point. The problem is solved by putting a decimal point after such a zero. Thus 250. means that the zero reflects a measurement; 250 means that the zero shows only the magnitude of the number. Similarly, 480,000 means the same as 4.8×10^5, but 480,000. means 4.80000×10^5.

28. WHAT ARE RULES FOR ASSIGNING SIGNIFICANT FIGURES?

The rules for assigning significant figures are based on deducing which digits could not have been the result of rounding off another number. When we round a number, we replace digits with zeroes, so zeroes are what we have to scrutinize. The following rules are a systematic scrutiny to determine if the digits in a number reflect actual measurements with their associated degrees of uncertainty.

1. Non-zero digits are significant.
2. Zeroes that lie between two non-zero digits cannot be the result of a rounding operation, so these are significant.
3. Zeroes that serve only to hold the decimal place are not significant, for example none of the zeroes in the numbers 550, 0.003 or 10,000,000 are significant.
4. Zeroes to the left of non-zero digits and left of the decimal point are significant.
5. The rules for determining 'if a zero is significant' do not apply if one knows for certain that the number was measured as a zero rather than rounded. We have ten fingers measured to the nearest whole finger (not nine or eleven rounded to ten). In that context, the number 10 has two significant figures, even though, by the rules, the number 10 has only one significant figure.
6. Even if a zero is measured yet it may appear to anyone reading the data to have been rounded, the data should be written in scientific notation. If a number is written in scientific notation, all zeroes are significant.

Some important numbers have values by definition. An example is that one minute is sixty seconds long. These numbers are not measurements and have never been rounded. They are known as exact numbers. When we decide where to round the result of a calculation, we ignore the exact numbers because there is no uncertainty associated with them.

***Note*:** The use of electronic calculators has greatly increased the importance of understanding significant figures and of observing the rules governing their use. A calculator with an eight-digit display capability may display eight digits in the answer to a calculation, regardless of the number of digits entered. For example, dividing 5.0 by 1.67 on a calculator may give the answer: $5.0 / 1.67 = 2.9940119$. The correct answer, 3.0, has only two significant figures, as in the least accurate number (5.0) in the problem. All other digits displayed by the calculator are insignificant.

29. HOW A NUMBER THAT RESULTS FROM ADDITION, SUBTRACTION, MULTIPLICATION OR DIVISION SHOULD BE ROUNDED?

When we add and subtract using pencil/pen and paper (rather than a calculator), we line up the decimals so that we are adding digits from the same decimal place to one another. Decimal places shared by all of the numbers contribute to the result. Higher powers of ten contribute, even if they aren't shared by all of the numbers. Lower powers of ten don't have much impact on the result. The number that results from addition or subtraction should be rounded at the lowest decimal place that the original numbers have in common. If we add the numbers 2.2 + 41.066 + 19.11 we get the result 62.376. This implies that we measured to the thousandths place, but one of the numbers was only measured to the tenths place. The sum should be rounded to 62.4 at the tenths place, which is the lowest decimal place that all three numbers share.

When a calculation is performed, the number of significant figures in the numerical answer is determined by the precision of the measurements used in the calculation. It is often necessary to round off the calculated result to the proper number of significant figures. The rules for rounding off are:

1. If the digit following the last one to be kept is less than 5, all unwanted digits are discarded. For example, to report the quantity 36.723 mL using four significant figures, write 36.72 mL; to report it using three significant figures, write 36.7 mL.
2. If the digit following the last one to be kept is 5 or greater, the digit to be kept is increased by 1. Thus, to report 38.785 mL using four significant figures, write 38.79 mL; for three significant figures, write 38.8 mL; for two significant figures, write 39 mL.

When we multiply or divide numbers, we change their scale. The decimal point may shift as a result, so it does not give us an indication of the uncertainty of the original measurements. The number that results from multiplication or division should be rounded so that it has the same number of significant figures as the original number with the fewest number of significant figures. For example, if we multiply the numbers 22 × 30.1 × 1.4 we get the result 927.08. The lowest digit is the one with uncertainty, so this implies that we made measurements that had only one in ten thousand parts uncertainty. We need to round the result to 930, which has two significant figures, since the original numbers had two and three significant figures. In calculations involving multiplication and division, the answer should contain the same number of significant figures as the measurement in the calculation that contains the fewest significant figures.

When calculations involve different kinds of operations, apply the rules sequentially. Suppose a student measures the volume and mass of a sample of metal pieces. She measures the mass on a balance using a beaker to contain the pieces. She measures the volume by adding the pieces to some water in a graduated cylinder and reading the volume displaced. She then uses the data to calculate the density of the metal using the formula D = M/V. In calculations involving addition or subtraction, the answer can show only as many decimal places as are common to all the measurements used in the calculation. Note that the location of the decimal point in each of the measurements, rather than the number of significant figures, determines the number of significant figures in the answer.

Table 18.9: Lab experiment data

	Description	*Measurement/Result*
1.	Mass of empty beaker	113.229 g
2.	Mass of beaker + metal	158.954 g
3.	Mass of metal (2–1)	46.725 g
4.	Volume of water	100.5 mL
5.	Volume of water + metal	117.5 mL
6.	Volume of metal (4–3)	1.70×10^1 mL
7.	Density	2.69 g/mL

The result of subtracting the mass of the beaker from the beaker + metal is rounded to the third decimal place by the addition/subtraction rule. The result of subtracting the volume of water from the volume of water + metal is rounded to the first decimal place by the subtraction rule. This gives 17.0 with three significant figures. The number is written in scientific notation to make it clear that the zero is significant. The density is the mass of the metal divided by the volume of the metal. When we divide a number with five significant figures by one with three significant figures, the answer is correctly rounded to three significant figures.

30. HOW IMPORTANT IS UNIT ANALYSIS TO A CHEMIST?

Unit analysis or dimensional analysis is a method for converting from one system of measurement to another. In this technique, all measurements are written with their units and conversion factors are written as ratios such that units cancel, top to bottom, like common factors in fractions. For example, to convert a person's height from English inches to metric meters, we need a conversion factor for English to metric length. The only one we know is 1 inch = 2.54 cm. But we also know that 100 cm = 1m. His height is 67 inches, so we start with that. Then we choose a conversion factor for inches and write it as a ratio with the inch unit in the denominator. That gives centimeters in the numerator, so we choose another conversion factor with centimeters and write it as a ratio with centimeters in the denominator. The inches and centimeters cancel out leaving us with the meters as the units. Multiply the numerators and divide by the denominators. In this case, the answer is rounded to two significant figures because the original measurement had two significant figures. The other values are exact numbers. The conversion between inches and centimeters is an unusual example of a definition between two totally different measurement systems.

Though chemistry students will seldom convert English to metric system, some topics require students to convert metric and SI derived units. For example, there are two values for the gas law constant, R: 0.08206 $^{\text{L atm}}/_{\text{mol K}}$ and 8.314 $^{\text{J}}/_{\text{mol K}}$. These two values are the same quantity but use different units. We can show this to be the case using unit analysis and some information about the derived units from the previous section. We know that a joule can be reduced to SI base units. That means we need to convert liters and atmospheres into base units; the moles and degrees Kelvin are fine. One milliliter is the same as a cubic centimeter. A cubic meter is a cube with each side 100 centimeters long, or 1×10^6 cm^3.

We have a new conversion factor: 1L = 0.001 m^3. If we convert atmospheres to Pascals, we can find SI base units. Using the conversion for atmospheres to kilopascals, knowing that there are 1000 Pascals in a kilopascal and using the definitions of the Pascal and the joule, we get: 1 atm = 1.01325×10^5 J/m^3. Note that the meters units do not cancel. We have cubic meters in the denominator. We can use these conversions to convert the gas law constant. This is close enough to the accepted value to prove the point. Now we need to think a bit about the significant figures. The original constant, 0.08206, has four significant figures. The atmosphere to kilopascals conversion has six significant figures. All of the other conversions are exact, because they are based on metric system definitions rather than measurements. The result is rounded to four significant figures by the multiplication/division rule.

31. WHY DO WE CONSIDER CONVERSION FACTORS?

Measurements made during a chemical experiment are often used to calculate some other property. Frequently it is necessary to change measurements from one unit to another, as from inches to feet, meters to centimeters, or hours to seconds. A relationship between two units that measure the same quantity is a conversion factor. For example, the conversion factor between feet and yards is: 3 ft = 1 yd. A conversion factor relates two measurements of the same sample. The measurements may be of the same property (in 3 ft = 1 yd, both measurements are of length) or of different properties of the same sample. In saying that 3 mL alcohol weighs 2.4 g, we are considering two different properties of the same sample, volume and mass. Together these measurements express a conversion factor, for they refer to the same sample and show a relationship between its volume and its mass.

Conversion factors are so named because they offer a way of converting a measurement made in one dimension to another dimension. They do not change the original property; rather they only indicate how it is measured. We have listed many conversion factors within the metric system and between the metric and English systems. Conversion factors that define relationships, such as 3 ft = 1 yd or 1 L = 1000 mL, are said to be infinitely significant. This statement means that the number of figures in these factors does not affect the number of significant figures in the answer to the problem.

32. HOW ARE PROBLEMS SOLVED BY UNIT ANALYSIS?

Problem solving by unit analysis (or dimensional analysis) is based on the premise that, in an arithmetic operation, units as well as numbers can be canceled. The idea may be new to us but the method is familiar. For example, if we were asked how many inches are there in 6 feet, we would reply without hesitation that 6 feet equals 72 inches. In doing this calculation, we would be using a familiar relationship, or conversion factor, between inches and feet, namely, 12 in. = 1 ft. This relationship can be written as a conversion factor or as a ratio (fraction). Both are equal to 1 because they measure the same distance:

$$\frac{12 \text{ in.}}{1 \text{ ft}} \text{ and } \frac{1 \text{ ft}}{12 \text{ in}}$$

To convert 6 feet to inches, we would use the conversion factor on the left because it allows us to cancel the unit (dimension) we do not want (feet) and arrive at an answer with the unit we do want (inches).

$$6 \cancel{\text{ft}} \times \frac{12 \text{ in.}}{1 \cancel{\text{ft}}} = 72 \text{ in.}$$

Note that only the units cancel; the numerical values remain.

Problem solving by unit analysis can be divided into the following steps:

1. Determine what quantity is wanted and in what units.
2. Determine what quantity is given and in what units.
3. Determine what conversion factor (or factors) relates the units given to the units wanted.
4. Determine how the quantity and units given and the appropriate conversion factors can be combined into a mathematical equation in which the unwanted units cancel and only the wanted units remain.
5. Perform the mathematical calculations and express the answer using the proper number of significant figures. Next look at the equation again and estimate the answer. If our estimate is close to the calculated answer, all is probably well; if it is quite different, we should check our calculations and be sure that we performed all operations correctly and that we properly placed the decimal point.

Many problems require two or more conversion factors to arrive at the units wanted. Problem solving by unit analysis is a straight forward method by which we can organize our thinking and approach problems in a systematic way.

33. WHAT IS CORRELATION BETWEEN DENSITY AND SPECIFIC GRAVITY?

We know that chemists determine the properties of matter, particularly those properties that help identify the composition of a sample. We can measure the mass and volume of a sample, as has been done for several samples of iron with results shown in following table. However, neither their masses nor their volumes show that all the samples are iron, but all the samples do have the same ratio of mass to volume, as is shown in the far right column. This ratio is called density.

Mass, volume, and density of iron samples

Sample	*Volume (cm^3)*	*Mass (g)*	*Density (g/cm^3)*
A	1.05	8.25	7.86
B	25.63	201.5	7.862
C	90.7	713	7.86
D	0.02471	0.1942	7.859

$$\text{Density} = \text{mass/volume}$$

All samples of the same kind of matter under the same conditions have the same density. Density is a physical property that characterizes and identifies a particular kind of matter. The densities of solids are usually given in grams per cubic centimeter (g/cm^3), the densities of gases in grams per liter (g/L) and the densities of liquids in grams per milliliter (g/mL). We know that 1 mL = 1 cm^3. Using these units, the density of water is given as 1.000 g/mL at 4°C. The densities of most metals are greater than that of water. The densities of liquids vary; some are less dense than water, whereas others are denser. For example, the density of gasoline is about 30% less than that of water and the density of chloroform is about 50% greater. Densities vary with temperature. For example, the density of water at 4°C is 1.000 g/mL and at 80°C is 0.9718 g/mL; the density of oxygen is 1.43 g/L at 0°C and 1.10 g/L at 80°C. Except for water, the densities in Table 18.9 are given at 0°C.

Table 18.9: Densities of some common solids and liquids under normal conditions

Metals (g/cm^3)		*Other solids (g/cm^3)*		*Liquids (g/mL)*	
Aluminum	2.70	Bone	1.85	Chloroform	1.49
Gold	19.32	Butter	0.86	Ethyl alcohol	0.791
Magnesium	1.74	Cork	0.24	Gasoline	0.67
Mercury	13.59	Diamond	3.51	Water (4°C)	1.000
Sodium	0.97	Sugar	1.59		

Density is a conversion factor that relates mass to volume. If we know two of the three quantities (mass, volume and density), we can calculate the third one. Often, particularly in discussing fluids, specific gravity is reported rather than density. The specific gravity (sp gr) of a substance is the ratio of its density to that of a reference substance: Specific gravity = density of a substance/density of a reference substance.

Generally, water is the reference substance for comparing solids and liquids and air is the reference substance for comparing gases. A value of specific gravity must state the temperature at which the densities were measured. Specific gravity has no units, because the density units cancel in its calculation. For example, we calculate the specific gravity of benzene at 20°C as follows:

$$\text{sp gr}\ {}^{20}_{4} = \frac{\text{density of benzene at 20°C}}{\text{density of water at 4°C}}$$

$$= \frac{0.8784\ \text{g/mL}}{1.0000\ \text{g/mL}}$$

$$= 0.8784$$

34. WHAT FACTORS DETERMINE THE AMOUNT OF HEAT ENERGY ASSOCIATED WITH A PARTICULAR SAMPLE?

We have learned that chemistry is concerned with the properties of matter and with the energy changes that matter undergoes. We have discussed properties related to the mass and volume of a sample of matter. In this section we examine properties related to energy. Energy is measured either in joules (J) or in calories (cal), where the conversion factor relating the two units is: 4.184 J = 1 cal. The terms kilojoules (kJ), 1000 J and kilocalorie (kcal), 1000 cal, are also commonly used. The large calorie (Calorie) used in nutrition is equal to one kilocalorie. The amount of heat energy associated with a particular sample is dependent on its temperature, its mass and its composition. Let us consider temperature before discussing its relationship to the energy of a sample. Temperature measures how hot or cold a sample is relative to something else, usually an arbitrary standard.

Temperature is measured with a thermometer and is most commonly reported using one of three different scales: Fahrenheit (F), Celsius (C) (sometimes called centigrade) and Kelvin (sometimes called absolute). The relationship between temperatures on these three scales is straightforward if we understand how a thermometer is constructed and calibrated. Two essential features of a thermometer are: (1) it contains a substance that expands when heated and contracts when cooled and (2) it has some means to measure the expansion and contraction. In the thermometer with which we are be most familiar, the substance that expands and contracts is mercury. In order to measure its expansion or contraction, the mercury is confined within a small, thin-walled glass bulb connected to a very narrow or capillary tube. When the temperature increases, the mercury expands and its level in the capillary tube rises. This increase in height is proportional to the increase in temperature.

35. HOW IS A THERMOMETER CALIBRATED?

In order to calibrated a new thermometer first, the mercury bulb of the thermometer is immersed in a mixture of ice and water. When the height of the mercury in the column because constant, a mark is made. This mark is one reference point. The ice-water mixture is then heated to boiling and kept at that temperature while the height of the mercury in the column rises to a new constant level. Another mark is made on the column at this level; this mark is a second reference point.

Further steps depend on whether this thermometer will measure temperature on the Celsius, Fahrenheit, or Kelvin scale. If the Celsius scale is to be used, the reference point for the ice-water mixture is labeled 0°C and that for boiling water is labeled 100°C. The distance between these two reference points is divided into 100 equal segments. If the thermometer is to measure temperature on the Fahrenheit scale, the reference point for the ice-water mixture is labeled 32°F and that for boiling water is labeled 212°F. The distance between these two points is divided into 180 equal segments. If the thermometer is to measure temperature on the Kelvin scale, the ice-water reference point is labeled 273.15 K, the boiling-water reference point is labeled 373.15 K and the distance between these two marks is divided equally into 100 segments. Notice that K does not use a degree symbol. The symbol K means "degrees Kelvin." As we can see, the temperatures measured by any of these thermometers do not differ; the difference is in the units with which each temperature is reported. The relationships between the three temperature scales are illustrated in Fig. 18.3.

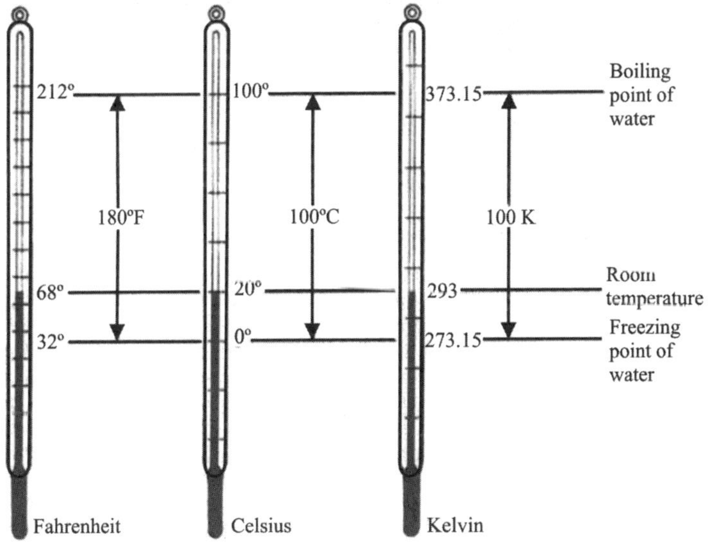

Fig. 18.3: Fahrenheit, Celsius and Kelvin thermometers

36. HOW ARE THE THREE SCALES OF TEMPERATURE INTERRELATED AND WHAT IS IMPORTANCE OF TEMPERATURE?

A temperature reading on any one of the three scales can be converted to a reading on any other. First, consider a conversion from degrees Celsius to degrees Fahrenheit. Figure 18.3 shows that, between the temperature readings of the ice-water and boiling-water marks, there are 180 Fahrenheit degrees but only 100 Celsius degrees. This relationship can be written as a conversion factor:

$$180^\circ \text{ F} = 100^\circ \text{ C} \quad \text{or} \quad \frac{180^\circ \text{ F}}{100^\circ \text{ C}} = \frac{9^\circ \text{ F}}{5^\circ \text{ C}} = \frac{1.8^\circ \text{ F}}{1^\circ \text{ C}}$$

In other words, a temperature increase of 9 Fahrenheit degrees is equivalent to an increase of 5 Celsius degrees. The above said figure also shows that the numerical values assigned to the two ice-water reference points differ by 32 degrees; a reading of 0° on the Celsius scale corresponds to a reading of 32° on the Fahrenheit scale. Combining these facts in an equation, we get:

$$^\circ\text{F} = \tfrac{9}{3}(^\circ \text{ C}) + 32 \quad \text{or} \quad ^\circ\text{F} = 1.8(^\circ \text{ C}) + 32$$

This equation can be rearranged to give the Fahrenheit to Celsius conversion equation:

$$^\circ\text{C} = \tfrac{9}{5}(^\circ\text{F}) - 32 \quad \text{or} \quad ^\circ\text{C} = \frac{(^\circ\text{F} - 32)}{1.8}$$

What is the relationship between the Celsius and Kelvin scales? Because each scale has exactly 100 divisions, or degrees, between the ice-water temperature and the boiling-water temperature, the temperature change represented by a Celsius degree is the same as that represented by a Kelvin. The scales differ in the readings at the ice-water reference point; the reading is 0° on the Celsius scale and 273.15 on the Kelvin scale. Therefore, to convert a Celsius temperature to a Kelvin temperature, simply add 273.15.

$$\text{K} = ^\circ\text{C} + 273.15 \qquad \text{or} \qquad ^\circ\text{C} = \text{K} - 273.15$$

Remember that K is not preceded by the degree symbol (°). The symbols for Fahrenheit and Celsius do require the degree symbol; for example we write 212°F and 100°C, but 373.15 K. Among the data used to identify a substance are the temperatures at which it changes state. The melting point, (mp) of a substance is the temperature at which it changes from a solid to a liquid (or from a liquid to a solid, in which case it may be called the freezing point). The boiling point (bp) of a substance is the temperature at which under normal conditions the substance changes from a liquid to a gas. The melting points and boiling points of several substances are shown in Table 18.3.

The far right column in Table 18.3 shows the physical state of several substances under normal conditions. The physical state is predictable from the melting and boiling

Table 18.10: The melting point boiling point and physical state at 20°C of several substances

Substance	*Melting*	*Boiling point, °C*	*Physical state point, °C at 20°C*
Propane	–190	–42	Gas
Chloroform	–64	62	Liquid
Sodium chloride (table salt)	801	1413	Solid
Quartz	1610	2230	Solid

points of a substance. A substance that boils below room temperature, 20°C, will be a gas under normal conditions, one that melts below room temperature and boils above it will be a liquid and one that melts above room temperature will be a solid.

37. WHY A PIECE OF IRON LEFT IN BRIGHT SUNSHINE QUICKLY BECOMES TOO HOT TO TOUCH, WHEREAS A SAMPLE OF WATER WITH THE SAME MASS LEFT IN THE SAME LOCATION FOR THE SAME LENGTH OF TIME BECOMES ONLY PLEASANTLY WARM?

When energy in the form of heat is added to a sample, the resulting temperature change depends on the sample's mass and composition. We become aware of this dependence on composition when we notice that a piece of iron left in bright sunshine quickly becomes too hot to touch, whereas a sample of water with the same mass left in the same location for the same length of time becomes only pleasantly warm. The difference is due to the difference in composition and is expressed quantitatively in the specific heats of the two materials. The specific heat, of a substance is the amount of energy required to raise the temperature of a 1-g sample by 1 degree Celsius. Typically specific heat has units of joules per gram °C (J/g°C). The specific heat of iron is 0.4525 J/g°C; that is, 0.4525 J is required to raise the temperature of 1 g iron by 1°C. The specific heat of water is 4.184 J/g°C, so 4.184 J are required to raise the temperature of 1 g water by 1°C. Each kind of matter has a unique specific heat. Several are listed in Table 18.11. Specific heat is a conversion factor that relates energy input to sample mass, composition and temperature change.

Table 18.11: Specific heat of some common substances

Metal	*J/g°C*	*Liquid*	*J/g°C*
Silver	0.238	Benzene	1.70
Copper	0.385	Ether	2.21
Gold	0.130	Ethyl alcohol	2.43
Iron	0.453	Ice (solid)	2.06
Lead	0.159	Water	4.18

The specific heat of a substance can be calculated if we know the amount of energy required to cause a measured temperature change in a sample of known mass.

38. STUDENTS IN CHEMISTRY LAB ARE WARNED TO USE THE SAME BALANCE FOR ALL THE MASS MEASUREMENTS FOR AN EXPERIMENT. WHAT KIND OF ERROR MIGHT OCCUR IF A STUDENT USES DIFFERENT BALANCES DURING THE COURSE OF A SINGLE EXPERIMENT?

All balances are sensitive to their environments. Dozens of factors contribute to the accuracy of a balance: minor things like air currents and the slope of the countertop and major things like spilled chemicals on the pan. Any given balance will read slightly high or slightly low on any given day. However, these differences tend to be fairly repeatable. If a student uses the same balance throughout an experiment, the errors tend to "subtract out" when weighing by difference. If the student changes balances, some readings will by high and others low by random amounts. This will cause a random error.

39. WHAT WILL HAPPEN TO THE PRECISION AND ACCURACY OF THE RESULTS FOR THE STUDENT WHO CHANGES BALANCES DURING AN EXPERIMENT?

The precision should be poor if a student randomly changes balances when repeating the same procedures. Suppose balance A reads high by 0.1 gram and balance B reads low by 0.1 gram. If the student uses balance A to do the first trial and balance B to do the second trial, the two results should be different by roughly 0.2 gram. We can't predict how this will affect the accuracy of the student's result because we don't know how many other errors he or she might have made and what impact those errors had. All of the errors could cancel each other out and lead to a perfect result, but most of us aren't that lucky.

40. HOW MANY SIGNIFICANT FIGURES ARE THERE IN THE NUMBER IN THE TITLE 1001 ARABIAN NIGHTS?

There are four significant figures in the number 1001. The zeroes between non-zero digits count as significant figures because they cannot be the result of a rounding operation. The author understood the ideas behind the rules for assigning significant figures. '1000 nights' implies a pretty long time, but the number '1001nights' implies that we know how many nights to the nearest night.

41. HOW CAN WE TRANSLATE THE TITLE 20,000 LEAGUES UNDER THE SEAINTO METRIC UNITS?

The metric title would be 100,000 Kilometers under the Sea. One league is three miles, so 20,000 leagues is 60,000 miles. 1.000 mile is 1.609 km, so 60,000 miles is 96,540 km which rounds to 100,000 km. The number 20,000 has one significant figure by the rules. Context can always over-turn the rules, but, in this case, the context is "really deep" rather than a measurement to the nearest league.

42. WHY AVOGADRO'S HYPOTHESIS IS AN INTRIGUING HYPOTHESIS?

Avogadro's number, also known as Avogadro's constant, is defined as the quantity of atoms in precisely 12 grams of Carbon-12. Amedeo Avogadro lived in the early-19th century and was an Italian savant known for his role in many different scientific disciplines. His most famous statement is known as Avogadro's Law and is a hypothesis that states: "Equal volumes of ideal or perfect gasses, at the same temperature and pressure, contain the same number of particles, or molecules." The designation is recognition of Amedeo Avogadro, who was the first to state that volume of a gas is proportional to the number of atoms it has. Avogadro's number is given as: $6.02214179 \times 10^{23}$ mol^{-1}.

Avogadro's hypothesis is an intriguing hypothesis, because it says that quite different elements, such as nitrogen and hydrogen, still have the same number of molecules in the same volume of an ideal gas. While in real world settings this is not strictly true, it is statistically quite close and so the ideal model still has a great deal of value. The constant can be expressed as: $(p_1)(V_1)/(T_1)(n_1) = (p_2)(V_2)/(T_2)(n_2) = \text{constant}$, where p is the pressure the gas is at, T is the temperature it is at, V is the volume of gas and n is the number of molar units. Part of Avogadro's genius and while Avogadro's number was named after him, is that he was able to see this fundamental relationship long before the experimental evidence was available to validate it. His innate understanding of the nature of ideal gasses was astounding and it wasn't until decades later that experimental evidence finally supported his hypothesis.

43. WHO CALCULATED AVOGADRO'S NUMBER AND WHEN WAS ITS IMPORTANCE RECOGNIZED?

In the 1860s, more than 50 years after Avogadro first made his hypothesis, the Austrian high school teacher Josef Loschmidt calculated how many molecules were in a single cubic centimeter of a gas under typical pressure and temperature. He determined this to be approximately 2.6×10^{19} molecules, a number now known as Loschmidt's Constant and which has since been expanded to $2.68677725 \times 10^{25}$ m^{-3}. Throughout the early years of the 20th century a search was undertaken to discover the precise value of Avogadro's number. Molecules were still largely theoretical entities to many scientists until the early part of the 20th century and so actually determining the value through experiment was not feasible. Once it became feasible, however, it was immediately apparent that the value of Avogadro's number was important, as it reflected on the fundamental nature of ideal gasses.

The term Avogadro's number was first used in a paper from 1909, by the scientist Jean Baptiste Jean Perrin, who later went on to win the Nobel Prize in Physics in 1926. He stated in the paper that, "The invariable number N is a universal constant, which may be appropriately designated 'Avogadro's Constant'. For years leading up to the 1960s there was some dispute as to the actual value of Avogadro's number. Some factions used oxygen-16 to base their calculation on, while others used a naturally occurring isotope of oxygen, leading to slightly different values. In 1960 the constant was changed to be based on carbon-12, making Avogadro's number much more regular.

44. HOW ARE ATOMS AND MOLECULES QUANTIFIED?

As the nucleus of an atom is about 10^{-13} cm in diameter and the diameter of the entire atom is about 10^{-8} cm, about 0.0000003937 inch, it is obvious that special units are required for describing the size of atoms and usually measurements are provided in terms of the angstrom that is equal to 10^{-10} m, or 10^{-8} cm. To put this on some sort of imaginable scale we can say that there are 10 million angstroms in a millimeter. Measuring the spatial dimensions of an atom, however, is not as important as measuring its mass and naturally, the mass of an atom is also almost inconceivably small. For instance, it takes about 5.0×10^{23} carbon atoms to equal just one gram of its mass. Again, the numbers boggle the mind, but the following consideration may put this into perspective. We have already established just how tiny an angstrom is; now we may consider the fact that if 5.0×10^{23} angstrom lengths were laid end to end, they would stretch for a total of about 107,765 round trips from Earth to the Sun.

It is obvious, then, that an entirely different unit should be used for measuring the mass of an atom and for this purpose, chemists and other scientists use an atomic mass unit (abbreviated amu) which is equal to 1.66×10^{-24} g. Even so, scientists can hardly be expected to be constantly measuring the mass of individual atoms; rather, they rely on figures determined for the average atomic mass of a particular element. Average atomic mass figures range from 1.008 amu for hydrogen to over 250 amu for elements of very high atomic numbers. Figures for average atomic mass can be used to determine the average mass of a molecule as well, simply by combining the average atomic mass figures for each atom the molecule contains. A water molecule, for instance, has an average mass equal to the average atomic mass of hydrogen multiplied by two and added to the average atomic mass of oxygen.

45. WHAT IS A MOLE?

The quantitative relationship between chemical substances in a reaction is known as stoichiometry. Avogadro, who was a pioneer in this field of chemistry, hypothesized that there was a specific number that would represent the number of atoms or molecules in a mole of that atom or molecule. The weight of that unit known as a mole would be equivalent to the atomic or molecular weight of the atom or molecule in grams. According to this theory, one mole of carbon-12 would have a mass of 12 grams because carbon-12 has an atomic weight of 12 (6 neutrons and 6 protons). One mole of hydrogen would weigh one gram and would contain the same number of atoms as one mole of carbon. The magical number was, in fact, discovered to be 6.024×10^{23}. It was named the Avogadro Number in honor of the father of stoichiometry even though he did not actually determine the exact number. The modern definition of a mole is as follows:

Exactly one mole represents the number of carbon atoms in exactly 12 grams of the carbon-12 isotope. For an atom a mole represents 6.02×10^{23} atoms. For a molecular compound, one mole represents 6.02×10^{23} molecules. One could even take it to extremes and say that one mole of bananas is equal to 6.02×10^{23} bananas. Now one might ask, "What relevance does the Avogadro number have for us?" The answer is, just about everything when we are working in a chemistry lab. On the other hand, it is totally useless as a unit of measure for bananas.

One atom or even ten atoms are too small for an individual to measure out in a lab. A mole of a substance equals the gram-formula mass or the gram-molar mass. This equals the sum of all of the masses of all the elements in the formula of the substance. Basically, if one were to count all of the carbon atoms in one mole of carbon-12, there would be 6.02×10^{23} atoms and it would weigh 12 grams (remember the atomic weight of carbon is 12). Using the information above, it is possible to calculate concentrations of solutions and make up solutions of desired concentration. It is also possible to use this information to determine how much of a given base would be needed in order to neutralize a specific acid and reach a pH of 7.

46. WHAT IS CORRELATION BETWEEN AVOGADRO NUMBER AND THE MOLE?

Just as using average atomic mass is much more efficient than measuring the mass of individual atoms or molecules, scientists need a useful means for comparing atoms or molecules of different substances and for doing so in such a way that they know they are analyzing equal numbers of particles. This cannot be done in terms of mass, because the number of atoms in each sample would vary: a gram of hydrogen, for instance, would contain about 12 times as many atoms as a gram of carbon, which has an average atomic mass of 12.01 amu. What is needed, instead, is a way to designate a certain number of atoms or molecules, such that accurate comparisons are possible.

In order to do this, scientists make use of a figure known as Avogadro's Number which is named after Italian physicist Amedeo Avogadro (1776-1856) and is equal to 6.022137×10^{23}. Earlier, we established the almost inconceivable scale represented by the figure 5.0×10^{23}; here we are confronted with a number 20% larger. But Avogadro's number, which is equal to 6,022,137 followed by 17 zeroes, is more than simply a mind-boggling series of digits. In general terms, Avogadro's number designates the quantity of molecules (and sometimes atoms, if the substance in question is an element that, unlike oxygen, appears as single atoms or even atomic and ionic species) in a mole (abbreviated mol). A mole is the SI unit for "amount of substance," and is defined precisely as the number of carbon atoms in 12.01 g of carbon. It is here that the value of Avogadro's number becomes clear: as noted, carbon has an average atomic mass of 12.01 amu and multiplication of the average atomic mass by Avogadro's number yields a figure in grams equal to the value of the average atomic mass in atomic mass units.

Chapter 19

ENVIRONMENTAL CHEMISTRY

1. HOW DID LIFE BEGIN ON EARTH?

Today we see innumerable varieties of insects, birds and animals besides human beings on the earth. Similarly there are various kinds of trees, plants and bushes. All of them come under the category of living beings. To study the origin of life on the earth, the scientists study the fossils of those animals, trees and plants which lived on the earth millions of years ago. Excavations carried out at various places on the earth have made available fossils, rocks and bones which throw a significant light on the existence of living beings of ancient times.

There are many theories about how life began on earth. While almost all religions teach that all forms of life were created by some divine or supernatural being or force, some theories state that life originally came to earth from outer space through micro-organisms. Many biologists and biochemists believe that life developed from non-living materials. However, scientific studies have revealed that life on the earth started with the unicellular organism called amoeba. This happened some 570 million years ago. From this very unicellular amoeba evolved the multicellular life on the earth. The first 345 million years saw the development of marine life. The subsequent 160 million years had the development of reptiles. In the last 65 million years, the evolution of mammals took place. Humans' birth is a phenomenon of only one million years.

It is observed from the study of the cells of plants and animals that they are made of proteins, carbohydrates, fats and many other substances. All these things must have been existing on the earth much before the origin of the unicellular life. Now the question arises: how did these substances which are so essential for life come. Recent studies have shown that these formed by electric discharge. Around 3500 million years ago, the earth's atmosphere was full of ammonia, methane, water vapour etc. At that time the earth was very hot and the sky was overcast with clouds. Lightning and thunder were common occurrences. The discharge during the lightning led the chemical combination of the molecules of these gases resulting in the formation of proteins, carbohydrates etc. which finally gave birth to the unicellular amoeba. Even though we have been able to solve many mysteries relating to the origin of life on the earth, there are many more which are still needed to be looked into. For example, when atoms and molecules are lifeless, how come their combination creates life? What is it which enables a unicellular amoeba to produce a multicellular creature? We still have to find answers to many such complex questions.

2. WHAT IS DARWIN'S THEORY OF EVOLUTION?

For a long time it was believed that each form of life on the earth appeared separately and that none had ever changed its form. Charles Darwin was the first scientist who disproved this belief and laid the foundation of the modern theory of evolution. According to this theory, evolution is the process by which all living things on earth today descended from common ancestors that lived millions of years ago. All plants and animals have evolved in an orderly way and continue to change even now.

Born in 1809, Charles Darwin received his education in Edinburgh University. He was deeply interested in the study of biology. When he was only 22, he got an opportunity to go around the world on a voyage on the Beagle ship of England. The voyage lasted five years and during this period Darwin could see many new countries and study different creatures. It was on this voyage that Darwin formed many ideas on evolution and adaptation. After his return to England, Darwin received a letter from the naturalist Alfred Russel Wallace, who had formed similar conclusion about evolution. A joint paper by Darwin and Wallace was read in 1858. Darwin then wrote an account of his voyage and published it in the form of a book. The title of the book was 'The Origin of Species by Natural Selection'. Its first edition was published in 1859. Setting a record, it was sold out on the very first day, it appeared in the market.

The 'Origin of Species' attempted to prove on the basis of Darwin's study that the living creatures had acquired their present shape and form after generations of evolution; and all of them had a common ancestry. In addition to this, his theory on natural selection explains that in the struggle for life only the fittest creature survives. Each organism differs slightly from its predecessor and in the course of time they acquired their present form. Though the book caused uproar when it was first published, today, Darwin's theories are widely accepted. Darwin devoted his entire life to the study of complex problems of biological science. He avoided entering the controversies surrounding his work and left it to others to debate the supposed consequences of "Darwinism". The great naturalist died in 1882.

3. HOW IS SOIL FORMED?

Soil is the upper most surface layer of the earth in which plants grow and on which directly or indirectly, all life depends. It is composed of rocks, minerals, organic matter, water and air. It did not take one or two days for the soil to form but millions of years. Millions of years ago, soil existed in the form of rocks. Since then weather and other natural forces like wind, rain, snow, water, heat etc. had been interacting with the rocks to break them into smaller fragments. These forces further converted the rock-pieces into sand. Bacteria, carbonic acid and some micro-organisms converted this sand into soil. In course of time bacteria converted the bodies of dead plants and animals also into soil. Soil is classified according to its colour, texture, chemical properties and climate. Colours of soil range from yellow and red to black. Soils are also categorized as acidic, alkaline and neutral. Extremely acidic or alkaline soil does not support plants.

There are three main types of soil: a) Pedalfers, associated with temperature and humid climate, contain iron and aluminium salts, b) pedocalo, associated with low rainfall regions, contain solvable substances such as calcium carbonate and other salts and c) Laterites, tropical red or yellow soils are heavily leached and rich in iron and aluminium. Based upon soil classifications, modern farming and plantation are made to yield maximum returns. An organic matter called humus is constantly being added to the parent soil. Humus makes the soil more fertile. Deep, well developed soil is divided into four layers called horizons. The top layer, where most plants grow, is called the A- horizon or top soil. This layer is rich in organic material and contains some minerals. The next layer is called B-horizon or sub soil. It is rich in minerals, especially in clay, but with little or no organic matter. Farmers often mix the A-and B-horizons by ploughing. The third is called C-horizon and it is a layer of weathered and shattered rocks. It is called the fragmented rock; it is quite similar to parent soil. The last layer is D-horizon or bed rocks.

4. WHAT IS MEANT BY SOIL CONSERVATION?

Throughout human history, soil has been repeatedly ruined and destroyed. This is highly dangerous to a civilization, because soil is an essential resource that is useful in many ways. Some societies have even collapsed due to mistreatment of their soil resources, so modern people must be cautious to avoid a similar fate. Soil is composed of both inorganic materials and living organisms. It provides the basis for life, giving nutrients to plants, which allow animal life to exist. Soil is also used extensively in human production. It contains natural fibers and produces lumber and crops that people consume. Crops are usually grown on mollisols, grassland soils. Soils found in other places, such as tropical rain forests, usually lack the nutrients necessary to support the growing of crops. When bedrock breaks down, it is converted into soil. Some soil also comes from sediments that are left by the wind, a volcano, glaciers, or other sources. Soil resources can be lost through erosion. Wind and rain can uplift soil that is made easy to transport by various human processes and take it somewhere else. Soil conservation is being practiced in various forms in many places around the world. Conservation efforts aim to prevent erosion and maintain life-giving nutrients that are found in the soil. In the future, soil will have to be preserved if food is to be continued to be grown.

5. WHAT ARE CAUSES AND EFFECTS OF SOIL-EROSION?

The process of soil erosion occurs when the surface of the land is worn away and lost. Soil erosion can have highly dangerous effects. To prevent erosion, many human societies throughout the world history have used devices such as terraces. Had they simply allowed erosion to occur, their food supplies would have decreased severely. In some civilizations, erosion caused so much damage that the civilization declined and fell apart. There are many specific types of erosion. In gully erosion, thin water columns quickly remove the soil from an area. Rill erosion occurs when many tiny water channels are formed and interrill erosion happens when a layer of soil is removed. Saltation erosion is the removal of soil and minerals by wind, water, or gravity. The soil bounces and moves away to another location. Surface- creep takes place when the wind blows small particles along the ground, picking up soil particles and blowing them away through the air. Erosion has many causes, both natural and human-based. Water, wind, ice and other natural elements can start the erosion process. Many human processes also cause the soil to be eroded, with negative results for the inhabitants of the area. Erosion can ruin farmland, wetlands, habitats, streams and even whole ecosystems. Measurements of total erosion have been made over time. Using the Universal Soil Loss Equation and the Wind Erosion Equation, statistics can be standardized now, unlike in the past when studies were commonly performed in different ways.

6. HOW DO NATURAL ECOSYSTEMS REACT TO VOLCANIC ERUPTIONS?

We know that Mount St. Helens erupted on May 18, 1980. The explosion was tremendous, equivalent to 500 atomic bombs of the type dropped on Hiroshima. The mountain is located in the state of Washington in the United States. It is a part of the Cascade mountain range, a group of subduction volcanoes. The Cascades enter California on the South and continue into Washington. The USGS Volcano Hazards Program received a boost as the result of the Mt. St. Helens explosion. Scientists now know considerably more about volcanoes as a result of research conducted at the Cascades Volcano Observatory in Vancouver, Washington. Ecologists have also learned a great deal from studying the area near Mt. St. Helens. For example about how the natural ecosystems react to a volcanic eruption. Large areas previously disturbed by human inhabitants can regain their natural state after an eruption. Regions can recover their biodiversity and the ecosystem that previously prevailed can return. The public has also become more worried by the threat of a volcanic explosion. Large mountains covered with snow are extremely dangerous, because the snow can easily melt and cause great amounts of mud to gush down from the mountain. These mudslides can create dams that in later years will not be strong enough to stand, causing problems long after the initial eruption. The Mt. St. Helens eruption provided an opportunity for scientists from several fields to increase their knowledge and improve their theories.

7. WHAT IS A NATURAL RESOURCE?

A natural resource is something that exists naturally and is not produced by human beings. Some people think of natural resources as a gift to humanity. They are used now in countless ways and have been of benefit to people throughout history. Natural resources are often limited, or non-renewable. As a result, their excessive use can result in the depletion of the resources and could cause supplies to dry up. For example, some environmentalists argue that present rates of fossil fuel use are unsustainable because fossil fuel resources will run out in coming years. A "fund resource" is one that can be exhausted through human consumption. Fossil fuels fit this category, as do many other resources. Metals, for example, can be depleted if they are not recycled and used again. Many natural resources can renew themselves over time. However, there are limitations to this ability. First, if a resource is completely depleted, it may not be able to replenish itself. For example, forests that are chopped down or animals that are made extinct will not be able to return. Second, human consumption may occur at a rate faster than that at which the natural resource can renew itself. For example, fossil fuel consumption is far outpacing the rate at which fossil fuel supplies can be replenished. Natural resources are now being used in increasing quantities. Higher living standards and industrialization are increasing the use of natural resources in industry and population growth is also causing demand for resources to rise. Today, the United States alone accounts for 30% of the world's natural resource consumption, though it has just 5% of the world's population. Many other industrialized nations are also consuming great quantities of natural resources, often obtaining them from less developed countries.

8. WHAT IS THE MEANING OF A HABITAT AND HOW LIVING WAYS OF HABITATS ARE RESPONSIBLE FOR DEFORESTATION?

A habitat is a more permanent ecosystem that has stabilized and reached dynamic equilibrium. The types of living organisms in a habitat are likely to remain there in the future and the habitat will probably maintain some degree of constancy. Groups of species living in a unique environment constitute a habitat. Habitats are classified based on their physical and chemical properties, inhabitants and other features. The environment may be classified based on its topography, soil, temperature, rainfall, moisture and many other physical characteristics. Chemical properties, such as level of acidity, types of nutrients and pollutants present and oxidation - reduction status, also help to identify individual habitats. The living organisms found within a habitat also help to define it. Occasionally, those habitats with treed areas (forests) are classified as one group, while areas that instead have grasses (prairies) are grouped separately. Most major types of habitats have many unique subtypes. In forests, for example, different types of soil often lead to entirely different habitats.

Deforestation is the process of destroying or removing a forest ecosystem. In its place, a different form of ecosystem will come about. This is different from the process of clear-cutting a forest. Clear-cutting only removes the current trees, but allows new ones to grow in their place because it does not fundamentally change the ecosystem. It is often hard to measure deforestation. The definitions of forests and deforestation are continually changing and data acquired by scientists is not always accurate and reliable for analysis. Over the course of history, humans have often destroyed forests. Settlers often burned trees down to have land to plant crops and at many places people burned forests to develop grasslands full of large game animals. A few major causes are responsible for deforestation. Expanding agriculture and animal herds can cause trees to be removed, as can the harvesting of wood to be used as fuel. Other important factors leading to deforestation are timber harvest, grazing and random fires. Besides killing trees and ruining forest ecosystems, deforestation presents many other problems as well. Species may become extinct due to the loss of their habitat, erosion and flooding can occur and desertification and decreased land productivity may also result. The general climate of an area can also change and more carbon dioxide can be released. Deforestation has become a major issue around the world. Some countries have developed programs aimed at reducing deforestation by creating incentives to adopt less wasteful ways of living.

9. WHEN AND WHERE DID ENVIRONMENTAL ISSUES APPEAR FIRST?

Ancient Greek and Roman thoughts began the environmental sciences. Major philosophers advanced differing views that affected others for years to come. The early Greeks searched for a reasonable understanding of nature. The ideas they developed would influence Europe in the future. Thales of Miletus (early 6th c. BC), the founder of the Ionian school, was the initiator of Greek philosophy on the environmental sciences. The Ionian school hoped to discover the nature of the universe and how it came about. According to Thales, everything was composed of water. The combination of the sun and water was the source of life. Empedocles (c.493–c.433 BC) argued that the universe was composed of four elements, earth, air, fire and water, which combined in different ways to produce different results. In his Memorabilia, the Greek writer Xenophon (c.428–c.354 BC) contends that the planet exists for people. The climate of the earth and the living creatures on it, are for the benefit of humanity. Xenophon says that the earth was created by a God for people. His work would later influence Christianity. In Rome, Cicero (106–43 BC) wrote that nature should be controlled by humans. The earth was designed for humanity, as was each different species.

To the theory of the four elements, the Greek philosopher Aristotle (384–322 BC) added a fifth: the heavens. After breaking with Plato (429–347 BC), Aristotle thoroughly investigated the natural world. He thought that the earth could be divided into sections defined by the amount of heat each received and that people could only live in the more inhabitable areas. Aristotle accepted the theory that underground wind caused volcanoes and earthquakes. He understood the origins of rivers and believed in a pattern of rainfall. Teleology, which explained things in terms of their purpose, was a major part of Aristotle's theory. It caused him to heavily research reproduction. Aristotle's Historia Animalium details over 500 species, examining common characteristics between the species. Throughout the ancient periods, interest in botany grew because of its relation to medicine. Many philosophers advanced varying theories that would affect Europe in later years.

10. HOW IMPORTANT ARE FOSSIL FUELS?

Fossil fuels are a non-renewable resource. They are used in the production of energy and have been consumed at increasing rates in recent history. Fossil fuels include coal, natural gas and oil. All three of these are available in a limited supply and are being rapidly depleted. The first major use of fossil fuels began during the Industrial Revolution in the 18th century. However, it took until the beginning of the 20th century for coal to replace wood as the dominant source of fuel for the new industrial economy. Coal took over because it became much easier and cheaper to mine and it was a better source of energy than wood.

Within few years, oil and natural gas started replacing coal. These fossil fuels are cleaner than coal and easier to transport. Also, oil can be used when liquid fuel is needed. In the 1970s, the world and especially the US experienced serious fossil fuel shortages. When OPEC put an embargo against the United States, many people began to conserve energy and realize the effects of a lack of fossil fuels. In 1984, the three major fossil fuels accounted for 82% of the world's commercial energy production. In the US, 91% of the energy supply came from burning those fossil fuels. Throughout the history of industrialization, major changes have occurred. In the beginning, the industrializing countries used substances such as wood that were available locally to generate energy. Now, by contrast, the developed countries rely on fossil fuels that are transported to them.

11. WHAT ARE POSITIVE AND NEGATIVE ASPECTS OF INDUSTRIAL REVOLUTION?

The Industrial Revolution began in England in the middle of the 18th century. A variety of ingenious inventions paved the way for machines to be used in production. The industrial societies used machines that were able to make use of energy to do work. Earlier, this work had to be done by humans or animals. The fact that burning coal could replace the need for human labor deeply altered human and environmental history. While the Industrial Revolution meant that more goods could be produced for human consumption, it also meant that more pollution would be emitted into the sky and more natural resources would have to be exploited in the production process. Industrialization has resulted in pollution of water, air and soil. Harmful waste has increased astronomically. All of these problems pose significant threats to both the environment and to human life.

Populations in industrialized countries moved from rural areas to urban areas. This exacerbated the effects of rising pollution. The concentrated population and a higher standard of living also produced larger amounts of waste. Agriculture was also affected by industrialization. More advanced machines and techniques for farming became available. These new methods caused the soil to degenerate and increased the amount of land used in farming - ruining animal habitats. As with the urbanization of thousands of years before, the Industrial Revolution changed the way people thought about themselves in relation to nature. Industrialization told many people that they had mastered nature and were now apart from and above it.

12. WHAT IS FIRE CALLED MAN'S BEST FRIEND BUT WORST ENEMY?

A material that burns, combines quickly with oxygen in the air and is said to produce fire. Burning, then, is a rapid process of oxidation. Most materials that burn are made up of mainly carbon and hydrogen. Combustion is another name for burning. In 17th century combustion was explained by other theory. Antoine Lavoisier, a French chemist, discovered in 1 774, that oxygen is essential for combustion. Some substances combine with oxygen extremely slowly. In that case no light and very little heat is produced. Rusting of iron is an example of such a slow oxidation. For a fuel to burn, certain conditions must be met. First, a substance that can burn must be present; second, a good supply of oxygen must be present and third, the fuel must be treated until it reaches its ignition temperature. The ignition temperature is also called the kindling point. Some substances like paper have low kindling points while fuels like coal must be heated considerably before they can burn. Some substances like white phosphorus have a very low ignition point that is why it is kept under water.

It would be almost impossible to count the number of ways in which fire is used. Fire provides light and cooks the food. Fire is used to heat water to make steam. Steam is used to run umpteen kinds of engines. Fire is used in industries for separating metals from their ores. Fire is also used in automobiles, boats and airplanes. When fire is under control it might be called man's best friend. But once it gets out of control it is undoubtedly his worst enemy. Every year fires kill thousands of lives. It can destroy buildings and forests in large a proportion if it gets out of control.

13. WHAT ARE ADVANTAGES AND DISADVANTAGES OF SMOKE?

Whenever wood, coal or some other substance burns, smoke comes out. Smoke is produced due to the incomplete combustion of fuels. If complete combustion of fuels takes place, no smoke will be produced. In fact smoke is a colloid of solid particles in a gaseous form. Most of the fuels contain carbon, hydrogen, oxygen, nitrogen and small quantities of sulphur. When any fuel burns, we get carbon dioxide, water vapour, nitrogen and traces of sulphur dioxide. For the complete combustion of fuels a large quantity of oxygen is required because burning is a process of oxidation. Due to the lack of oxygen, fuels do not burn completely. As a result of this smoke is produced. Smoke mainly consists of carbon dioxide, water vapour and carbon particles. When the number of carbon particles present in the smoke is large, it appears gray or dark. These very carbon particles get deposited on the walls of the chimneys. This deposit is called soot.

Smoke is the biggest pollutant of air and is very harmful from the health point of view. Today it has become a major problem in the cities. If it does not get scattered by wind, the atmosphere of the cities becomes highly foggy. Smoke is particularly harmful for the heart and lungs and is the cause of many diseases. It is also harmful to the eyes. However, in certain respects, smoke is useful also. It is used to protect the orchards from cold. In wars, smoke helps in camouflaging. It is also helpful in rain formation by enabling the water vapour to condense on its particles.

14. WHAT ARE CAUSES AND EFFECTS OF AIR POLLUTION?

There are many different chemical substances that contribute to air pollution. These chemicals come from a variety of sources. Among the many types of air pollutants are nitrogen oxides, carbon monoxides and organic compounds that can evaporate and enter the atmosphere. Air pollutants have sources that are both natural and human. Now, humans contribute substantially more to the air pollution problem. Forest fires, volcanic eruptions, wind erosion, pollen dispersal, evaporation of organic compounds and natural radioactivity are all among the natural causes of air pollution. Usually, natural air pollution does not occur in abundance in particular locations. The pollution is spread around throughout the world and as a result, poses little threat to the health of people and ecosystems. Though some pollution comes from these natural sources, most pollution is the result of human activity. The biggest causes are the operation of fossil fuel-burning power plants and automobiles that combust fuel. Combined, these two sources are responsible for about 90% of all air pollution. Some cities suffer severely because of heavy industrial use of chemicals that cause air pollution. Places like Mexico City and Sao Paulo have some of the most deadly pollution levels in the world.

Air pollution is responsible for major health effects. Every year, the health of countless people is ruined or endangered by air pollution. Many different chemicals in the air affect the human body in negative ways. Just how sick people will get depends on what chemicals they are exposed to, in what concentrations and for how long. Studies have estimated that the number of people killed annually in the US alone could be over 50,000. Older people are highly vulnerable to diseases induced by air pollution. Those with heart or lung disorders are under additional risk. Children and infants are also at serious risk. Because people are exposed to so many potentially dangerous pollutants, it is often hard to know exactly which pollutants are responsible for causing sickness. Also, because a mixture of different pollutants can intensify sickness, it is often difficult to isolate those pollutants that are at fault. Many diseases could be caused by air pollution without their becoming apparent for a long time. Diseases such as bronchitis, lung cancer and heart disease may all eventually appear in people exposed to air pollution. Air pollutants such as ozone, nitrogen oxides and sulfur dioxide also have harmful effects on natural ecosystems. They can kill plants and trees by destroying their leaves and can kill animals, especially fish in highly polluted rivers.

15. WHAT ARE PROBABLE SOLUTIONS TO THE AIR POLLUTION PROBLEMS?

Air pollution has many disastrous effects that need to be curbed. In order to accomplish this, governments, scientists and environmentalists are using or testing a variety of methods aimed at reducing pollution. There are two main types of pollution control. Input control involves preventing a problem before it occurs, or at least limiting the effects the process will produce. Five major input control methods exist. People may try to restrict population growth, use less energy, improve energy efficiency, reduce waste and move to non-polluting renewable forms of energy production. Also, automobile-produced pollution can be decreased with highly beneficial results.

Output control, the opposite method, seeks to fix the problems caused by air pollution. This usually means cleaning up an area that has been damaged by pollution. Input controls are usually more effective than output controls. Output controls are also more expensive, making them less desirable to tax payers and polluting industries. Current air pollution control efforts are not all highly effective. In wealthier countries, industries are often able to shift to methods that decrease air pollution. In the United States, for example, air pollution control laws have been successful in stopping air pollution levels from rising. However, in developing countries and even in countries where pollution is strictly regulated, much more needs to be done.

16. WHAT IS GREEN HOUSE EFFECT?

On June 23, 1988, James Hansen, the director of the Goddard Institute at NASA, told the Senate Committee on Energy and Natural Resources that global warming was a reality and that is was extremely dangerous. Global warming, also known as the greenhouse effect, immediately received international attention. Scientists, environmentalists and governments around the world took an interest in the subject. Global warming is called the greenhouse effect because the gases that are gathering above the earth make the planet comparable to a greenhouse. By trapping heat near the surface of the earth, the greenhouse effect is warming the planet and threatening the environment.

Many scientists criticized Mr. Hansen's report and the debate over global warming continues today. Current fears stem largely from the fact that global warming is occurring at such a rapid pace. Models are predicting that over the next century, the global temperature will rise by several degrees. Some scientists still do not think that the effects of global warming are as severe as some people say. They think that droughts, hurricanes and floods often blamed on global warming might actually have other causes. One major difficulty in studying global warming is the fact that weather data exists only for the last century and a half. As a result, understanding the present and predicting the future are very difficult.

17. WHAT ARE CAUSES OF GLOBAL WARMING?

Global warming has a variety of causes. One of the largest factors contributing to global warming is the general problem of overpopulation and its many effects. The greater number of people consumes more items which take more energy to make; they drive more cars and create larger amounts of garbage. These factors all increase the global warming problem. Many different gases can increase the planet's temperature. The number of different products and human activities that contribute to global warming are so numerous that finding solutions to the problem is very difficult. Using a refrigerator releases dangerous gases, turning on the lights requires energy from a power plant and driving to work causes gas emissions from the car. Countless other normal activities lead to global warming. Though having an atmosphere is important, the greenhouse effect may be making it excessively thick. The levels of gases covering the Earth have soared with industrialization and developed countries now produce about 75% of greenhouse gases. The most common gas is carbon dioxide, accounting for about 50% of all greenhouse gases. Other gases, including methane, CFCs, nitrogen oxides and ozone, also contribute to forming the greenhouse layer. Because these gases are produced by so many important and common processes, limiting their production to prevent global warming will be difficult. As population increases and Third World countries begin to use greater amounts of energy, the problem may expand rather than contract.

18. SHOULD WE RESTRICT EMISSIONS?

When humans or natural forces release chemicals or other substances into the environment, the process is known as emission. Emissions often cause pollution. Air pollution, for example, is caused when chemicals are released into the air. However, because natural sources as well as humans cause pollution, it is difficult to define that term exactly. Many natural processes cause emissions. When a volcano erupts, it lets out acid, ash and many toxic gases. When fires break out in forests, they release smoke, soot, carcinogenic hydrocarbons, dioxins and carbon dioxide. Of course, human processes generate tremendous emissions as well. Driving a car, operating a factory, or using a fireplace all cause emissions to be put out into the air. It would be unfair to say that all pollution comes from human beings, because nature itself produces much of it. As a result, some people think it is either unfair or not worthwhile to restrict human pollution production. These people argue that there will be large natural emissions anyway. It is important to discover how much pollution nature can handle. If restricting human pollution will reduce aggregate pollution to a level at which favorable results occur, it may be desirable to restrict pollution. However, if many of the bad effects of pollution are going to occur anyway, then perhaps pollution should not be restricted as much.

19. WHAT ENVIRONMENTAL PROBLEMS ARE ATTRIBUTED TO SMOG?

In many areas around the word, smog has reached extraordinary levels. Some governments have quickly reacted with adequate measures in response to the problem. The word smog is a combination of the words; smoke and fog. The term was invented by a Glasgow public health official, Des Voeux. Smog causes a smoky dark atmosphere to arise over cities. It decreases visibility and creates a haze throughout the area.

Numerous studies have monitored smog throughout the world. Some of the world's dirtiest cities have millions of inhabitants, all of whom are threatened by the smog. Modern Los Angeles suffers severely from smog, as London did in the 19th century. These two areas released certain chemicals into the air and created a foggy atmosphere. In London, where fog levels have now fallen far below those of years ago, people were often unable to see their hands and sometimes could not walk around. It took a long time for governments to act to control smog. The Clean Air Act of 1970 in the United States limited legal smog levels. The Environmental Protection Agency now measures levels of smog and regulates smog producers. Despite government action to reduce them, smog levels remain very high in many cities. Even those areas that do successfully reduce smog may be the victims of smog blown in from other locations.

20. WHAT ARE CAUSES OF SMOG?

Smog is caused by many factors. The smog that enters the atmosphere consists of over 100 chemicals, many coming from different sources. Particulates present in smog include carbon monoxide, dirt, soot, dust and ozone. To really create the smog effect, sunlight, hydrocarbons and nitrogen oxides have to mix together. Major producers of smog include automobiles, fires, waste treatment, oil production, industrial solvents, paints and coatings. Car engines, especially diesel engines, as well as gas stations that allow gas to be leaked out are huge contributors to the smog problem. Gas vapor that gets away from gas pumps contributes to the hydrocarbons needed to form smog. Diesel engines emit particles of soot that enter the atmosphere. Lead is also a major problem, especially when found in gasoline that is combusted in automobiles. Though the US and many of his countries now use unleaded gas, Third World nations remain dependent upon cheaper leaded gas. Today, the smog problems created by cars are becoming increasingly severe. As gas prices decline, consumers are buying cars that use more gas and hence pollute the atmosphere to a greater extent.

21. WHAT ARE EFFECTS OF SMOG?

When an area becomes covered in smog, the people feel the effects immediately. Unlike other problems, which may be harder to understand and visualize, smog creates immediate problems that everyone will experience. Smog creates many harmful health effects. It can cause anything from minor pain to deadly diseases such as lung cancer. Smog slowly ruins people's lungs to an extent as great as that of cigarettes. The human body has difficulty defending itself against the harms of smog. Smog can irritate and inflame pulmonary membranes, causing chest pains, coughing and throat irritation. Other illnesses such as colds and pneumonia can also be brought on by exposure to smog.

People with asthma problems are under an even greater threat. Even minor exposure to smog may cause these people to get asthma attacks. Mexico City has the world's worst levels of smog. As a result, children and the elderly are advised not to live in the city. The heavy smog levels have had serious health effects in the city, though many poorer people who need city jobs have to live there and suffer from the smog. Smog is not only a city problem. As smog levels increase, winds are carrying smog away from urban areas and harming people and ecosystems far away. Agriculture is also hurt by smog. Soybeans, wheat, tomatoes, peanuts, lettuce and cotton are all subject to infection when exposed to smog.

22. HOW SMOG FORMATION CAN BE AVOIDED?

Smog for nation can be decreased by limiting those processes that create it. In the United States, where huge amounts of smog are produced, action is being taken to reduce smog levels. Governments are combating smog in several ways. Laws are encouraging producers of automobiles to develop cars that produce less smog and chemical companies are being watched over and restricted from producing certain harmful substances. In addition to national solutions which often deal with large plants that produce smog, local and individual efforts are underway as well. Many communities and their governments are attempting to restrict usage of dangerous products such as barbecues, house paint and kitchen cleaners. In Denver, Colorado, the second most polluted US city after Los Angeles, several innovative measures have been taken in an attempt to decrease smog. The government has requested that people not drive to work at least one day a week, encouraged use of oxygenated fuels and experimented with fuels that create less smog. Not to be outdone, Los Angeles' city council came up with a long list of measures aimed at reducing smog. They decided to require carpooling, offer free bus services and create many environmentally-friendly regulations for homes. Research is being done to find less harmful alternatives for many smog-producing processes. Scientists and corporations are testing new fuels from new sources and trying to replace other products that create problems as well. In the future, many cities and countries will realize that they need to follow the examples set by cities like Denver and Los Angeles. Hopefully, they will not have to acquire those cities' high smog levels before taking action.

23. WHAT AIR POLLUTANTS ARE POTENTIAL CARCINOGENS?

A carcinogen is defined as any substance that causes cancer. Such substances affect the metabolic process in DNA. Many elements in the environment act as carcinogens. One of the first links observed by researchers between the environment and cancer was radioactivity, a link made evident by the bombing of Hiroshima, Japan, during World War II. Following this event, the cancer rate in surrounding areas of Japan skyrocketed. Subsequently, many other environmental agents have been found to be carcinogenic as well. These include synthetic chemicals, sunlight, air pollutants, heavy metals, chemical pesticides and cigarette smoke. Heavily industrialized areas often exhibit a higher cancer rate. Cigarettes are the leading cause of lung cancer, a dominant type of cancer in the United States. Not only smoking, but even secondhand inhalation of smoke has been found to be carcinogenic. Air pollution is a cause of cancer as well. Inhabitants of urban areas have a 33 percent higher risk of developing lung cancer. Occupational exposure to various chemicals is also a cause of cancer. Before the Toxic Substances Control Act of 1976, potentially hazardous chemicals were often not tested before being put on the market and carcinogenic substances, such as asbestos, led to many workers' deaths. Vinylchloride, used in plastic production and anesthetic gases from operating rooms, have been known to cause cancer among workers exposed to them. In addition, overexposure to ultra-violet radiation, as a result of ozone layer depletion, is a leading cause of skin cancer.

24. WHY IS AIR POLLUTION A GLOBAL PROBLEM?

Every day, the average person inhales about 20,000 liters of air. Every time we breathe, we risk inhaling dangerous chemicals that have found their way into the air. Air pollution includes all contaminants found in the atmosphere. These dangerous substances can be either in the form of gases or particles. Air pollution can be found both outdoors and indoors. Pollutants can be trapped inside buildings, causing indoor pollution that lasts for a long time. The sources of air pollution are both natural and human-based. As one might expect, humans have been producing increasing amounts of pollution as time has progressed and they now account for the majority of pollutants released into the air. Air pollution has been a problem throughout history. Even in Ancient Rome people complained about smoke put into the atmosphere. The effects of air pollution are diverse and numerous. Air pollution can have serious consequences for the health of human beings and also severely affects natural ecosystems. Because it is located in the atmosphere, air pollution is able to travel easily. As a result, air pollution is a global problem and has been the subject of global cooperation and conflict. Some areas now suffer more than others from air pollution. Cities with large numbers of automobiles or those that use great quantities of coal often suffer most severely from problems of air pollution.

25. HOW CAN WE SOLVE THE GLOBAL WARMING PROBLEMS?

The threat of global warming is among the most important of all modern environmental problems. There are a variety of ways of dealing with it, each attempting to combat one of the many causes of global warming. The problems that cause global warming include overpopulation, deforestation, ozone depletion, garbage dumping and many others. These all have unique solutions which are now being promoted by environmentalists. Certain laws and treaties are aimed at reducing the emission of pollutants that result in global warming. In 1988, the International Conference on the Changing Atmosphere drew scientists and decision makers from 48 countries. Some policies could successfully reduce global warming. Raising fossil fuel prices, taxing emissions and encouraging people to take environmentally friendly action through such activities as planting trees will all help. Because many problems leading to global warming are caused or contributed to by overpopulation, people are beginning to work to reduce family sizes. Family planning services actually help in the fight against global warming. Education is a key method of reducing the greenhouse effect. By teaching people about such things as deforestation, environmental activists hope to prevent the problems that ultimately lead to global warming. Widespread media attention to the global warming problem is also increasing awareness. This is causing both individuals and governments to act more responsibly towards the environment.

26. HOW USE OF ENERGY EFFICIENTLY CAN DECREASE POLLUTION?

Over the past years, human energy consumption has risen dramatically. The rise in use of energy resources has meant increasing costs and the depletion of non-renewable supplies. These and other factors have led people try to use energy in a more efficient manner. The First Law of Thermodynamics says that energy can be transferred between forms, but not created or destroyed. The Second Law of Thermodynamics says that usage of heat to perform work inevitably causes some heat to be lost to the surrounding environment. As a result, perfect efficiency in energy use is impossible. For many things that use energy, a maximum efficiency limit exists. A four-cycle internal combustion engine can achieve at most 54% efficiency, while a diesel engine can reach 56%. Steam engines perform only at 32% efficiency. In reality, engines are considerably worse than this due to designs that prevent achievement of maximum efficiency. If one includes all devices that use energy, about 85% of energy is lost due to inefficient usage. Many low-efficiency devices exist because when they were developed, energy costs were very low. Only later, when energy costs rose, did it make sense to be more energy efficient.

Devices that increase efficiency often are priced higher than those that perform inefficiently. However, because less energy is used and the devices often last longer (such as fluorescent light bulbs), the efficient devices are sometimes cheaper in the long run. Governments are trying to promote efficient energy use for the sake of the future. Laws such as the Clean Air Act in the US and other laws elsewhere encourage companies to develop more efficient methods of energy consumption.

27. WHAT THREATS HAVE BEEN POSED BY CHLOROFLUOROCARBONS?

Chlorofluorocarbons (CFCs) are the family of synthetic chemicals that are compounds of the elements: chlorine, fluorine and carbon. CFCs are stable, nonflammable, noncorrosive, relatively nontoxic chemicals and are easy and inexpensive to produce. Scientists developed the first CFCs during the late 1920s. The compounds subsequently became used in a wide range of industrial products in the United States, Europe and Japan. Manufacturers used CFCs as refrigerants in refrigerators, freezers, air conditioners and heat pumps and as propellants in aerosols and medical inhalers. CFCs also served as insulating foams in packaging materials, furniture, bedding and car seats. Cleaning agents for electronic circuit boards, metal parts and dry cleaning processes also used CFCs.

During the 1970s, scientists linked CFCs to the destruction of Earth's ozone layer. The manufacture of CFCs has since been banned in most countries. CFCs damage the ozone layer when they escape from sources such as leaky car air conditioners, discarded plastic-foam egg cartons and old home air conditioners crushed in a landfill. The CFCs drift up to the stratosphere, an upper layer of the atmosphere where strong ultraviolet radiation from the Sun breaks them down. As they break down, they release chlorine, which depletes the protective ozone layer. A single chlorine atom can destroy over 100,000 ozone molecules. Ozone absorbs ultraviolet radiation. When the ozone layer thins, more of a harmful type of ultraviolet radiation called UVB reaches Earth's surface. Studies show that UVB radiation can cause skin cancer. UVB has also been linked to cataracts of the eye and to suppression of the immune system. Fish, shrimp, crabs, amphibians and other animals that live in or around water, including phytoplankton, microscopic organisms that form the foundation of aquatic food webs, are all at risk from UVB radiation.

Scientists have worked hard to find alternatives to CFCs that spare the ozone layer. Alternative refrigerants have completely replaced CFCs in many countries and substantial progress has been made finding alternatives for foam products and air conditioning. Many of the alternatives have the added bonus of being highly energy efficient, helping the United States and other countries meet goals of reduced global warming emissions. Despite these important steps, previously released CFCs continue to damage the ozone layer because they can remain in the atmosphere for more than 100 years. Ozone depletion is particularly bad over the Antarctic, seasonally producing an ozone hole, a large area of the atmosphere that contains extremely low amounts of ozone. Low levels of ozone have also been observed over the United States, Canada, Japan, Russia and Europe. Scientists predict that the highest levels of CFCs in the atmosphere will occur during the first decade of the 21st century. If international agreements hold, the ozone layer is expected to recover around 2050.

To combat the problem of ozone thinning, in 1978 the U.S. Congress banned CFCs used in aerosols in the United States. In 1987, 27 nations signed an international agreement called the Montréal Protocol, calling for a 50 percent reduction in CFC production by 1998 and an eventual end to all manufacture of CFCs. Later amendments accelerated the timetable for eliminating CFC production. The Montréal Protocol has been remarkably successful in curtailing production of CFCs. More than 175 nations have ratified the protocol.

28. WHY 'DDT' HAS BEEN PLACED UNDER STRICT CONTROL?

DDT, dichlorodiphenyltrichloroethane, is a colorless chemical pesticide, used to eradicate disease-carrying and crop-eating insects. It was first isolated in Germany in 1874, but got prominence in 1939 when the Swiss Nobel Prize-winning chemist Paul

Müller recognized it as a potent nerve poison on insects. First used heavily in World War II for preinvasion spraying, DDT was disseminated in great quantities thereafter throughout the world to combat yellow fever, typhus, elephantiasis and other insect-vectored diseases. In India, DDT reduced malaria from 75 million cases to fewer than 5 million cases in a decade. Crops and livestock sprayed with DDT sometimes as much as doubled their yields. With the publication of the American marine biologist Rachel Carson's Silent Spring in 1962, suspicion grew that DDT, by entering the food chain and eventually concentrating in higher animals, caused reproductive dysfunctions, such as thin eggshells in some birds. Some insect pests also gradually developed DDT-resistant strains whose populations grew unchecked while their natural predators, such as wasps, were being eradicated by spraying. In 1973 DDT was banned in the U.S. except for use in extreme health emergencies. Many other nations have also banned it or placed it under strict control.

29. WHAT ARE REQUIREMENTS OF LANDFILLS?

The landfills accept garbage and charge the groups that bring it to the landfills. These groups include businesses as well as municipalities. Every day, the landfill receives a huge tonnage of garbage. When the garbage is placed on the landfill, it is compacted three to five times. Compacting the garbage helps to prevent fires from starting in the garbage and it keeps rats away. The landfill also has only a limited amount of space, so compacting garbage also allows more garbage to be sent there. After compacting the garbage, the operators of the landfill cover up all the new garbage in a five-inch layer of dirt. This helps to keep birds, seagulls and rats away. It also reduces the odor that the landfill spreads out to the surrounding community. The landfill is built in the shape of a bathtub. It has a four to five foot layer of clay and a 60-milliliter layer of plastic lining. This ensures that no contaminated water from the landfill will enter the water supply. To further protect against this, the landfill operators remove water from the landfill every day and send it to a treatment center. Eventually, the landfill will reach its capacity and it will not be possible to place any more garbage on it. At that point, the garbage will be covered up with grass and the area will be used as a skiing area, golf course, model airplane flying field, or bike trail.

30. WHAT DANGERS ARE ASSOCIATED WITH HAZARDOUS WASTES?

In addition to releasing gases and particles into the atmosphere, humans produce waste that is dumped on the environment. Often, this waste is hazardous and dangerous to both nature and human life. The levels of dangerous wastes continue to grow. Industries and individuals continue to be largely unaware of this major environmental problem. As a result, many people and industries are failing to prevent the creation of hazardous waste or to limit the negative effects it produces. Individuals often throw out goods without realizing that they are headed for a landfill and could be dangerous for the environment. No matter where people put these hazardous waste materials, there is always a chance that they could find their way into the ground and eventually into our bodies. Corporations usually want to avoid the costs associated with having to limit creation of hazardous waste. Consequently, they build landfills on site and fill them with waste, or sometimes pay to have their waste removed. Often, hazardous materials are transported to areas that accept money to take the waste. It may prove very difficult to reduce hazardous waste in the future. Unlike many other environmental problems, waste creation is something people do not often think about. In the future, people may have to reduce not only their generation of hazardous waste, but also their consumption of many products that end up in landfills.

31. WHAT IS THE BEST METHOD OF REDUCING THE HAZARDOUS WASTE PROBLEM?

Hazardous waste is produced both on a huge scale by major industries and on a relatively small scale by individuals. No matter where it comes from, waste can be dangerous. One of the main causes of the abundance of hazardous waste is that people do not realize how large a problem it is. Because it can be simply removed and sent to a landfill, it is often assumed that the problem ends there. Industries have often displayed an unwillingness to find ways to deal with hazardous waste because of the expenses associated with it. Many industries and governments create crude landfills to store waste and often just dump waste chemicals into nearby bodies of water.

Chemicals used for industrial processes often create dangerous forms of waste. The amount of these chemicals has risen heavily in the past, as more areas of the world industrialize and new products are produced. Over 80,000 different chemicals are used in industries worldwide. Often, it is difficult and expensive to get rid of these chemicals and to store them in a way that does not endanger human life or the environment. Obviously, not all of these chemicals are dangerous, but many are and they do create serious problems. Around the world, hundreds of millions of tons of hazardous waste are produced annually. Rather than cleaning up or storing waste more carefully, one method of reducing the hazardous waste problem may be to simply stop producing so much of it.

32. WHAT ARE EFFECTS OF HAZARDOUS WASTES?

Every year, major health problems result from hazardous wastes. Increasing amounts of hazardous waste have caused increasing health problems. There are over 80,000 chemicals in existence and many are used commonly in industrial processes. Often, these chemicals find themselves in places where they are able to harm human health. Insufficient research has been done to provide data on the effects of every chemical. Because waste chemicals often mix together, it will also be necessary to learn how combinations of these chemicals affect human health. To compound the problem created by a lack of knowledge, 1,500 new chemicals are invented every year and many are introduced into industrial processes. Though the health problems being created are uncertain, they definitely do exist. In 1989, a school in New Jersey had to be closed because students there had suffered excessive exposure to chromium. It was later learned that large amounts of chromium had been dumped nearby and had blown over to the school area. Often, the lack of knowledge and certainty about what chemicals are dangerous reduces the possibilities of restricting their use. Sadly, it is often only after someone has died or become seriously ill that governments will intervene and reduce levels of dumped hazardous waste.

33. HOW HAZARDOUS WASTE PROBLEMS CAN BE SOLVED?

Hazardous waste has created many problems and dangers that have not gone unnoticed. Despite the fact that the problem receives less attention than many other environmental threats, some successful solutions to the problem of hazardous have been suggested and used. As with air pollution and many other environmental problems, hazardous waste can be controlled through input and output controls. The government can increase regulations on the disposal of hazardous waste to ensure that problems do not occur. It could also limit the amount of waste industries are allowed to produce, or provide incentives to create less waste. After hazardous waste has been created, there are several actions that can be taken. Industries can break down chemical compounds into less dangerous forms, or store waste in ways that protect the environment from being exposed to the waste. Not only major industries but individuals as well must become part of the solution. They can choose not to buy those products which require the production of hazardous waste, attempt to influence policymakers and produce less hazardous waste themselves. Many scientists think that waste production can be cut. Experts say that waste can be reduced by at least one-third using existing technologies and methods. Some countries, including many in Europe and Asia, are working to reduce waste levels. A great deal of success is being achieved in eliminating this serious problem.

34. WHAT ARE CHARACTERISTICS OF WATER?

Water is the common name applied to the liquid state of the hydrogen-oxygen compound H_2O. The ancient philosophers regarded water as a basic element typifying all liquid substances. Scientists did not discard that view until the latter half of the 18th century. In 1781 the British chemist Henry Cavendish synthesized water by detonating a mixture of hydrogen and air. However, the results of his experiments were not clearly interpreted until two years later, when the French chemist Antoine Laurent Lavoisier proved that water was not an element but a compound of oxygen and hydrogen. In a scientific paper presented in 1804, the French chemist Joseph Louis Gay-Lussac and the German naturalist Alexander von Humboldt demonstrated jointly that water consisted of two volumes of hydrogen to one of oxygen, as expressed by the present-day formula H_2O. Almost all the hydrogen in water has an atomic weight of 1. The American chemist Harold Clayton Urey discovered in 1932 the presence in water of a small amount (1 part in 6000) of so-called heavy water, or deuterium oxide (D_2O); deuterium is the hydrogen isotope with an atomic weight of 2. In 1951 the American chemist Aristid Grosse discovered that naturally occurring water contains also minute traces of tritium oxide (T_2O); tritium is the hydrogen isotope with an atomic weight of 3.

Pure water is an odorless, tasteless liquid. It has a bluish tint, which may be detected, however, only in layers of considerable depth. Under standard atmospheric pressure (760 mm of mercury, or 760 torr); the freezing point of water is 0°C (32°F) and its boiling point is 100°C (212°F). Water attains its maximum density at a temperature of 4°C (39°F) and expands upon freezing. Like most other liquids, water can exist in a super cooled state; that is, it may remain a liquid although its temperature is below its freezing point. Water can easily be cooled to about –25°C without freezing, either under laboratory conditions or in the atmosphere itself. Super cooled water will freeze if it is disturbed, if the temperature is lowered further, or if an ice crystal or other particle is added to it. Its physical properties are used as standards to define the calorie and specific and latent heat and in the metric system for the original definition of the unit of mass, the gram. Water is one of the best-known ionizing agents. Because most substances are somewhat soluble in water, it is frequently called the universal solvent. Water combines with certain salts to form hydrates. It reacts with metal oxides to form acids. It acts as a catalyst in many important chemical reactions.

35. WHAT IS NATURAL WATER CYCLE?

Hydrology is the science concerned with the distribution of water on the earth, its physical and chemical reactions with other naturally occurring substances and its relation to life on earth; the continuous movement of water between the earth and the atmosphere is known as the hydrological cycle. Under several influences, of which heat is predominant, water is evaporated from both water and land surfaces and is transpired from living cells. This vapor circulates through the atmosphere and is precipitated in the form of rain or snow.

On striking the surface of the earth, the water follows two paths. In amounts determined by the intensity of the rain and the porosity, permeability, thickness and previous moisture content of the soil, one part of the water, termed surface runoff, flows directly into rills and streams and thence into oceans or landlocked bodies of water; the remainder infiltrates into the soil. A part of the infiltrated water becomes soil moisture, which may be evaporated directly or may move upward through the roots of vegetation to be transpired from leaves. The portion of the water that overcomes the forces of cohesion and adhesion in the soil profile percolates downward, accumulating in the so-called zone of saturation to form the groundwater reservoir, the surface of which is known as the water table. Under natural conditions, the water table rises intermittently in response to replenishment, or recharge and then declines as a result of continuous drainage into natural outlets such as springs.

36. WHY PURE WATER RARELY OCCURS IN NATURE?

Because of its capacity to dissolve numerous substances in large amounts, pure water rarely occurs in nature. During condensation and precipitation, rain or snow absorbs from the atmosphere varying amounts of carbon dioxide and other gases, as well as traces of organic and inorganic material. In addition, precipitation carries radioactive fallout to the earth's surface. In its movement on and through the earth's crust, water reacts with minerals in the soil and rocks. The principal dissolved constituents of surface and groundwater are sulfates, chlorides and bicarbonates of sodium and potassium and the oxides of calcium and magnesium. Surface waters may also contain domestic sewage and industrial wastes. Ground waters from shallow wells may contain large quantities of nitrogen compounds and chlorides derived from human and animal wastes. Waters from deep wells generally contain only minerals in solution. Almost all supplies of natural drinking water contain fluorides in varying amounts. The proper proportion of fluorides in drinking water has been found to reduce tooth decay. Seawater contains, in addition to concentrated amounts of sodium chloride, table salt, many other soluble compounds, as the impure waters of rivers and streams are constantly feeding the oceans. At the same time, pure water is continually lost by the process of evaporation and as a result the proportion of the impurities that give the oceans their saline character is increased.

37. HOW IS HARDNESS OF WATER CAUSED AND REMOVED?

Suspended and dissolved impurities present in naturally occurring water make it unsuitable for many purposes. Objectionable organic and inorganic materials are removed by such methods as screening and sedimentation to eliminate suspended materials; treatment with such compounds as activated carbon to remove tastes and odors; filtration; and chlorination or irradiation to kill infective microorganisms. In aeration, or the saturation of water with air, water is brought into contact with air in such a manner as to produce maximum diffusion, usually by spraying water into the air in fountains. Aeration removes odors and taste caused by decomposing organic matter and also industrial wastes such as phenols and volatile gases such as chlorine. It also converts dissolved iron and manganese compounds into insoluble hydrated oxides of the metals which may then be readily settled out.

Hardness of natural waters is caused largely by calcium and magnesium salts and to a small extent by iron, aluminum and other metals. Hardness resulting from the bicarbonates and carbonates of calcium and magnesium is called temporary hardness and can be removed by boiling, which also sterilizes the water. The residual hardness is known as noncarbonate, or permanent, hardness. The methods of softening noncarbonate hardness include the addition of sodium carbonate and lime and filtration through natural or artificial zeolites which absorb the hardness-producing metallic ions and release sodium ions to the water. Sequestering agents in detergents serve to inactivate the substances that make water hard. Iron, which causes an unpleasant taste in drinking water, may be removed by aeration and sedimentation or by passing the water through iron-removing zeolite filters, or the iron may be stabilized by addition of such salts as polyphosphates. For use in laboratory applications, water is either distilled or demineralized by passing it through ion-absorbing compounds.

38. HOW IS WATER DESALINIZED?

To meet the ever-increasing demands for fresh water, especially in arid and semiarid areas, much research has gone into finding efficient methods of removing salt from seawater and brackish waters. In the U.S., desalinization research is directed by the Bureau of Reclamation, Department of the Interior. Several processes are being developed to produce fresh water cheaply. Three of the processes involve evaporation followed by condensation of the resultant steam and are known as multiple-effect evaporation, vapor-compression distillation and flash evaporation. The last-named method, the most widely used, involves heating seawater and pumping it into lower-pressure tanks, where the water abruptly vaporizes (flashes) into steam. The steam then condenses and is drawn off as pure water. In 1967, Key West, Florida, opened a flash-evaporation plant and thus became the first city in the U.S. to draw its fresh water from the sea.

Freezing is an alternate method, based on the different freezing points of fresh and salt water. The ice crystals are separated from the brine, washed free of salt and melted into fresh water. In another process, called reverse osmosis, pressure is used to force fresh water through a thin membrane that does not allow the minerals to pass. Reverse osmosis is still undergoing intensive development. Electro dialysis is being used to desalt brackish waters. When salt dissolves in water, it splits into positive and negative ions, which are then removed by electric current through anion and cation membranes, thus depleting the salt in the product water. Although developmental work on electro dialysis is continuing, a number of commercial plants are in operation. In 1962 Buckeye, Arizona, became the first town to have all its water supplied by its own electro dialysis-desalting plant, which provides about 2,460,000 liters (about 650,000 gallons) of water daily at a cost of about $1 per 6300 liters (1670 gallons).

One major problem in desalinization projects is the cost of producing fresh water. Water from conventional sources, such as wells and reservoirs, is sold for less than 30 cents per 3800 liters delivered to the home and water for irrigation is usually priced at less than 5 cents per 3800 liters. The dual-purpose atomic power and water-desalting plants now being planned are designed to produce fresh water for between 20 and 30 cents per 3800 liters. Most experts expect more immediate results from efforts to purify mildly brackish water that contains between 1000 and 4500 parts per million of minerals, compared to 35,000 parts per million for ocean water. Because water is potable if it contains fewer than 500 parts per million of salt, the cost of desalting brackish water is correspondingly less than it is for desalting seawater.

39. HOW DOES SNOWFALL OCCUR AND WHAT ARE ITS BENEFITS?

Snow is formed of tiny crystals of frozen water that fall from the sky. Snow flakes have an infinite variety of shapes. They are usually hexagonal and may be flat, needle shaped or star shaped. No two snowflakes are exactly alike. Freshly fallen snow reflects about 95% of the sun's heat back into the space. Due to sun's heat there is a continuous evaporation of water from the seas, rivers, lakes, ponds etc. Since water vapor is lighter than air, it goes up in the atmosphere and turns into cloud. We know that the temperature decreases as we go higher in the atmosphere. Since the capacity of air to hold water vapour decreases with the fall in its temperature, at a certain height a large quantity of water vapour gets accumulated in the air. The air having excess of water vapours is said to be supersaturated. Under this condition the water vapours condense on the dust and smoke particles present in the air, on further cooling, they get converted, into snow particles. These particles combine with each other to form crystals of snow. When the air cannot bear their weight, they fall down as snow flakes and go on accumulating on the mountains.

Now the question arises: why does it happen on the mountains only and why does it not fall in the plains too? In fact the possibility of snowfall at any place depends on two factors, the altitude of the place from the sea-level and its distance from the equator. The higher the altitude of the place, the greater is the probability of snowfall there. Similarly the larger its distance from the equator, more are chances of snowfall there. Even if the amount of snow formed in the atmosphere is very large, only a very small fraction of it falls down on the mountains in the form of snow. The remaining portion comes down as rains because while passing through hotter regions, it melts and turns into water. However, the temperature being low at the mountains, the snow does no melt into water there. It goes on piling up as snow and under the increasing weight of its accumulation layer after layer, becomes harder and harder.

Snowfall is very useful for us. When snow melts in summer, the melted water flows to rivers and is used for irrigation purposes. Fresh snow is very light and a good insulator, protecting the underlying plants from severe cold. Snow is a bad conductor of heat because of the air trapped between its particles. As such it acts as a blanket for the earth. In colder regions, people make houses of snow as it provides excellent protection against cold weather. In hilly areas, the roofs of the houses are made in a slanting shape, so that snow does not accumulate there. It saves the inhabitants from the adverse-effects of snow and also prevents the roofs from collapsing.

40. WHY ARE DETERGENTS CONSIDERED AS PUBLIC NUISANCE?

The term Detergent is applied to materials, the solutions of which aid in the removal of dirt or other foreign matter from contaminated surfaces. Until the 1940s soap was the only important detergent. Today soap is but one of a great many detergent products. The ingredients of detergents are often called surface-active agents, or surfactants, because they act upon a surface. A common feature of detergents is that they are made up of comparatively large molecules (molecular weight over 200). One part of the molecule is soluble in organic material and the other part is soluble in water. During World War II the shortage of fats, from which soap is made, spurred the development of soapless or synthetic detergents, primarily in the United States. After the war the need for new types of detergents for automatic washing machines accelerated the trend.

Detergents are made by treating an aromatic, benzene-type, compound with sulfuric acid, followed by neutralization with alkali to convert the product to its sodium salt. The detergent products of these reactions came into wide use in the late 1940s and early '50s and proved to be effective in hard water and cold water, whereas soap is often wholly ineffective under both conditions. These detergents, however, became a public nuisance because, unlike soaps, they were neither soluble nor biodegradable; that is, once put into water they tended to remain there, resisting conversion into less complex and more soluble substances. The detergents tended to create foam in cesspools and in sewage-disposal plants as well. They even appeared in naturally occurring ground and surface waters.

Replacing the aromatic compound with a so-called linear alkyl-type compound in the process, described above, led to a more desirable product. It was as effective as the former kind in its detergent action but was more biodegradable and soluble. The new linear alkylate sulfonate is changed to harmless products by microorganisms in cesspools, sewage-treatment plants and ordinary soils. Manufacturers of detergents in the U.S. have changed their processes to produce only biodegradable or soft, detergents. The changeover was done voluntarily between 1963 and 1965.

41. HOW GRAVE IS THE RAIN FOREST DESTRUCTION PROBLEM?

The atmosphere and oceans are not the only parts of the environment being damaged. Rain forests are being quickly destroyed as well and their survival is questionable. E.O. Wilson, a biologist at Harvard, called the depletion of rain forest areas "the greatest extinction since the end of the age of dinosaurs." Unlike some environmental issues, rain forest depletion has fortunately received significant public and media attention. Despite the opposition to the cutting down of rain forests, the problem continues. Every year, Brazil chops down an area of forest the size of the state of Nebraska. In addition to the Amazon's rain forests, many other forests are being cut down as well. In Indonesia, Zaire, Papua-New Guinea, Malaysia, Burma, the Philippines, Peru, Colombia, Bolivia and Venezuela, rain forests that were once great have been lost. According to some estimates, 50 million acres of rain forest are cut down every year. The United Nations says the figure is closer to 17 million acres. The World Wildlife Fund says that every minute, 25 to 50 acres are cut or burned to the ground. The world's growing population has been a primary cause of rain forest destruction. More people need land to live on and wood products to consume. Limiting population growth may be the first in a series of steps that would limit the destruction of the rain forests.

42. WHAT IS MEANT BY WATER POLLUTION?

Attention for water pollution exploded in the 1980s. The oil spill of the Exxon Valdez showed many around the world just how horrible the effects of water pollution could be. However, even the Exxon Valdez spill barely touched the surface of the problem of water pollution. The ship spilt only 5% of the oil spilt that year and oil is just one of many pollutants that people dump into the water every year. Every year, 14 billion pounds of sewage, sludge and garbage are dumped into the world's oceans. 19 trillion gallons of waste also enter the water annually. The problem of ocean pollution affects every nation around the world. This is especially true because water is able to transport pollution from one location to another. For many years, chemicals were dumped into bodies of water without concern. While many countries have now banned such behavior, it continues to go on today. As the world has industrialized and its population has grown, the problem of water pollution has intensified. The simple fact that millions of people live along coastlines and near rivers, means that these bodies of water are likely candidates for heavy and destructive pollution. It is hard to know now what our oceans will look like in the future. Just how damaged they will be by pollution is uncertain.

43. WHAT ARE CAUSES OF WATER POLLUTION?

One of the reasons that the water pollution problem is so severe is that it is not actually illegal to dump pollutants into water bodies. Sewage, sludge, garbage and even toxic pollutants are all dumped into the water. Often, governments either do not care or simply look the other way. Across the world, about half of all sewage is dumped into water bodies in its original form. No efforts are made to disinfect the sewage or to remove especially harmful pollutants. Even if sewage is treated, problems still arise. Treated sewage forms sludge, which is sent out into the sea and dumped. Many cities and countries dump sewage out at sea. Often, they place it not far from their own coastline, often killing all the sea wildlife in the dumping area. In addition to sewage, chemicals dumped by industries and governments are another major source of water pollution. Oil, such as that spilled by transport ships, has been dumped into the water since the US Civil War. Every year, between 1 and 10 billion tons of oil are spilt, killing many species and destroying the ecosystem in the area. Cleanup efforts have been weak, as only about 10% of the oil is removed by the most successful efforts.

44. WHAT ARE EFFECTS OF WATER POLLUTION?

The effects of water pollution are varied and depend on what chemicals are dumped and in what locations. Boston Harbor is a strong example of how badly pollution can damage bodies of water. The water is filled with toxic waste and sewage and routinely receives more waste when rainfall pushes it into the harbor. Many bodies of water near urban areas are highly polluted which is the result of both garbage dumped by individuals and dangerous chemicals legally or illegally dumped by industries. The main problem caused by water pollution is that it kills life that inhabits water-based ecosystems. Dead fish, birds, dolphins and many other animals often wind up on beaches, killed by pollutants in their habitat. Pollution disrupts the natural food chain as well. Pollutants such as lead and cadmium are eaten by tiny animals. Later, these animals are consumed by fish and shellfish and the food chain continues to be disrupted at all higher levels. Eventually, humans are affected by this process as well. People can get diseases such as hepatitis by eating seafood that has been poisoned. Ecosystems can be severely changed or destroyed by water pollution. Many areas are now being affected by careless human pollution and this pollution is coming back to hurt humans.

45. WHAT ARE SOLUTIONS TO WATER POLLUTION?

Many laws have been made to restrict industries from dumping materials into the water. However, many laws remain weak and many countries do not restrict water pollution. In the United States, the Clean Water Act was written to completely put an end to all dumping of pollutants into water. The law has not been that effective in many areas, but in other locations, it has achieved its goals. Since the Clean Water Act, other legislation has been enacted as well. The world has spent tremendous sums of money trying to clean up water. From 1972-1990, the US spent over $250 billion. Many non-governmental projects are also being carried out in an effort to clean up the water. Industries are beginning to reduce the amount of chemicals they dump into water and environmental groups are participating in cleanup projects. The plastics industry, blamed for some of the worst pollution of the water, is making its products degradable. However, many environmentalists think this is hardly enough. Public reaction to the water pollution problem has also been influential. Unfortunately governments respond when public anger rids.

46. WHAT FACTORS INTENSIFY THE WATER SUPPLY PROBLEMS?

Almost 71% of the surface of the Earth is covered by Water. The substance is what makes the Earth unique; without it and its unique properties; life could not survive on the planet. Water also serves to maintain the climate and make pollutants less dangerous to living creatures. Approximately 97% of the earth's water resources are salt water, the remaining 3% being fresh water. Living things need water to survive. The weight of a human being is about 70% water and plants and animals are composed of anywhere from 50% to 97% water. Unfortunately for some people in the world, fresh water supplies are distributed in a highly unequal manner. While fresh water is abundant in some locations, other areas are extremely dry and in constant need of water supplies. What intensifies the problem of access to water resources is that only about .003% of the water in the world comes in surface water and groundwater. Groundwater is stored in aquifers, which can be accessed by people. Currently, water consumption rates are rapidly increasing. In some processes, such as irrigation, water is consumed, which means it cannot be reclaimed locally as surface water. Water shortage problems can be partially solved through the creation of reservoirs to store water, transferring water to areas that need it and gaining access to underground supplies of water. However, many problems related to water still exist. Some areas experience shortages of water, some are adversely affected by flooding, sometimes water is located in undesirable places, sometimes drinking water is unavailable, salinization of water occurs and desertification occurs where there are insufficient amounts of water.

47. WHAT DO PEOPLE MEAN BY ACID RAIN?

The term acid rain refers to what scientists generally call acid deposition. It is caused by airborne acidic pollutants and has highly destructive results. Scientists first discovered acid rain in 1852, when the English chemist Robert Agnus dubbed the term. From then until now, acid rain has been an issue of intense debate among scientists and policy makers. Though acid rain, one of the most important environmental problems of we all, cannot be seen, the invisible gases that cause this problem usually come from automobiles or coal-burning power plants. Acid rain moves easily, affecting locations far beyond those that let out the pollution. As a result, this global pollution issue causes great debates between countries that fight over polluting each other's environments.

While scientists have been studying the true causes of acid rain for years, majority of environmental scientists have concluded that production of polluting gases by humans was the primary cause, but many others have included natural causes as well. Recently, more intensive research has been done and countries and agencies concerned have the information that they need to prevent acid rain and its dangerous effects. The levels of acid rain vary from region to region. In Third World nations without pollution restrictions, acid rain tends to be very high. In Eastern Europe, China and the Soviet Union, acid rain levels have also risen greatly. However, because acid rain can move about so easily, the problem is definitely a global one.

48. WHAT ARE CAUSES OF ACID RAIN?

For many years, there was considerable debate and disagreement over what caused acid rain. Recent scientific work, however, has helped to clarify this. The primary causes of acid rain are sulfur dioxide and nitrogen oxides. These chemicals are released by certain industrial processes and as a result, the more industrialized nations of Europe as well as the US suffer severely from acid rain. Most sulfur dioxide comes from power plants that use coal as their fuel. Automobiles produce about half of the world's nitrogen oxides. As the number of automobiles in use increases, so does the amount of acid rain. Power plants that burn fossil fuels, also contribute significantly to emission of oxides of nitrogen.

Though human causes are primarily responsible for acid rain, natural causes exist as well as fires; volcanic eruptions, bacterial decomposition and lightening also greatly increase the amount of nitrogen oxide on the planet. However, even the gigantic explosion of Mt. St. Helens released only about what one coal power plant emits in a year. Once the tiny pollutant molecules enter the atmosphere, they can travel for thousands of miles. Eventually, the particles will combine with other compounds to produce new, often harmful, chemicals. Acid rain comes down to the earth in the form of rain, snow, hail, fog, frost, or dew. Once it reaches the ground, the acidity in the substance can harm and even destroy both natural ecosystems and man-made products, such as car finishes.

49. WHAT ARE EFFECTS OF ACID RAIN?

Acid rain is having harmful effects both on people and on the natural ecosystems of the world. Scientists today are convinced that acid rain is severe in many areas and that it is having an adverse effect on the environments of those locations. The problem of acid rain is rapidly spreading. Because it is mainly caused by industrial processes, automobiles and power plants, those countries that are developed have the most severe acid rain problems. However, as the undeveloped nations begin to industrialize, acid rain will increase greatly. Determining just how much the planet is being hurt by acid rain is very difficult because the ecosystems that it affects are so diverse and complex.

Many ecosystems are affected by acid rain. Bodies of water, such as lakes and rivers, see many of their inhabitants die off due to rising acidity levels. Acidic water also ruins plant nutrients, hurting plants' ability to survive and to give life to other organisms. Human-made products are also experiencing degradation from acid rain. Cars can lose their finishes and outdoor statues are beginning to rust that has ruined many historical artifacts and monuments. Acid rain's effects are destructive and long lasting. Though scientists have studied lakes, streams and many other natural ecosystems to prove its negative effects, acid rain continues to be produced and is increasing in many parts of the world.

50. WHAT ARE PROBABLE SOLUTIONS TO ACID RAIN PROBLEM?

As scientific studies have proved that acid rain is a dangerous and highly destructive problem, various ways to limit acid rain have been invented and some are now being used. Debate over acid rain and ways of preventing it have raged between environmentalists and corporations. Businesses such as power companies and car makers oppose controlling acid rain because they fear the effects on their profits. But in some cases, industries have attempted to curb acid rain production. The Northern States Power Company of U.S. began working to reduce acid rain in the 1980s and has invested over a billion dollars to that end.

There are many ways that power plant companies can adopt to reduce acid rain creation. They can use coal with low sulfur content, they can remove the sulfur from smoke their plants release and can limit processes known to generate high levels of acid rain. Policy makers and environmental experts are now looking into the best methods to limit acid rain. Environmentalists advocate the installation of sulfur cleaning scrubbers in factories, washing sulfur out of coal and finding new methods of burning coal. Power plant operators are looking for less expensive solutions to the problem. Individuals can help by conserving energy or driving their cars less. Governments can pass laws restricting pollution levels, or can use a variety of methods such as tradable emission permits to reduce acid rain. Whatever way it is done, acid rain will certainly have to be limited in the future.

51. WHAT ARE POSSIBLE SOLUTIONS TO THE RISKS ASSOCIATED WITH IRRIGATION?

While irrigation techniques have greatly enhanced agricultural output, such practices can sometime create more problems than they solve. There are several environmental as well as health problems relating to irrigation. Irrigation consumes a great deal of freshwater, thus reducing the drinking water supply for various populations. Irrigation sites also attract and encourage the breeding of mosquitoes, a situation that can result in the transmission of the diseases that these insects carry. Another very large concern is the use of wastewater for irrigation; this practice increases the risk of many sewage related diseases, including cholera and hepatitis. There are, however, a number of possible solutions to the risks associated with irrigation. Taking precautions with irrigation-scheduling and other simple improvements can result in at least short-term water and health conservation. Other possible solutions include strategic work with dams and channel-digging to regulate water flow.

Collaboration between water development planners and health authorities can also greatly lessen disease risks; by weighing economic costs against health consequences, compromises that are effective and fairly inexpensive can be reached that entail long-term benefits for all parties involved. If preventive steps such as these are not taken, other alternatives will be needed, such as developing chemical control strategies and treatment centers, to cope with already existing problems. There are several anti-risk strategies involving the use of wastewater. Using wastewater in the planting process, for example, has proved much less damaging than its use during growth cycles. Risks are also greatly reduced by discontinuing the use of wastewater some time before harvest. Finally, wastewater is obviously much less harmful when used with non-food products.

52. WHAT ARE CONSEQUENCES OF DESERTIFICATION?

Desertification is the expansion of desert lands into previously non-desert areas. The process occurs due to both natural and human causes. Deserts occur anywhere where there is less than 10 inches of rainfall each year and if an area receives only slightly more than this, its ecosystem is fragile and is a candidate for desertification. Long droughts can cause desertification to take place by removing the soil. If humans cause the loss of the soil, such as by overgrazing, over-cultivation, deforestation, irrigation and excessive plowing, desertification can follow. In many developing countries, problems such as high population growth and poor land use are causing desertification. Presently, desertification is occurring with greatest rapidity in the African Sahel. The desertification process can be a catastrophe for the inhabitants of an area. By preventing the growing of food, desertification can lead to famine and poverty. Already desertification has ruined huge chunks of land. Across the world, the total amount of land that has been lost is about equal to the size of Brazil (2 billion acres or 810 million hectares). Annually, 15 million acres (6 million hectares) are lost to desertification. Today, in most developed counties, trees are being replanted faster than they can be cut down, so desertification is not as severe a threat. In developing countries, by contrast, poverty has created a threat to trees, which can be sold or used as fuel. In these areas erosion is also common and along with deforestation, it is causing heavy desertification.

53. WHAT ARE ILLS OF OVERPOPULATION?

The world's population has been booming for years. The population is now threatening to reach the stage where there are simply too many people for the planet to support. Around 1850, the world population has reached one billion. By 1987, it was at five billion and still rising rapidly. Third World nations are responsible for a great deal of the population growth. In 1989, about 90% of the people being born were in developing countries. The populations of Third World countries are expected to continue to boom. The United Nations Population Fund predicts that by the middle of the 21st century, the world's population will stabilize at about 14 billion people. If fertility rates were decreased to 2.1 births per woman, population stabilization could be achieved sooner. In 1968, Paul Ehrlich published 'The Population Bomb'. The book described how the world population had risen, citing the discovery of agriculture as the cause. The book predicted that population growth would result in widespread famine and even nuclear war. While some of his predictions turned out to be false, famine did occur at high levels in later years. Overpopulation has been disastrous for the planet. Greater populations have polluted and consumed more, ruining the environment and creating or intensifying a variety of problems. Also, with the food supply limited, increase in population make shortages in many parts of the world even worse.

54. WHAT ARE CAUSES OF FAMINES?

Famine is defined as the starvation of a population of people. It can result in massive losses of lives. Famine can occur as a result of many factors. One frequent explanation is overpopulation. As the world grows at a rate of 250,000 people each day, serious strain is placed on the natural environment and the amount of food that it can produce. The growing world population and resulting need for food production can also cause depletion of topsoil, vegetation clearing and overgrazing. Natural disasters can also cause widespread famine. Situations of flood or drought greatly damage crops. These results in food shortages and increased prices for what little food is available. Also, crop destruction is accompanied by a shortage of employment and thus of the income needed to buy food, leaving many families no alternatives but to starve. A variety of other environmental conditions, including the depletion of natural resources, can cause widespread hunger and starvation.

55. HOW FOOD RESOURCES CAN BE INCREASED?

Though there is more than enough food available to feed everyone in the world at present and food supplies should not be a problem, many people do not have enough to eat. Every year, between 12 and 20 million people die from under nutrition, malnutrition, starvation and diseases caused by lack of food. Food problems come from food quality, quantity, storage and distribution, as well as other factors such as poverty. Rapidly rising population levels are increasing the need for food. To worsen the problem, the added people will have to live on land that could otherwise be used for food production. Food is grown differently in rich and poor nations. The developed world uses industrialized methods of food production, while less developed countries use methods more reliant on land and labor. Industrialized countries have learned how to increase food output from creating hybrid crops. In poorer countries, some governments have redistributed land on a more equal basis, decreasing starvation. Some people think that if dry lands and topical forests were converted into crop-growing land, food production could be greatly increased. However, other people disagree, arguing that these areas could not support farming. Neo-Malthusians advocate a sustainable-earth agricultural system in the developing countries. That system would mean land would be redistributed and human labor and natural fertilizers would be used in place of chemicals and industrial processes.

56. WHAT ENVIRONMENTAL ISSUES ARE RELATED TO MALNUTRITION?

Malnutrition is an enormous problem throughout the world. Although it is often a result of poverty, frequently it also results from environmental problems. Such problems can, for example, impair a family's capacity to produce food as a result of land degradation, water salinity, over-irrigation, soil erosion, droughts and flooding. These conditions may lead to widespread malnutrition. Malnutrition can take a serious toll on human health. In fact, it is estimated that malnutrition from all causes was responsible for twelve percent of all deaths worldwide in 1990. There are many sorts of deficiencies resulting from malnutrition. These include iron, iodine and vitamin A deficiencies, as well as the specific protein-energy malnutrition. Soil characteristics can play a large part in the development of these deficiencies by influencing which nutrients are present in the food of a region. Mountainous regions are known for iodine deficiencies; areas with weather extremes, such as arid, infertile, or wet, humid land, can be deficient in vitamin A; and many parts of Africa, South America and Asia suffer from protein-energy malnutrition, as well as all three deficiency types, due to various environmental conditions.

Each deficiency causes its own range of health problems. Iron deficiencies are known as anemia. This can cause hemorrhaging in instances such as childbirth, as well as physical productivity reductions. They can also undermine children's' learning capacity. Vitamin A deficiencies can cause loss of eyesight, ranging from deteriorating night vision to complete blindness. Many persons blinded die within the following year. Lack of iodine is a leading worldwide cause of brain damage and mental retardation. Iodine deficiencies can also lead to iodine deficiency disorders and goiter, enlargement of the thyroid gland. In pregnant women, this deficiency can cause brain damage in the developing fetus. Protein-energy malnutrition can do harm to the immune system, making it difficult for affected children to combat potentially damaging diseases, such as tuberculosis and malaria. Such diseases can worsen the deficiencies. Malnutrition can also heighten the effect of toxic chemicals on humans and various deficiencies can influence the body's absorption of substances such as lead, leading to more dangerous health hazards.

57. WHY ARE RENEWABLE ENERGY SOURCES DEPLETED?

There are three types of resources in the world: renewable, non-renewable and perpetual. Perpetual resources are not affected by human use of them. Examples are sunlight and wind. Renewable resources are those that are replenished through biogeochemical and physical cycles. By contrast, non-renewable resources do not replenish themselves, or, as in the case of fossil fuels, do so only at a very slow rate. Renewable resources are seldom perfectly renewable. If their levels are heavily decreased, they may not be able to completely replenish themselves. For example, if a species is nearly eradicated, it may not be able to avoid extinction.

Renewable resources can also be lost through pollution. Though water renews itself, if it is polluted, it is no longer useful for human use. Urban sprawl, cultivation, irrigation, grazing, deforestation, fishing, hunting and habitat destruction can all be causes of the destruction of an otherwise renewable resource. There are both organic and inorganic types of renewable resources. Examples of renewable organic substances include plant and animal species. Inorganic examples include water and certain gases like oxygen. During different periods of time, different things are thought of as resources. As the usefulness of certain resources like wind has varied, their classification has changed too. There have been numerous efforts to prevent the mistakes that lead to the depletion of renewable resources.

58. HOW IS SOLAR ENERGY GENERATED?

Solar power is an extremely clean and renewable form of energy. It can be used both in automobiles and in generating power for consumption in other ways. Many people credit Sun Day, May 3, 1978, as the beginning of the movement into the Solar Age. On that day, 20 million people worldwide learned about ways to utilize solar energy to generate power. Immediately, governments responded by writing laws favorable to the development of solar energy. However, hope that solar energy would be used highly in the future decreased in the 1980s as US President Ronald Reagan decreased support for solar energy projects.

The generation of solar energy is relatively simple. A box with a heat-trapping dark bottom and glass on top holds the sun's heat. The devices can be easily deployed anywhere. One major advantage of solar energy is that it is available to everyone and can be harnessed by individuals everywhere. Anyone can set up a solar collector to obtain solar energy, making power distribution across large areas unnecessary. Though solar energy might save money in the long run, initial costs are a major obstacle. However, many people have found solar energy use to be profitable overall. Many solar developments have been made throughout the world. Solar water heaters are used in 90% of houses in Cyprus, 65% in Israel and are also common in Japan and Australia.

59. WHAT ARE BENEFITS AND DRAWBACKS OF HARNESSING OF WIND ENERGY?

Harnessing the power of the wind was first accomplished long ago. In Persia in 600 BC, wind was used to lift water. Later, ships were powered by the wind and windmills were used throughout Europe. In the 1890s, wind was used to create electricity. Farms used windmills for obtaining water from wells deep in the ground. However, as supplies of gas and oil became abundant and cheap, people switched to using those resources for their power needs. Interest in wind came about because, like solar, it is a very clean form of energy. The oil shocks of the 1970s furthered interest in wind and other alternative energy sources. However, wind research also suffered a major setback in the early 1980s. Despite research and development cuts, there is considerable wind research and usage today. From 1973-1983, over 10,000 wind machines were constructed throughout the world. Countries such as Denmark, Sweden, the US and many other industrialized nations have deployed wind technology. In total, 95 countries are now using windmills. Often, power is gained through wind farms, large groups of wind turbines. China and India are also working to develop and use wind equipment to gain energy. A few people have objected to wind energy. Some say the wind farms are too noisy and cause traffic congestion nearby. Others complain that windmills kill birds.

60. WHAT IS CORRELATION BETWEEN RENEWABLE ENERGY AND SUSTAINABLE ECONOMICS?

Some environmentalists and economists argue in favor of a sustainable economy. This system has a variety of implications, but basically means that the population and quantity of goods would remain constant over time. First, it means that the society would desire a constant GNP, with some sectors growing and others declining at an equal and opposite rate. Overall, the economy would continue to achieve the same level of output each year. Those sectors of the economy that increased the sustainability of the environment, such as renewable energy or the production of long-lasting goods, would be encouraged to grow. If new technology made growth possible without decreasing sustainability, growth would be allowed. Second, the country would try to maximize the use of renewable resources. This would mean relying on wind, water, the sun, or another renewable resource for power production, rather than burning fossil fuels, of which there are limited supplies. Third, the sustainable economy would seek to achieve economic and ecological sustainability. It would view biological capital as being of equal importance to financial capital.

Fourth, an incentive system would be used to reward those who minimize the toll they exert on the environment and produce long-lasting items for human use. Materialism would be discouraged and people would be encouraged to use only essential goods and services. Rather than consuming our natural resources, the sustainable economy would seek to use what we have already taken and leave natural resources as a backup supply. Finally, a sustainable economy would attempt to build a more egalitarian society. While trying to minimize the number of goods people used, the economy would also try to equalize what goods people had.

61. WHY IS NORTH AMERICA A LEADING CONTRIBUTOR TO POLLUTION?

North America includes the United States and Canada, two of the world's most modern and wealthiest countries. Together, the United States and Canada account for 14% of the world's land, covering 18.8 million square miles. Canada and the US happen to be tremendously rich in natural resources. Vast supplies of resources of all types have contributed heavily to economic growth in the region. Economic growth, industrialization and major production of goods have all meant that North America is a leading contributor to waste and pollution. Now, however, the area is working towards limiting degradation of the environment. In both Canada and the United States, a great deal of public attention has turned towards the environment. Many environmental organizations have sprung up and are educating people about the possible dangers of current ways. The governments of these nations have also taken a strong interest in the environment. For many years the United States wrote environmental laws such as the Clean Air Act and the two nations restricted pollution and other harmful processes. Now, the countries are beginning to work together to save the environment. When NAFTA came into force in 1994, the Commission for Environmental Cooperation (CEC) was established. The CEC works on environmental conservation, protecting human health and the environment, trade and economy, enforcement co-operation and environmental law and information and public outreach.

62. WHAT ENVIRONMENTAL ISSUES ARE SPECIFIC TO THE THIRD WORLD NATIONS?

The Third World refers to the poorer and undeveloped countries of the world. Often, these countries have extremely poor environmental situations. In many Third World nations, pollution is unrestricted. Countless other environmental problems are also not addressed by the government. Usually, creating and enforcing environmental regulations would be economically disastrous for a poor country. As a result, it is forced to choose between buying food and having a clean environment. The more pressing concern is almost always the former. Often, rich Western countries take advantage of the dilemma of Third World countries. They dump garbage and hazardous waste in developing countries. First World companies might also build plants, which emit considerable pollution, in Third World nations to avoid the regulations these companies would face at home. Some transnational corporations that produce chemicals deemed overly dangerous in the First World find a market in the Third World. There, governments cannot restrict usage of these chemicals because it would be too costly to citizens trying to make a living. Countries in Latin America, Eastern Europe, Africa and Asia are the greatest victims of this environmental inequality. In addition to problems created by development and industrialization, poorer nations also suffer environmental difficulties caused by poverty and war, among other causes. Many environmental problems arise in the Third World. Air pollution, water pollution , deforestation, desertification, soil erosion and poisoning of the environment are among the largest of these. Third World nations are aware of these problems and are working to solve them. The United Nations and other international organizations have looked into preventing foreign companies from dumping waste in developing nations, making transnational corporations pay for the pollution they produce and enlisting the First World in helping to clean up the Third World's environment.

63. WHY A VARIETY OF ENVIRONMENTAL PROBLEMS ARE THRIVING IN EASTERN ASIA?

Eastern Asia and the Pacific areas are some of the most highly populated parts of the world. With about 1.2 billion people, China is the world's most populous country. Other highly populated countries such as India and Indonesia give Eastern Asia a massive population and one that is responsible for many environmental problems associated with overpopulation. In total, Eastern Asia contains about 58% of the global population, though it has only 23% of the world's land. The region includes East Asia, Southeast Asia and the Pacific Rim. It stretches from Mongolia in the north down to New Zealand in the south. The Pacific Ocean, the largest in the world, is located in Eastern Asia. The area also contains the Indian Ocean, the world's third largest. Eastern Asia tends to be very poor, with a large number of people living in poverty. Hundreds of millions of people suffer through poverty in the area, resulting in many environmental difficulties. In the early years of the 1990s, there was great economic growth across the region. Countries like China mounted impressive growth figures and began to industrialize. Now, industrialization has brought the harmful environmental affects commonly associated with it. A variety of environmental problems are especially threatening Eastern Asia. Destruction of forests, land degradation, deforestation, marine pollution and limited water supplies are all having a negative affect on the environment and its human inhabitants.

64. WHY IS IT SAID THAT THE EXTINCTION PROBLEM REMAINS SEVEREST TODAY?

When the last members of a species are killed, it becomes extinct. Natural causes have always been a source of extinction, but many human processes have intensified the rate at which extinction occurs. Mass extinctions have happened about every 50 million years. 250 million years ago, as many as 96% of all species inhabiting the earth died off, the greatest mass extinction ever. Of all the creatures that have inhabited the earth throughout history, 99.9% have gone extinct. Long ago, species went extinct due to volcanic eruptions, habitat destruction, food supply loss, or asteroids and meteors that hit the earth.

Though extinctions have always occurred, the rate at which the process is occurring in modern times is astounding. According to one estimate, in the last 400 years the earth has lost 100 types of mammals and 258 bird species. In the course of history, many animal species have been killed off by humans. In 80 AD, the European lion became the first species in recorded history to become extinct. Early North American settlers wiped out tens of millions of passenger pigeons, destroying the species by 1914. These settlers also wiped out species of grizzly bears and wolves and one type of fox and cougar. Since the arrival of the Pilgrims, more than 500 plant and animals species in North America have become extinct. In 1988, Harvard professor Edward O. Wilson estimated that 17,500 species were going extinct annually. It is hard to know exactly how extinction will affect humans. Though different scientists and environmental activists have different opinions on the issue, they agree to the fact that the extinction problem remains severest today.

Chapter 20

SOME MATHEMATICAL CONCEPTS

1. WHAT IS SIGNIFICANCE OF 'PI(π)?

The unique number 'pi' is defined as the ratio of a circle's circumference to its diameter. This number is independent of the size of a circle and for all practical purposes its approximate value is taken as 22/7 or 3.1416. In fact, the fraction 22/7 is slightly greater in value than 'pi'. For many centuries mathematicians have been fascinated by its unique characteristic. The strangest thing about this number is that nobody has been able to calculate its exact value. Computer scientists have now computed pi to over one million decimal places.

At one time the scientists tried to prove that 'pi' was a fraction. When any fraction is written in a decimal number, the same digits always appear over and over again in a special pattern. If 'pi' were a fraction, there would be a repeating pattern to its digits. But strangely enough a repeating pattern in 'pi' could not be found. Finally in 1761, a Swiss mathematician named Johann Heinrich Lambert settled the matter once and for all. He proved that pi is not a fraction. Pi is the sixteenth letter of the Greek alphabet and the abbreviation used to name a relationship, the ratio between the circumference of a circle and that circle's diameter. As such, it is a mathematical constant. Pi allows us to find the circumference of the circle if we know the radius or diameter and allows us to find the radius and diameter if we only know the circumference. The Greek form, ð, is often used in mathematics and scientific settings.

Pi is given as different figures, both for ease and because the calculation of pi has improved over time. In middle school or junior high school, when it is first introduced to students, a value of 3.14 may be used. As students mature, they may be given the value 3.14159 and then 3.14159265. The fact is that pi had been calculated to 707 places in the nineteenth century and in the twenty-first century, we can claim that it has been calculated to 200 billion digits. Pi is an irrational number, which means that it is a real number with a non-repeating decimal expansion: it cannot be represented by an integer ratio.

In 1882, British physicist Ferdinand Lindemann showed that ð is a transcendental number, which means that it cannot be the root of any algebraic equation with rational coefficients. The problem of trying to construct a square that has equal area to a given circle had been proposed in 1800 B.C. and Lindemann used his discovery to prove what had been suspected: that "squaring the circle" is impossible using only straight edge and a compass. As an acronym, PI can also refer to a number of other concepts. On the list are: private investigator, politically incorrect, protease inhibitor, polynomial identity and many other terms.

Now the question arises what is the significance of 'pi' in our daily life? Suppose we have an automobile tyre that has diameter of one meter. If we want to find its circumference, we can find it out by measuring with a tape. Another way of finding the circumference is to multiply the diameter by this strange number 'pi'. This number is used to calculate the circumference of all the circular objects. The mathematicians are still engaged in research in this direction to see if the digits are arranged in any special way.

2. WHAT IS AN INTEGER?

An integer is what is more commonly known as a whole number. It may be positive, negative, or the number zero, but it must be whole. In some cases the definition of whole number will exclude the number zero, or even the set of negative numbers, but this is not as common as the more inclusive use of the term. Integers are the numbers people are most familiar with and they serve a crucial role in virtually all mathematics. To understand what an integer is, that is to say, why it is different than simply a 'number', we must look at the other sets of numbers that can exist. Many of these sets overlap with the integer set in some areas and some are virtually identical. Others have very little in common with any integer; these sorts of numbers tend to be much less familiar to most people.

The subset of positive integers is probably the oldest set of numbers. This group is often referred to as the set of counting numbers, since these are the numbers used to count things and ideas. The numbers in the positive integer set are all of the whole numbers above zero. So the set would be listed as {1, 2, 3, 4 ...} and so on, forever. Like the set of integers itself, the positive integers are infinite. Since people have been counting as far back as we know of, this set has also existed for a very long time. Although it may not have been known to be infinite, the set was still essentially the same.

A very closely related set is the set of all non-negative integers. This set is identical to the set of positive integers, except that it also includes the integer zero. Historically, the number zero was an innovation that came about quite a bit after the counting numbers had been in wide-spread use. Both of these sets may be referred to as the set of natural numbers. Some mathematicians prefer to exclude the integer zero from the natural numbers, while others find it useful to include it. If we consider the more inclusive definition, we can then define an integer as any member of the set of natural numbers, as well as their negative counterparts.

Beyond the integer we find other sets that are more complicated. The next logical progression is the set of all rational numbers. A rational number is any number that can be discussed as a ratio between two integers. This means that an integer itself would be rational, 2/2 is a ratio, but is also simply equal to 1, while 8/2 is also a ratio and also equal to the integer 4. It also means that fractions are rational numbers, 3/4 is not an integer, but it is a rational number.

The next step out would be the set of real numbers. These could most easily be described as any number which could be placed on a number line. This would include any integer, as they can obviously be placed on a number line as well. It would also include any rational number, as fractions can be placed on a number line. It further includes numbers which cannot be expressed simply as the ratio between two numbers, for example, the square root of two produces a string of digits after the decimal place which go on infinitely, so it can never be adequately described as a rational number, but it is a real number. The final set of numbers commonly dealt with is the set of complex numbers, sometimes described as irrational numbers. These numbers have no actual place on a number line, but have a use in many mathematical processes nonetheless. Complex numbers include an imaginary component, usually given as i, where i^2 is equal to -1.

As we can see, there are many different types of numbers and each have their place in the world of mathematics and the many disciplines in which it is used. An integer can best be described both by what it is and by what it is not. An integer is any whole, positive number, from one to an infinitely-large number. An integer is the number zero. An integer is any whole, negative number, from negative one to an infinitely-large negative number. An integer is not any number which has a remainder beyond the decimal place. An integer is not a special real number, such as pi or e. And an integer is not a complex or irrational number too.

3. WHAT IS UNIQUE WITH WHOLE NUMBERS?

The term ‘whole number’ is one, we find often in mathematics. We may also see the term integer or natural number used in the same manner. Essentially the definition of a whole number is based around what it doesn’t contain. A whole number can’t be a fraction of a number, a percentage, or have a decimal. If we have a number like 21.32, it has a whole number portion (21). But in itself, this number is not a whole number because it contains a fraction (.32).

We also hear natural numbers referred to as counting numbers. They’re really the first numbers children learn. They include zero, though it has some unique properties and every number that is not fractional below and above zero. The sequence 1,2,3,4… is a whole number sequence. Similarly 0, -1, -2, -3, -4 are also natural numbers. The definition for whole number may seem unnecessary to some, but in fact in early math we soon begin teaching children the properties of integers. We teach, for instance the associative property in addition, which means that when we add real numbers together, it doesn’t matter if they’re grouped differently into parentheses. (1 + 2) + 3 is equal to

1 + (2 +3). Other properties applying to whole numbers with addition and multiplication include the following:

Identity Property: Any whole number added to zero equals that number (0 + 23 = 23).

Commutative Property: Order doesn't matter when multiplying or adding two integers (3 × 4 = 4 × 3) (3 + 4 = 4 + 3).

Integers also matter when we ask students to round answers. At some point in our lives, we all need to figure out a little math in our minds, especially when we buy things. If we look at a number with a fraction, for instance the price Rs. 29.95 for lunch at a restaurant, we may need to figure out how to tip. While some people may just want to tip to the exact penny, others merely round up or down to the nearest counting number to figure tip. Thus Rs. 29.95 is rounded up to Rs. 30. If we tip 20%, we tip Rs. 6. It's sometimes more practical and simpler to use whole numbers in everyday applications of math.

Even organizations like the Internal Revenue Service (IRS) would prefer to work with integers instead of decimal amounts. We can round up or down when our deductions, income and et cetera aren't whole numbers. We might also discover that some people employ rounding up only when it comes to disbursements and only round down to a counting number when estimating their total taxable income.

While many children contend that mathematics is a worthless pursuit, we find ourselves lost without a basic understanding of the counting number. We use these constantly, no matter what our occupation. Real numbers answer such basic questions as how many people are coming to dinner, or how many hours are there in a day, days in a year, or minutes in an hour. We consider whole numbers as they relate to how many apartment complexes are in a building, the number of available parking spaces, or the cartons of milk we need to buy. People often laugh at the use of decimal numbers in various statistical reports, such as in the following: the average family has 2.5 children. Our minds immediately convert this to a counting number, since there is no such thing as a .5 child. We read this in whole numbers too, as the following: the average family has two to three kids.

4. WHAT IS A NATURAL NUMBER?

A natural number, which can also be called a counting number, is represented by the digits from 0, 1, 2, 3... through to infinity. In math law, there must be an infinite number of natural number digits, since each natural number is defined in part by having a number that follows it. These numbers are also whole numbers, not fractions or decimals and can be used for counting or ordering. The main distinction between a natural number and an integer is that natural numbers are only positive. There is no number below zero and a natural number can't be followed by zero, such as in the case with -1, 0. Essentially this defines natural numbers as anything zero or above that is whole and not fractional.

The concept of zero evolved long after civilizations began using counting numbers. Earliest records of counting numbers (from 1-10) date back to more than 4000 years ago when the use of specific written code to signify place were used by the Babylonians. The Egyptians wrote hieroglyphs for each digit, but it wasn't until about 1000 BC that the concept of zero was created by the Mayan and Olmec civilizations. Though the Olmec and Mayan groups show the first records of the use of zero, the concept of zero also developed in India, in the 7th century BCE. It was the Indian use, rather than Mesoamerican use that was adopted by civilizations like the Greeks.

There are many ways in which natural numbers can be used in math applications. They can limit problems by suggesting that the answer must be a natural number. They are also studied in specific application in set theory, mathematics that evaluates sets of things. Number theory may evaluate natural numbers as part of the set of integers or independently to see if they behave in certain ways or exhibit certain properties. Perhaps one of the widest uses of natural numbers comes to us very "naturally." When we are young we learn to count from 0 onward. Even young children can easily begin to learn the difference between one and two, or explain how old they are. This study continues as children start going to school and learn to manipulate natural numbers, how to multiply, divide, add and subtract them. Only after the concept of natural numbers is learned, is the concept of integers introduced and the possibility of negative numbers, which can confound some kids at first, is usually learned in fourth or fifth grade at earliest.

5. WHAT IS UNIQUENESS OF ZERO?

Zero is a fascinating little number and it has some very distinctive properties. Ever since zero was invented, mathematicians have struggled to define it and to use it in their work, with the properties of zero being arrived at through the use of mathematical proofs which are intended to illustrate those properties at work. Even with proofs to support the rationale behind some of the properties of zero, this number can be quite slippery. People haven't always used zero. A crude form of zero as a placeholder appears to have been used by Babylonian mathematicians, but Indian mathematicians are usually credited with coming up with the idea of zero as a number, rather than just a placeholder. Almost immediately, people struggled to define the number and learn how it worked and explorations into the properties of zero got quite complex.

Numbers can be classified as positive or negative, depending on whether they are greater or less than zero, but zero itself is neither of them. Zero is also even, something which comes as a surprise to some people when they learn about the properties of zero, as they often assume that it is either odd or outside of the even/odd dichotomy. In fact, extensive math could be used to show us how zero is classified as even, but the simplest way to show how zero is even is to think about what happens when we have a multiple digit number which ends in an even number. 1002 ends in a 2, an even number, so it is considered even. Likewise the same is true with 368, 426 and so forth. Numbers which end in zero are also treated as even, illustrating that zero is itself even.

The Addition Property of Zero states that adding 0 to a number does not change that number. 37+0 equals 37, for example. In the Multiplication Property of Zero, mathematicians state that multiplying a number by zero always ends in zero: if we multiply six oranges zero times, we end up with no oranges. Some other properties of zero are with addition and subtraction. While subtraction of a positive number from zero ends in a negative number, subtraction of a negative number from zero ends in a positive.

Zero has another property which is familiar to anyone who has tried to divide a number by zero with a graphing calculator. Division by zero is simply not allowed in mathematics and if we attempt it, a calculator usually returns the message "undefined," "not allowed," or simply "error." The Indians actually tried very hard to prove that one could divide by zero, but they were unsuccessful. However, we can divide zero by other numbers (although not by zero), although the result is always 0. 0/6, for example, equals 0.

6. WHAT IS AN IMAGINARY NUMBER?

An imaginary number is a mathematical term for a number whose square is a negative real number. Imaginary numbers are represented with the letter *i*, which stands for the square root of -1. This definition can be represented by the equation: $i^2 = -1$. Any imaginary number can be represented by using *i*. For example, the square root of -4 is $2i$. When imaginary numbers were first defined by Rafael Bombelli in 1572, mathematicians believed that they did not really exist, hence their name. Decartes coined the term imaginary in reference to these numbers in his 1637 book, La Geometrie. However, imaginary numbers are as real as any other numbers and have gradually come to be accepted by the mathematical community and the world at large. The work of mathematicians Leonhard Euler and Carl Friedrich Gauss in the 18th and 19th centuries was instrumental in this change.

While imaginary numbers are meaningless in the "real world" of most individuals, they are indispensable in such fields as quantum mechanics, electrical engineering, computer programming, signal processing and cartography. For perspective, consider that negative numbers were also once considered fictitious and that such concepts as fractions and square roots might be considered meaningless to a person who does not need them in everyday life, though they are quite real to others.

To better understand imaginary numbers, geometry may be helpful. Picture a standard number line: zero is in the center, positive numbers are found to the right of zero and negative numbers are found to the left. At the zero point, visualize another line perpendicular to the first, stretching up and down rather than right and left. This is the axis of imaginary numbers, also known as the y-axis in geometry, while the "standard number line" is the x-axis. Positive imaginary numbers extend up from the zero point and negative imaginary numbers extend down. Zero is the only number that is considered both real and imaginary.

7. WHAT IS MEANT BY 'SUBTRACTIVE NOTATION' IN ROMAN NUMERALS?

Roman numerals are a numbering system which was widespread throughout Europe until approximately 900 CE, when Arabic numerals supplanted Roman numerals for most applications. Roman numerals can still be seen on formal documents to mark the date or being used to iterate simple numerals, such as those used on an outline. For mathematical purposes, however, Roman numerals have long since been discarded, because they are clumsy and difficult to work with, in comparison to Arabic numerals.

The inspiration for Roman numerals can be found in Attic numerals, which were used in Greece from around the seventh century BCE. Attic numerals used symbols to represent the numbers: 1, 5, 10, 100, 1,000 and 10,000. The symbol representations probably came from tally sticks which would have been used to mark down goods as they were counted. The Etruscans located in what is now Italy picked up the Attic numeral system and adapted it for their own use, creating symbols to represent 1, 5, 10, 50 and 100.

The Romans changed the symbols used for Etruscan numerals in addition to adding a few. Under the system of Roman numerals, I represents one, V stands for five, X means 10, L is used for 50, C marks 100, D indicates 500 and M is 1000. A bar placed over a symbol multiplies its value by 1,000. All numbers in Roman times would have been written out using these symbols. 17, for example, would have been written as XVII. As can be seen, Roman numerals require the reader to add the symbols together in order to discern the number being represented. This can be quite time consuming, especially with large numbers such as MMMDCXIII, which reads as 1,000 + 1,000 + 1,000 + 500 + 100 + 10 + 1 + 1 + 1, or 3,613.

However, numbers such as four are not, by convention, written out in Roman numerals as IIII. Instead, a system called subtractive notation is used. Subtractive notation means that if a symbol of lesser value is placed in front of a symbol of higher value, the lesser symbol should be subtracted from the greater one. Hence, IV means four, just as MCMLXXXIV means 1984. When using subtractive notation, only multiples of 10 are used, so VC for 95 would be incorrect and XCV would be proper. It is also considered improper to skip orders of magnitude when using subtractive notation, meaning that XM would not be used to represent 990, but CM could be used to indicate 900.

8. WHAT IS DIFEERENCE BETWEEN PRECISION AND ACCURACY?

Accuracy is how close an experimental reading or calculation is to the true value. Lack of accuracy may be due to error or due to approximation. The less is the total error in an experiment or calculation, the more accurate the results will be. Error analysis can provide information about the accuracy of a result. Errors in experiments stem from incorrect design, inexact equipment and approximations in measurement. Imperfections in equipment are a fact of life and sometimes design imperfections are unavoidable as well. Approximation is also unavoidable and depends on the fineness and correctness of the measuring equipment. For example, if the finest marks on a ruler are in centimeters, then the final measurement is not likely to have an accuracy of more than half a centimeter. But if the ruler includes millimeter markings, then the measurements can be an order of magnitude more accurate.

Accuracy differs from precision. Precision deals with how repeatable measurements are, if multiple measurements return numbers that are close to each other, then the experimental results are precise. The results may be far from accurate, but they will be precise. Approximations are also unavoidable in calculations. Neither people nor computers can provide a totally accurate number for 1/3 or pi or any of several other numbers and often the desired accuracy does not require it. The person calculating does, however make sure that the approximations are sufficiently small that they do not endanger the useful accuracy of the result. Accuracy becomes an issue in computations because rounding errors accumulate, series expansions are attenuated and other methods that are not analytical tend to include errors.

If we buy several items, all of which are subject to a sales tax, then we can calculate the total tax by summing the tax on each item. However, for the total tax to be accurate to the penny, we must do all the calculations to an accuracy of tenths of a penny (in other words, to three significant digits), then round the sum to the nearest penny (two significant digits). If we calculated only to the penny, then each measurement might be off by as much as half a penny and the total possible error would be this amount multiplied by the number of items bought. If we bought three items, then the error could be as large as 1.5 cents; if we bought 10 items then our total tax could be off by as much as five cents.

As another example, if we want to know the value of pi to an accuracy of two decimal places, then we could express it as 3.14. This could also be expressed as 3.14 +/- 0.005 since any number from 3.135 to 3.145 could be expressed the same way, to two significant decimal points. Any calculations using a number accurate to two decimal places are only accurate to one decimal place. In a similar example, the accuracy of a table can either refer to the number of significant digits of the numbers in a table or the number of significant digits in computations made from the table.

9. WHAT IS IMPORTANCE OF APPROXIMATION?

In mathematics, making an approximation is the act or process of finding a number acceptably close to an exact value; that number is then called an approximation or approximate value. Approximating has always been an important process in the experimental sciences and engineering, in part because it is impossible to make perfectly accurate measurements. Approximation also arises because some numbers can never be expressed completely in decimal notation. In these cases approximations are used. For example, irrational numbers, such as pi (π), are nonterminating and nonrepeating decimals. Every irrational number can be approximated by a rational number, simply by truncating it. Thus, π can be approximated by 3.14, or 3.1416, or 3.141593 and so on, until the desired accuracy is obtained.

Another application of the approximation process occurs in iterative procedures. Iteration is the process of solving equations by finding an approximate solution, then using that approximation to find successively better approximations, until a solution of adequate accuracy is found. Iterative methods, or formulas, exist for finding square roots, solving higher order polynomial equations, solving differential equations and evaluating integrals.

The limiting process of making successively better approximations is also an important ingredient in defining some very important operations in mathematics. For instance, both the derivative and the definite integral come about as natural extensions of the approximation process. The derivative arises from the process of approximating the instantaneous rate of change of a curve by using short line segments. The shorter the segment is the more accurate the approximation will be, until, in the limit that the length approaches zero, an exact value is reached. Similarly, the definite integral is the result of approximating the area under a curve using a series of rectangles. As the number of rectangles increases, the area of each rectangle decreases and the sum of the areas becomes a better approximation of the total area under investigation. As in the case of the derivative, in the limit that the area of each rectangle approaches zero, the sum becomes an exact result.

Every function can be expressed as a series, the indicated sum of an infinite sequence. For instance, the sine function is equal to the sum: $\sin(x) = x — x^3/3! + x^5/5! — x^7/7! + ...$ In this series the symbol (!) is read "factorial" and means to take the product of all positive integers up to and including the number preceding the symbol ($3! = 1 \times 2 \times 3$ and so on). Thus, the value of the sine function for any value of x can be approximated by keeping as many terms in the series as required to obtain the desired degree of accuracy. With the growing popularity of digital computers, the use of approximating procedures has become increasingly important. In fact, series like this one for the sine function are often the basis upon which handheld scientific calculators operate.

An approximation is often indicated by showing the limits within which the actual value will fall, such as 25 ± 3, which means the actual value is in the interval from 22 to 28. Scientific notation is used to show the degree of approximation also. For example, 1.5×10^6 means that the approximation 1,500,000 has been measured to the nearest hundred thousand; the actual value is between 1,450,000 and 1,550,000. But 1.500×10^6 means 1,500,000 measured to the nearest thousand. The true value is between 1,499,500 and 1,500,500.

10. HOW IMPORTANT ARE ERRORS AND WHAT IS CORRELATION BETWEEN ERROR AND UNCERTAINTY?

Error is the amount of deviation in a physical quantity that arises as a result of the process of measurement or approximation. Another term for error is uncertainty. Physical quantities such as weight, volume, temperature, speed, or time must all be measured by an instrument of one sort or another. No matter how accurate the measuring tool is, be it an atomic clock that determines time based on atomic oscillation or a laser interferometer that measures distance to a fraction of a wavelength of light some finite amount of uncertainty is involved in the measurement. Thus, a measured quantity is only as accurate as the error involved in the measuring process. In other words, the error, or uncertainty, of a measurement is as important as the measurement itself.

As an example, imagine trying to measure the volume of water in a bathtub. Using a gallon bucket as a measuring tool, it would only be possible to measure the volume accurately to the nearest full bucket, or gallon. Any fractional gallon of water remaining would be added as an estimated volume. Thus, the value given for the volume would have a potential error or uncertainty of something less than a bucket.

Now suppose the bucket were scribed with lines dividing it into quarters. Given the resolving power of the human eye, it is possible to make a good guess of the measurement to the nearest quarter gallon, but the guess could be affected by factors such as viewing angle, accuracy of the scribing, tilts in the surface holding the bucket, etc. Thus, a measurement that appeared to be 6.5 gal (24.6 l) could be in error by as much as one quarter of a gallon and might actually be closer to 6.25 gal (23.6 l) or 6.75 gal (25.5 l). To express this uncertainty in the measurement process, one would write the volume as 6.5 gallons +/-0.25 gallons.

As the resolution of the measurement increases, the accuracy increases and the error decreases. For example, if the measurement were performed again using a cup as the unit of measure, the resultant volume would be more accurate because the fractional unit of water remaining, less than a cup, would be a smaller volume than the fractional gallon. If a teaspoon were used as a measuring unit, the volume measurement would be even more accurate and so on.

As the example above shows, error is expressed in terms of the difference between the true value of a quantity and its approximation. A positive error is one in which the observed value is larger than the true value; in a negative error, the observed value is smaller. Error is most often given in terms of positive and negative error. For example, the volume of water in the bathtub could be given as 6 gallons +/-0.5 gallon, or 96 cups +/-0.5 cup, or 1056 teaspoons +/-0.5 teaspoons. Again, as the uncertainty of the measurement decreases, the value becomes more accurate. An error can also be expressed as a ratio of the error of the measurement and the true value of the measurement. If the approximation were 25 and the true value were 20, the relative error would be 5/20. The relative error can also be expressed as a percent. In this case, the percent error is 25%.

Measurement error can be generated by many sources. In the bathtub example, error could be introduced by poor procedure such as not completely filling the bucket or measuring it on a tilted surface. Error could also be introduced by environmental factors such as evaporation of the water during the measurement process. The most common and most critical source of error lies within the measurement tool itself, however. Errors would be introduced if the bucket were not manufactured to hold a full gallon, if the lines indicating quarter gallons were incorrectly scribed, or if the bucket incurred a dent that decreased the amount of water it could hold to less than a gallon.

In electronic measurement equipment, various electromagnetic interactions can create electronic interference, or noise. Any measurement with a value below that of the electronic noise is invalid, because it is not possible to determine how much of the measured quantity is real and how much is generated by instrument noise. The noise level determines the uncertainty of the measurement. Engineers will thus speak of the noise floor of an instrument and will talk about measurements as being below the noise floor, or "in the noise."

Measurement and measurement error are so important that considerable effort is devoted to ensure the accuracy of instruments by a process known as calibration. Instruments are checked against a known, precision standard and adjusted to be as accurate as possible. Even gas pumps and supermarket scales are checked periodically to ensure that they measure to within a predetermined error. Nearly every country has established a government agency responsible for maintaining accurate measurement standards.

11. WHAT IS A 'SYMMETRICAL POLYHEDRON'?

A polyhedron is a three dimensional shape bounded by a series of flat planes that are known as faces. A simple example of a polyhedron is a cube, a six-faced polyhedron made with square faces. There is actually some dispute in the math world about what constitutes a polyhedron, exactly and some mathematical disciplines use the term "polyhedron" to describe a very specific type of shape. For the most part, however, the definition of a three dimensional shape characterized by having flat faces is reasonably accurate. Each face of a polyhedron is a polygon, another type of geometric figure. Polygons are flat figures made with a series of connecting line segments which form a closed shape. A rectangle is a simple polygon, with a star being a more complex example of a polygon. The polygons in a polyhedron can vary in size, shape and arrangement considerably and when all of the polygons are identical, the resulting polyhedron is known as a "symmetrical polyhedron."

As a general rule, a polyhedron is named according to the number of faces it has. An octahedron has eight faces; a dodecahedron has 10 and so forth. Sometimes, descriptive terms about the shape of the polyhedron will be added as well. A pyramid, for example, is a special type of four or five sided polyhedron. Stellated polyhedrons have numerous pyramidal outcroppings called stellations which create a series of points, causing the polyhedron to take on a three-dimensional star shape. Polyhedra can be convex, which means that a line drawn between any two points on the shape will pass through the shape without straying beyond its boundaries, or non-convex, in which the line passes outside the polyhedron. A rectangular block is an example of a convex polyhedron: if we draw an imaginary line between any two points on the block, it will pass through the block. By contrast, a stellated polyhedron is nonconvex, because points which go outside the body of the polyhedron can be drawn between the stellations.

People interact with polyhedrons every day and the study of polyhedra is critical to a wide variety of industries. For example, many products are packaged in polyhedral containers, with the containers being designed for optimum manufacturing efficiency and stackability. Many homes are polyhedra, created with a series of flat planes which have been used to create an enclosed space. The study of polyhedra tends to be introduced to children at a young age, since understanding polyhedra can be a key to grasping other mathematical concepts.

12. WHAT IS SPECIAL WITH A TETRAHEDRON?

A tetrahedron is a type of polyhedron which has four faces, making it the smallest possible type of polyhedron. This geometric figure is the basis for a wide variety of geometry problems and examples of tetrahedra can be seen in architecture, the arts and even daily life. In fact, chances are very good that there's a tetrahedron in our vicinity. To understand the tetrahedron, it is desirably necessary to discuss a few key terms in geometry. A polygon is a flat or "planar" shape created with a series of connecting line segments: a triangle, for example, is a polygon. A polyhedron is a three-dimensional object composed of multiple polygons which meet to form straight edges. A well-known example of a polyhedron is a cube, a six-sided polyhedron. If the edges are curved, as in the case of a cylinder, the shape is no longer a polyhedron.

In the case of a tetrahedron, the polygons are all triangles by default, because in order to create a three-dimensional object with four polygons, each polygon must have three sides to connect with the other three polygons. The triangles can come in a variety of styles: when equilateral triangles are used, a tetrahedron is known as a "regular tetrahedron." Tetrahedra are also sometimes called triangular pyramids, because they include a flat base and an apex. There are lots of ways to play with this shape in mathematics. Triangles themselves are very interesting shapes from a mathematical standpoint, so an assortment of triangles is all the more interesting. Tetrahedra can also be joined together to create numerous other polyhedra, especially in the case of regular tetrahedra.

The tetrahedron is an example of a convex polyhedron. This means that if we randomly select any two points on the tetrahedron and connect them with a line, the line will pass through the tetrahedron and not stray outside of it. By contrast, in a non-convex polyhedron, the line would at some point travel outside the polyhedron. Generally, the more faces a polyhedron has, the harder it is to make it convex and at a certain point, it must become non-convex to accommodate all the faces. Some architects like to use this shape to add visual interest to their designs. Some cultures have also historically attached religious significance to this shape, or to collections of tetrahedra. The star tetrahedron, for example, is a polygon created by merging two tetrahedra which face in opposite directions, creating an eight-pointed star.

13. WHAT ARE MATHEMATICAL AND ARCHITECTURAL PYRAMIDS AND WHAT IS MEANT BY CHEERLEADING?

Pyramids are one of the fundamental solids used in Euclidean geometry. There are three elements required to make a pyramid. It is a three-dimensional figure with triangular side faces that meet at a common vertex, with a polygon base. Pyramids are classified by the shape of their base. A triangular pyramid has a triangular base and three triangular side faces. A square pyramid has a square base and four triangular sides. A pentagonal pyramid has a pentagonal base and five triangular sides. A hexagonal pyramid has a hexagonal base and six triangular sides and so on. Pyramids are referred to as n-sided, where n is the number of sides of the polygon that forms the base. An n-sided pyramid has n triangular faces.

Pyramids are also a shape favored for architectural construction, both ancient and modern. The ancient cultures that built pyramids include Ancient Egypt, Mesopotamia, several Mesoamerican cultures, China and Rome. In Egypt, where pyramids were used in funeral complexes, two types, step pyramids and "true" pyramids, were both built. In Mesopotamia, the type of pyramid built was the ziggurat, a temple with terraced stories. Many Chinese pyramids, used as burial monuments, have flat tops.

The pyramid is also one of the fundamental cheerleading structures and allows for endless variations. The World Cheerleading Association (WCA), which includes pyramids/basket tosses as one of the six technical skills scoring categories, issues General Safety Guidelines with rules for the use of pyramids in competition. There are six competition levels and special rules for what kinds of pyramids are allowed at each level. Pyramids may be 2, 2½, or 3 persons high, but for WCA competition, the limit is 2½ persons high at Level Six. There are three types of roles in cheerleading pyramids: bases, spotters and flyers. Bases form the foundation for the flyers to mount on. Spotters may be simply standing by to assist, support, or catch someone in case of need, or may also brace the stunt.

14. WHAT ARE TRIANGLES?

Triangles are one of the fundamental figures used in Euclidean geometry. There are three elements required to make a triangle. It is a three-sided plane or two-dimensional figure, in which the sum of the interior angles equals exactly 180 degrees. There are two common systems of triangle classification. One focuses on the sides and designates three types of triangle. Equilateral means "equal sides," and in an equilateral triangle, all three sides are the same length. This means that the angles will also be equal, all 60°, making the triangle equiangular as well. Isosceles means "equal legs," and an isosceles triangle has two sides that are equal in length. This also means that the two angles formed where the equal sides meet the third sides are equal. Scalene comes from a word meaning "uneven," and a scalene triangle has three unequal sides. As we might suspect, then, the three angles are unequal as well.

The other triangle classification schema approaches triangles from the point of view of the measurements of the internal angles. It, too, designates three types of triangle. In an acute triangle, the largest internal angle is acute, less than 90°. This means that all the angles are acute. In a right triangle, there is one right angle, an angle of exactly 90°. This means that the other two angles will be acute. In an obtuse triangle, one internal angle is obtuse. This means, again, that the other two angles will be acute.

The word triangle is also used for things that bear some resemblance to a triangle, either in having the approximate shape of a triangle or in having three points that are connected in some way. The musical instrument called a triangle is a metal rod shaped like a triangle, except that one end is open. It is hung by a thin wire so that it can vibrate when struck by a triangle beater. It is used in both orchestral and band music. A so-called love triangle is a romantic situation that involves three people and generally implies that they would each prefer a twosome.

Triangle is also used in reference to places that have a more-or-less triangular shape. The Bermuda Triangle, the area roughly marked out by Bermuda, Fort Lauderdale and Puerto Rico, is a well-known bit of the planet because it has a reputation for unexplained disappearances. There are a number of Golden Triangles, including the area of Southeast Asia noted for its opium production, however, a golden triangle is also an isosceles triangle constructed so that the ration of the hypotenuse to the base is equal to the so-called golden ratio, so we come full circle back to mathematics.

15. HOW DO PRISMS WORK?

In geometry, prism refers to a certain three-dimensional shape: a polyhedron or star polygon connected to an identical shape by rectangular sides. A cube or rectangular box is one example of this type of prism, but instead of having a rectangle on the top and bottom, as in this example, a geometrical prism may also have a triangle, a six-pointed star, or any other many-sided shape. In optics, a prism is a transparent object with smooth, flat sides that refracts light. Many optical prisms are also geometrical prisms, often triangular prisms, but they may also be other shapes. An optical prism is commonly made of glass or clear plastic, but any material may be used as long as it is transparent to the light source that is to be refracted.

When light is refracted, its course is bent due to an obstacle in its pathway. We can see an example of refracted light by placing a pencil in a clear glass of water and observing how it appears to split where the water begins. This happens because water has a higher refractive index than air, so light moves more slowly through water. An optical prism also has a higher refractive index than air, so light passing through it is refracted.

Light of different frequencies or wavelengths travels at different speeds and refracts at different angles as a result. In the visible spectrum, the portion of the electromagnetic (EM) spectrum visible to the human eye, different frequencies of light appear as different colors. Pure white light is a combination of light of many frequencies, so when pure white light shines through a prism, it breaks up into its constituent wavelengths, creating a rainbow effect.

Red, the lowest frequency visible to the human eye, refracts less than violet, the highest frequency visible to the human eye, so red bends at a less severe angle than violet; all the colors in between bend at different degrees, so the prism makes a rainbow out of white light. This is also how rainbows arise in nature, as they occur when the air is full of water vapor that refracts light in a similar way to a prism. Other types of prisms have different effects on light. In addition to the refractive prism, a prism may be reflective or polarizing. Reflective prisms reflect light, like a mirror and are used in binoculars, among other applications. A polarizing prism breaks light up into different polarizations, or electromagnetic charges, rather than different frequencies.

Subject Index

B

C

D

E

F

G

H

M

N

O

P

T